Waterproof and Water Repellent Textiles and Clothing

The Textile Institute Book Series

Incorporated by Royal Charter in 1925, The Textile Institute was established as the professional body for the textile industry to provide support to businesses, practitioners and academics involved with textiles and to provide routes to professional qualifications through which Institute Members can demonstrate their professional competence. The Institute's aim is to encourage learning, recognise achievement, reward excellence and disseminate information about the textiles, clothing and footwear industries and the associated science, design and technology; it has a global reach with individual and corporate members in over 80 countries.

The Textile Institute Book Series supersedes the former 'Woodhead Publishing Series in Textiles', and represents a collaboration between The Textile Institute and Elsevier aimed at ensuring that Institute Members and the textile industry continue to have access to high calibre titles on textile science and technology.

Books published in The Textile Institute Book Series are offered on the Elsevier web site at: store.elsevier.com and are available to Textile Institute Members at a substantial discount. Textile Institute books still in print are also available directly from the Institute's web site at: www.textileinstitute.org

To place an order, or if you are interested in writing a book for this series, please contact Matthew Deans, Senior Publisher: m.deans@elsevier.com

Recently Published and Upcoming Titles in The Textile Institute Book Series

Antimicrobial Textiles, Gang Sun, 9780081005767

Active Coatings for Smart Textiles, Jinlian Hu, 9780081002636

Advances in Women's Intimate Apparel Technology, Winnie Yu, 9781782423690

Smart Textiles and Their Applications, Vladan Koncar, 9780081005743

Advances in Technical Nonwovens, George Kellie, 9780081005750

Activated Carbon Fiber and Textiles, Jonathan Chen, 9780081006603

Performance Testing of Textiles, Lijing Wang, 9780081005705

Principles of Textile Finishing, Asim Kumar Roy Choudhury, 9780081006467

Forensic Textile Science, Debra Carr, 9780081018729

Crazing Technology for Polyester Fibers, Victor Goldade and Nataly Vinidiktova, 9780081012710

Natural Dyes for Textiles, Padma Vankar, 9780081012741

Colour Design, Second Edition, Janet Best, 9780081012703

High-Performance Apparel, John McLoughlin and Tasneem Sabir, 9780081009048

Sustainability in Denim, Subramanian Muthu, 9780081020432

Fibrous Filter Media, Philip Brown and Christopher Cox, 9780081005736

The Textile Institute Book Series

Waterproof and Water Repellent Textiles and Clothing

Edited by

John Williams

The Textile Institute

ELSEVIER

WP
WOODHEAD
PUBLISHING
An imprint of Elsevier

Woodhead Publishing is an imprint of Elsevier
The Officers' Mess Business Centre, Royston Road, Duxford, CB22 4QH, United Kingdom
50 Hampshire Street, 5th Floor, Cambridge, MA 02139, United States
The Boulevard, Langford Lane, Kidlington, OX5 1GB, United Kingdom

Library of Congress Cataloging-in-Publication Data
A catalog record for this book is available from the Library of Congress

British Library Cataloguing-in-Publication Data
A catalogue record for this book is available from the British Library

ISBN: 978-0-08-101212-3 (print)
ISBN: 978-0-08-101134-8 (online)

For information on all Woodhead publications
visit our website at https://www.elsevier.com/books-and-journals

Publisher: Matthew Deans
Acquisition Editor: David Jackson
Editorial Project Manager: Edward Payne
Production Project Manager: Omer Mukthar
Cover Designer: Victoria Pearson

Typeset by SPi Global, India

Contents

Contributors

Riza Atav University of Namık Kemal, Tekirdağ, Turkey

Jeni Bougourd Consultant, London, United Kingdom

Seong-O Choi Kansas State University, Manhattan, KS, United States

Luminţa Ciobanu "Gheorghe Asachi" Technical University of Iaşi, Iaşi, Romania

Irina Cristian "Gheorghe Asachi" Technical University of Iaşi, Iaşi, Romania

Prince Dabhi Institute of Chemical Technology, Matunga, India

Alice J. Davies University of Leeds, Leeds, United Kingdom

Angela Davies De Montfort University, Leicester, United Kingdom

Miyu Du University of Leeds, Leeds, United Kingdom

Nicholas W.M. Edwards University of Leeds, Leeds, United Kingdom

Parikshit Goswami University of Leeds, Leeds, United Kingdom

Lauren Heine Northwest Green Chemistry, Spokane, WA, United States

Dorin Ionesi "Gheorghe Asachi" Technical University of Iaşi, Iaşi, Romania

Zehra Evrim Kanat Namık Kemal University, Tekirdağ, Turkey

Veronika Kapsali University of the Arts London, London, United Kingdom

Asimananda Khandual College of Engineering & Technology, Bhubaneswar, India

Jooyoun Kim Seoul National University, Seoul, Republic of Korea

Carmen Loghin "Gheorghe Asachi" Technical University of Iaşi, Iaşi, Romania

Emil Loghin "Gheorghe Asachi" Technical University of Iaşi, Iaşi, Romania

Ameersing Luximon The Hong Kong Polytechnic University, China

Ningtao Mao University of Leeds, Leeds, United Kingdom

Jan Marek INOTEX Ltd, Dvůr Králové n.L., Czech Republic

Lenka Martinková INOTEX Ltd, Dvůr Králové n.L., Czech Republic

Jane McCann Design Consultant, Northern Ireland, United Kingdom

Roshan Paul University of Beira Interior, Covilhã, Portugal

Silvia Pavlidou Materials Industrial Research and Technology Center, Athens, Greece

Natalie Pomerantz US Army Natick Soldier Research, Development and Engineering Center, Natick, MA, United States

Usha Sayed Institute of Chemical Technology, Matunga, India

Quoc T. Truong US Army Natick Soldier Research, Development and Engineering Center, Natick, MA, United States

Margaret H. Whittaker ToxServices LLC, Washington, DC, United States

Hikmet Ziya Özek University of Namik Kemal, Tekirdağ, Turkey

Part One

Principles of waterproofing and water repellency in textiles

Introduction to waterproof and water repellent textiles

1

Carmen Loghin, Luminţa Ciobanu, Dorin Ionesi, Emil Loghin, Irina Cristian
"Gheorghe Asachi" Technical University of Iaşi, Iaşi, Romania

1.1 Introduction

Protection against environmental factors is the initial function of clothing. In a wet environment, the basic requirement for garments is to keep the wearer dry by being waterproof and or water repellent. The difference between the two terms is essential when characterizing the behaviour of textile materials in reference to liquid water. In contact with water, water repellent materials form drops that can be easily removed from the fabric surface but for longer contact with water or with a higher pressure difference, the material will absorb water. Water repellent textiles are often high density woven materials made of very fine yarns or common materials with hydrophobic surface treatment. Waterproofing is defined as the property of a material not to be penetrated by fluids. The waterproofness of a fabric can be measured using two testing methods: one that simulates raining and the other (more common) that subjects the fabric to hydrostatic pressure. The minimum value for the hydrostatic pressure without leaking at its surface, at which a fabric is considered rainproof is 5000 mm water column, while for waterproof materials the hydrostatic pressure can reach 10,000–15,000 mm water column (Loghin, 2003). For high quality waterproof materials designed for aggressive conditions, the hydrostatic pressure varies between 15,000 and 30,000 mm water column. Such fabrics are completely waterproof even under very high pressure.

First historical mentions regarding the hydrophobization of textiles are in the 15th century, when sailors tried to obtain sea water protective clothing by impregnating it with linseed oil, animal fat or wax. The first bio-inspired waterproof clothing product (*kamleika*) belongs to Aleut American Indians who used dried seal or whale intestines; the seams have been sealed with animal glues to make the product totally waterproof (Lynch and Strauss, 2015).

The first waterproof fabric was produced and patented by Charles Macintosh in 1823 in England (Shephard, 2012). The process to produce waterproof materials involves the spreading of a rubber layer between two woven fabrics. The problems related to the use of garments made of this material, caused by the unstable rubber, were eliminated by the process of rubber vulcanization that led to a textile material more stable in environmental conditions. The process was patented in 1844 by Charles Goodyear in the United States, and Thomas Hancock in England.

For a long period, rubberized textile fabrics were the raw material for waterproof garments. The main problem with these garments is reduced comfort due to the

Waterproof and Water Repellent Textiles and Clothing. https://doi.org/10.1016/B978-0-08-101212-3.00001-0

overheating of the wearer's body and high resistance to vapour passing out through the clothing layers. The sweat vapours condense in contact with the interior surface of the clothing, humidifying the textile layers in direct contact with the skin and causing increased discomfort. Subsequent researches conducted in the production of waterproof textiles led to a new type of material, waterproof-breathable fabrics.

Ventile fabrics are waterproof, breathable, densely woven materials developed in the UK during WWII to replace flax in garments for outdoor, military, medical and work wear applications. The first microporous membrane (polytetrafluoroethylene PTFE, also known as Teflon) was created in 1969 by W. L. Gore and Associates. The first GORE-TEX materials appeared on the market in 1976, starting a revolution in the concept of waterproof-breathable garments.

Water repellent textiles are obtained using specific finishing surface treatments. A review by Schuyten et al. (1948) shows that these hydrophobic treatments were developed significantly starting with 1920s.

Waterproof-breathable textiles represent a significant global market, with major players from the US, Europe and Asia. A press release for a report from Grand View Research Inc. (2016) indicates the value of the waterproof breathable textiles market in 2014 was \$1.43 billion. Membrane waterproof-breathable products account for 71% of the overall demand, while garments remain the main application. The report anticipates a constant growth of this market, stimulated by the need for comfortable multifunctional products, the use of innovative technologies to produce biomimetic and smart textiles, and the focus on recyclable and eco-friendly products. With an estimated compound annual growth rate (CAGR) over 5% per annum, the market of waterproof-breathable textiles is expected to reach \$2.18 billion by 2020.

1.2 Areas of application of waterproof and water repellent textiles

Waterproof and water repellent materials are currently used in the three major textile areas (clothing, home and outdoor products and technical textiles). There are a large number of possible applications, from rain garments to medical and military equipment (Singha, 2012). Regardless of the applications for which waterproofness is the determinant function, the complexity of the conditions during use requires the multicriterial design of the fabric structure and its testing to ensure a high number of functional characteristics such as: vapour permeability, tensile strength, abrasion resistance, flexural strength (repeated cycles), resistance to low and high temperatures, resistance to light, chemical resistance and more.

Several standards are used for the evaluation of waterproof-breathable and water repellent textiles.

Waterproofness is measured as the hydrostatic pressure needed to penetrate the waterproof-breathable fabrics. The standards used for determining waterproofness are:

- ASTM D 3393-91 Standard Specification for Coated Fabrics—Waterproofness.
- AATCC TM 127-water resistance: hydrostatic pressure test.

- ISO 811 Textile fabrics—Determination of resistance to water penetration—Hydrostatic pressure test.
- BS 3424-26 Testing coated fabrics. Methods 29A, 29B, 29C and 29D. Methods for determination of resistance to water penetration and surface wetting.

Breathability is evaluated based on water vapour transmission (WVT). There are several test methods, applicable to both coated and laminated fabrics:

(1) The upright cup test—JIS L 1099, JIS Z 0208, ISO 2528, Desiccant Method of ASTM E96, JIS K 6328.
(2) The inverted cup method—JIS L 1099, similar to ASTM E96-BW test method.
(3) The sweating hot plate method (evaporative resistance)—ISO 11092, ASTM F 1868.
(4) The dynamic moisture permeation cell—ASTM F 2298.

Water repellency is tested using the following standards:

- AATCC TM 22-water repellency: spray test.
- ISO 9865-water repellency: Bundesmann rain shower test.
- AATCC TM 35-water resistance: rain test.
- ISO 22958:2005 Textiles—Water resistance—Rain tests: exposure to a horizontal water spray.
- EN 14360-rain test (test method for ready-made garments).
- AATCC TM 42-water resistance: impact penetration test.

1.2.1 Clothing design specifics according to the end use

Garments remain the most frequent use of waterproof and/or water repellent fabrics. Due to the complex requirements of the users (protection, comfort, aesthetic, identity, etc.), waterproof materials must have a sum of properties that ensure the multifunctional characteristics of the garment. The level of performance of the waterproof and water repellent materials used for garments is determined by two groups of factors: (i) subjective variables, defined by the requirements and the level of comfort of the final user and (ii) objective variables, defined by the environmental conditions, risk factors and specifics of the activities carried out by the user.

A first level of classification for waterproof clothing contains:

- conventional wet-weather clothing;
- work clothing and uniforms (including military);
- clothing for sport and leisure; and
- personal protective equipment (PPEs) (for risk conditions).

The main problem when using waterproof fabrics for garments is the comfort of the wearer. Usually, waterproof technologies consisted in covering and blocking the pores of the textile substrate so water absorption and transfer from the exterior towards the body are no longer possible. This way, the material acts like a barrier for the humidity in the environment, apparently satisfying the protection function of the garment. In the relationship between the human body, the garment and environment, the transfer of humidity must be analysed in both directions from and towards the body. Human skin sweats continuously, both at rest (insensible perspiration or *perspiratio insensibilis*), or in activity (sensible perspiration). For example, during intense effort the average human body (with a 1.8 m^2 skin surface) produces approx. 1000 cm^3 perspiration per hour to reach its thermal balance (Holmes, 2000). The humidity produced through perspiration must be eliminated as vapour in a process of mass transfer through the layers of the garment. This problem was solved in the 1980s by the production of waterproof-breathable textiles, which are materials waterproof for liquid water in the environment but permeable to sweat vapours from the body passing through clothing layers (Section 1.3.3.3).

1.2.1.1 Conventional wet-weather clothing

Conventional garments are used in wet environments (rain and snow) and their waterproofness is determined by special characteristics of the materials. The basic material can be water repellent for short periods of rain or snow, or waterproof and water repellent for long exposures to rain or snow. This type of garment includes:

- waterproof garments (raincoats, jackets, trousers);
- waterproof garments with high thermal insulating characteristics and low weight (e.g. equipment for winter sports);
- waterproof bioactive garments (insect repellent, antiallergic, antibacterial) to be used in outdoor activities in insect-infested environments, and in medical applications where the risk for infections is very high;
- waterproof UV garments for outdoor sports (fishing, camping, hiking, etc.), as well as work wear for workers exposed to UV; and
- low maintenance products with increased cleaning characteristics (e.g. work wear, clothing for children, etc.).

Work clothing, including uniforms, can be considered conventional waterproof garments, used in similar conditions but only in the working period.

The fundamental aspects that have to be taken into consideration when designing a waterproof garment and subsequently the selection criteria (Loghin and Ciobanu, 2008; Mukhopadhyay and Midha, 2008a; Chinta and Satis, 2014) are:

1. waterproofness level;
2. fabric weight;
3. level of thermal physiological comfort;
4. the capacity to allow the transfer of sweat vapour;
5. aesthetics;
6. durability (strength to repeated flexural cycles, tear, tensile, friction strength);
7. launderability (washing/dry cleaning/tumble drying);
8. resistance of the water repellent treatment to repeated laundering and cleaning;
9. visibility (for work wear);
10. identification/identity (for uniforms); and
11. flame retardant (for work wear and uniforms), etc.

Waterproofness and air (wind) resistance/proofness of the materials are the most important issues in the case of weather protective outer garments. For such garments, water/airproofness must be considered at constructive, structural and technological levels, as illustrated in Fig. 1.1.

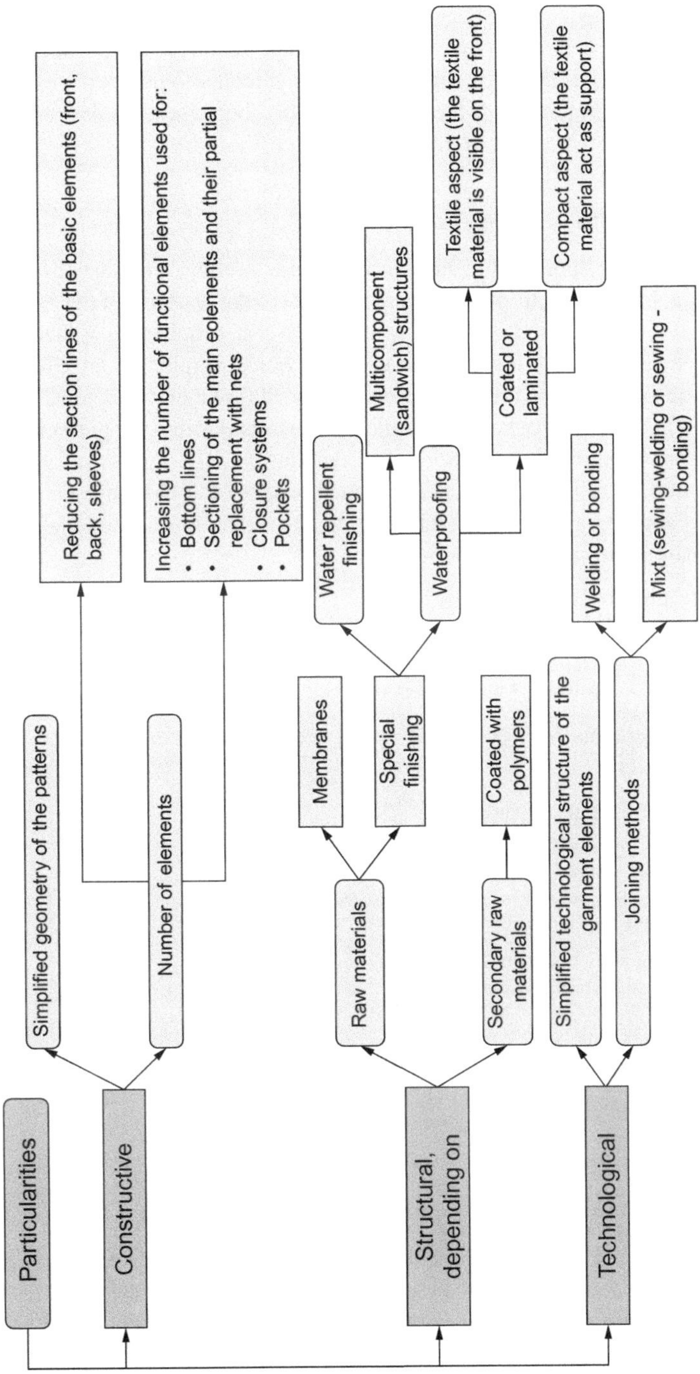

Fig. 1.1 Design characteristics for waterproof clothing.

Designing a waterproof garment requires certain constructive solutions (geometrical pieces, few cut lines, most sealed constructive variants for closures and/or pockets) and technological processing (welding or bonding/ sealing, sewing and welding, sewing and sealing). Replacing sewing with welding technologies or the use of seam sealing are decisions essential for the garment quality, because the holes produced by the needles during sewing lead to diminished waterproofness.

1.2.1.2 Sport and leisure garments

The requirements for sport and leisure garments are similar to the ones for conventional clothing, but the functional performance level of the waterproof materials needs to be higher due to increased sweating in intense effort, requiring a higher moisture transfer rate from the body towards the exterior. For example, for an effort rate of 500 watts, the perspiration rate is approximately 800 g/hour (Mukhopadhyay and Midha, 2008b). The main part of this humidity transfers through the garment, the rest representing losses through ventilation and respiration. The results of athletes are often influenced decisively by the performance level of the clothing. The potential of waterproof and/or water repellent materials to be used for multifunctional sport garments is illustrated in the examples below:

- waterproof garments for winter sports;
- self-ventilating waterproof garments;
- waterproof garments with moisture control; and
- antimicrobial and antifungal waterproof garments.

The basic requirements for waterproof sportswear and leisurewear are as follows, ranked in this order:

1. good heat and mass transfer capacity;
2. good vapour permeability in relation to air proofness; and
3. moisture control, assured by:
   - humidity absorption and its transport towards the exterior (environment);
   - keeping the skin dry; and
   - quick drying after humidity absorption;
4. good dimensional stability in wet state;
5. aesthetic aspect;
6. low weight;
7. pleasant touch;
8. low maintenance; and
9. durability, etc.

1.2.1.3 Personal protective equipment

PPE represent a set of individual protective means (clothing, footwear, head protection, gloves, masks, etc.). Protective clothing must ensure complete or almost complete insulation from the environmental factors (weather, hazards), some of these factors are harmful to human health. In the conditions of hazardous environments, waterproofness is a major requirement that most of the time must be correlated with other requirements for the protective garment in specific working conditions. The complex protection requirements for the working and protective equipment and/or working environment can be as follows:

1. For work that involves fluid flow: the possibility for self starting or self blocking, spill, immersion-waterproof fabric, seam sealing, resistance to dynamic loads.
2. For work that involves microbiological cultures, bacteria, physiological fluids: impervious fabric, seam sealing, mechanical strength, decontamination capacity.
3. Outdoor work environment with low temperatures: thermal insulation, waterproof fabric, seam sealing.
4. Work environments with high humidity, precipitations, air currents: waterproof fabric, seam sealing.
5. Work environments with dangerous powders or suspensions of microorganisms: impervious fabric, seam sealing, decontamination capacity.
6. Gaseous work environment, toxic vapours, aerosols: impervious fabric, seam sealing, specific chemical resistance.
7. Gaseous work environment, inflammable vapours or explosives: impervious fabric, seam sealing, specific chemical resistance, antistatic properties, flame retardant characteristics.
8. Sterile work environment, clean rooms: impervious fabric, seam sealing, antistatic properties.

In the case of conventional clothing, waterproofness is perceived as the property of the materials to oppose the passing of air and water and therefore a measure of protection against atmospheric factors while ensuring the wearer's comfort. It therefore needs to be breathable, not impermeable. In the case of protective clothing, waterproofness (imperviousness or impermeability) has many aspects, depending on the system of factors and leading to a specialization of the waterproof materials. The following types of impervious textiles can be listed:

- impervious to water/air and permeable to water vapours;
- impervious to chemical agents;
- impervious to biological agents (microorganisms, physiological fluids);
- impervious to radioactive contaminants; and
- impervious to micro-particles generated by the human body, e.g. for clean rooms.

1.2.2 Home and outdoor textiles

Waterproof and water repellent textiles become more and more common in products used at home, as well as for the outside.

For home textiles, main applications are pillow protectors, bed covers, bed sheets and mattress covers. Shower curtains can also be made from waterproof textile materials. Decorative mobile walls can also be made from this type of material.

If needed, further treatments can be applied to the textile-coated materials, like antifungal, antidust mites and antibacterial, halogen-free fire retardant treatments (Sen, 2008).

Home outdoor applications include small shade structures and other decorative outside elements, covers for chairs and tables in the garden, etc. Apart from clothing for outdoor activities, waterproof/water repellent materials are also used for specific equipment such as tents, backpacks, hiking gear, insect repellent curtains, etc.

1.2.3 Technical applications

There is a wide range of technical/functional applications that use waterproof or water repellent textiles including agricultural, civil engineering, medical, industrial and packing applications.

In agriculture, waterproof-breathable textiles are used for different cultures (breathable ground cover for weed control, waterproof sheeting, root protective bags for transporting, greenhouse covers, tree shelters), as well as structures with agricultural use (leak-proof sheeting for water and liquid fertilizer tanks and flexible water tanks) and packing for product transport.

Architectural textiles are lightweight, flexible materials that can be used for temporary and permanent structures. Waterproof coated or impregnated textiles can save energy and decrease costs, while allowing for innovative creative approaches to architecture. Water repellent treatments are also applied to such materials, especially for outdoor applications, to improve their behaviour in wet weather.

Temporary applications using textile membranes (woven fabrics with PVC) include tents, clear-span structures, tension fabric structures and air structures and commonly built for exhibition spaces, structure for leisure activities, short-term commercial spaces, social gatherings, storage facilities, etc. Such materials are high strength woven fabrics made of glass, fibres, PES or polyethylene coated with PVC, silicone, PTFE ref (Houtman, 2015).

Textile membranes are suitable for roofs due to their lower weight, controlled mechanical strength including impact, resistance to weather, controlled translucence, sound insulation capacity, fire retardant characteristics and resistance to UV (Zerdzicki, 2015). To increase their behaviour, a hydrophobic top coating can be added to the materials, while titanium dioxide (TiO_2) photo-catalyst provide self-cleaning properties. Another advantage is that textile membrane roofs can be fixed or retractable.

Another domain of application refers to decorative waterproof textiles like canopies, awnings, marquees, shading structures and advertising structures that are placed on buildings or are in the immediate public space.

Waterproof textiles are widely used in medical applications, for nonimplantable and healthcare and hygiene products. Literature presents examples of waterproof breathable textiles (woven, knitted) used for orthopaedic orthoses to improve the level of comfort for patients, replacing the neoprene commonly used. Another orthopedic end-use is a knitted breathable cast for upper or lower limbs (Sherif and Roedel, 2011). Modern multilayer wound dressings have an outer layer (mostly nonwoven) that is waterproof-breathable.

Healthcare products include wheelchair cushions, mattress covers, pillow protectors, bed-stretchers, stretchers and hospital cases. Another application is for surgical gowns and drapes. Apart from their characteristics, these products must also ensure a clean environment around the patients and medical staff, so the coating must include antifungal and antibacterial substances. The coating can also be designed to ensure viral protection, an important issue in hospitals.

1.3 Basic aspects regarding waterproof and water repellent textiles

1.3.1 Textile and water interaction mechanism

Because waterproofness is a requirement imposed mainly by the environment (weather conditions), the behaviour of textile materials towards liquid water must be commented upon. From this point of view, textile materials can be divided into:

1. materials with water absorption and retaining characteristics—*hydrophilic materials*;
2. materials that repel water—*hydrophobic materials*.

The capacity of the textile surface to absorb or repel liquid water is explained through the surface tension developed at the interface between the water drop and textile surface (see Fig. 1.2). The surface tension γ_{12} generated at the interface depends on the fibrous composition of the textile material, the structural parameters of the yarn and material and the microstructure of the contact area (smooth, micro rough, continuous, discontinuous, etc.) (Park et al., 2016).

Another factor influencing the balance of the superficial tensions is the material's porosity, namely the state of the transversal pores after water repellent finishes (open, blocked uni- or bilaterally). Fig. 1.3 presents the simplified structure of a textile material, emphasizing the transversal (P_T), longitudinal (P_L) and superficial (P_S) pores.

The behaviour of a textile material towards liquid water is evaluated based on the value of the contact angle (θ), with the following formula (Young's Equation):

$$\cos\theta = \frac{\gamma_{13} - \lambda_{12}}{\gamma_{23}} \quad (1.1)$$

where γ_{12}, γ_{13} and γ_{23} represents the surface (interfacial) tensions of the fabric-water (γ_{12}), fabric-air (γ_{13}), and water-air (γ_{23}) contact.

Theoretically, the value of the contact angle is placed in the interval 0 angle and 180 angle. The textile materials can be classified accordingly into (Gugliuzza and Drioli, 2013; Zimmermann et al., 2009):

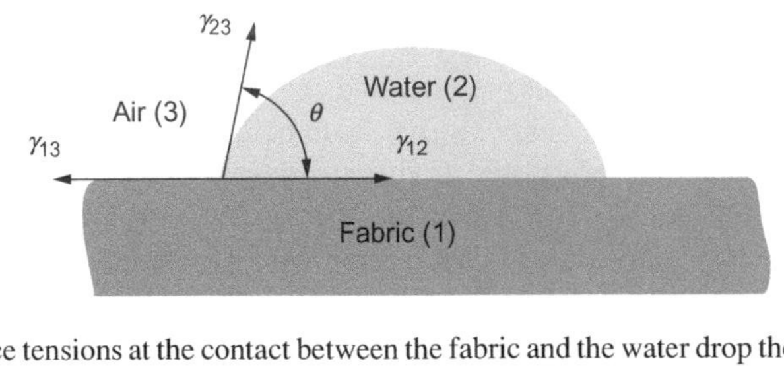

Fig. 1.2 Surface tensions at the contact between the fabric and the water drop theoretical model. From Hoblea, Z., 1999. Structuri textile—Structura şi proiectarea îmbrăcămintei (Textile Structures—Garment Structure and Design). Gh.Asachi Publishing House, Iasi, pp. 50–60, ISBN 973-99209-4-2. Published with the author's permission.

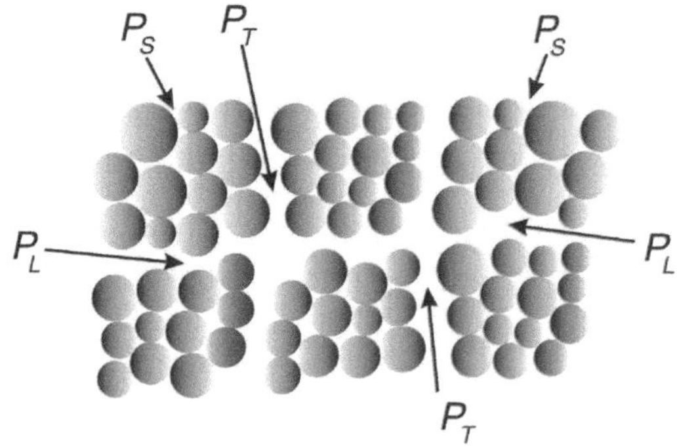

Fig. 1.3 The porous structure of a textile surface (simplified model).
From Hoblea, Z., 1999. Structuri textile—Structura şi proiectarea îmbrăcămintei (Textile Structures—Garment Structure and Design). Gh.Asachi Publishing House, Iasi, pp. 50–60, ISBN 973-99209-4-2. Published with the author's permission.

- superhydrophilic materials, $\theta \rightarrow 0$ angle;
- hydrophilic materials (including textiles), 0 angle $< \theta <$ 90 angle;
- hydrophobic materials (including textiles), 90 angle $\leq \theta <$ 150 angle;
- superhydrophobic materials (including textiles), 150 angle $\leq \theta \leq$ 180 angle.

1.3.2 Water repellent textiles

Hydrophobic textiles present the advantage of air permeability but offer less protection against water, being generally used for conventional garments or as an exterior layer for waterproof clothing. Based on the resistance to cleaning agents, the hydrophobicity can be permanent (durable water repellent, DWR) or temporary (Gibson, 2008).

Depending on the way the water repellent effect is obtained, there are two groups of textile materials:

1. inherent water repellent textile materials;
2. textile materials with water repellent finishing.

Water repellent characteristics are specific to compact textile structures. Inherent water repellent materials are (i) high density woven fabrics, made of very fine yarns and filaments and (ii) nonwoven materials.

Microfibres and microfilaments present a high technological potential, with a practically unlimited area of applications (garments, household, medical and technical textiles) due to their special surface properties. In this group are included fibres with fineness 0.3–1 dtex, with the interval 0.3–0.1 dtex for super-microfibres. These fibres are made of synthetic polymers—PTE, PA, PP, PAN or cellulose (Purane and Panigrahi, 2007).

The specific properties and implicitly the end-use are determined by the morphological structure of the microfibres and their specific technology. The properties of the woven fabrics controlled through the particular characteristics of the microfibres refer to:

- water repellency and air impermeability due to the high density of the fabrics made of microfilaments (e.g. a fabric made of 0.2 dtex microfilaments has a thread density of 7000 microfilaments/cm^2) (Kaynak and Babaarslan, 2012);
- the vapour permeability varies in acceptable limits due to the interstitial porosity of the textile surface;
- increased filtering/absorption capacity of solid particles in reference to other fibres determined by the significantly higher specific surface, given by the number of microfibres per unit area and the cross section geometry (segmented, cross or island type structure); and
- increased liquid absorption capacity concurrent with an increased drying rate due to the same bigger specific surface, intensifying the capillary activities at textile surface level.

The hydrophobization of the textile materials is carried out with different chemicals that ensure high superficial tension in relation to water. These substances orient their hydrophobic groups towards the textile fibres thus forming a protective brush against water. The water hydrophobization agent forces are null, facilitating the water drop to maintain its spherical shape without spreading onto the fibres. In general, the limitations of the water repellent treatments refer to low surface energy and extended surface porosity.

The technological variants for hydrophobization include:

1. Hydrophobization with additives (aluminium organic salts, aluminium soaps, paraffin emulsions with aluminium salts).
2. Hydrophobization with resin type reactive agents (perfluoro ester-aziridine , zirconium compounds, radical crosslinking of methyl or cyanoethyl silicones (Indu Shekar et al., 2001) fluorocarbon (FC) resin (Kuhr et al., 2016)).
3. Hydrophobization through chemical modification of the fibres (esterification or etherification reactions).
4. Textile finishing with nanoparticles that create an ultrahydrophobic surface with self-cleaning characteristics (lotus effect). Oleophobization techniques give textile materials the property of repelling oils and thus creating a protection against dirt and smudges, while increasing the hydrophobization effect. FC resins are used as oleophobization agents—water emulsions or solutions in solvents (Kuhr et al., 2016).
5. Plasma treatment of the textile materials (Colleoni et al., 2015), plasma polymerization or plasma depositing of organic-silicone polymers (Kale and Palaskar, 2010) can give a hydrophobic character to materials that are typically not hydrophobic, like 100% cotton.

1.3.3 Waterproof textiles

An initial example of this type of materials is the high-density woven fabric VENTILE, made of 100% cotton, thread density up to 95 yarns/cm and waterproofness corresponding to 500–750 mm water column hydrostatic pressure (Mukhopadhyay and Midha, 2008a, 2008b). Due to the vapour permeability given by its structural porosity (transversal pores), VENTILE can be considered the first textile breathable-waterproof material.

Generally, conventional waterproofing treatments lead to the impossibility of fluids passing through textile materials due to the closing of the pores by covering them with a layer of polymer or a membrane (Ahn et al., 2010).

Considering their morphological structure and/or their technology, waterproof materials can be classified as follows:

1. inherent waterproof materials;
2. textile materials with waterproofing finishing treatments (coated and laminated).

W.L. Gore's expanded PTFE membrane is often regarded as the starting point of commercially available high performance waterproof breathable membranes. Initially an expanded PTFE membrane claiming 90% void volume was laminated to a support fabric, however, the pores became contaminated by sweat or detergents thus reducing the overall waterproofness (now used in windstopper fabrics). To overcome this drawback a thin hydrophilic polyurethane coating was applied to the body side of the membrane to prevent contamination (know as 2nd generation GoreTex).

1.3.3.1 Inherent waterproof materials

Inherent waterproof materials include materials with compact, nonporous structures that are completely impermeable to liquid or vapour water, namely:

1. polymeric foils;
2. textile materials laminated with polymeric foils.

Polymeric foils present a continuous compact and nonporous structure. With the development of plastics, polymeric foils became raw materials for weather protective garments, the so-called raincoats. Low density polyethylene (LDPE) and polyvinyl chloride (PVC) are the thermoplastic polymers most used for foils and films, the production technology requiring the planar extrusion of the melted polymers. Polymeric foils are suited for waterproof clothing due to their isotropic and compact structure, the low thickness (0.25–0.5 mm) and specific mass, the pieces being joined using adequate welding techniques. The use of PE and PVC foils presents the following major disadvantages:

- low mechanical strength, limiting the possibilities of using these waterproof materials to common applications, even if they exhibit good resistance to chemical or biological agents that recommend such materials for protective garments and
- high resistance to vapour transfer caused by the compact structure, generating discomfort due to the lack of ventilation within the microenvironment of the clothing system (overheating, perspiration at skin level, etc.).

The mechanical strength of the foils can be improved by laminating them to textile substrates (frequently nonwoven materials) extending their applications to technical and decorative textiles (Uludag et al., 2011).

1.3.3.2 Textile materials with waterproofing finishes

Waterproof materials are generally obtained using covering techniques that are considered surface finishing treatments. Covering is a general term referring to the placement on one or both sides of a textile material of one or more layers of adherent polymeric products that in the end form a film.

There are two technologies for this type of material that have different ways for the application of the polymer:

- Coating technology, where the polymer is applied by direct layering and superficial impregnation, usually in the final stage of obtaining the waterproof material. The polymer can be applied as a paste or a high viscosity liquid. Such coatings are superthin, in the range of 10–100 μm.
- Laminating technology that involves in a first stage the formation of a laminating layer (membrane or foam) that is subsequently spread on the surface/surfaces of the textile material. The membrane is extremely thin (e.g. around 10 μm for PTFE) so the final thickness of the film remains also in the range of 10–100 μm.

Coated waterproof textiles

Impregnation is a particular case where the polymer is deposited uniformly on the entire textile surface as a solution or a low- or high-viscosity dispersion using different processes that require the following technological phases: impregnation, drying and consolidation of the pellicle. A general characteristic of the impregnated materials is that the components cannot be clearly separated because the polymer is dispersed among the structural elements of the textile surface (see Fig. 1.4). The finishing technologies used cover either sides of the material (total impregnation) or just one side.

Laminated waterproof textiles

The general characteristic of laminated materials is that the components are clearly delimited and in some cases they can even be detached (see Fig. 1.5).

Such multicomponent products (two or more layers, one of which being the textile fabric) require bonding by the use of:

- a special adhesive added to the polymer (solutions in organic solvents, powders, granules, fibres) and
- the adhesive properties of one or more component layers (membranes, foams, expanded foils).

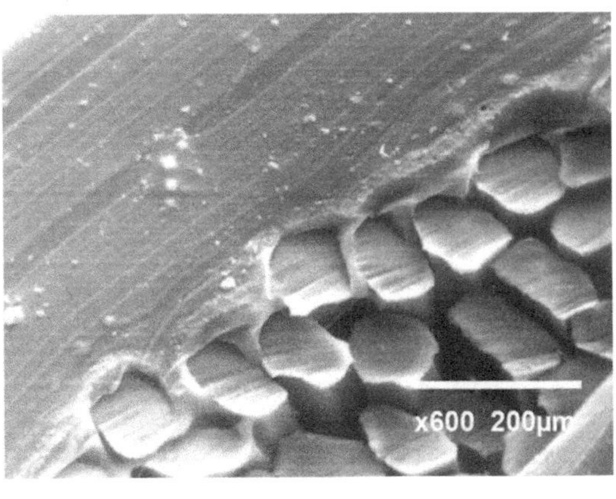

Fig. 1.4 Coated waterproof textile (SEM image ×600) (Loghin, 1998).

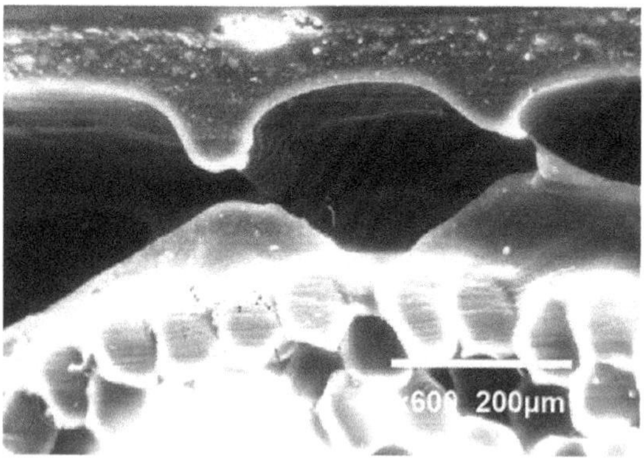

Fig. 1.5 Laminate waterproof textile (SEM image ×600) (Loghin, 1998).

Natural or synthetic polymers are suitable for laminating textile materials. They present layering and adhesive characteristics, as well as properties determined by the application. Rubber is the only suitable natural polymer, while the range of synthetic polymers is much wider.

An analysis of the consumption of synthetic polymers shows that 90% represent polyurethanes. Table 1.1 shows the most common polymers used for coating textile materials and the pellicle/film characteristics (Loghin, 2003).

The morphological structure of the coated and laminated materials and the nature of the polymers are important from the garment manufacturing point of view, as they are key factors in obtaining a perfectly sealed waterproof product.

Table 1.1 Coating polymers used for waterproof textiles

No.	Polymers	Pellicle/film properties	Observations
1.	Synthetic rubbers	- abrasion resistance - flexibility and elasticity - waterproofness - resistance to chemical agents	Are applied as dispersions or solutions in organic solvents, require vulcanization
2.	Polyolefins	- flexibility - waterproofness - resistance to frosting - resistance to chemical agents	Low pressure PE and PTFE are used, especially as compact membrane or foam
3.	Polyvinyl chloride	- stability to chemical agents - stability to light	Is applied as solutions in solvents and thermoplastic films

Table 1.1 **Continued**

No.	Polymers	Pellicle/film properties	Observations
4.	Acrylic derivatives	- stability to chemical agents - waterproofness - flexibility - elasticity - resistance to ageing	Are applied as solutions/aqueous dispersions or solutions in organic solvents
5.	Polyurethanes	- mechanical strength - resistance to ageing - elasticity - flexibility - waterproofness - noncreasing	Are applied as solutions in organic solvents or as foams

The morphological structure of these materials contains the components of the solid–gas ensemble that define the covering polymer and of the polymer-textile substrate system that characterizes the product at macroscopic level. When presenting the components, most authors recommend the macroscopic towards microscopic system. This way, the morphological structure includes:

1. The number of layers that make the coated or laminated waterproof material and their relative position in the garment.
2. The type of solid–gas system for the covering layer, meaning the absence or the presence of the pores (compact or porous layer) and the absence or the presence of other added substances.
3. The structure of the textile substrate; woven, knitted or nonwoven fabrics that can have different finishing treatments.

Considering their position in the garment, the coated or laminated waterproof materials can be:

- With the covering layer towards the exterior; most representative are materials covered with elastomers and some materials laminated with compact foils. These materials have a high decontamination and/or cleaning capacity, being recommended for chemical protection and protection against particles.
- With the covering layer towards the interior—especially used for wet weather protection. This variant is used for laminated materials (e.g. Gore-Tex) with a membrane or film with low mechanical strength. For increased durability, the polymeric film is covered with a hydrophilic polyurethane (PU) layer and/or a thin textile fabric (2.5L and 3L).

Based on the number of layers, there are:

- materials with two layers (made of two layers, one of which is the textile substrate). The group includes most coated materials and 2L laminated materials. Fig. 1.6 illustrates graphically some structural variants for garments made of 2L laminated materials. The laminated

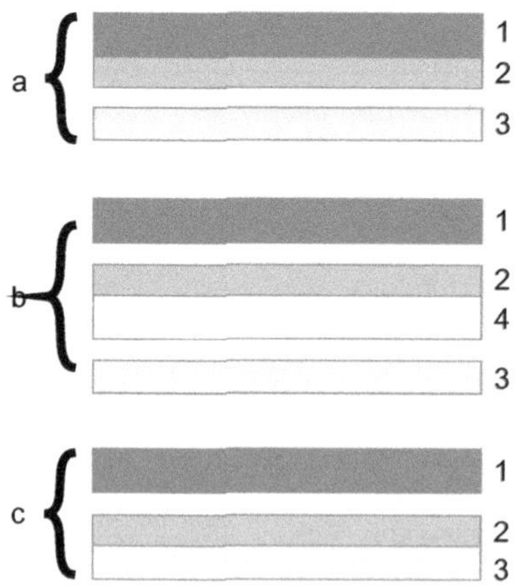

Fig. 1.6 Structural variants of laminated materials. *1*, outer fabric; *2*, laminated polymer; *3*, lining; and *4*, support material (knitted or nonwoven fabric).

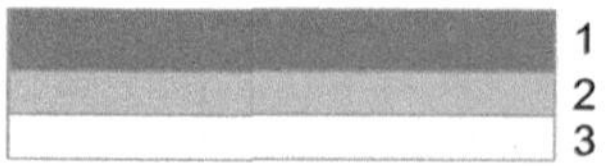

Fig. 1.7 Laminated material 3L. *1*, textile fabric (exterior layer); *2*, laminating polymer; and *3*, lining.

structures can be used as outer materials (A), as an intermediary layer between the outer material and the lining (B) or as lining (C). The presented structural variants are typical for garments for weather protection for which the outer layer presents hydrophobic and/or oleophobic characteristics.

- multilayer materials are made of at least 2.5 and 3 different layers, most representative being the sandwich 3L laminated materials (Fig. 1.7) used especially for protective clothing (protective equipment for fire fighters) and work clothing (industrial, police or military uniforms).

Another characterization of waterproof materials based on the solid–gas system of the covering layer refers to the presence or absence of the pores (compact or porous layer):

- waterproof textiles with compact coating polymer (nonporous structure) applied as solution, dispersion (Fig. 1.8) or laminated with compact foils, that are in the same measure impermeable to water liquid and vapour water and
- laminated textiles with microporous layers, considered to belong to the group of breathable materials, characterized by waterproofness and vapour permeability and used at large scale for manufacturing waterproof garments.

Ease of sealing and durability determine use of 2, 2.5 or 3 layer materials. The film is the weakest point

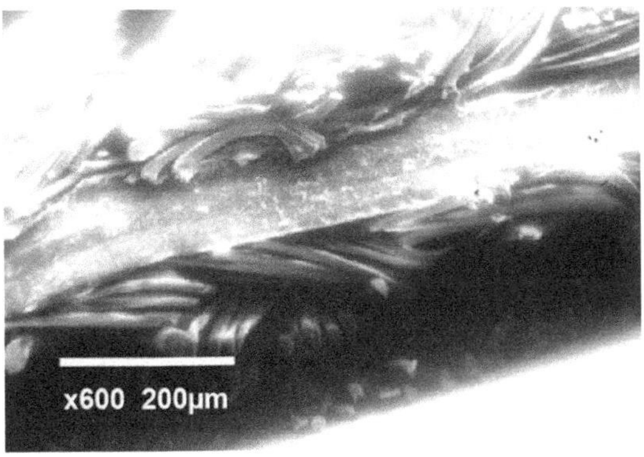

Fig. 1.8 Compact coated waterproof textile (SEM image ×600) (Loghin, 1998).

1.3.3.3 Breathable-waterproof textiles

As mentioned before, the main problem in using waterproof textile materials for garment manufacturing is the need to ensure to a satisfactory degree the wearer's comfort, as this function is affected by the decrease in ventilation in the clothing layers. Humidity absorption and transfer towards the external environment are determined by five physical mechanisms, as follow (Loghin and Ciobanu, 2008):

- diffusion of the water vapour through the pores of the textile materials, determined by the partial pressure difference created between the outer and inner surface;
- vapour adsorption and migration at fibre or yarn surface, determined by the surface tension generated in contact with water;
- humidity adsorption and desorption in gas and/or liquid state in and from the fibres, generally with hysteresis that sometimes can lead to fibre swelling;
- condensation followed by evaporation through free spaces, determined by simultaneously reaching the negativity condition for the difference in partial pressure, respectively temperature; and
- convection from the internal microclimate (caused by the movement of the clothed human body) is mainly a phenomenon of thermal exchange, but it also involves the displacement of humid air through the fabric structure/layers. Garment endings (like cuffs, collar, bottom end) intensify the convection rate.

To produce coated and laminated waterproof-breathable textiles, the vapour transfer through the clothing layers towards the external environment is ensured by:

1. Materials with compact hydrophilic outer layer (Fig. 1.9). This layer is made of polymers with hydrophilic groups (—OH, —COOH, —NH_2, —COO—, —CONH) (Loghin et al., 2009), in which case the vapours are eliminated through an absorption–desorption mechanism. An example is the SYMPATEX membrane (Akzo Nobel) made of a co-polyester obtained by grafting the polyester with polyether (Loghin et al., 2008).
2. Materials with porous outer layer, in which case the vapours are transferred by diffusion. This porous outer layer (hydrophilic PU film) is added to prevent the PTFE membrane becoming contaminated with different agents (oil, dirt, detergents, etc.). The hydrophilic PU layer insures the mechanical strength of the material.

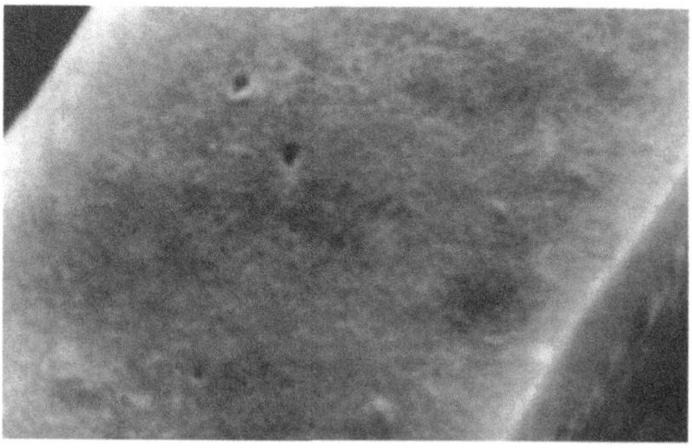

Fig. 1.9 Compact hydrophilic layer (SYMPATEX membrane) (Loghin, 1998).

To describe the different behaviour in relation to liquid water and vapours (the principle of membranes waterproof-breathable), there are certain considerations regarding the type, average diameter, form and distribution of the pores.

Considering their type, the pores can be (i) individual; closed, partially closed or open (see Fig. 1.10); (ii) distributed in a net, communicating between them and with both sides of the membrane, this being the typical structure of materials coated with microporous PU.

Considering the average diameter, the pores can be divided into (Colleoni et al., 2015): (i) macropores, with diameter over 50 nm and specific surface 0.5–2 $m^2\ g^{-1}$; (ii) mesopores, with diameter between 2 nm and 50 nm and specific surface 20–150 $m^2\ g^{-1}$; and (iii) micropores, with diameter under 2 nm (comparable to small molecules) and specific surface 400–900 $m^2\ g^{-1}$. The structures with pore dimensions less than 10^3 nm are considered microporous, while the ones with dimensions exceeding 10^3 nm are considered foams.

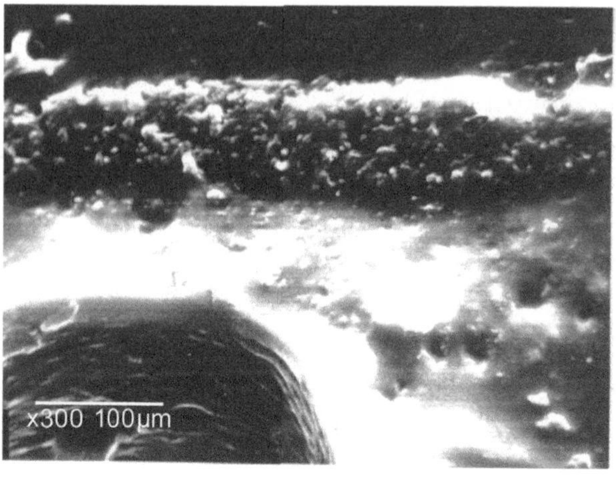

Fig. 1.10 Porous coating polymer (SEM image ×600) (Loghin, 1998).

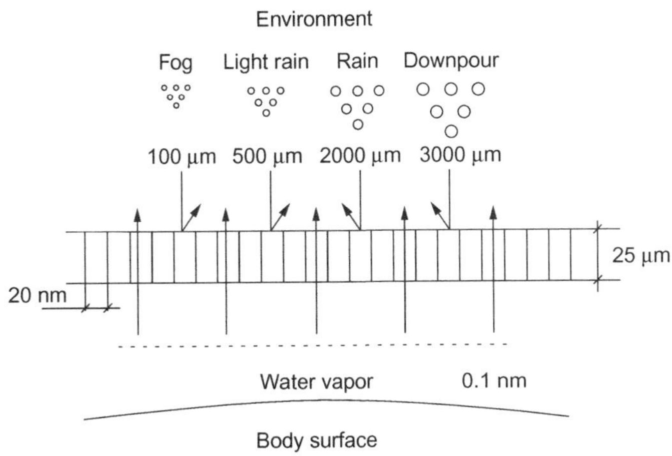

Fig. 1.11 The principle of microporous membranes.

Fig. 1.11 presents the principle of microporous membranes (Loghin, 1998). Laminated and coated microporous materials have pores with much lower dimensions (2–50 nm) than the smallest rain drop (fog – 100 μm), but bigger than the water molecules (0.1–10 nm) (Ahn et al., 2010).

Taking into consideration the dimensions and density, the pore distribution can be uniform or nonuniform. Analysis of the porosity of the outer layer is determinant in evaluating the level to which the garment will respond to the requirements during use.

1.3.3.4 *Multifunctional waterproof fabrics*

Recent developments in the field of multifunctional materials refer to waterproof textiles with multiple functions. Apart from common repellent finishes (often c8-fluorocarbon giving both oil and water repellency while maintaining high breathability), such materials require other specific treatments to create multifunctionality.

The outer layers of the coated and laminated waterproof textiles can be produced as a matrix-dispersed phase composite, the additional substances enlarging the range of possible applications where multiple functionality is required:

- Waterproof-breathable and electrostatic shielding. The multifunctional material is manufactured by coating or laminating a multilayer knitted fabric made of carbon core filaments and PES/stainless steel spun yarns with microporous PU. The resulting material is waterproof-breathable, and water and oil repellent, while ensuring electrostatic shielding (Varnaite-Zuravliova et al., 2016). The antistatic characteristics of the waterproof-breathable material are given by adding Ag nanoparticles in the polymer mass (Shyr et al., 2011).
- Water repellent and flame retardant. Cellulosic materials used for tent manufacturing are covered with hexagonal boron nitrite nanosheets; this treatment creates hydrophobic ($\theta > 90$ angle) and flame retardant characteristics (Yaras et al., 2016).

- Waterproof-breathable and UV barrier, antimicrobial and antistatic properties. The anti-UV, antimicrobial and antistatic properties of the waterproof materials are obtained by adding to the polymer mass nanoparticles like nano-silver, TiO_2, ZnO, SiO_2, Al_2O_3 and UV blockers (Gowri et al., 2010).
- Waterproof-breathable with enhanced vapour transfer properties obtained by laminating with layers of electrospun nanowebs. The resulting materials have waterproof characteristics similar to materials laminated with Polytetrafluorethylene (PTFE, GoreTex) but also improved vapour permeability, making these materials suited for outdoor clothing (Ahn et al., 2010).

1.3.3.5 Ecological issues regarding waterproof-breathable and water repellent fabrics

A significant issue is the recycling and environmental impact of these materials. Some of the polymers or finishing agents are not biodegradable, certain chemicals used are toxic, and the manufacturing processes are energy-intensive. There are materials that have components of a common nature (like SYMPATEX that has PES substrate and hydrophilic polyester membrane) and can be recycled simultaneously. In most cases, the components have different chemical structures and their recycling requires separation and selection. Extracting, collecting and/or removing chemical substances from the macrostructure of the waterproof and water repellent materials represents a great challenge for the future.

1.4 Conclusions

In a wet environment, the basic requirement for garments is to keep the wearer dry by being waterproof and/or water repellent. Waterproof and water repellent materials have a large range of applications, being widely used for garment manufacturing in conventional garments for weather protection, uniforms and work wear, and clothing for sport and leisure.

In contact with water, water repellent materials form drops that can be easily removed from the fabric surface. A water repellent fabric is resistant to wetting by water droplets and to the spreading of water over its surface. The water repellency of a fabric prevents the water absorption into the macrostructure of the fabric, with good influence on garment weight and fabric breathability.

Waterproof materials for clothing must also ensure the wearer's comfort, presenting the capacity to transfer water vapour from the microclimate through the garment system. Waterproof-breathable materials coated or laminated with microporous polymers or hydrophilic membranes are commonly used for this basic function. The materials with a compact outer layer are used for technical applications, and home and outdoor textiles, because their design includes specific requirements such as mechanical strength, biodegradability, resistance to biological agents or low weight.

References

Ahn, H.W., Park, C.H., Chung, S.E., 2010. Waterproof and breathable properties of nanoweb applied clothing. Text. Res. J. 81 (14), 1438–1447.

Chinta, S.K., Satis, D., 2014. Studies in waterproof breathable textiles. Int. J. Rec. Dev. Eng. Technol. 3 (2), 16–20.

Colleoni, C., Guido, E., Migani, V., Rosace, G., 2015. Hydrophobic behavior of non-fluorinated sol-gel based cotton and polyester fabric coatings. J. Ind. Text. 44 (6), 815–834.

Gibson, P., 2008. Water-repellent treatment on military uniform fabrics: physiological and comfort implications. J. Ind. Text. 38 (1), 43–53.

Gowri, S.S., Almeida, L., Amorim, T., et al., 2010. Polymer nanocomposites for multifunctional finishing of textiles: a review. Text. Res. J. 80 (13), 1290–1306.

Grand View Research Inc., 2016. Waterproof Breathable Textiles Market Size, WBT Industry Report 2024, 250 pp., Report ID: 978-1-68038-316-4.

Gugliuzza, A., Drioli, E., 2013. A review on membrane engineering for innovation in wearable fabrics and protective textiles. J. Membr. Sci. 446, 350–375.

Holmes, D., 2000. Performance characteristics of waterproof breathable fabrics. J. Ind. Text. 29 (4), 306–316.

Houtman, R., 2015. Materials used for architectural fabric structures. In: Llorens, J. (Ed.), Fabric Structures in Architecture. first ed. Woodhead Publishing Series in Textiles. ISBN: 9781782422402, pp. 101–121.

Indu Shekar, R., Kasturiya, N., Raj, H., Mathur, G.N., 2001. Studies on effect of water repellent treatment on flame retardant properties of fabric. J. Ind. Text. 30 (3), 222–254.

Kale, K.H., Palaskar, S., 2010. Atmospheric pressure plasma polymerization of hexamethyldisiloxane for imparting water repellency to cotton fabric. Text. Res. J. 81 (6), 608–620.

Kaynak, H.K., Babaarslan, O., 2012. Polyester microfilament woven fabrics. In: Jeon, H.-Y. (Ed.), Woven Fabrics. InTech Publisher, ISBN: 978-953-51-0607-4, pp. 155–178.

Kuhr, M., Aibibu, D., Cherif, C., 2016. Targeted partial finishing of barrier textiles with microparticles, and their effects on barrier properties and comfort. J. Ind. Text. 45 (5), 853–878.

Loghin, C., 2003. Sudarea materialelor textile (Welding of Textile Materials). Performantica Publishing House, Iaşi, ISBN: 973-7994-35-3.

Loghin, C., Ciobanu, L. (Eds.), 2008. Îmbrăcăminte funcţională—Modelarea şi simularea funcţiilor de protecţie. PIM Publishing House, Iasi, ISBN: 978-606-520-128-6, pp. 45–60.

Loghin, C., Nicolaiov, P., Ionescu, I., Hoblea, Z., 2009. Functional design of equipments for individual protection. In: Proceedings of International Conference Management of Technological Changes. 2, ISBN: 978-960-89832-8-1, pp. 693–696.

Loghin, C., 1998. Cercetări privind aplicarea procedeelor neconvenţionale de asamblare la realizarea subansamblelor şi produselor vestimentare (Research on the application of unconventional assembly processes for garments producing). Ph.D. Thesis, "Gheorghe Asachi" Technical University of Iasi, Doctoral School.

Lynch, A., Strauss, M.D. (Eds.), 2015. Ethnic Dress in the United Stated: A Cultural Encyclopedia. Rowman & Littlefield Publishing Group Inc, USA, ISBN: 978-0-7591-2148-5.

Mukhopadhyay, A., Midha, V.K., 2008a. A review on designing the waterproof breathable fabrics. Part I. Fundamental principles and designing aspects of breathable fabrics. J. Ind. Text. 37 (3), 225–262.

Mukhopadhyay, A., Midha, V.K., 2008b. A review on designing the waterproof breathable fabrics. Part II. Construction and suitability of breathable fabrics for different uses. J. Ind. Text. 38 (1), 17–40.

Park, S., Kim, J., Park, C.H., 2016. Influence of micro and nano-scale roughness on hydrophobicity of a plasma-treated woven fabric. Text. Res. J. (00), 1–15. https://doi.org/10.1177/0040517515627169.

Purane, S.V., Panigrahi, N.R., 2007. Microfibres, microfilaments & their applications. AUTEX Res. J. 7 (3), 148–158.

Sen, A.K., 2008. Coated Textiles: Principles and Applications. CRC Press, Taylor & Francis, ISBN: 9781420053456. pp. 175–200.

Schuyten, H.A., David Reid, J., Weaver, J.W., 1948. Imparting water-repellency to textiles by chemical methods—a review of the literature. Text. Res. J. 18 (7), 396–398.

Shephard, A., 2012. Waterproof dress: patents as evidence of design and function from 1880 through 1895. Cloth. Text. Res. J. 30 (3), 183–199.

Sherif, F., Roedel, H., 2011. Technical textiles as a new route to enhance orthopedic casts' properties. Int. J. Cloth. Sci. Technol. 23 (1), 25–33. https://doi.org/10.1108/09556221111096714.

Singha, K., 2012. A review on coating & lamination in textiles: processes and applications. Am. J. Polym. Sci. 2 (3), 39–49. https://doi.org/10.5923/j.ajps.20120203.04.

Shyr, T.W., Lien, C.H., Lin, A.J., 2011. Coexisting antistatic and water-repellent properties of polyester fabric. Text. Res. J. 81 (3), 254–263.

Uludag, M.O., Ozcan, G., Unal, H., 2011. Disposable hydrophilic antimicrobial laminated non-woven bed sheet. Int. J. Cloth. Sci. Technol. 23 (4), 222–231.

Varnaite-Zuravliova, S., Sankauskaite, A., Stygiene, L., et al., 2016. The investigation of barrier and comfort properties of multifunctional coated conductive knitted fabrics. J. Ind. Text. 45 (4), 585–610.

Yaras, A., Er, E., et al., 2016. Cellulosic tent fabric coated with boron nitride nanosheets. J. Ind. Text. 45 (6), 1689–1700.

Zerdzicki, K., 2015. Durability evaluation of textile hanging roofs materials. Mechanical engineering [physics.class-ph]. Ph.D. Thesis, Universite d'Orleans.

Zimmermann, J., Seeger, S., Reifler, F.A., 2009. Water shedding angle: a new technique to evaluate the water-repellent properties of superhydrophobic surfaces. Text. Res. J. 79 (17), 1565–1570.

Further reading

Global Market Insights, 2012. Waterproof breathable textiles market, https://www.gminsights.com/pressrelease/waterproof-breathable-textiles-wbt-market.

Hes, L., Loghin, C., 2009. Heat, moisture and air transfer properties of selected woven fabrics in wet state. J. Fiber Bioeng. Inform. 2 (3), 141–149.

Hoblea, Z., 1999. Structuri textile—Structura şi proiectarea îmbrăcămintei (Textile Structures—Garment Structure and Design). Gh. Asachi Publishing House, Iasi, ISBN: 973-99209-4-2, pp. 50–60.

Lee, K., Cho, G., 2014. The optimum coating condition by response surface methodology for maximizing vapor-permeable water resistance and minimizing frictional sound of combat uniform fabric. Text. Res. J. 84 (7), 684–693.

Loghin, C., 2013. Some aspects regarding the radio frequency welding of textile composites, bulletin of the polytechnic institute. Mech. Eng. 59 (1), 125–133.

Loghin, C., Ursache, M., Mureşan, R., Mureşan, A., 2010. Surface treatments applied to textile materials and implications on their behaviour in wet conditions. Ind. Textilă 61 (6), 284–290.

Development of waterproof breathable coatings and laminates

2

Hikmet Ziya Özek
University of Namik Kemal, Tekirdağ, Turkey

2.1 Introduction

A waterproof and breathable fabric incorporates two distinct functions of 'waterproofness' and 'breathability'. It should basically provide protection from the rain, wind and cold but also maintain a comfortable microclimate just below the fabric layer. The idea of a waterproof fabric is not new; in very old times, people were also in need of such fabrics (Mierzinski, 1903). Waterproof clothes and covering were needed for outdoor endeavours of all kinds from farming to sailing, for riding and for the military, as well as for various sports.

The term 'waterproof' literally means something that is impervious to water. A waterproof material is expected to fully protect from the weather conditions like rain, sleet and wind. Waterproof fabrics should entirely prevent the penetration and absorption of liquid water, in contrast to water repellent or rain resistant fabric, which only retards the penetration of water into the fabric's structure. The first record of the word 'waterproof' goes back to the 18th century (Etymology Dictionary, n.d.). It was, therefore, universally used in earlier publications and marketing together with the term waterproofing. In comparison with rainproof or showerproof, waterproof is more widely used and refers to a rather general case of being unaffected by water. The term 'water repellent' is a relative term because it is taken to be the relative degree of resistance displayed by a fabric to surface wetting, water penetration, water absorption or any combination of these properties. The differences between them are tabulated in Table 2.1.

Traditionally, fabric can be made waterproof by coating or binding with an uninterrupted layer of impervious flexible material. This fabric, however, does not necessarily provide the function of breathability. The most widely used method of producing a waterproof fabric is by coating the fabric with a solid polymeric coating such as neoprene, polyurethane or polyvinyl chloride (Mierzinski, 1903; Lister, 1963). Since such coatings are generally not porous, they form a continuous solid barrier to liquid water and other liquids, such as oils. On the other hand, solid nonporous continuous layers are impermeable to both the passage of air and of water vapour. Such coated fabrics exhibit waterproof quality but not water vapour permeability. In other words, they are not 'breathable' (Lomax, 1989; Holme, 2003).

In general, the term 'breathability' implies that the material is actively ventilated. In the case of textiles, a breathable fabric should passively allow water vapour to

Waterproof and Water Repellent Textiles and Clothing. https://doi.org/10.1016/B978-0-08-101212-3.00002-2

Table 2.1 **Comparisons of certain features of water repellent, waterproof and waterproof and breathable fabrics**

Features	Water repellent fabric	Waterproof fabric (nonbreathable)	Waterproof and breathable fabric
Pores	Relatively open	Filled	Partly filled
Resistance to water droplets	Resistant to wetting	Highly resistant to wetting	Highly resistant to wetting
Resistance to water penetration	Permits water passage under external hydrostatic pressure	Highly resistant even under external hydrostatic pressure	Highly resistant even under external hydrostatic pressure
Air permeability	Usually high	Zero	High
Water vapour permeability	Medium to high	Zero	Sufficient to high
Comfort	Sufficient to high	Very low	Medium to high
Cost	Low	Low to medium	High
Terminological class	Permeable	Impermeable	Semipermeable

Adapted by Holme, I., 2003. Water repellency and waterproofing. In:Heywood, D. (Ed.), Textile Finishing. Society of Dyers and Colourists, West Yorkshire, pp.137–213; Rowen, J.W., Gagliardi, D., 1947, Properties of water repellent fabrics, Research Paper RP1762, Part of the Journal of Research of the National Bureau of Standards, 38, January, 103–117.

diffuse through itself, but at the same time, the fabric should restrict the entry of liquid water. This particular function is not relevant to the waterproof character. A fabric can be waterproof but not breathable because of the nature of the waterproofing treatment. Such fabrics are referred to as nonbreathable waterproof fabrics and do not allow any internal moisture vapour to escape, which makes them a poor choice for garments used in activities like skiing and ice climbing. Heavy-duty, PVC-coated rainsuits for commercial fishermen and dock workers are a good example of this kind of fabric.

In addition to some low-activity apparel, for many nonapparel end uses such as technical textiles, industrial fabrics, and outdoor awnings, this lack of breathability presents no problem. Nevertheless, during various levels of physical activity, the human body provides cooling partly by producing insensible perspiration. If the water vapour generated by sweating cannot escape to the surrounding atmosphere, the relative humidity of the microclimate inside the clothing increases. As a result, corresponding thermal conductivity of the insulating air increases making the clothing uncomfortable, possibly unbearable in certain cases. Breathability of garments and fabrics is, therefore, crucial with regard to garment comfort and, more importantly, to maintaining a steady body temperature.

Major differences between water repellent, waterproof and waterproof and breathable fabrics are given in Table 2.1 (Holme, 2003; Rowen and Gagliardi, 1947). The striking difference between the nonbreathable and breathable waterproof fabric is water vapour permeability and hence, increased comfort. It should also be noted that with reference to this characteristic, water repellent fabrics are also superior to non-breathable waterproof fabrics. Water repellent finishes are often incorporated into the outer fabric surface to reduce ingress of liquid into interyarn spaces, which adversely reduces breathability. On the other hand, the absence of water repellent finishes is expected to result in soiling or staining, which will, in turn, require the fabric to be laundered more often. Therefore, both the durability and performance efficiency of such finishes are important.

2.2 History of waterproof and breathable fabrics

Waterproof clothes and coverings were needed for outdoor work and activities of all kinds, from hunting and farming to sailing and sports, as well as for shelter. The materials with which indigenous people kept themselves warm and dry are the early versions of waterproof textiles. These may even be some of the earliest functional textiles capable of stopping water from passing through the fabric. It is worth remembering that the need for breathable fabrics arose from the vulnerable impermeability of non-breathable waterproof fabrics.

The utilization of waterproof fabrics is believed to have begun in ancient civilizations with natural textiles, including silk and wool, and continued with cotton and linen in the 19th and 20th centuries. Oiled silk, being strong, waterproof, windproof and extremely light, was one of the first high-performance fabrics. It was first used in umbrellas by the Chinese over 1000 years ago, and vegetable oil was used on silk through the 19th century (Kovacevic et al., 2010; Parsons and Rose, 2011). The application of coated textiles for shelters, covers, protective clothing, liquid containers, etc., dates back to antiquity. Historically, the earliest recorded use of a coated textile was by the natives of Central and South America, who applied latex to a fabric to render it waterproof. The Aleut tribe of Native American Indians (living in the Aleutian Islands between Alaska and Russian Siberia) used dried seal or whale intestines and sealed the seams with animal-derived glues to make a poncho-like garment called 'kamleika' (Parsons and Rose, 2011). It served as a totally waterproof jacket for hunting and was made of a natural membrane. This may be considered as one of the early uses of membranes for waterproofing. Semi-permeable natural membranes like stomachs and intestines were used in clothing by indigenous people through the beginning of the 20th century.

Many of these early solutions came through trial and error, using materials close at hand. As in the case of sailors who treated heavy-duty sailcloth with linseed oil and a mix of other waxes (and sometimes other, more dubious additives) to make weather-proof capes in the 15th century. These were the origin of so-called 'oil skins'. Manually oiling sailcloth (first linen and later cotton) with linseed oil from flax seeds evolved to become an industrial process by the end of the 18th century in Scotland. This was the origin of the processes used for waxed clothing in the 19th century. Over time, the linseed oil methods were replaced by nonsticky paraffin waxes and (tightly) woven fabrics. These were the first procedures to apply coats of several agents onto the textile substrate, and the resulting textiles can be considered the predecessors of multilayered materials.

Early attempts at waterproofing fabrics sometimes involved rubber; however, these were not particularly successful because clothes treated with rubber were not easy to wear. If the weather was hot, the clothing became supple and tacky, and it was hard and inflexible when it was cold. This problem was solved in the early 19th century by Charles Macintosh, a Scottish chemist and chemical manufacturer. Through experiments, Macintosh discovered a better way to use rubber in clothing. He managed to dissolve rubber in naphtha, making a liquid. Macintosh then brushed this liquid onto fabric, making the fabric waterproof. In 1823, Macintosh patented his process for making waterproof fabric. This process involved sandwiching a layer of moulded rubber between two layers of fabric treated with the rubber-naphtha liquid. It took some time to develop the industrial process for spreading the rubber-naphtha mixture onto the cloth. The patented waterproof fabric was produced in factories around Manchester in 1824. The first customer was the British military, and Macintosh's discovery led to other innovative uses of rubber, including tires (Schoeffler and Gale, 1973; Levitt, 1986; Shephard, 2012).

The process for vulcanizing rubber was developed by Charles Goodyear in Philadelphia, Pennsylvania, USA, in 1839. Rubber was made more elastic and more easily meltable by heating with the addition of sulphur through the vulcanization process. The extensive utilization of rubber thanks to vulcanization was regarded as one of the major achievements of the 19th century. In 1843, Thomas Hancock took the waterproof fabric invented by Charles Macintosh and made it better using vulcanized rubber.

In 1903, one of the earliest publications (Mierzinski, 1903) focusing on waterproof fabrics presented a definition of waterproof fabric as fabrics that are protected from the destructive effect of dew, rain, snow, etc., by covering over the fibres or the filling-up of the interstices between them with a substance insoluble in water. It also emphasized that objects covered by these fabrics were also expected to be protected from these atmospheric influences. This was probably one of the first specifications for durability of waterproofing. Waterproofing trials were reportedly done with fats, oils, varnishes, pigments, gutta-percha, India rubber and other media. Various alternative methods were also introduced and tried for the treatment of woven fabrics by chemical means, so as to precipitate insoluble oxides in the fibre, or to impregnate the fibre with organic substances such as paraffin, wax, etc., to make them impervious to water. One hundred and five British patents on waterproofing methods were taken between 1877 and 1902, proving both the importance of waterproofing and the existence of various efforts for developing such fabrics in those years.

Shephard (2009, 2012) also confirms these extensive efforts in his academic work and publication. Research revealed the register of eighty patents in the United States from 1880 to 1895 that incorporated waterproofing compounds or techniques. Based on the waterproofing technique used, rubber was the most common waterproofing technique reported, with 30 patents; oil or paraffin finishes were second, and chemical compounds were third. There were also eight patents combining multiple techniques to achieve waterproof capabilities.

The capability of all these methods to produce nonbreathable waterproof fabrics was limited. It is interesting to note that Mierzinski (1903) remarked in his early book that absolute impermeability could never be attained.

Chemically treated fabrics gradually began to predominate by the early 20th century. For World War I, Thomas Burberry created the all-weather trench coat (Parsons and Rose, 2011; Schoeffler and Gale, 1973). The coat was made of a yarn-dyed fine twill cotton gabardine, chemically processed to repel rain. Despite the fact that these trench coats were first made for soldiers, after the war, their popularity quickly spread. They were also much cooler than those made of Macintosh's fabric.

A waterproof fabric acquires the quality of being impervious to water, even under pressure, by means of various treatments. Waterproof fabrics and products were warmly welcomed in the beginning, until the public realized they were not sufficiently comfortable or pleasurable to use. Staying dry is not just about preventing water from getting in, it's also about letting sweat out. Being 'waterproof' means that moisture does not pass through the fabric; water doesn't come in, and sweat doesn't go out. The solution to this particular problem led to the introduction of waterproof and breathable fabrics.

More than any other country, the United Kingdom has been the origin of these innovations, probably because of its significant textile industry as well as its climate. Ventile clothing resulted from British statesman Winston Churchill's demand for a material to protect British military pilots during combat conditions. After many trials, scientists at the British Cotton Industry Research Association (Shirley Institute) in Manchester, England, developed Ventile (Parsons and Rose, 2011; Schoeffler and Gale, 1973). The original Ventile immersion suits proved impenetrable to arctic winds and icy water, extending survival time tenfold and saving 80% of ejected pilots. This fabric was made of cotton due to the shortage of flax during the World War II. The fabric Ventile was not impermeable (like rubber- or plastic-coated cloths), so it was breathable, making it ideal for a wide range of active pursuits. It may be considered as the first application of waterproof, breathable fabrics.

The very first commercial introduction of polymer-based waterproof, windproof and breathable fabric (Gore-Tex) was introduced in 1976 by W.L. Gore. Development and commercialization of Gore-Tex took 8 years after Dr. R.W. Gore discovered how to turn PTFE into a microporous material in 1969. Having discovered that the polymer could be 'expanded', he attempted to stretch heated rods of PTFE by about 10%. As it turned out, the right conditions for stretching PTFE were the opposite of what Dr. Gore believed, and instead of slowly stretching the heated material, a sudden, accelerating yank unexpectedly caused it to stretch by about 800% (Chemical Heritage Foundation, 2015). Eventually, he achieved the transformation of solid PTFE into a microporous structure that incorporated about 70% air (Chemical Heritage Foundation, 2015; Gore, n.d.). Expanded PTFE was initially used to produce the waterproof and breathable membrane used in Gore-Tex. Gore-Tex was a byproduct of the 1920s and 1930s development of synthetic polymers as strategic alternatives to rubber and silk for specialized clothing. It was claimed that the thin film of expanded (ePTFE) polymer contained 1.4 billion tiny holes per square centimetre (Holmes, 2000). In its earliest form, this structure was 'poisoned/contaminated' by sweat and detergents, thus reducing the pressure at which water could pass through it. This led to the introduction of second generation, 'hybrid' Gore-Tex with a hydrophilic 'pu' coating on the body side of the membrane. The original product still finds application today as 'Windstopper'. The discovery of ePTFE is considered a revolutionary step for textiles and relevant industries and has led to new inventions. New emerging technologies are just that, 'emerging', and need time for successive generations to yield improvements, modifications and ever-crucial cost decreases, to appear and perform in commercial products.

2.3 Basics of waterproof breathable fabrics for comfort

Fabrics are highly porous materials, as they are composed of fibres and entrapped air. The porosity of most fabrics ranges from 50% to 95%, depending on the fibre fineness, the amount of twist in the yarns, the yarn count and the tightness of the fabric (Morris, 1953). The term 'breathable fabric' implies that it is capable of allowing moisture transfer through its cross-section. Naturally, all textile fabrics are expected to be capable of breathing, unless they are specially treated in some way. A typical treatment is the addition of a waterproof layer/coating which functions to stop water (liquid) from coming in or going out. In the case of developing waterproof and breathable fabrics, the waterproofing treatment should not cause elimination of a fabric's breathing ability. Hence, the characteristics of a waterproof shell should be adjusted to allow the transmission of water vapour to make it breathable as well as waterproof.

Garments made of both waterproof and breathable fabrics provide the wearer with a greater level of comfort in many adverse situations. While the mild discomfort caused by a modest accumulation of condensation between the fabric and body skin can be tolerated somewhat, the consequences of wearing relatively impermeable inner or outer clothing may be more serious in extreme conditions. Thus, comfort is generally an important consideration for the consumer in purchasing performance or protective apparel. The aim of waterproofness and breathability for a clothing fabric application is illustrated in Fig. 2.1. Such fabric stops water droplets from the environment from penetrating into the fabric, while allowing the water vapour resulting from body

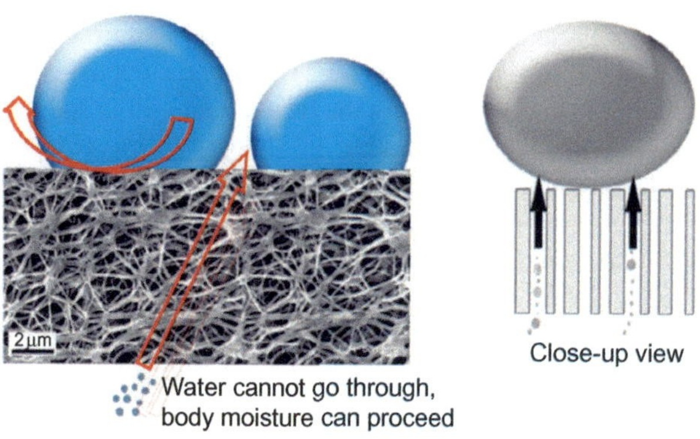

Fig. 2.1 The functions of 'waterproof' and 'breathable' character in clothing fabrics and a schematic close-up view. (The figures are informative and not based on real scale.)

moisture to go through the fabric. Currently, the usage of waterproof breathable fabrics is enormous, and many alternative techniques have been developed to produce super-hydrophilic or even smart versions of waterproof breathable textiles. Waterproof garments which are also breathable were previously distinguished by significantly high cost. However, such items are generally more affordable today than they were in the past (Mukhopadhyay and Midha, 2008).

The inescapable demand for breathable fabric in garments is due to human physiology. The body attempts to maintain a constant core temperature through a balance of heat loss and heat gain. Body heat is normally gained through activities such as exercise and shivering, and also with the application of external heat sources such as heat packs. On the other hand, the body loses heat by conduction, convection, radiation or evaporation. The latter is carried out by the sweat glands in response to external factors such as ambient temperature and humidity. Clothing is used outside the skin to extend the body's range of thermoregulatory control by imparting a barrier to heat exchange. There are two forms of perspiration: insensible perspiration and sensible perspiration (Arens and Zhang, 2006). Insensible perspiration occurs from both the skin (trans-epithelial) and respiratory tract. It is not under regulatory control and accounts for the main source of daily heat loss from the body. Sensible perspiration is defined as the evaporation of moisture through the outer layers of skin, or from the skin's surface when wetted by sweat (perspiration) or some other external agency. During physical activity, the body facilitates some cooling action by sensible perspiration.

If the water vapour generated by the body cannot be released to the surrounding atmosphere, the relative humidity of the microclimate inside the clothing increases causing a corresponding increased thermal conductivity of the insulating air, and the clothing becomes uncomfortable. The normal body core temperature is 37°C, while skin temperature varies between 33°C and 35°C, depending on conditions. If the core temperature goes beyond critical limits of about 34°C and 42°C, then disastrous consequences become unavoidable. In cases where perspiration cannot evaporate while liquid sweat (perspiration) is produced, the body is prevented from cooling at the same rate as heat is generated. This may happen during physical activity, and hyperthermia might occur because the body core temperature increases (Vander et al., 1998). On the other hand, if heat loss exceeds the ability of the body to generate heat, body temperature decreases below normal levels. A decrease in body temperature to 35°C or below is likely to stimulate hypothermia. This usually results from prolonged exposure to cold environments.

It is therefore very important for outdoor garments to allow both the transmission of moisture vapour by diffusion (so as to facilitate evaporative cooling) and to provide sufficient insulation for hindering heat loss. The heat energy produced during various activities and the perspiration required to provide adequate body temperature control have been studied and published (Basavanthappa, 2008; Holmes, 2000; Sen, 2008). Table 2.2 shows the perspiration rates for activities ranging from sleeping to maximum work rate.

For a person with no activity, energy expended is 50–60 W/m^2 per hour. The work rate of a person may increase to 100 W while engaged in normal routine indoor

Table 2.2 Heat energy produced by various activities and corresponding perspiration rates

Activity	Work rate (W/m^2 h)	Perspiration rate (mL/m^2 h)	Daily perspiration rate (g/m^2 per day)
Sleeping	40–60	20.00–99.08	450–2280
Sitting	95–100	165.14	3750–3800
Gentle walking	185–200	330.28	7550–7600
Active walking with no pack	285–300	499.77	11,480–11,500
Active walking with light pack	385–400	660.56	15,180–15,200
Active walking with heavy pack	485–500	825.70	18,900–19,000
Mountain walking with heavy pack	650–810	986.50–1321.13	22,700–30,400
Maximum work rate	1100–1200	1664,45–1981,70	38,300–45,600

activity. The metabolic heat generated is readily dissipated through the clothing as sweat. At rest, a body will produce about 60–90 mL of water vapour per hour at ambient conditions. Moderate exertion like gentle walking will increase that amount to about 330–350 mL per hour. During sporting activity, for example, tennis or cycling, metabolic heat increases six times, and the perspiration rate reaches 840 mL (Chinta and Gujar, 2013). When the body evaporates 1 kg of sweat from skin into vapour, 2538 kJ (604 kilocalories) of energy is lost. In other words, evaporation of 1 g/min of sweat is equivalent to 42 W. Evaporation of 1 L of sweat per hour means heat loss of 700 W (Desai, 2016). Consequently, the breathability of a garment is really important for allowing moisture due to sweating to pass through the fabric. During sweating, the human body humidity is more or less absorbed by the clothing. If the humidity remains in the fabric and is not transported to the surface for evaporation, cooling cannot occur. The body warms up, and even more sweat is produced. In 1991, it was reported that modern breathable waterproof fabrics were capable of transmitting more than 5000 g/m^2 per day. In 1998, the transmission rate reached 10,000 g/m^2 per day, and recently, levels over 30,000 g/m^2 per day are often claimed (Lomax, 1991; Holmes, 2000; Sierra Trading Post, n.d.). However, various test methods use different size fabrics under differing temperature or humidity conditions, and therefore, can give vastly differing values (Williams, 1997).

From the thermal comfort point of view, moisture transmission through textile materials in both liquid and vapour forms are equally important.

During the diffusion of moisture vapour through clothing textiles materials, resistance to moisture diffusion varies in different layers. In simple modeling of this phenomenon, the first layer next to the skin is evaporating fluid media which remains full of saturated water vapour, then a conventional fabric including confined air segments and additional functional layer or layers attached to the fabric. The boundary air layer in proximity of the outer face of the garment is surrounded by ambient air. The phenomenon of water vapour transfer through three different situations with reference to clothing application is given in Fig. 2.2. As seen, in regard of the transmission of water vapour, the addition of a waterproof and breathable layer makes no difference in contrast to waterproof layer. The ability of breathing of the breathable layer allows the water vapour pass through the clothing and surrounding layers into the ambient air. On the other hand, water droplets cannot go through waterproof layers while easily passes through the fabric with no extra layer.

It is also worth mentioning that the function of breathability only matters in the case of waterproof textiles since most textile structures are generally permeable due to the nature of the textile material and construction. With reference to waterproof fabrics, this nature is removed by the addition of a waterproof layer which prevents the water molecules passing through as well as limiting air ventilation. Breathable fabrics passively allow water vapour to diffuse through them while preventing the penetration of liquid water because of this waterproof character. To meet upmarket demands, waterproof breathable materials are expected to have a certain level of waterproofness. In general, breathability is meaningless without a high standard of waterproofness and initial hydrostatic head pressure level of 500 cm of water (50 kPa) for high quality products to 130 cm of water (13 kPa) for lower grade products (Mukhopadhyay and Midha, 2008; Fung, 2002). A breathable layer or cover facilitates minimizing discomfort or corrosion by removing moisture from beneath the layer. On the other hand, it should prevent penetration by liquid phase water, while allowing water vapour to diffuse outward, as shown in Fig. 2.3. This particular response is enabled by the breathing mechanism of the waterproof and breathable segment. Alternative solutions

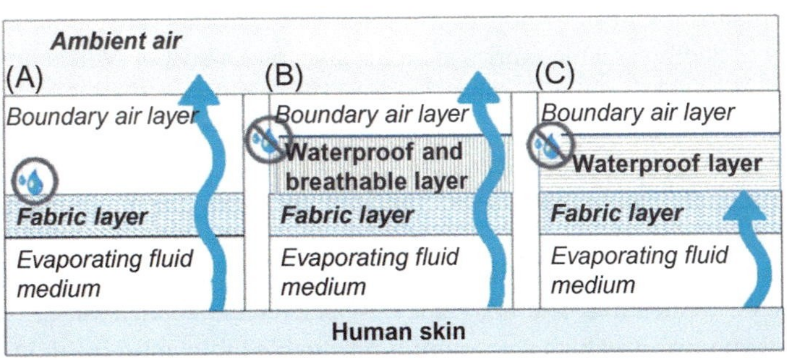

Fig. 2.2 Comparison of the schematic water vapour transfer on clothing application of conventional textile with no extra layers (A) and with extra segments of waterproof and breathable (B) and only waterproof layers (C).

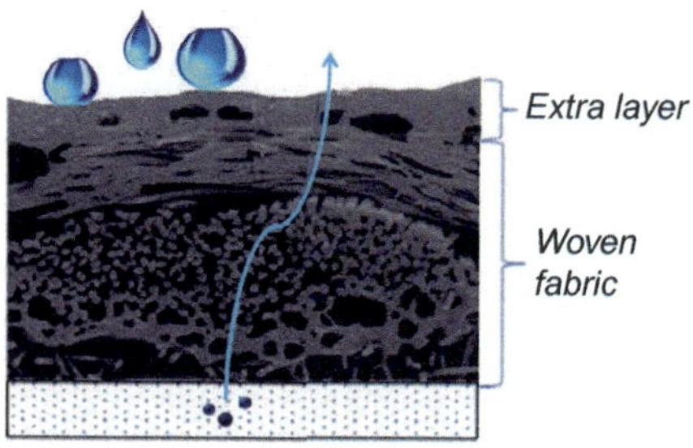

Fig. 2.3 The principle of a waterproof breathable fabric. (Water drops cannot go through the extra layer, while moisture vapour can pass through both layers.)

exist for this mechanism. The use of a microporous structure for the segment is one early solution. The pores of the breathable layer are roughly 20,000 times smaller than a drop of liquid water, but 700 times larger than a water molecule, the approximate size of which is measured as 2.75 Å. Thus, they are too small to allow liquid water to pass through, but large enough to allow the passage of molecules of water vapour, which are fairly separated.

The volume of waterproof breathable fabric trade worldwide is also increasing every year. The demand for the development of a wide variety of waterproof breathable fabrics drives researchers to explore new techniques to impart waterproofness and breathability in addition to traditional waterproof breathable fabrics on the market.

The usage areas of waterproof breathable fabric trade are quite extensive and gradually increasing parallel to development and emerging applications. A great scope for development also arises with the development of new materials and innovations in the field. Commercial focus on the research and development is targeted on cost effective production methods and better functional characteristics combined with environmentally friendly and sustainable solutions.

These types of fabrics are called technical textiles, an umbrella term describing high performance textiles developed in the last few decades. The application areas for waterproof and breathable fabrics are extensive, and a host of typical applications have been reviewed by various authors (Lomax, 1991; Holmes, 2000; Mukhopadhyay and Midha, 2008; Dhanabalan et al., 2013). A revised list of applications arranged on the basis of technical textile categories is given in Table 2.3. Although the range of applications appears to be very diversified, the sporttech group has far and away the largest share of products.

The global waterproof textiles market has been segmented on the basis of raw material, type, applications and geography. Whatever the application or type, there are certain requirements which a waterproof breathable fabric must fulfil. In fact, wetting, wicking and moisture vapour transmission (MVT) properties are critical aspects for assessing the functional and comfort performance of textile products.

Table 2.3 **Typical usage areas of waterproof and breathable fabrics**

Main category of technical textiles	Specific products/application examples
Clothtech	Interlinings Umbrella cloth Waterproof fasteners and sealings
Sporttech	Leisurewear: jackets, trousers, overtrousers, anoraks, raincoats, etc. Active sportswear garments Swimwear Rainwear, skiwear, Mountaineering garments and gaiters Hats, gloves Sport footwear (lining, panel or insert) Trekking and mountaineering shoes Camping boots (lining, panel and insert) Tents, sleeping bag covers Rucksack and backpack cloth
Protech	Special protective military clothing Heavy duty clothing, functional workwear Survival suits Clean room clothing Fireman garments, farmer clothing
Hometech	Outdoor fabrics Mattress cover and bedding Outdoor upholstery
Mobiltech	Car covers, car seating Cargo wraps in aircrafts Special curtains for ship
Medtech	Surgical garments Hospital bedware, wound dressing Special hygienic products
Packtech	Specialized tarpaulins, special packaging
Indutech	Specific filtration

2.4 Behaviour of waterproof breathable fabrics

2.4.1 *Fabric wetting*

A waterproof textile surface is required to resist wetting by applied liquid water. The surface is expected to be completely covered with a layer which is impermeable to water, water vapour and air. On the other hand, water repellent fabrics are often limited in applications because they may not perform well under hydrostatic pressure, such as the forces that a textile would experience during a rain shower. Both types of fabrics are expected to display a nonwettable character.

Wetting is a dynamic process and spontaneous wetting is the migration of a liquid over a solid surface towards thermodynamic equilibrium (Kissa, 1996). It is the displacement of a solid–air (vapour) interface with a solid–liquid interface. In the case of waterproof fabrics, this replacement is prevented. Wettability of a surface is usually determined by measuring the static contact angle of a liquid droplet contacting the surface. Where a drop of liquid is placed upon a solid surface and does not spread, the drop shape appears to be constant, and the angle between the surface and the liquid water meniscus near the line of contact, measured through the droplet, gives an indication of the wettability or nonwettability of the solid surface by the liquid (Kissa, 1996). The wetting of a fibre is defined as the condition resulting from its contact with a specific liquid under specified conditions. Wettability is the potential of a surface to interact with liquids having specified characteristics. Nonwettability describes a specific condition where the liquid is not allowed to spread over the fabric surface during interaction with liquid. A picture presented in Fig. 2.4A displays the contact angle with typical forces occurring during wetting process. The three-phase contact line is the line where three phases (liquid, solid and vapour) meet. Forces in the tangential direction at the three-phase contact line are defined by the well-known Young's

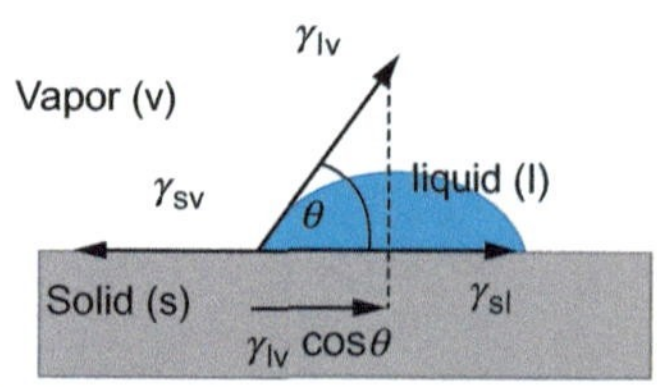

(A)

Level of wetting	Contact angle range	Energetic relationship	Graphic representation
Non wettability	$\theta = 180°$	$\gamma_{sl} - \gamma_{sv} > \gamma_{lv}$	
Low wettability	$90° \leq \theta \leq 180°$	$\gamma_{sv} - \gamma_{sl} < 0$	
High wettability	$0° \leq \theta \leq 90°$	$\gamma_{sv} - \gamma_{sl} > 0$	
Complete wetting	$\theta = 0°$	$\gamma_{sv} - \gamma_{sl} > \gamma_{lv}$	

(B)

Fig. 2.4 Interfacial forces acting at equilibrium on a liquid drop–solid system and variation of the parameters depending on the condition of wetting.

equation in spreading and wetting dynamics (Zisman, 1964). The equation connects three interfacial tensions, γ_{sl}, γ_{sv} and γ_{lv}, with the value of the equilibrium contact angle, θ, where γ_{sl}, γ_{sv} and γ_{lv} are solid–liquid, solid–vapour and liquid–vapour interfacial tensions, respectively:

$$\gamma_{sv} = \gamma_{sl} + \gamma_{lv} \cos\theta \tag{2.1}$$

$$\cos\theta = (\gamma_{sv} - \gamma_{sl})/\gamma_{lv} \tag{2.2}$$

Further work by Dupre, as described in an early study (Rowen and Gagliardi, 1947), defined the equation relating the free surface energies with the work of adhesion. The interfacial tension of liquid–vapour (or air) γ_{lv} is usually expressed as γ.

$$W_A = \gamma(1 + \cos\theta) \tag{2.3}$$

There are primary three processes involved when a textile surface is exposed to water: immersion, adhesion and spreading. A liquid drop placed on a solid (or another immiscible liquid) surface may start to spread to a greater or lesser extent. Its extent depends on the intermolecular forces interacting between the two phases. Quantitative measure of the spreading wetting is the work of spreading, W_s, which is also called the 'spreading coefficient', S_s, defined as follows:

$$W_s = S_s = \gamma_{sv} - (\gamma + \gamma_{sl}) \qquad 2.4$$

Its value can be positive or negative, depending on the surface free energy of the solid and the liquid involved in the wetting process.

$$W_s > 0 \ \text{ positive if } \ \gamma_{sv} > (\gamma + \gamma_{sl})$$
$$W_s < 0 \ \text{ negative if } \ \gamma_{sv} < (\gamma + \gamma_{sl})$$

In case of positive value, the liquid spreads spontaneously over the solid surface, even up to its monomolecular layer if the solid surface is sufficiently large. This is called 'the wetting process'. If the work of spreading is negative, the liquid drop will not spread but will remain on the surface and form a definite wetting contact angle. That results in occurrence of beads in small separate droplets since the water sticks to itself. This is the nonwetting process as it takes place in the case of waterproof or water repellent fabrics. Hydrophilic (water loving) materials have low contact angles, and as a result, the water bead spreads into a film. Hydrophobic (water hating) materials have a high contact angle, so the water sticks to itself, forming beads in small separate droplets. Variations in bead formation and other parameters for basic wetting conditions are presented in Fig. 2.4B.

Due to the complexity of contact angle phenomena, the experimentally observed contact angle might or might not be equal to Young's equilibrium contact angle, θ (Miller, 1977). Theoretically, the equilibrium contact angle of a completely nonwetting liquid on a perfectly smooth, homogeneous, impermeable and nondeformable surface should be 180 degrees, but in practice, none of these conditions are obtained, and indeed, the action of gravity can deform the shape of the drop. Thus, values less than 180 degrees are always observed (Miller, 1977; Holme, 2003).

In cases in which surface heterogeneity or physical roughness generates hysteresis, the actual microscopic variations of slope on the surface create barriers that pin the motion of the contact line and alter the macroscopic contact angles. Interpretation of such contact angle data in terms of Young's equation can be misleading because the equation fails to consider surface topography (Miller, 1977). This is the case for most textiles' fibres and for almost all textile fabrics with no treatment.

Textile fabrics are naturally macroporous structures. The sizes of the pores may vary from 100 μm and up. The treatment for waterproofness either decreases the pore diameters, making the pore 'nonwettable', or covers the surface with an extra layer with nonporous or microporous structure. Decreasing the pore diameters results in increasing the angle of contact which the surface of the water droplets makes with the fibre around the pores. Surface tension then prevents the water molecules from spreading and penetrating into the fabric. Various materials such as waxes or silicones may be applied for this treatment. Another technique is to integrate a nonporous or microporous outer shell or layer over the fabric surface, so that it would keep out even the smallest water droplets because of either no pores or pore size of the order of 1 μm.

The main parameters that determine a fabric's resistance to wetting are (Zisman, 1964; Kissa, 1984; Holme, 2003; Patnaik et al., 2006):

- The chemical nature of the fibre surfaces, that is, the presence of polar or nonpolar groups.
- The geometry and roughness of the fibre surfaces (e.g. longitudinal striations, fissures, crenulations, etc., and modified cross-sections that promote wicking).
- The nature of the capillary spacing in the fabric such as interfiber and interyarn capillary spaces.
- The existence of residual surfactants or impurities.
- The temperature of the liquid.

2.4.2 Moisture transfer

Breathability does not, as the term might imply, relate to an exchange of air or that the fabric is actively ventilated. Instead it defines the ability of a fabric to allow moisture vapour to pass through it. It should be remembered that a fabric may achieve this in different ways, and exchange of air is only one possible option.

2.4.2.1 Assessment of moisture transfer

The rate at which water vapour passes through a fabric is measured in terms of moisture vapour transport rate (MVTR). MVT or moisture vapour permeability (MVP) appears to be a more technical and explanative term than breathability, noting that the fabric transports water vapour from the body. No standard is defined for breathability, but MVP and MVT are both defined in test standards. According to EN 31092:1993, water vapour permeability is 'a characteristic of a textile material or composite depending on water vapour resistance and temperature. Water vapour permeability is expressed in grams per square meter hour pascal'. According to BS 3546:2001, water vapour permeability is the 'ability of a coated fabric to transmit water vapour above a specified level while maintaining a high degree of water penetration resistance'. Differing methods are expressed in different units such as %, mm still air, g/m^2 per day, Ret, etc. Many of these cannot be easily compared, though some are additive, so predictions about the effects of different layers can be made (Williams, 1997).

The water vapour permeability index is the ratio of thermal and water vapour resistances, in accordance with the definition in standard EN 31092:1993. Water vapour resistance, Ret, can be thought of as the 'opposite' of breathability and is, according to BS 3546:2001, the 'water vapour pressure difference between the two faces of a material divided by the resultant evaporative heat flux per unit area in the direction of the gradient'. Air permeability is intrinsically linked to breathability. Air permeability is 'the velocity of an air flow passing perpendicularly through a test specimen under specified conditions of test area, pressure drop, and time' (according to BS EN ISO 9237: 1995). All air-permeable fabrics are breathable to some extent, though not all breathable fabrics are air permeable.

It has already been proved that most issues of clothing comfort involve the mechanisms by which clothing materials influence heat and moisture transfer from the skin to the environment (Rees, 1971; Haghi, 2011). Heat transfer by convection, conduction, radiation and moisture transfer by vapour diffusion are the most important mechanisms in very cool or warm environments. Vapour transmission obviously plays an important role in garment comfort. More specifically, it concerns the diffusion of moisture vapour through the thickness of the fabric. It is mostly an important issue when used as a garment, but also in various special applications such as footwear, outdoor covers (like a tent) and protective clothing, wrapping or packaging, where high resistance to liquid water is required together with considerable permeability of water vapour.

Water may transfer through textile materials in vapour and in liquid form. Within the scope of this section, only transfer in the vapour form will be discussed. A considerable amount of work has been carried out on mechanisms of water vapour transfer through fibres (Fourt et al., 1957; Hong et al., 1988; Weiner, 1970; Watt and Kabir, 1975; Benisek et al., 1987; Ito and Muraoka, 1993), textile fabrics and textile systems (Peirce et al., 1945; Fourt and Harris, 1947; Whelan et al., 1955; Woodcock, 1962a,b; Crow, 1974; Saxena, 1999; Das et al., 2007; Huang and Chen, 2010; Das and Kothari, 2012; Vivekanandan and Sreenivasan, 2012), layered fabrics (Wang and Yasuda, 1991; Yasuda and Miyama, 1992; Yasuda et al., 1994; Rossi et al., 2004; Barauskas and Abraitiene, 2011; Huang, 2016) and clothing assemblies (Farnworth, 1986; Gibson, 1993; Wang et al., 2007; Guo et al., 2008).

However, for the most part, investigations were carried out only under steady state conditions, and limited work has been done with waterproof fabrics. Water vapour transfer is regarded as being a particularly important factor in the manufacture of waterproof breathable fabrics. These fabrics are used in clothing for outdoor occupations (e.g. farming and construction industries) and clothing for arduous pursuits (e.g. cycling and mountaineering). Such fabrics are also used in the offshore oil and fishing industries and uniforms for the armed services. This protective clothing is, in general, exposed to low temperatures, wind and rain, and occasionally to severe conditions like wind-driven rain.

2.4.2.2 Mechanism of moisture transfer

The behaviour of water vapour transmission in textiles varies in accordance with the construction and optional treatment types (Lomax, 1985; Ruckman, 1997a). Water vapour and liquid water are transmitted through elementary untreated textiles by the following mechanisms (Lomax, 1985; Das et al., 2007):

(1) Simple diffusion through the interyarn spaces. This process is controlled by the water vapour pressure gradient across the inner and outer faces of the fabric. The resistance to diffusion is governed by the fabric construction, that is, the size and concentration of interyarn spaces and the fabric thickness.

(2) Capillary transfer through fibre bundles. In this mechanism, liquid water is wicked through the yarns and desorbed or evaporated at the outer surface. The efficiency of yarn wicking depends on the surface tension, that is, wettability of the fibre surfaces, and on the size, volume and number of capillary spaces within the fibre bundle. The nature of these interfiber spaces is determined by the choice of yarn and fabric construction.

(3) Diffusion through individual fibres. This mechanism involves absorption of water vapour into the fibres at the inner surface of the fabric, diffusion through the fibre structure, and desorption at the outer surface. The ability of fibres to undergo water vapour diffusion is obviously enacted by the hydrophilic or hydrophobic nature of the fibre.

(4) Transmission of water vapour by forced convection. This is a mode of moisture transfer caused by convection which takes place as the air flows over a moisture layer. This is known as the forced convection method. The mass transfer in this process is controlled by the difference in moisture concentration between the ambient air and the moisture source.

The water vapour transfer rate of a fabric is essentially conducted by interyarn or interfiber spaces. The vapour diffuses through the air spaces between the fibrous materials. It is obvious that a relatively open fabric structure promotes the diffusion process. As was already discussed with reference to Fig. 2.2, during the diffusion of moisture vapour through textile materials, the resistance to the moisture vapour diffusion comes in different layers. These different layers which are positioned successively are (i) the evaporating fluid layer (which remains full of water saturated vapour), (ii) the confined air layer (between the skin and fabric), (iii) the boundary air layer and (iv) the ambient air layer. Moisture vapour resistance mainly depends on the air permeability of the fabric and represents its ability to transfer the perspiration coming out of human skin. The resistance provided by the fabric is lower than that of the external boundary layer and often much lower than the inner confined air layer between skin and fabric (Das and Alagirusam, 2010).

In the diffusion process, the vapour pressure gradient acts as a driving force in the transfer of moisture from one side of a textile layer to the other. The relation between the flux of the diffusing substance and the concentration gradient was first described by Fick (Morton and Hearle, 2008; Haghi, 2011).

$$F_A = -D_{AB} \cdot \frac{dC_A}{dx} \tag{2.5}$$

where, F_A is the rate of moisture flux or substance diffusing across A, a unit area per unit time, 'dC_A/dx' is the concentration gradient, and D_{AB} is the diffusion coefficient or mass diffusivity of one component A, diffusing through media, B. Water vapour in the porous media can move by molecular or Fickian diffusion if the pores are large enough. Water vapour can diffuse through a textile structure in two ways: simple diffusion through the air spaces between the fibres and yarns, and along the fibre itself (Fohr, 2002). In the case of diffusion along the fibre, water vapour diffuses from the inner surface of the fabric to the fibres' surface and then passes along the interior of the fibres and its surface, reaching the outer fabric surface.

The transfer of moisture vapour takes place from the wetter environment to the dryer environment until an equilibrium is reached. At a specific concentration gradient, the diffusion rate along the textile material depends on the porosity of the material and on the water vapour diffusivity of the fibre. The diffusion coefficient of water vapour through air is 0.239 cm^2/s, while it is around 10^{-7} cm^2/s through cotton fabric. The moisture diffusion through the air portion of the fabric is almost spontaneous, whereas it is limited through a fabric system by the rate at which moisture can diffuse into and out of the fibres, due to the lower moisture diffusivity of the textile material (Kothari, 2000; Das et al., 2007).

High-density woven fabrics and microporous polymers transmit water vapour according to Fick's law of diffusion. The diffusion constant does not alter with changes in water vapour concentration within the polymer or with changes in temperature; therefore, water vapour transfer through microporous coated fabrics depends on the vapour pressure gradient across the fabric (Gretton et al., 1996). This law can only be applied to diffusion under steady-state conditions (Finn et al., 2000).

In the case of hydrophilic fibrous assemblies, vapour diffusion does not obey Fick's law. Several researchers concentrating on the mechanism of water vapour transfer in waterproof breathable fabrics have reported that water vapour transfer in such fabrics does not obey Fick's law, but it is governed by a non-Fickian, anomalous diffusion (Osczevski and Dolhan, 1989; Gibson, 1993; Gretton et al., 1996; Ren and Ruckman, 2003). The non-Fickian, anomalous diffusion takes place in two stages. The first stage corresponds to a Fickian diffusion, but the second stage is much slower than the first, following an exponential relationship between the concentration gradient and the vapour flux (Das et al., 2007; Li and Luo, 2000).

Hydrophilic polymer coatings also do not obey Fick's law; on the contrary, they show non-Fickian behaviour (Gibson, 1999; Kannekens, 1994). In this case, vapour transmission under steady-state conditions is given by the following equation:

$$WVT = DS(p_1 - p_2)/l \tag{2.6}$$

where WVT is the water vapour transfer rate, $(p_1–p_2)$ is the partial pressure gradient between the two fabric faces, l is the thickness of polymer coating, D is the diffusion constant and S is the solubility coefficient (Gretton et al., 1996).

Since garments made of waterproof breathable fabrics are normally worn in wet conditions, and because the clinging of such garments decreases comfort for the wearer, some researchers also tried to investigate the effect of moisture and condensation on water vapour transfer in waterproof breathable fabrics. Keighley (1985) noted that the breathability of certain rainwear fabrics appeared to increase in wet conditions. Therefore, some microporous products are finished with a thin hydrophilic polyurethane top coat, which in itself has a relatively high water vapour permeability, increasing WVP efficiency as seen in Fig. 2.5 (Scott, 1995).

Ruckman (1997a,b,c) carried out a series of studies to observe the water vapour transfer rate in waterproof breathable fabrics under steady-state conditions, under windy conditions and under rainy and windy conditions. From the experimental work, the following results were obtained:

(1) Water vapour transfer in waterproof breathable fabrics decreased as rain temperature increased.
(2) Waterproof breathable fabrics did breathe under rainy conditions; however, the breathability of most of them ultimately ceased after long exposure to prolonged severe rainy conditions.
(3) More condensation was observed on all fabrics under rainy conditions than under dry conditions except for PTFE-laminated fabrics, which formed the least condensation.
(4) The water vapour transfer rate was reduced under wind-driven rainy conditions compared to that under rainy conditions for all fabrics due to the disturbance of both rain and condensation.

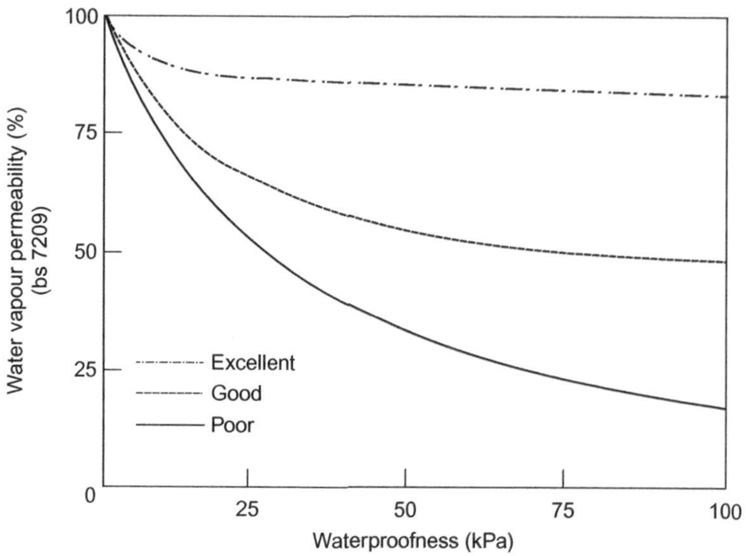

Fig. 2.5 Typical relationship between WVP versus waterproofness. Reproduced from Scott (1995).

It appears that the water vapour transfer depends very much on atmospheric conditions. It can generally be concluded that wind increases and rain decreases the water vapour transfer rate of a fabric, giving in descending order of water vapour transfer performance: windy, dry, wind-driven rainy and rainy.

2.4.3 Properties of waterproof breathable fabrics

The most important features of waterproof breathable fabrics are obviously the prime functions of waterproofness and breathability. There are various evaluation methods to measure these properties.

The quality of waterproofness: Hydrostatic head is the measurement of a material's waterproofness. It is measured in mm and represents the height of a 1 square inch wide of water column to make an opening in the material. Currently, the lowest grade of waterproof rating of 5000 mm is required to be considered as truly waterproof. On the other hand, the minimum level may be referred to as 1000 mm by some manufacturers. The current classifications accepted by many outdoor manufacturers are given in Table 2.4.

Acceptable MVTR: This value represents the rate at which moisture passes through a square metre (m^2) of the fabric from the inside to the outside in a 24-hour period and values vary depending on the test protocol employed. An MVTR rating of 10,000 or higher are required for more active pursuits.

Resistance to evaporative heat transfer (R_{et}) measures how breathable a garment is in units m^2 Pa/W, with lower being better for more aerobic activities. The method developed by the Hohenstein Institute using the Skin Model (DIN EN 31092/ISO 11092 and ASTM F 1868-02) differs from the other methods (Hohenstein Institute, 2014). The values of RET are widely used to compare various products. The values of resistance to water vapour transmission and expected performance are given in Table 2.5.

The other desirable properties of waterproof breathable fabrics are listed below:

Table 2.4 The classes of waterproofness level

Waterproof rating hydrostatic head (mm)	Resistance provided	Weather conditions
0–5000 mm	No resistance to some resistance to moisture	Light rain, dry snow, no pressure
6000–10,000 mm	Rainproof and waterproof under light pressure	Light rain, average snow, light pressure
11,000–15,000 mm	Rainproof and waterproof except under high pressure	Moderate rain, average snow, light pressure
16,000–20,000 mm	Rainproof and waterproof under high pressure	Heavy rain, wet snow, some pressure
20,000 mm over	Rainproof and waterproof under very high pressure	Heavy rain, wet snow, high pressure

Table 2.5 **The evaluation of the resistance to water vapour transmission 'R_{et}' values**

Range of R_{et} (m^2 Pa/W)	Expected performance
0–6	Very good or extremely breathable. Comfortable at higher activity
6–13	Good or very breathable. Comfortable at moderate activity rate
13–20	Satisfactory or breathable. Uncomfortable at high activity rate
20–30	Unsatisfactory or slightly breathable. Moderate comfort at low activity rate
30 and over	Unsatisfactory or not breathable. Uncomfortable and short tolerance time

- Optimum heat and moisture regulation (thermo-regularity effect).
- Absorption of surplus heat.
- Rapid moisture absorption and conveyance capacity.
- Rapid drying to prevent catching cold.
- Durability (against tear, peel, abrasion resistance).
- No/minimum water absorption of the layer of clothing just positioned to the skin.
- Dimensionally stable even when comes in contact to water.
- Easy care/launderability.
- Lightweight.
- Soft and pleasant touch.

2.5 Classification of waterproof breathable fabrics

Waterproof and breathable fabrics are a kind of functional fabric promoted in highly competitive markets. They are developed to maintain comfort of clothing or sustainment of certain substances or surrounding environments, in case of apparel and technical usage, respectively. The classification of waterproof and breathable (WB) fabrics is made through various research, as tabulated in Table 2.6. As seen in the table, the classification is fundamentally based on two criteria:

- The structure of the WB layer.
- The technique employed for developing WB fabric.

Classifications by both criteria are equally acceptable and correct, but, in this chapter, the classification is made in accordance with the technique employed for developing WB fabric. From the structure point of view, these fabrics are basically of three types. The earliest type consists of a tightly woven cotton fabric, suitably proofed, in which the fibres swell when wetted to further close up the fabric structure and resist water penetration. The second type, microporous, consists of a laminated or coated fabric in which the layer has very small gaps throughout, which allows water vapour to diffuse through and reduces liquid water penetration. The third kind of WB is composed of nonporous structure and has been developed as hydrophilic films in which water vapour is absorbed on one side of the film and re-evaporates from the other side whilst

Table 2.6 Classification of waterproof and breathable fabrics on the basis of structure and treatment technique with reference to earlier works

Source of work	Classification by structure				
	Type 1 High density wovens	Type 2 Microporous	Type 3 Solid polymer	Type 4 Bi-component	Other types
Lomax (1985)	Tightly woven	*Pol. coatings* *Laminated films*	*Coatings* *Laminated films*		
Mukhopadhyay and Midha (2008)	Closely woven	Microporous membranes and coatings	Hydrophilic membranes and coating	Combination of microporous and hydrophilic membranes and coating	Use of retroreflective microbeads Smart breathable fabrics Biomimetics
Sen (2008)	Closely woven	Microporous coatings and laminates	Hydrophilic coatings and films		
Dhanabalan et al. (2013)	Closely woven	Microporous membranes and coatings	Hydrophilic membranes and coating	Combination of microporous and hydrophilic membranes and coating	Smart breathable fabrics
Mukhopadhyay and Midha (2016)	Closely woven	Microporous membranes and coatings	Hydrophilic membranes and coating	Combination of microporous and hydrophilic membranes and coating	Use of retroreflective microbeads Smart breathable fabrics

Continued

Table 2.6 **Continued**

Source of work	Classification by treatment technique				
	Type 1 **High density wovens**	**Type 2** **Laminates**	**Type 3** **Coating**	**Type 4** **Combined techniques**	**Smart techniques**
Roey (1991)	High-density fabrics	Melt-blown or cast film • Microporous • Hydrophilic	Coating • Microporous • Hydrophilic		
Holmes (2000)	Densely woven	Membranes	Coating		Biomimetics
Holme (2003)	Densely woven	Membranes • Microporous • Hydrophilic	Coating • Microporous • Hydrophilic		Biomimetics
Wood (2016)	(Woven)	Laminates	Coatings		

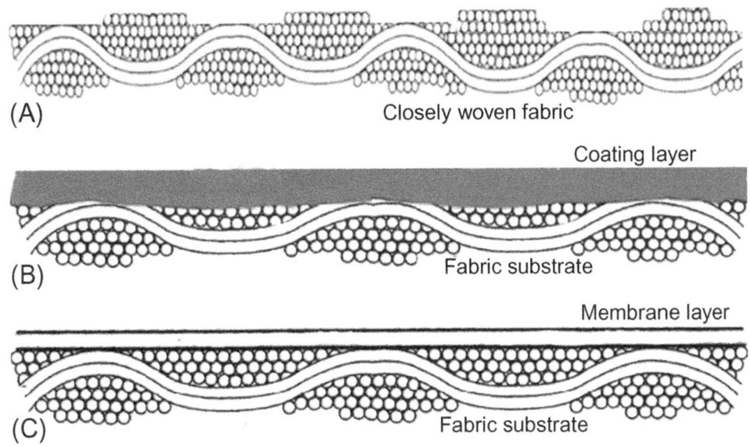

Fig. 2.6 Major types of waterproof breathable fabrics: (A) typical view of tightly woven fabric (pore size:10–3 μm), (B) coated woven fabric and (C) laminated woven fabric.

preventing liquid water penetration. The other types are a derivation or combination of these basic structures. A combination of the last two types can be called bi-component or fused structures.

This special group of functional fabrics can be produced by tight weaving, coating, laminating, smart techniques or combinations of these. The schematic views of all three techniques are given in Fig. 2.6. On the basis of treatment technique, and in consideration of earlier works listed in Table 2.6, WB fabrics can be divided into six groups as follows:

- *Tight weaving*: Densely woven fabrics
- *Coating*: Microporous and solid hydrophilic coating
- *Laminating*: Microporous and solid hydrophilic membranes
- *Combination of coating and laminates*: Bi-component structures (microporous and solid)
- *Smart techniques*: Smart breathable fabrics
- WB fabric based on biomimetics

2.6 Waterproof breathable fabric structures

2.6.1 High-density woven porous structures

The high-density porous structure is obtained by weaving specially designed dense yarns at particularly dense weave constructions. The very first type of this structure is the famous Ventile fabric by the Shirley Research Institute in England. The woven fabric is obtained with the minimum possible pores. When this fabric is inserted into water, the pore size is reduced due to the swelling of cellulosic fibres. Naturally, very high water pressure is required to penetrate such fabric. When the fabric is dry, the spacing (gap) between warp and weft yarn is larger, about 10 μm, and it can provide a highly permeable structure (Lomax, 1985). In case of rain or wetting of the fabric, the expansion of cotton fibre reduces the spacing between the yarns to 3–4 μm like an obturator mechanism. Combinations of water repellent finishing ensures that the fabric will not be exposed to further penetration. The choice of repellent treatment is critical, because it should still allow absorption of water by the cotton yarn to swell and constrict the interyarn pores.

2.6.2 Microporous structures

Microporous structures for a waterproof breathable layer can be obtained through both coating and lamination techniques such as coatings and membranes, respectively. These types of membranes and coatings are hydrophobic in nature. Microporous coatings may be produced by wet coagulation, thermo-coagulation or foam-coating methods, or by mechanical fibrillation (Lomax, 1989; Holme, 1986; Kramar, 1998). The dimensions of microholes occurring in the structure are smaller than raindrops but much larger than water vapour molecules, as seen in Fig. 2.1 (Holmes, 2000).

The most widely used polymers for microporous fabrics are PTFE (which offers an inert, extremely hydrophobic polymer with very open porous structures) and polyurethanes (which offer toughness, flexibility and wide variation in formulation, allowing tailor-made polymers). Acrylics, polyamino acids and polyolefins are also used. Microporous membranes are defined as having a narrow pore-size distribution, usually in the submicrometer range, although they can range from 0.1 to 10 μm. It is common for microporous films to have 1.2–1.4 billion pores per square centimetre in a film 10–50 μm thick. Microporous arrangement enables water vapour to pass through permanent pores of various sizes by diffusion, as seen in Fig. 2.7. It can be applied to a fabric substrate either by a coating of porous paste or by laminating a porous membrane film. The pore sizes of this microporous structure may vary from 0.02 to 50 μm depending on the manufacturing technique. The actual structure of porous membrane is seen in Fig. 2.8 with a surface and cross-section view.

The first commercial microporous layer in textiles was introduced by W.L. Gore. This solid ePTFE material consisted of a microporous structure that was about 70% air (Chemical Heritage Foundation, 2015). The expanded PTFE film is known to contain a network of micropores that is 82% by volume (Mukhopadhyay and Midha, 2008). It is claimed to have around 1.4 billion pores per square centimetre (9 billion pores per square inch) with sizes ranging from 0.1 to 50 μm. In comparison to water molecules, these pores are 20,000 times smaller, but 700 times larger than that of water vapour molecules (Gohlke and Tanner, 1976; Shekar et al., 2003). Since the film is mechanically weak, it is expected to be laminated to a textile fabric substrate.

As shown in Fig. 2.9A the expanded, amorphously locked, porous material (10) of this film comprises a large plurality of nodes (11) which are oriented perpendicularly to the direction in which the expansion was effected. These nodes, on the average about 50 μm in size and fairly irregular in shape, lie closely together and in many instances, appear to touch at points. A given node is connected to adjacent or nearby

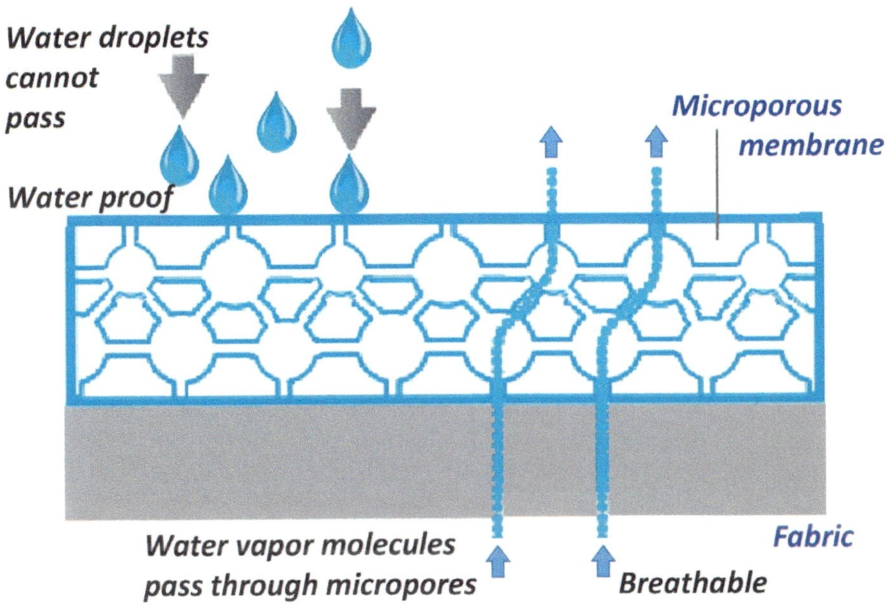

Fig. 2.7 Working principles of microporous structure in water vapour transfer.

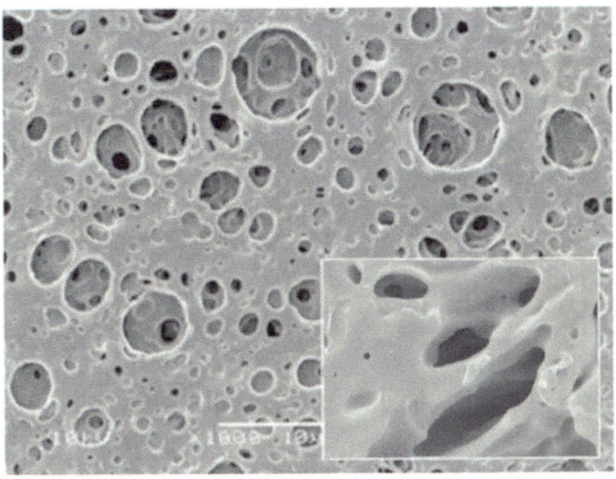

Fig. 2.8 SEM image of the surface of the polyurethane microporous membrane magnification ×1000, small pictures cross-section view magnification ×5000.

nodes by fibrils (12). These fibrils interconnecting the nodes are oriented parallel to the direction of expansion and appear to be characteristically wide and thin in cross-section, the fibrils with a typical width of around 0.1 μm (1000 Å) vary in length from 5 to 500 μm, depending upon the amount of expansion. While Fig. 2.9A shows a uni-axial expansion effect, it will be appreciated that with expansion biaxially and with

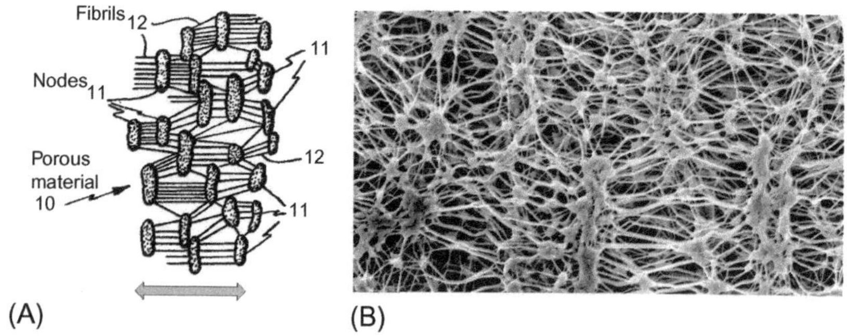

Fig. 2.9 The schematic and SEM image of expanded PTFE. (A) Direction of uniaxial expansion; (B) Gore-Tex expanded PTFE membrane.

expansion in all directions, similar fibril formation occurs in said directions with the production of spider-web-like or cross-linked configurations and attendant increases in strength. As expected, the porosity increases while the voids or spaces between the polymeric nodes and fibrils become more numerous and larger in size (Gore, 1976).

Porosity also increases as the voids between the nodes and fibrils become more numerous and larger in size. The following factors affect the porosity and strength of the microporous layer or film:

1. Crystallinity. The polymer should have high crystallinity, preferably >98%.
2. Temperature and rate of stretching. Higher temperature and higher rate of stretch lead to a more homogeneous structure with smaller, closely spaced nodes, interconnected with a greater number of fibrils, increasing the strength of the polymer matrix.
3. Temperature and duration of heat treatment. The role of PTFE is crucial in the case of Gore-Tex fabric containing bi-component microporous membrane. In PTFE, all the carbon atoms in the backbone of the polymer chain are fully bonded to fluorine atoms. This C—F bond strength is responsible for chemical inertness, UV resistance and good ageing property. Further, the low coefficient of friction results in a smooth finish and can withstand temperatures from −24°C to 280°C. This fully fluorinated polymer film has a significantly low surface energy compared to water, and hence is not wetted by water (Shekar et al., 2003).

A number of new microporous coatings and laminate techniques have also been developed over the past decade, particularly in Japan where there is a vast home market and export capacity for synthetic-based leisurewear. Most of these lightweight fabrics are based on porous polyurethane direct coatings or laminated films which have been produced by wet coagulation. Microporous aluminium layers are included in some grades of fabric, and are claimed to improve the warmth of a garment.

A number of diverse methods are available to generate an interconnecting pore structure in solid polymer films and coatings (Scott, 1995; Holmes, 2000; Mukhopadhyay and Midha, 2008; Sen, 2008). Various methods of obtaining microporous membranes and coatings are as follows:

For both techniques	Only for coating	Only for laminate
Wet coagulation	Thermo-coagulation	Mechanical fibrillation
Solvent extraction	Solubilizing one component in the mixture	
Radio frequency (RF)/ion/UV or E beam radiation Melt-blown technology Point-bonding technology	Foam coating	

2.6.2.1 Wet coagulation process

In 1951, a coagulation process for a polyurethane was published for the first time; since 1954, additional process steps have been developed to produce microporous polyurethanes (Traubel, 1999).

The utilization of the chemical ability of PU to coagulate into a fine interconnecting porous structure by means of a solvent exchange process with dimethyl formamide (DMF) and water (Krishnan, 1992) has resulted in the generation of a PU-based microporous film. A polyurethane polymer, usually thermoplastic, is dissolved in an organic solvent which is also water-soluble, and then the resulting viscous solution is directly coated onto a fabric, or cast as a thin film on a release paper. While still liquid, the film is exposed to water vapour or steam in a controlled manner. The water vapour dissolves in the film, causing phase separation of the polymer, producing a coagulated gel with microporous voids, channels and blowholes. The film is then washed and dried. The process requires careful control and is relatively slow. It produces coatings or films with pores in the range of 0.1–20 μm. Entrant, Porvair and Porelle are examples of commercial products made in this manner (Scott, 1995; Sen, 2008; Mukhopadhyay and Midha, 2008).

The structure of the soft segments of polyurethanes remarkably influences polyurethanes' coagulation. Soft segments with a higher hydrophilic property result in polyurethanes with a better coagulation behaviour. The structure of the hard segments is of minor importance. The process produces a very fine interconnecting structure in which the micropores are small enough to keep droplets of water out, but large enough to let the small water vapour molecules in. It appears that a considerable proportion of lightweight fabrics are based on porous PU-direct coatings or laminated films which have been produced by wet coagulation.

2.6.2.2 Thermo-coagulation

A coating polymer can also be implemented to a fabric substrate from a mixture of a relatively volatile solvent mixed with a proportion of a higher boiling nonsolvent. The polymer coating precipitates out as a microporous layer when the true solvent evaporates faster than the nonsolvent during drying operation. This technique is also known as the phase seperation method for forming a microporous coating which was developed for the Ucecoat 2000 (Lomax, 1985). The process appears to be advantageous over the wet coagulation process, in which immersion and washing baths are not required (Lomax, 1985; Sen, 2008; Shekar et al., 2003).

In a PU-based coating process by thermo-coagulation technique, PU is dissolved in a solvent mixture of methyl ethyl ketone, toluene and water, incorporating 15%–20% solids and coated onto the fabric. The low-boiling solvent evaporates, leading to precipitation of PU in the nonsolvent. As the concentration of the nonsolvent reaches a critical concentration level, the PU precipitates out in a highly porous form, and the leavings of the solvent and the nonsolvent evaporate from the coating as the fabric passes through the oven.

2.6.2.3 Foam coating

A typical foam formulation consists of polymer emulsion, a foaming agent such as ammonium stearate, a thickening agent, filler and, possibly, a cross-linker. In foam coating, a low level of penetration is expected, and the coated fabric has much softer handle and better drape. This process is rather useful for fabric of open construction. The coating also has a certain degree of permeability. The foam is generated by blowing air into the emulsion. The density of the foam depends on the volume of air blown. Foam density is adjusted in accordance with the structure of the fabric (Sen, 2008; Mukhopadhyay and Midha, 2008).

In the development of a microporous coating, a mixture of PU and PU/polyacrylic acid esters are dispersed in water and then foamed. The foam is stabilized by means of additives and then coated on to one side of the fabric. The coated fabric is dried to create a microporous coating. The fabric is finally calendered under low pressure to compress the coatings. Because the foam cells are relatively large, a fluorocarbon (FC) polymer water repellent finish has traditionally been applied to improve the water-resistant properties (Mukhopadhyay and Midha, 2008).

2.6.2.4 Solvent extraction

Solvent extraction is a process in which a polymer dissolved in a water-miscible solvent is coated directly onto the fabric. The microporous structure of the coating is developed by passing the coated fabric through a coagulation bath, where the solvent is displaced by water (an older process) whereby finely divided water-soluble salts are incorporated into the coating formulation. The salt particles are subsequently extracted from the dried and cured coating by passing the fabric through a water bath.

2.6.2.5 Solubilizing one component in the mixture

In this method, a moisture-permeable waterproof coating is formed on a fabric. The first step is applying to a fabric a water-based coating containing a film-forming polymer, which forms a water-insoluble polymeric layer upon being dried or heat treated and a water-soluble enzyme-degradable polymer, then drying or heat treating the fabric to form a layer on it. After drying, the fabric is treated with an aqueous solution of an enzyme capable of selectively degrading the water-soluble polymer, extracting the degraded water-soluble polymer from the layer, rendering the layer on the fabric microporous (Mukhopadhyay and Midha, 2008).

2.6.2.6 Radio frequency/ion/UV or E beam radiation

In this method, specific coating materials are deposited onto textile substrate by ion beam sputtering under different conditions. A fast method of producing microporous film was developed (Gregor et al., 1988) by Gelman Science in the United States. The polymer is based on acrylates, and monomers and oligomers are crosslinked under a radiation source (UV/E beam) and cured in milliseconds. The pore sizes are about 0.2 μm, and the film can be very thin, the order of a few angstroms.

2.6.2.7 Melt-blown/hot melt technology

Breathable fabric produced with melt-blown technology may consist of at least one layer of coarse, melt-spun, thermoplastic filaments and at least one layer of fine, melt-blown, thermoplastic microfibres. The layers are thermally bonded together at intermittent points and, while being heated, are subjected to a force in at least one direction without tearing. The coarse filaments are elongated in the direction of force, and the fine microfibres are straightened in the direction of force, in the absence of drawing, to form a denser array of the microfibres having a lesser thickness within the resulting fabric (Etzold, 1999).

2.6.2.8 Point bonding technology

Point-in-Point (PiP) patented technology (Krings et al., 2003) is complementary to the continually increasing hot melt applications in the areas of bonding and laminating. In the PiP process, the adhesive points on two sides of a three-ply system are brought into alignment by means of electronic control. Lamination and bonding occurs in one production step. With this process, adhesive points on one side are congruent with the points on the second side of a film or textile. Advantages of the PiP technology include the possibility of greater water vapour transmission, as there is a higher amount of free surface between the adhesive points, leading to greater breathability (of up to about 40%) and comfort for the wearer (Mukhopadhyay and Midha, 2008). However, the gap between the adhesive points is small for water particles. Application areas include climate-membranes used for protective weather and work garments.

2.6.2.9 Mechanical fibrillation

A well-known example of the interconnecting pore structure is very thin PTFE films used in the Gore-Tex laminates. Certain polymer films can be stretched in both directions and then annealed to generate microscopic rips and tears throughout the membrane. Biaxial stretching of certain polymers results in microscopic tears throughout the membrane so as to imparts an acceptable microporous structure. Like ePTFE membranes used in the Gore-Tex laminates, a thin microporous membrane is produced from solid PTFE sheet through a novel drawing and annealing process as discovered by R. Gore. In drawn form, the tensile strength is increased threefold (Gohlke and Tanner, 1976).

Required porosity develops thanks to void formation between nodes and fibrils. When the films are biaxially stretched, similar fibril formation occurs in the other direction with the production of cobweb-like or crosslinked configurations with an increase in strength. As the voids between the nodes and fibrils become more numerous and larger in size, porosity increases. Factors affecting the porosity and strength of the film can be summarized as (Sen, 2008):

- The crystallinity of the polymer should be high, preferably >98%.
- Regarding the degrees of the temperature and rate of stretching, the higher the temperature and rate of stretch, the more homogeneous the microporous structure. The occurrence of smaller and closely spaced nodes which are interconnected with a greater number of fibrils increases the strength of the polymer matrix.
- As for the temperature and duration of heat treatment, a heat treatment above the melting point of the polymer increases amorphous content of the polymer formation. This amorphous region reinforces the crystalline region, thus enhancing the strength without substantially altering the microstructure.

2.6.3 Solid polymer structures

Materials adopting this technique have a continuous solid polymer-based film or coating that has a high resistance to ingress of liquids, provided that they are free from pinholes and that the film has a certain minimum thickness. Those solid membranes and coatings are usually thin hydrophilic films because water vapour permeability rates through hydrophilic films are inversely proportional to their thickness (Lomax, 1990).

The transmission of moisture vapour is achieved thanks to molecular diffusion as the moisture is absorbed by the membrane or coating and released on the other side. They consist of modified polymers and diffuse moisture by molecular diffusion or by adsorption diffusion-desorption process (Kramar, 1998; Fan and Hunter, 2009). The phenomenon of water vapour transmission is illustrated in Fig. 2.10. From the hydrophilic group containing polyurethane solution the solvent will be evaporated leaving a compact film behind (Gottwald, 1996). Solid film incorporates hydrophilic functional groups along the molecular chain. Hydrophilic groups built into the polymer chains can absorb, diffuse and desorb water vapour molecules through the film (Lomax, 1990). Such groups can be —O—, —CO, —OH or —NH2, all of which can form reversible hydrogen bonds with water molecules. These bonds are comparatively weak and can readily be broken down by thermal motion. Water molecules diffusing through the film pass stepwise along the molecular chains, a process which is facilitated by introducing pendant side groups, which prevents close-packing of adjacent chains (Scott, 1995). Sympatex and Excepor-U are examples of these type of structures.

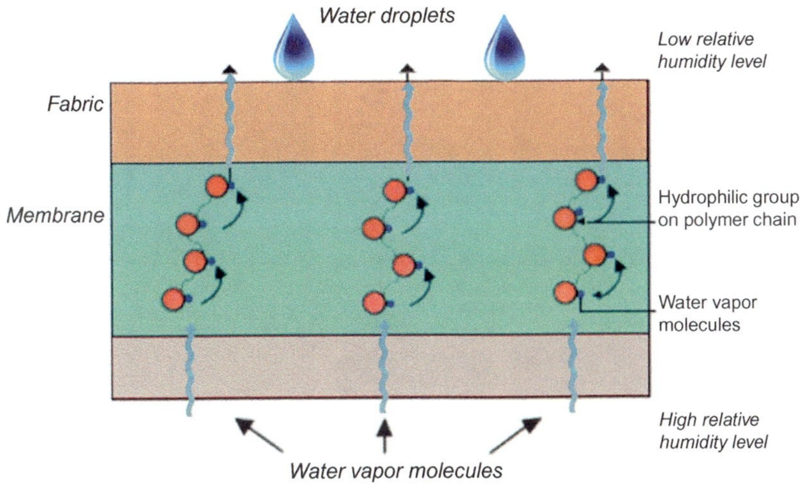

Fig. 2.10 The mechanism of moisture transmission at hydrophilic structures.

2.6.4 Bi-component structures

Another method of developing a waterproof breathable structure is to combine microporous and hydrophilic membranes and coatings. It is claimed that this combination results in obtaining better properties and performance (Lomax, 1985). Krishnan (1992) suggested that hydrophilic coatings and membranes can be developed using a combination of hydrophilic and hydrophobic urethane components to obtain better properties while maintaining other physical properties.

A microporous coating or film is further coated with a hydrophilic layer to increase waterproofness and to seal the surface pores, reducing the possibility of contamination of the microporous layer by dust particles, detergents, pesticides, etc. A hydrophilic finish on a microporous structure is used to upgrade the water-resistance of microporous coatings. Care has to be taken to select a hydrophilic finish which does not cause an unacceptable decrease in breathability. In the case of membranes, the microporous mesh or material is imbued with a hydrophilic material like polyurethane. In the case of coatings, hydrophilic finishes are applied over microporous films that have been attached to the fabric. This ensures enhanced waterproofing capacity without hampering breathability to a large extent (Roey, 1991).

The SEM images of some of the WB fabric structures are given in Fig. 2.11.

2.6.5 Smart waterproof breathable fabrics and biomimetics

Smart coating material is used to develop breathable textiles based on passive and active hydrophilic coatings (Jassal and Agrawal, 2010). A more durable method may be to use a waterproof breathable film or 'membrane' which lies inside the garment (such as socks and gloves) incorporating a thin film inside.

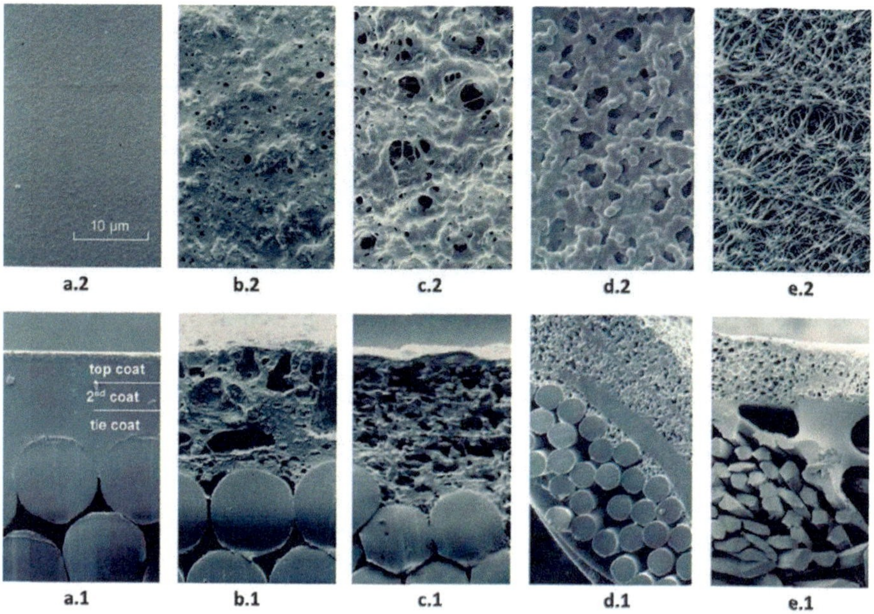

Fig. 2.11 SEM images of various types of membranes used in waterproof breathable fabrics (WBFs). Top row shows surfaces of membranes; bottom row, sections through coatings, laminated to a polyamide fabric. a.1 and a.2, solid hydrophilic coating; b.1 and b.2, microporous PU produced by wet coagulation; c.1 and c.2, microporous PU produced by phase separation (c.2) with additional very thin pore-sealing topcoat; d.1 and d.2, mechanically foamed PU dispersion with mechanically-foamed acrylic coating; e.2, ePTFE membrane; e.1, PU film produced by solvent extraction laminated to polyester with discontinuous adhesive. Reproduced from Lomax (2007).

In general, breathable coatings are made up of hydrophobic microporous coatings and laminates. Due to their hydrophobic character, these types of membranes tend to limit WVTR and lowers the performance. Therefore, the use of hydrophilic coatings has been tried. They are copolymers composed of hydrophilic and hydrophobic segments in which hydrophobic moieties provide the water resistance property and facilitate adherence of the coating to the substrate, while the hydrophilic part, which is a soft segment, allows water-vapour permeability (Kannekens, 1994; Schledjewski et al., 1997). The absence of open spaces and macro-pores in the coatings/laminates provide protection against wind and water.

Highly hydrophilic poly(acrylamide) is used to develop breathable fabrics (Save et al., 2002). This polymer coating was chemically integrated onto a cotton fabric using polycarboxylic acid compounds as crosslinkers and reaction catalysts. This coating is reported to provide a high degree of water vapour transmission along with high levels of integration with the cotton substrate. However, only a moderate level of hydrostatic water head protection could be obtained in addition to the stiffness of the fabric.

High molecular weight PVA is also used as an alternate hydrophilic polymer system that could overcome the drawbacks of poly(acrylamide). It is integrated onto cotton substrate using a low molecular weight PAAc as a crosslinker (Bunyakan and

Hunkeler, 1999). By this method, the development of a hydrophilic-coated textile with a high level of breathability is achieved. The approach of hydrophilic breathable coatings appears to be applicable for particular end uses. These coatings are passive, and they do not adapt to changing environmental conditions such as temperature. A breathable fabric with temperature-dependent response may be desirable for certain specialized applications where high variations of temperature are encountered over a short period. The use of stimuli-sensitive polymers appears to a potential solution to develop breathable coatings. Temperature-responsive polymers (or smart polymers) that show a reversible transformation from one state to another as a response to temperature stimulus from the environment provide a big opportunity for creating intelligent materials (Jassal et al., 2006; Jassal and Agrawal, 2010).

Water vapour transmission through the use of shape memory polyurethane (SMP) is another example of smart techniques. The experimental trials and discussion of water vapour permeability of SMP films prove that a large change in water vapour permeability occurs at the transition point (glass transition temperature/crystal melting point temperature). This is due to the morphological change of the polymer membrane and the presence of hydrophilic groups in the polymer structure, also enhances permeability, thanks to the increasing solubility of water vapour molecules in the membrane (Hu and Mondal, 2006).

Phase change material (Chung and Cho, 2004; Li and Zhu, 2004) is another class of smart material applicable for developing breathable fabrics. Although its role as a class of material is not to directly regulate the passage of water vapour through the fabric, its incorporation leads to improvement in thermal and moisture management when worn close to the skin.

Biomimetic studies are primarily focused to transfer ideas inspired by biology into practical solutions to everyday problems. This concept used in textiles has many ways for specific applications such as waterproof breathable textiles. A British team investigated the ability of penguins to survive in extreme cold conditions, and it was revealed that this ability is the result of their coat, a combination of feather and skin. They have the ability to switch their skin from insulating to waterproof using muscles connected to their feathers. As these connecting muscles contract, the skin becomes waterproof, and when it is released, it changes into an air-filled windproof coat. This phenomenon is called adaptive insulation.

Attempts have been made to extrapolate the phenomenon of adaptive insulation into garments. The fabric structure developed is two-layered, the layers of which are joined together using strips of textile material at some angle to the plane of these two layers. Skewing these parallel layers decreases the volume of air trapped between them, which results in decreased insulation. A jacket (namely Airvantage) has been produced using an ePTFE membrane along with a polyester structure (24% PTFE, 76% PE), which allows the user to control the quantity of air for suitable insulation (Kapsali, 2009).

Teijin Co., Japan developed Super-Microft, a fabric with high water repellency, by emulating the structure of a lotus leaf (Hongu and Phillips, 2001). Water rolls like mercury from the lotus leaf, whose surface is microscopically rough and covered with a wax-like substance with low surface tension. It was reported that Super-Microft exhibits good water repellent durability and a high wear resistance, and at the same time it possesses moisture permeability and water proof characteristics.

A fabric system exploiting a lightweight woven structure laminated onto a nonporous membrane based on Neoprene was developed by Stomatex Limited (Stomatex (2004)). Small u-shaped perforations were opened into the surface of the composite textile. Any increase in relative humidity caused the loose section of fabric (created by incisions) to curl back, thus increasing the system's permeability to air. When the microclimate becomes damp, the textile material alters its porosity to allow the renewal of saturated air. The system returns to its original position when the skin conditions are dry, and ventilation is no longer needed. However, the pores would also open in the rain, negating its use in rainwear. Nike implemented such a concept into a clothing system. It is based on a double-knit fabric, incorporating a hygroscopic layer next to the skin and a less hygroscopic layer on the outer surface of the fabric. The clothing has a fish scale pattern on the back panel that opens up as the wearer perspires to increase local ventilation and maintain comfort. It was reported that this sports clothing was worn by Maria Sharapova at the US Open in 2006.

2.7 Methods of developing waterproof breathable fabrics

2.7.1 High density wovens

The use of tight weaving stands out as the first method of developing a WB fabric. The tightly woven fabric created by using specially designed, dense yarns results in a very dense or tight construction and increases coverage of the fabric. This requires a balance between optimum coverage and porosity of the fabric structure.

The first effective waterproof breathable fabric is the famous Ventile fabric. It was introduced during the World War II in the 1940s by the Shirley Research Institute to be utilized by British Air Force pilots as a cold-immersion-resistant suit (Ventile n.d.). A selection of Egyptian extra-long staple cotton fibres was used to produce high count and low twist combed cotton yarn. By means of these yarns, very high density heavy flat fabric was woven employing the Oxford weave (Lomax, 1985). This ensures that there are minimal pores in the fabric. When this fabric is inserted into water, the pore size is reduced due to the swelling of cellulosic fibres as shown in Fig. 2.12.

The new versions of Ventile fabrics are produced by the Stotz & Co AG in Switzerland. Under a new brand name, 'etaProof', tightly woven fabrics are produced in various constructions, with varying weights of 170–300 g/m^2 depending upon the end use application. Those fabrics have warp and weft densities in the range of 66–95 ends/cm and 26–35 picks/cm, respectively. The count range of combed yarns are varied between Ne 30 and Ne 50 (Odermatt, n.d.). The natural product offers a high level of comfort due to its drape and breathability, while being durable and quiet in use. It also provides good resistance to tearing and burning. It is not as light in weight as synthetic fabrics. The Oxford construction provides a flat surface with good abrasion resistance and maximizes the closeness of the weave without greatly stiffening the fabric.

An experimental work was carried out by Shekar and colleagues (2005) to develop breathable waterproof linen fabric using dry-spun linen yarns. The principles of tight weaving are based on fabric engineering principles taking into account the cloth cover, using different twist levels, mass, and weave. It was observed that the fabric woven in

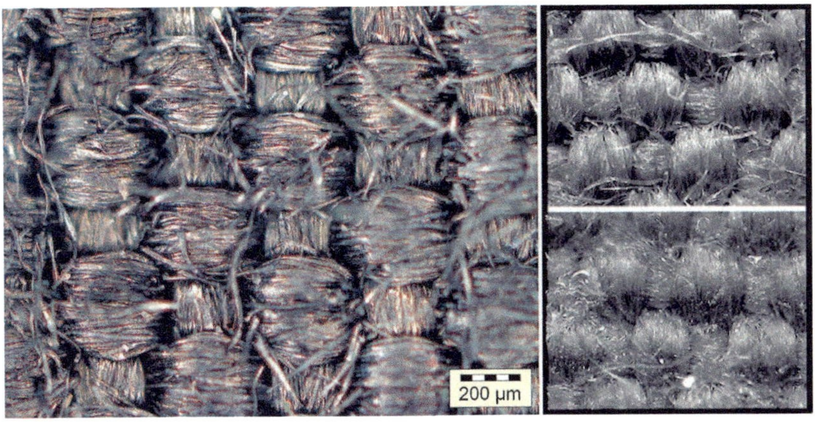

Fig. 2.12 SEM image of a tightly woven WB fabric and its dry and wet state (right).

Oxford weave with twist levels varying from 160–180 t/m exhibited superior performance in terms of waterproofness, water-vapour-permeability and water-retaining capability. This was relatively coarse fabric woven with a yarn count of Nm 3.85 at 11 ends/cm and 8 picks/cm. This study proved the importance of using fabric engineering concepts to achieve optimum levels of functional requirements with regard to WB fabric.

Synthetic filament yarns can also be used in a similar way by using inherently water repellent fibres. However, they do not swell when inserted in water, and hence, further coatings are required to obtain desirable results (Holmes, 2000). The waterproofness of these fabrics is low, and they can withstand only moderate hydrostatic pressure.

It is possible to use new generation microfibres of 0.05 to 1 d polyester, polyamide, viscose or acrylics to develop high-density, waterproof, breathable fabrics with improved functional properties. Those microfibres are obtained by melt-spinning two incompatible polymers into a single fine fibre, then one of the polymers is separated by dissolving in a specific solvent, leaving behind microfibre. The yarn may be woven into various dense fabric constructions like Oxford, taffeta or twill, and given a repellent finish of silicones or fluorochemicals. These fabrics have better water repellency than the cotton Ventile fabrics, and have very soft handle.

In consideration of previous studies, fundamental characteristics of microfibre based WB fabrics are reported by Kaynak and Babaarslan (2012) as follows:

- The yarns made from microdenier fibre contain many more filaments than regular Yarns, producing fabrics with water tightness and windproofness, while also improving breathability.
- Greater fibre surface area makes deeper, richer and brighter colours possible.
- Microfibres have quick stress relief, so microfibre fabrics resist wrinkling and retain shape.
- Microfibre fabrics insulate well against wind, rain and cold and are more breathable and more comfortable to wear.
- Microfibres are superabsorbent, absorbing over seven times their weight in water.
- Microfibre dries in one-third of the time of ordinary fibres.

The interspaces of the weave construction should be just large enough to allow water vapour to pass through, while water molecules are blocked out. On the other hand, as a foul weather clothing, there is a contradictory requirement that the interyarns are required to be as small as possible for maximum protection against wind and rain, while the fabric's outer surface should be nonabsorptive and hydrophobic to minimize wetting by rain or snow. Therefore, an inevitable design optimization conflict is unavoidable between waterproofness, windproofness and water vapour transmission performances of closely woven fabrics (Lomax, 1985).

2.7.2 Coatings for waterproof breathable fabrics

Coating techniques are simply the application of polymeric add-ons onto the fabric surface and the control of add-ons to preserve fabric aesthetics and properties as much as possible. This is not easy, and fabric handling is always a challenge. Coated fabric handling may be especially difficult because, after even just one layer is applied, the total fabric weight increases, and the surface nature of one side of the material changes significantly.

In coatings for waterproof breathable fabrics, the function of the polymeric add-on substance is to provide WB characteristics. Coated textiles usually consist of a textile substrate which has been combined with a thin, flexible film of a natural or synthetic polymeric substance. A coated fabric is shown in Fig. 2.6B incorporating a woven textile substrate onto which a specific paste including polymer substance is applied directly as a viscous liquid. The thickness of the film is controlled by means of a blade or similar aperture during the application of the coating substance. Control of polymer add-on is one of the fundamental principles of fabric coating, and various methods achieve this in different ways. Resin solids content or foam density and viscosity are polymer properties that are used to control add-ons, but machine characteristics and geometry are also relevant. Moreover, control must be achieved without adversely affecting fabric properties.

There is a wide range of coating techniques in industrial use, but they all require similar basic approaches (Scott, 1995).

(a) The textile fabric to be coated is supplied full-width on a roll.
(b) The fabric is fed under careful tension control to a coating or laminating head or zone.
(c) After application, the coated fabric is passed through an oven to cure the composite and remove volatile solvents before cooling and rolling up.
(d) The quantity of coating can range in weight from about 10 g/m^2 up to several hundred g/m^2 according to usage.

For developing waterproof breathable fabrics, microporous coatings and hydrophilic coatings are used. Coating is a joining process, and all the rules governing adhesion forces should be considered. The methods of coatings are various and selected in accordance with the preparation techniques of polymeric coating substance. The choice of a coating method depends on several factors (Sen, 2008):

- Nature of the substrate.
- Form and preparation of the resin and viscosity of the coating fluid.
- End product and accuracy of coating desired.
- Economics of the process.

2.7.3 Laminates for waterproof breathable fabrics

Laminated textiles consist of one or more layers of textile and component. The Textile Institute defines a laminated or combined fabric as, 'a material composed of two or more layers, at least one of which is a textile fabric, bonded closely together by means of an added adhesive, or by the adhesive properties of one or more of the component layers'. It results in modification of physical properties based on the individual characteristics of the separate components. Laminates are a relatively more recent development. The end result of the lamination technique is a fabric that is lighter, more flexible, softer and much more comfortable to wear, with added functions such as waterproofness and breathability.

If one component has limited stretch only in the warp, and the other, limited stretch only in the weft, the combined material will have limited stretch in both warp and weft. Lamination of any fabric invariably produces a laminate which is stiffer than either of the two starting materials, although this can be minimized through choice of the most suitable lamination method and adhesive. The adhesive used to join the components may reduce stretch further and cause additional stiffening; an objective in adhesive choice is to choose the adhesive which provides the strongest bond with the least amount (Fung, 2002).

The suitability of fabrics for lamination should be considered when products are designed; clearly, the more stretch, the better the handle and drape of the laminated product. Usually, a third material is used as the adhesive, but sometimes one of the materials being joined can itself act as the adhesive, as in flame lamination of polyurethane foam.

For many years, lamination was carried out by the application of both solvent- and water-based adhesives by back licking rollers or by some other coating device, and then drying off the solvent, sometimes on a Palmer unit, a large oil or steam-heated cylindrical drum with an endless blanket. If crosslinking was required, this was done separately on heated rollers or in a heated chamber. Hot-melt adhesive methods are cleaner and less energy-intensive with far fewer health and safety considerations. Hot-melt lamination is accomplished by two separate processes: first by means of applying the actual adhesive, and second, by bringing the two substrates together to form the actual bond under the action of heat and pressure. The latter process is often referred to as fusing. The method of application depends on the form of the adhesive, that is, film, web, powder or liquid (Fung, 2002).

2.8 Arrangements of layers at waterproof breathable clothing construction and review of commercial products

In search of more comfort and durability, the construction of waterproof-breathable clothing has evolved over the years. Instead of single layers, multifunctional additional layers are developed and arranged. In general, breathable films are typically

10–25 μm thick, and therefore, are too fragile to be used alone within a garment. As such, they are coated onto or laminated to a support fabric in a variety of forms.

The first layer is usually the face fabric, which is also the layer that incorporates the design, colour or texture of the fabric to make it stylish or fashionable. This layer is expected to be durable because it stays outside and is exposed to the surroundings. The second layer where the functional magic takes place is the waterproof, breathable part of the fabric. The final layer which is usually positioned on the inner side is the layer that protects the waterproof coating or membrane from abrasion and pore-clogging contaminants and helps the garment's form and look. The methods of construction are named for how many of these elements are fused together in the final fabric, and there are three main types as shown in Fig. 2.13. An extra layer as insert fabric attached to the waterproof breathable layer or for insulation purposes may also be added into these three layers. One example of this is a drop-liner, where the film is attached to a thin inner fabric and protected from the outside by a loose outer fabric.

2.8.1 2 Layer construction

In this construction type, only the face fabric and the waterproof breathable layer are laminated together. The inner protective layer is attached by a separate fabric or mesh liner that hangs on the inside of the garment. This was the original waterproof-breathable construction. The principle benefit is comfort, as that hanging liner forms a nice, drape-able interface with the skin. It is quite flexible, and is the cheapest type. The drawbacks are the extra weight of the added layer, the bulkiness of the hanging liner, and breathability concerns as the liner is one more layer moisture needs to go through. Durability is also a serious concern. This type of construction is commonly used in light-to-medium-weight garments such as casual and fashion jackets, and some rain pants.

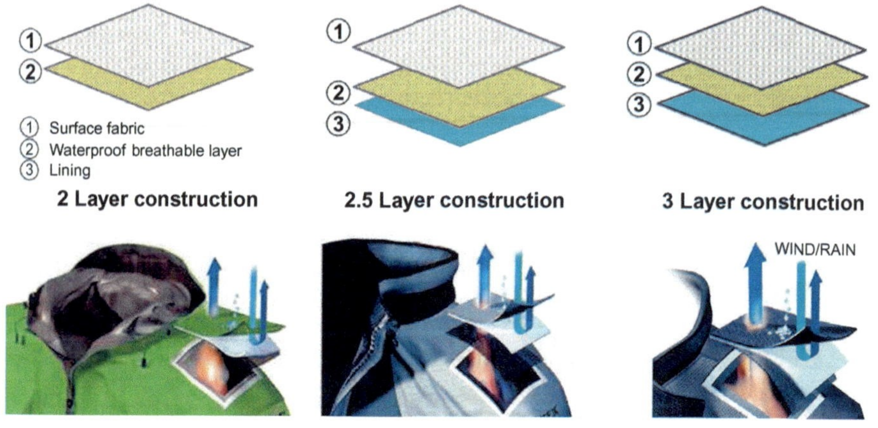

Fig. 2.13 Arrangement of individual layers in WB fabrics and application in outdoor jackets.

2.8.2 2.5 Layer construction

This version was introduced in the mid-1990s with the development of the featherweight Gore-Tex Paclite fabric. The 2.5L type is currently the most common type of construction for lightweight rain clothing. The face fabric and waterproof-breathable layer are almost similar to the previous type, but the protective fabric or scrim gets eliminated in exchange for a printed or sprayed-on partial protective layer. This version is definitely the lightest construction, and it is also claimed to be the most flexible, the most packable, and relatively less expensive construction available. The main drawbacks are durability, because reducing the protective layer increases wear on the waterproof-breathable layer, and breathability, because the spray-on protective layers do not reliably breathe as well as a hanging mesh or laminated scrim. The 2.5L construction appears to be customary in lightweight and ultralight garments.

2.8.3 3 Layer construction

This type is composed of three layers combined in one composite system. It is developed for the needs of alpinists and stubborn users who wanted to overcome the drawbacks of 2L construction. A very lightweight scrim fabric is laminated on the back side of the waterproof-breathable layer, effectively making a sandwich between the scrim and face fabric. This makes the fabric more durable, because the waterproof breathable layer is never exposed, even to the inside of a hanging liner. It is the most expensive solution, but offers the highest level of protection to the membrane but with a lower breathability compared to the 2 layer construction. The drawbacks to three-layer construction are that losing the liner decreases comfort, and lamination of three layers together decreases flexibility and increases stiffness and bulkiness. This construction is widely used in the top range of outdoor clothing.

The design of a garment includes a number of pieces which are either fused together or stitched. Stitching punctures the film allows ingress of water, unless the punctures are sealed in some way, often using hot-melt tape. The use of different fabric forms gives the designer range to alter styles and costs.

2.8.4 Review of Commercial Products

There are various types of waterproof breathable outdoor garments and numerous manufacturers or brands on the market. On the other hand, the manufacturers of functional WB fabrics are far fewer in number. A few outdoor sport brands have their own branded fabrics, but most use brands available in the market. As usual, some claim to be better than others. Every product has its own merits, and will perform depending on the conditions in and for which the garment is used. Their waterproofness performance on the basis of hydrostatic head measurement varies within the range of 8000–28,000 mm water per 24 hours. The range of MVTR (breathability) is estimated to be between 5500 and 30,000 g/m^2 per 24 hours. On the basis of Resistance to Evaporative Heat Transfer (RET) value, most commercial products perform within the range of 3 to 6, with very few measuring 6 (Sierra Trading Post n.d.; Gugel, 2017).

GoreTex is one of the most well-known manufacturers with a wide range of product: GoreTex Classic 2L and 3L, GoreTex Paclite 2.5L, GoreTex Pro 3L and GoreTex

C-Knit 3L (which is their latest product). Classic GoreTex fabrics are made using ePTFE membrane. In addition to ePTFE membrane, an anticontaminant PU layer is coated, and the shell displays hydrophilic monolithic character.

DryVent (previously HyVent) is another branded fabric developed by The North Face. It is a hydrophilic polyurethane coated fabric. HyVent is made in 2L, 2.5L and 3L construction fabrics. DryVent DT (Dry Touch) is another product with 5-layer skim coat of PU.

The WB fabrics eVent DVAlpine, DVStorm ve DVElite are usually made in 3L and have a microporous structure. eVent fabrics use original expanded PTFE membrane covered with an additional PU layer. Some products are also lined with an oleophobic and hydrophobic chemical layer to remain air permeable.

The 2.5L H2No shell fabric is developed by Patagonia, and it combines a water repellent shell fabric with a WB membrane and protective top coat barrier. It is lightweight and quite compressible.

Marmot NanoPro by Marmots is 2.5L, microporous, coated fabric. It is claimed that the new coating technology improves breathability up to 43% more than the previous version, Precip 2.5L construction, and consists of a combination of both types of fabric coatings. Precip uses a protein coating on the inside of the fabric to make it feel drier and more comfortable next to the skin and has no protection scrim. MemBrain was Marmot's first WB technology, and it is a PU membrane with monolithic character. It has the added advantage that PU is slightly hydrophilic, which means it will absorb some of the moisture that would otherwise be liquid condensate on the inside of the garment. MemBrain Strata is the latest version with 2.5L construction.

Sympatex offers WB fabrics with a base layer with conductive, wicking properties, mid-layers with an insulating and heat-regulating function, and top layers with a weather protection function (waterproof, breathable and windproof). It is a compact hydrophilic membrane with a thickness range of 5–25 μm. Sympatec also produces 100% recyclable membrane.

Polartec NeoShell and Power Shield Pro are membrane-based WB shell products. They also have insulation types for special protective clothings. Mont Hydronaute & Hydronaut Pro, PERTEX Shield+, Colombia Omni-Tech Dry and Mini-Faille, Toray Entrant and Dermizax are other well-known WB fabric brands.

Water Vapour Diffusion Test by the Dynamic Moisture Permeation Cell method developed by Gibson and his colleagues (Gibson et al., 1995) was applied to some commercial outerwear fabrics. The results for water vapour diffusion resistance for various relative humidity levels are given in Fig. 2.14. It is reported that the ASTM F 2298 test standard was used at the Gas Flow Rate of 2000 cm^3/min (Gibson, 2006). Results are shown in terms of resistance to the diffusion of water vapour in units of (s/m). The resistance units make comparing results obtained at different environmental conditions much easier. The lower the diffusion resistance, the more water vapour gets through the material. It may be seen that some materials like Gore-Tex, Sympatex, etc., have much better water vapour transport properties when they are in a humid environment than when they are in a dry environment. Other materials such as most textiles or microporous membranes, have almost constant water vapour diffusion resistance regardless of the environmental conditions.

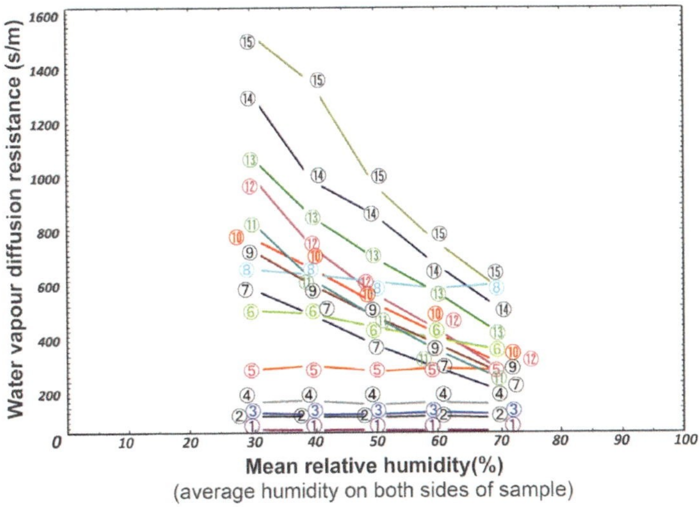

Fig. 2.14 Comparison of resistance to Water Vapour Diffusion of some commercial outerwear shell layers. The lower the resistance the more 'Breathable' the layer. ① Expanded PTFE Membrane, ② EVENT Laminate (BHA Technologies), ③ Schoeller Dryskin Extreme (Schoeller Textiles), ④ Entrant GII XT Laminate (Toray Industries), ⑤ Nextec (Patagonia 'Velocity'), ⑥ Lowe Alpine Triplepoint Ceramic, ⑦ Gore-Tex XCR, ⑧ Omni-Tech 'Titanium' Columbia Sportswear, ⑨ Entrant Dermizax (Toray Industries), ⑩ Membrain (Marmot), ⑪ Gore-Tex (Standard), ⑫ The North Face 'Hydroseal' (Burlington 'XALT'), ⑬ Sympatex Laminate, ⑭ Conduit (Mountain Hardwear), ⑮ Schoeller WB-Formula (Schoeller Textiles).
Data from Gibson, P.W., 2006. "Breathability" Comparison of Commercial Outerwear Shell Layers, U.S. Army Natick Soldier Research, Development, and Engineering Center.

2.9 Conclusions and future trends

There are various techniques and various types of WB fabrics available on the market. Each one offers different benefits, as well as their own unique drawbacks. Even though considerable progress has been made during the past decade, there is still significant potential for further technical and commercial developments. Laminated fabrics appear to perform better than coated fabrics under cold ambient conditions, transmitting more water vapour and preventing condensation from forming for a longer period of time. In general, the advantages of coatings, as compared to laminates, are lower-priced base materials and flexibility in the production process (Kannekens, 1994).

Based on the vapour permeability test, following BS 7209 (with some modification), Holmes et al., 1995 ranked breathable fabrics in order of decreasing performance as follows:

- Tightly woven fabric
  fabric made from synthetic filament
  fabric made from cotton fibre

- Membrane
  microporous membrane
  hydrophilic membrane
- Coatings
  hydrophilic coating

This may appear as contradictory with the findings of other studies because of the variance in effective weather conditions. It was found that both vapour pressure and natural convection within the air gap also affect the extent of water vapour transfer. The rates of water vapour transfer of different fabrics were ranked differently, depending on the presence of a temperature gradient. The ranking of fabrics in the presence of a temperature gradient was as follows: microfibre fabrics > PTFE laminated fabrics > cotton Ventile > hydrophilic laminated fabrics > poromeric polyurethane laminated fabrics > polyurethane coated fabrics.

In the presence of a temperature gradient, condensation is also a major factor, especially at air temperatures below 0°C. The amount of condensation is least on the inner surface of PTFE laminated fabrics, followed by cotton, Ventile, microfibre fabrics, hydrophilic laminated fabrics, poromeric polyurethane, laminated fabrics and polyurethane coated fabrics (Ruckman 1997a,b,c).

The technology behind designing breathable fabrics has continued to evolve for achieving improved functionality as well as cost effective manufacturing processes for a variety of applications. There are various essential development issues including improving material formulation for enhancing films' properties, controlling pore sizes and their distribution, developing improved monolithic films and coating materials. Trials on nanofibre webs show that it is possible to develop breathable fabrics made of electrospun nanofibre webs possessing the desired porous structure necessary to impart barrier and comfort characteristics. The unique combination of high specific surface area, flexibility, light weight and porous structure may perform better than conventional breathable fabrics. Smart breathable fabrics, in particular fabrics based on biomimetics, which are of recent origin, are also showing their potential.

Recent progress in the method of incorporation of the membrane, coating techniques, fabric substrates, lining materials and, above all, garment construction are leading to changes in market trends and play a vital role in designing breathable garments. In consideration of affordable solutions, application areas are also increasing. Recent developments in coating and laminating technology to manufacture textile fabrics provide more functionality and many possibilities, and have greatly expanded the use of textiles applications, not only for use in the production of apparel and shoes, but also for many technical textiles uses. The waterproof breathable textiles market size is projected to reach 2.18 billion dollars by 2022, growing 5.4% from 2015. It was valued at 1.20 billion dollars in 2013 (Grand View Research, 2014). Higher demands for active sportswear, along with development of eco-friendly products, should stimulate demand growth during the next 6 years. Because of the hazards of textiles made from synthetic raw materials, there is a tendency to adopt bio-based raw materials, which provide more comfort than the textiles produced from petroleum-derived synthetic materials. As seen in Fig. 2.15 with reference to the European market, based on a market share and forecast report by the Global Market Insights (2016), laminates continue to dominate.

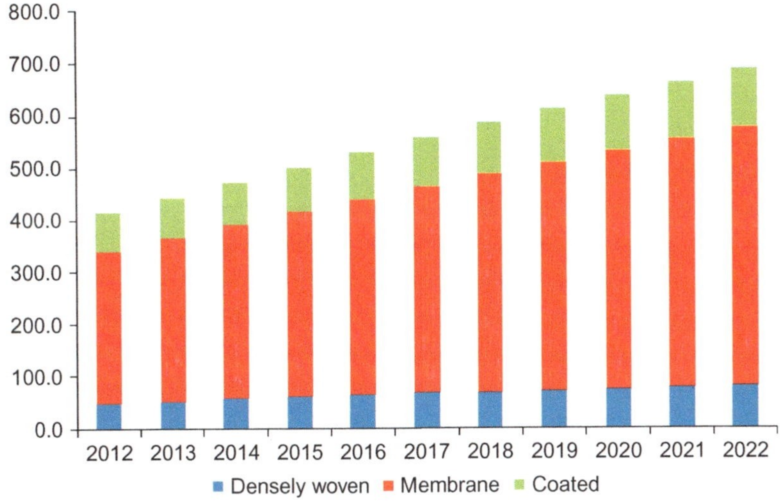

Fig. 2.15 Estimation of European waterproof breathable textiles market size by textile on the base of production techniques, 2012–22 (USD million).

References

Arens, E., Zhang, H., 2006. The skin's role in human thermoregulation and comfort. In: Pan, N., Gibson, P. (Eds.), Thermal and Moisture Transport in Fibrous Materials. The Textile Institute and Wood Head Publishing Ltd, Cambridge, pp. 560–602 (Chapter 16).

Barauskas, R., Abraitiene, A., 2011. A model for numerical simulation of heat and water vapor exchange in multilayer textile packages with three-dimensional spacer fabric ventilation layer. Text. Res. J. 81 (12), 1195–1215.

Basavanthappa, B.T., 2008. Community Health Nursing, second ed. Jaypee Brothers Medical Publisher Ltd, New Delhi.

Benisek, L., Harnett, P.R., Palin, M.J., 1987. The influence of fibre and fabric type on thermophysiological comfort. Melliand Textilber. 68, 878–888.

Bunyakan, C., Hunkeler, D., 1999. Precipitation polymerization of acrylic acid in toluene. I: synthesis, characterization and kinetics. Polymer 40 (23), 6213–6224.

Chemical Heritage Foundation, 2015. Robert W. Gore, August 2015. Available at: https://www.chemheritage.org/historicalprofile/robertwgore (accessed January 2017).

Chinta, S.K., Gujar, P.D., 2013. Significance of moisture management for high performance textile fabrics. Int. J. Innov. Res. Sci. Eng. Technol. 2 (3), 812–819. Available at: www.ijirset.com (accessed November 2016).

Chung, H., Cho, G., 2004. Thermal properties and physiological responses of vapor-permeable water repellent fabrics treated with microcapsule-containing PCMs. Text. Res. J. 74 (7), 571–575.

Crow, R.M., 1974. Heat and Moisture Transfer in Clothing Systems. Part 1. Transfer Through Materials, A Literature Review. NTIS, Ottawa. AD/A-003 837.

Das, A., Alagirusam, R., 2010. Moisture transmission. Science in Clothing Comfort, first ed. Woodhead Publishing India Pvt Ltd, New Delhi.

Das, S., Kothari, V.K., 2012. Moisture vapour transmission behaviour of cotton fabrics. Indian J. Fibre Text. Res. 37 (June), 151–156.

Das, B., Das, A., Kothari, V.K., Fanguiero, R., Araújo, M., 2007. Moisture transmission through textiles part I: processes involved in moisture transmission and the factors at play. AUTEX Res. J. 7 (2), 100–110. Available at: http://www.autexrj.org.

Desai, R., 2016. Sweating, An Educational Blog. Blog post, April, Available at:http://drrajivdesaimd.com/2016/04/14/sweating/ (accessed November 2016).

Dhanabalan, V., Rashmi, J.M., Laga, S.K., 2013. Waterproof Breathable Fabrics: Technologies and Practices. DKTES Textile & Engineering Institute, Ichalkaranji. Available at: http://www.textiletoday.com.bd/waterproof-breathable-fabrics-technologies-and-practices-2/ (accessed January 2017).

Etzold, S., 1999. Waterproof, Multilayered Nonwoven Fabric of Reduced Weight Having Good Vapour Permeability and Method for its Production. Corovin GmbH, Peine, Germany USP 5855992, Jan. 5.

Fan, J., Hunter, L., 2009. Engineering Apparel Fabrics and Garments, first ed. Woodhead Publishing Limited in Association with the Textile Institute, Cambridge.

Farnworth, B., 1986. A numerical model of the combined diffusion of heat and water vapor through clothing. Text. Res. J. 56 (11), 653–665.

Finn, J.T., Sagar, J.G., Mukhopadhyay, S.K., 2000. Effects of imposing a temperature gradient on moisture vapor transfer through water resistant breathable fabrics. Text. Res. J. 70 (5), 460–466.

Fohr, J.P., 2002. Dynamic heat and water transfer through layered fabrics. Text. Res. J. 72 (1), 1–12.

Fourt, L., Harris, M., 1947. Diffusion of water vapour through textiles. Text. Res. J. 17 (5), 256–263.

Fourt, L., Craig, R.A., Rutherford, B., 1957. Cotton fibres as means of transmitting water vapour. Text. Res. J. 27 (5), 362–368.

Fung, W., 2002. Testing, product evaluation and quality. Coated and Laminated Textiles. The Textile Institute, Woodhead, Cambridge, pp. 250–315 (Chapter 5).

Gibson, P.W., 1993. Factors influencing steady-state heat and water vapour transfer measurements for clothing materials. Text. Res. J. 63 (12), 749–764.

Gibson, P., 1999. Effect of temperature on water vapor transport through polymer membrane laminates, US Army Natick Research, Development and Engineering Center Technical Report, NATICKITR-99/015.

Gibson, P.W., 2006. Breathability" Comparison of Commercial Outerwear Shell Layers, U.S. Army Natick Soldier Research, Development, and Engineering Center. Available from: https://www.shelby.fi/images/support/breathability.pdf (accessed February 2).

Gibson, P., Kendrick, C., Rivin, D., Sicuranza, L., Charmchi, M., 1995. An automated water vapor diffusion test method for fabrics, laminates, and films. J. Coat. Fabr. 24 (4), 322–345.

Global Market Insights, 2016. Waterproof Breathable Textiles (WBT) Market Size By Textile (Membrane, Coated, Densely Woven), By Product (Footwear, Gloves, Garments), By Application (Active Sportswear), Industry Outlook Report, Regional Analysis, Application Potential, Price Trends, Competitive Market Share & Forecast, 2015–2022. Available at: https://www.gminsights.com/industry-analysis/waterproof-breathable-textiles-wbt-market-size (accessed January 2017).

Gohlke, D.J., Tanner, J.C., 1976. Gore-Tex® waterproof breathable laminates*. J. Coat. Fabr. 6 (July), 28–38.

Gore, R.W., 1976. Very Highly Stretched Polytetrafluoroethylene and Process Therefor. W. L. Gore & Associates, Newark, USP 3,962,153, June 8.

Gore, n.d., Gore Technologies. Available at: https://www.gore.com/about/technologies (accessed January 2017).

Gottwald, L., 1996. Water vapor permeable PUR membranes for weatherproof laminates. J. Coat. Fabr. 25 (January), 169–175.

Grand View Research, 2014. Waterproof Breathable Textiles (WBT) Market Analysis by Textile (Densely Woven, Membrane, Coated), by Product (Garments, Footwear, Gloves), by Application (Active Sportswear) and Segment Forecasts to 2020.

Gregor, E.C., Tanny, G.B., Shchori, E., Kenigsberg, Y., 1988. Sunbeam process microporous membranes: a high performance barrier for protective clothing. J. Coat. Fabr. 18 (1), 26–37.

Gretton, J.C., Brook, D.B., Dyson, H.M., Harlock, S.C., 1996. A correlation between test methods used to measure moisture vapour transmission through fabrics. J. Coat. Fabr. 25, 301–510.

Gugel, M., 2017. All about waterproof fabrics, Paddy Palling Blog. Available at: http://www.paddypallin.com.au/blog/all-about-waterproof-fabrics/ (accessed February 2017).

Guo, Y., Li, Y., Tokura, H., Wong, T., Chung, J., Wong, A.S.W., Gohel, M.D.I., Leung, P.H.M., 2008. Impact of fabric moisture transport properties on physiological responses when wearing protective clothing. Text. Res. J. 78 (12), 1057–1069.

Haghi, A.K. (Ed.), 2011. Heat & Mass Transfer in Textiles, second ed. WSEAS Press, Montreal.

Hohenstein Institute, 2014. Skin model vs. cup methods. Available at: http://www.hohenstein.de/en/inline/pressrelease_80768.xhtml (accessed January 2017).

Holme, I., 1986. Porous polymers & fusible films. J. Coat. Fabr. 15, 198.

Holme, I., 2003. Water repellency and waterproofing. In: Heywood, D. (Ed.), Textile Finishing. Society of Dyers and Colourists, West Yorkshire, pp. 137–213.

Holmes, D.A., 2000. Waterproof breathable fabrics. In: Horrocks, A.R., Anand, S.C. (Eds.), Handbook of Technical Textiles, first ed. Woodhead Publishing Limited in Association With the Textile Institute, Cambridge, pp. 282–315.

Holmes, D.A., Grundy, C., Rowe, H.D., 1995. The characteristics of waterproof breathable fabric. J. Cloth. Tech. Manag. 12 (3), 142.

Hong, K., Hollies, N.R.S., Spivak, S.M., 1988. Dynamic moisture vapor transfer through textiles: part 1: clothing hygrometry and the influence of fibre type. Text. Res. J. 58 (12), 697–706.

Hongu, T., Phillips, G.O., 2001. New Fibres, second ed. Woodhead Publishing Limited in Association With the Textile Institute, Cambridge.

Hu, J., Mondal, S., 2006. Study of shape memory polymer films for breathable textiles. In: Mattila, H.R. (Ed.), Intelligent Textiles and Clothing. Woodhead Publishing Limited in Association With the Textile Institute, Cambridge (Chapter 7).

Huang, J., 2016. Review of heat and water vapor transfer through multilayer fabrics. Text. Res. J. 86 (3), 325–336.

Huang, J., Chen, Y., 2010. Effects of air temperature, relative humidity, and wind speed on water vapor transmission rate of fabrics. Text. Res. J. 80 (5), 422–428.

Ito, H., Muraoka, Y., 1993. Water transport along textile fibres as measured by an electrical capacitance technique. Text. Res. J. 63 (7), 414–420.

Jassal, M., Agrawal, A.K., 2010. Intelligent breathable coatings and laminates for textile applications. In: Smith, W.C. (Ed.), Smart Textile Coatings and Laminates. Woodhead Publishing Limited in Association With the Textile Institute, Cambridge (Chapter 7).

Jassal, M., Agrawal, A.K., Save, N.S., 2006. Thermo responsive smart textile. Indian J. Fibre Textile Res. 31 (1), 52–65.

Kannekens, A., 1994. Breathable coatings and laminates. J. Coat. Fabr. 25, 41–59.

Kapsali, V., 2009. Biomimetic approach to the design of textiles for sportswear applications. In: Williams, J.T. (Ed.), Textiles for Cold Weather Apparel. Woodhead Publishing Limited in Association With the Textile Institute, Cambridge (Chapter 6).

Kaynak, H.K., Babaarslan, O., 2012. Polyester microfilament woven fabrics. In: Jeon, H.-Y. (Ed.), Woven Fabrics. InTech Textile Technology, Intech Open Science, Rijeka, Croatia, https://doi.org/10.5772/38483 Available at: https://www.intechopen.com/books/woven-fabrics/polyester-microfilament-woven-fabrics (accessed January 2017).

Keighley, J.H., 1985. Breathable fabrics and comfort in clothing. J. Coat. Fabr. 15, 89–105.

Kissa, E., 1984. Repellent finishes. In: Lewin, M., Sello, S.B. (Eds.), Handbook of Fiber Science and Technology, Vol. 2, Chemical Processing of Fibers and Fabrics, Part B Functional Finishes. Marcel Dekker, New York, NY, pp. 143–183.

Kissa, E., 1996. Wetting and wicking. Text. Res. J. 66 (10), 660–668.

Kothari, V.K., 2000. Quality Control: Fabric Comfort. Indian Ins. of Technology, Delhi.

Kovacevic, S., Ujevic, D., Brnada, S., 2010. Coated textile materials. In: Dubrovski, P.D. (Ed.), Woven Fabric Engineering. InTech, Intech Open Science, Rijeka, Croatiahttps://doi.org/10.5772/10468 Available at: http://www.intechopen.com/books/woven-fabric-engineering/coated-textile-materials. (Chapter 13).

Kramar, L., 1998. Recent and future trends for high performance fabrics providing breathability and waterproofness. J Coat. Fabr. 28 (October), 107–115.

Krings, W., Hansch, M., Krings, M., 2003. Lamination and bonding of three layer textile systems and breathable films. Melliand Int. 9 (2), 133–134.

Krishnan, S., 1992. Technology of breathable coatings. J. Coat. Fabr. 22, 71–74.

Levitt, S., 1986. Manchester Mackintoshes: a history of the rubberized garment trade in Manchester. Text. Hist. 17 (1), 51.

Li, Y., Luo, Z.X., 2000. Physical mechanisms of moisture diffusion into hygroscopic fabrics during humidity transients. J. Text. Inst. 91 (2), 302–316.

Li, Y., Zhu, Q., 2004. A model of heat and moisture transfer in porous textiles with phase change materials. Text. Res. J. 74 (5), 447–457.

Lister, W.N., 1963. Waterproof coated fabrics. In: Moilliet, J.L. (Ed.), Waterproofing and Water-Repellency. Elsevier, Amsterdam.

Lomax, G.R., 2007. Breathable polyurethane membranes for textile and related industries, J. Mater. Chem., Issue 27.

Lomax, G.R., 1985. The design of waterproof, water vapour permeable fabrics. J. Ind. Text. 15, 40–66.

Lomax, R., 1989. Ways of Water Proofing Breathable Fabrics. Textile Technology International. p. 305.

Lomax, G.R., 1990. Hydrophilic polyurethane coatings. J. Coat. Fabr. 20 (October), 88–107.

Lomax, G.R., 1991. Breathable, waterproof fabrics explained. Textiles 20 (4), 12–16.

Mierzinski, S., 1903. The Waterproofing of Fabrics. Greenwood and Son Co., London (translated from German).

Miller, B., 1977. The wetting of fibres. In: Schick, M.J. (Ed.), Surface Characteristics of Fibers and Textiles. Marcel Dekker, New York, NY, pp. 417–445.

Morris, G.J., 1953. Thermal properties of textile materials. J. Text. Inst. 44, 449–476.

Morton, D.H., Hearle, J.W.S., 2008. Physical Properties of Textile Fibres. Woodhead Publishing Limited in Association With the Textile Institute, Cambridge.

Mukhopadhyay, A., Midha, V.K., 2008. A review on designing the waterproof breathable fabrics part I: fundamental principles and designing aspects of breathable fabrics. J. Ind. Text. 37 (3), 225–262.

Mukhopadhyay, A., Midha, V.K., 2016. Waterproof breathable fabrics. In: Handbook of Technical Textiles, second ed., vol. 2: technical textile applications (Chapter 2).

Odermatt, D., n.d. EtaProof specifications, Technical Data. Available at: http://stotzfabrics.ch/en/outdoorfabrics/ (accessed January 2017).

Online Etymology Dictionary, Waterproof. Available at: http://www.etymonline.com/index.php?allowed_in_frame=0&search=waterproof (accessed December 2016).

Osczevski, R.J., Dolhan, P.A., 1989. Anomalous diffusion in a water vapour permeable, waterproof coating. J. Coat. Fabr. 18, 255–258.
Parsons, M., Rose, M.B., 2011. How Did the Waterproof Jacket Evolve? Part 1. Outdoor Gear Coach. Available at: http://www.outdoorgearcoach.co.uk/tools-resources/waterproof-jacket-evolution-part-1/ (accessed October 2016).
Patnaik, A., Rengasamy, R.S., Kothari, V.K., Ghosh, A., 2006. Wetting and wicking in fibrous materials. Text. Prog. 38 (1), 1–105.
Peirce, F.T., Rees, W.H., Ogden, L.W., 1945. Measurement of the water vapour permeability of textile fabrics. J. Text. Inst. 36, T169–T176.
Rees, W.H., 1971. Physical factors determining the comfort performance of textiles. In: Shirley Institute 3rd Seminar: Textiles in Comfort.
Ren, Y.J., Ruckman, J.E., 2003. Water vapour transfer in wet waterproof breathable fabrics. J. Ind. Text. 32 (3/1), 165–175.
Roey, M.V., 1991. Water-resistant breathable fabrics. J. Coat. Fabr. 21 (July), 21–31.
Rossi, R.M., Gross, R., May, H., 2004. Water vapor transfer and condensation effects in multilayer textile combinations. Text. Res. J. 74 (1), 1–6.
Rowen, J.W., Gagliardi, D., 1947. Properties of water-repellent fabrics. Research Paper RP1762, Part of the Journal of Research of the National Bureau of Standards, 38 (January), 103–117.
Ruckmann, J.E., 1997a. Water vapour transfer in waterproof breathable fabrics part 1: under steady-state conditions. Int. J. Cloth. Sci. Technol. 9 (1), 10–22.
Ruckman, J.E., 1997b. Water vapour transfer in waterproof breathable fabrics part 2: under windy conditions. Int. J. Cloth. Sci. Technol. 9 (1), 23–33.
Ruckman, J.E., 1997c. Water vapour transfer in waterproof breathable fabrics part 3: under rainy and windy conditions. Int. J. Cloth. Sci. Technol. 9 (2), 141–153.
Save, N.S., Jassal, M., Agrawal, A.K., 2002. Polyacrylamide based breathable coating for cotton fabric. J. Indust. Text. 32 (2), 191.
Saxena, R.K., 1999. Studies on water vapour transfer through nylon and cotton fabrics. Indian J. Fibre Text. Res. 24 (September), 188–192.
Schledjewski, R., Schultze, D., Imbach, K., 1997. Breathable protective clothing with hydrophilic thermoplastic elastomer membrane films. J. Coat. Fabr. 27, 105–114.
Schoeffler, O.E., Gale, W., 1973. Esquire's Encyclopedia of 20th Century Men's Fashion. McGrawHill Book Company, New York, NY.
Scott, R.A., 1995. Coated and laminated fabrics. In: Carr, C.M. (Ed.), Chemistry of the Textile Industry. Blackie Academic & Professional, London.
Sen, A.K., 2008. Coated Textiles: Principles and Applications, second ed. CRC Press, Taylor & Francis Group, Boca Raton, FL.
Shekar, R.I., Yadav, A.K., Kumar, K., Tripathi, V.S., 2003. Breathable apparel fabrics for defence applications. Man-Made Text. India 46 (12), 9–16.
Shekar, R.I., Kumar, K., Kotresh, T.M., 2005. Development of closely woven breathable linen fabric for water storage applications. Indian J. Fibre Text. Res. 30 (September), 335–339.
Shephard, A.J., 2009. Waterproof Dress: Patents as Evidence of Design and Function From 1880 Through 1895 (Ph.D thesis). The University of Missouri, Columbia, MO.
Shephard, A.J., 2012. Waterproof dress: patents as evidence of design and function from 1880 through 1895. Cloth. Text. Res. J. 30 (3), 183–199.
Sierra Trading Post, n.d. Waterproof Guide. Available at: http://www.sierratradingpost.com/lp2/waterproof-guide/ (accessed November 2016).
Stomatex, 2004. F.A.Q. What is Stomatex?. Available at: http://www.stomatex.com/faqs.htm (accessed February 2017).
Traubel, H., 1999. New Materials Permeable to Water Vapor. Springer-Verlag, Berlin, Heidelberg.

Vander, A., Sherman, J., Luciano, D., 1998. Human Physiology: The Mechanisms of Body Function, Available at:seventh ed. W.C. Brown Pub. Co, Dubuque. Available at: http://www.mhhe.com/biosci/ap/vander/student/olc/d-reading1.html (accessed October 2016).
Ventile Fabrics, n.d. Ventile® Fabrics the history. Available at: http://www.ventile.co.uk/history/ (accessed January 2017).
Vivekanandan, M.K., Sreenivasan, S., 2012. Dynamic transportation of water vapor through cotton and polyester-cotton blended fabrics part I: indices characterizing moisture buffering and their interrelationships. J. Eng. Fibers Fabr. 7 (4), 70–80.
Wang, J.H., Yasuda, H., 1991. Dynamic water vapor and heat transfer through layered fabrics: part 1: effect of surface modification. Text. Res. J. 61 (1), 10–20.
Wang, S.X., Li, Y., Tokura, H., Hu, J.J., Han, Y.X., Kwok, Y.L., Au, R.W., 2007. Effect of moisture management on functional performance of cold protective clothing. Text. Res. J. 77 (12), 968–980.
Watt, I.C., Kabir, M., 1975. Sorption of water vapor in jute fibers. Text. Res. J. 45 (1), 42–48.
Weiner, L.I., 1970. The relationship of moisture vapour transmission to the structure of textile fabrics. Text. Chem. Color. 2 (22), 378–385.
Whelan, M.E., MacHattie, L.E., Coodings, A.C., Turl, L.H., 1955. The diffusion of water vapour through laminae with particular reference to textile fabrics. Text. Res. J. 25 (3), 197–222.
Williams, J.T., 1997. A comparison of techniques used to assess the thermal burden of protective clothing. In: Stull, J.O., Schwope, A.D. (Eds.), Performance of Protective Clothing. vol. 6. American Society for Testing and Materials, Philadelphia, PA. ASTM STP 1273.
Wood, T.D., 2016. Rainworker: How It Works, Learn at REI. Available at: https://www.rei.com/learn/expertadvice/rainwearhowitworks.html (accessed February 2017).
Woodcock, A.H., 1962a. Moisture transfer in textile systems, part 1. Text. Res. J. 32 (8), 628–633.
Woodcock, A.H., 1962b. Moisture transfer in textile systems, part 2. Text. Res. J. 32 (9), 719–723.
Yasuda, H., Miyama, M., 1992. Dynamic water vapor and heat transfer through layered fabrics: part 2: effect of chemical nature of fibres. Text. Res. J. 62 (4), 227–235.
Yasuda, T., Miyama, M., Muramoto, A., Yasuda, H., 1994. Dynamic water vapor and heat transport through layered fabrics part III: surface temperature change. Text. Res. J. 64 (8), 457–461.
Zisman, W.A., 1964. Relation of the equilibrium contact angle to liquid and solid constitution. In: Gould, R.F. (Ed.), Contact Angle, Wettability, and Adhesion, Advances. In: Advances in Chemistry, vol. 43. American Chemical Society, Washington, DC.

Further reading

Adam, N.K., 1963. Principle of water repellency. In: Moilliet, J.L. (Ed.), Waterproofing and Water-Repellency. Elsevier Science, Amsterdam.
Bakshi, A.S., 2015. Development and Study of Waterproof Breathable Fabric Using Silicone Oil and Polyurethane Binder (M.Sc. thesis). Eastern Michigan University.
Gore, R.W., 1976. Process for Producing Porous Products. W.L. Gore & Associates, Newark, USP 3,953,566, April 27.
Save, N.S., Jassal, M., Agrawal, A.K., 2005. Smart breathable fabric. J. Indust. Text. 34 (3), 139–155.
Starov, V., Velarde, M., Radke, C., 2007. Wetting and Spreading Dynamics. CRC Press/Taylor & Francis Group, Boca Raton, FL.

Soil repellency and stain resistance through hydrophobic and oleophobic treatments

3

Silvia Pavlidou, Roshan Paul†*
*Materials Industrial Research and Technology Center, Athens, Greece, †University of Beira Interior, Covilhã, Portugal

3.1 Introduction

Textile soiling or staining resulting from the retention of soil or stain on a textile, originates from two main sources: the environment, and the body of the wearer. While the composition of soiling and staining varies according to source, it usually consists of two components: (1) a liquid component, usually an oil, fat or grease, and (2) a solid component made up of small particles. Oils and fats from food and cosmetics, secretions from human skin and lubricating oils and grease from household products are the main source of liquid components, while airborne dust particles constitute main sources of solid components (Baghaei and Mehmood, 2011).

Soiling occurs when the soil comes in contact with the fabric and is retained as a more or less stable unit. A fabric becomes soiled either by contact with airborne or liquid-borne substances or by direct contact with another soiled surface. Soil particles floating in the atmosphere settle down on the fabrics, while liquid-borne soils that come in contact with the fabric may evaporate or filter off, leaving undissolved or suspended particles on the fabric surface (Hashim, 1986).

With any fabric structure, pores are created during the interlacing of the yarns; yarns also have inter-fibre spaces. These porous cavities have the ability to trap potential stains as a result of capillary forces driving the liquid stains. In addition, fine particles from contact with solid materials can adhere to the fibre and fabric surface due to electrostatic charges, or become embedded into the yarn and/or fabric interstices, which also results in stains (Hashim, 1986).

It becomes obvious that soil repellency and stain resistance, referring to the ability of a fabric to prevent liquid absorption and fine particles adherence to its surface, are directly linked to oleophobic and hydrophobic functionalities (Textile Innovation Knowledge Platform, 2017). In fact, it is mostly for marketing reasons that the scientific terms hydrophobic and oleophobic are translated for the consumer into buzzwords, such as soil repellent, stain resistant or self-cleaning (Mahltig, 2014).

Since hydrophobicity and oleophobicity are surface properties of a material, they are usually achieved through specific surface treatments of a textile, involving, in most cases, the application of a so-called soil-repellent or stain-resistant finish. While

Waterproof and Water Repellent Textiles and Clothing. https://doi.org/10.1016/B978-0-08-101212-3.00003-4

water repellency can be achieved using different product groups, e.g. waxes, silicones and silanes, oil (and thus soil) repellency is normally attained only with fluorocarbon polymers. These can be tailored to fit the different demands of the users and the intended purposes (Schindler and Hauser, 2004; Souma, 2012).

It is worth noting that repellent finishes often bring along more functionalities to a textile, such as better durable press properties, more rapid drying and ironing and increased resistance to acids, bases and other chemicals. However, in addition to the desired repellency effects and functionalities, repellent finishes can also have adverse effects. These include problems with static electricity, poor soil removal and/or greying (soil re-deposition) in aqueous laundering, stiffer handle, reduced permeability and increased flammability (Schindler and Hauser, 2004).

Even more, the main issue is that the fluorinated compounds used to endow soil resistance to textiles have been shown to be hazardous for both human health and the environment. They have been found in locations far away from any source proving they are persistent. They have been detected in human blood across the world and have been found to be endocrine disrupting. Much effort is, therefore, concentrated on finding less harmful alternatives or new methods to reduce the use of these substances (Hedegard, 2014). Plasma treatments and use of nanotechnology have been proposed as alternatives, and will be discussed in the following sections.

The field of soil-repellent and stain-resistant textiles receives much attention and research efforts, since such products are important in all parts of the textile market; for clothing, home and technical applications (Schindler and Hauser, 2004). In fact, the use of soil-repellent and stain-resistant finishes gives added value to consumer products in the apparel sector, with this technology now being widely marketed on children's clothing and workwear, reducing maintenance requirements and extending the lifetime of the garments. Military clothing benefit from such finishes as the garments are required to perform in battlefield conditions with reduced care and laundering. In the interiors sector, stain-resistant finishes have widespread use in applications such as upholstery and floor coverings. Extreme-performance upholstery requires durability, stain resistance and ease of cleaning. Such upholstery and trim with stain-resistance characteristics is of added value in the automotive industry to lengthen a product's life and make day-to-day care easier. Finally, stain resistance of medical and hygiene textiles is mainly concentrated on uses such as interior textiles, carpets, curtains, bedding and uniforms, including protective clothing. Preventing staining or penetration of the textile can be helpful in controlling microbial activity and cross contamination (Textile Innovation Knowledge Platform, 2017). It follows that the consumer would not want to pay the same price for a more environmentally friendly fabric with reduced performance that requires cleaning or reproofing more often.

3.2 Soil repellency and stain-resistance mechanism

Soiling and soil resistance are complex phenomena, involving the interrelationship of the nature of the fibre surface, the fibre/fabric structure, as well as the nature of the soil

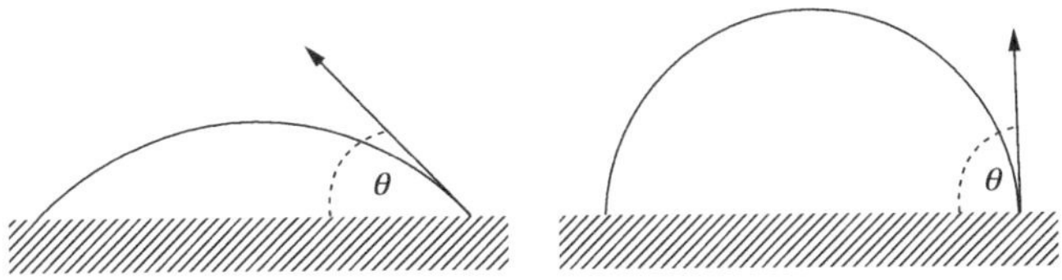

Fig. 3.1 Scheme of drops on solid surfaces, indicating the relevant contact angles.

and chemical finishes (Hashim, 1986). As already mentioned, for a surface to be stain resistant, it needs to be both water and oil repellent.

Usually, the wetting behaviour of water onto a surface is simply determined and classified by the contact angle of water deposited on the solid substrate (Fig. 3.1). The lower the contact angle the higher the wettability. Surfaces with contact angles against water of 90° or higher are usually considered hydrophobic surfaces. Similarly, to gain information on the wetting behaviour of surfaces by different oils and solvents, it is possible to determine the contact angles of all requested liquids on all requested solid materials (Mahltig, 2014).

However, for estimating wetting properties, besides using contact angles, the term surface tension or surface energy is used. The surface tension is typical for every material (solid or liquid) and is related to the chemical composition at the surface. In summary, a liquid does not wet a surface if the surface tension of the liquid is higher than the surface tension of the solid substrate (Mahltig, 2014).

Accordingly, repellent properties can be achieved by lowering the free energy at the fibre surface. The surfaces that perform low interactions with liquids are called low-energy surfaces. The critical surface energy or surface tension (γ_c) of these low-energy surfaces must be smaller than the surface tension of liquid water or of the oil that is to be repelled. The surface tension of oils (γ_L=20–30 mN m^{-1}) is two to three times less than the surface tension of water (73 mN m^{-1}). Therefore, when the low-energy surface is produced by treating the textile with fluorocarbons having a surface tension of 10–20 mN m^{-1}, the treated surface not only repels the oil but also exhibits repellency against water. Conversely, if the low-energy surface is produced by treating the textile with silicone or paraffin wax having surface tension of 24–30 mN m^{-1}, the treated textile will repel the water but not oil (Baghaei and Mehmood, 2011). Table 3.1 summarizes the surface tensions of polymers typically used for synthetic fibres production, along with the surface tensions of typical liquids and oils (Mahltig, 2014).

Finally, it is important to note that besides the chemical composition of the liquid and the solid surface, the surface structure of the solid substrate also plays a key role in determining the wetting behaviour (Mahltig, 2014). For example, introducing surface roughness on a hydrophobic substrate will increase the contact angle to water, since the shape of the droplet becomes more spherical to reduce the contact between water and solid (Hedegard, 2014). Further, in designing a super-repellent surface that is not wet even to liquids with a lower surface tension than water, micro- and nano-scale surface roughness have a great impact in addition to low surface energy

Table 3.1 Surface tensions of polymers typically used for synthetic fibres production and surface tensions of typical liquids and oils (Mahltig, 2014)

Polymer surface	Surface tension (mN m^{-1})	Liquid	Surface tension (mN m^{-1})
Technical fibres		Octane	22
Polytetrafluoroethylene (PTFE)	18.5	Decane	24
Polyvinylidenefluoride (PVLF)	25	Dodecane	25
Polyvinylfluoride (PVF)	28	Hexadecane	28
Polyethylene (PE)	31	Octanol	28
Polyvinylchloride (PVC)	39	Hexadecane/paraffin oil mix	29
		Paraffin oil	31
Common synthetic fibres		*p*-Xylene	31
Polyethylene terephthalate (PET)	43	Cyclohexane	33
Polyamide 66 (PA)	46	Chlorobenzene	36
		Diiodomethane	51
		Formamide	58
		Glycerol	63
		Water	72

(Park et al., 2016). A bi-modal structure in the micro- and nanometre range, e.g. the morphology of the lotus plant, seems to be ideal (Fig. 3.2). Air inclusions within the microstructure minimize the effective contact area of the interface to external materials such as liquids and soiling (Hegemann, 2015). However, this surface roughness may not be sufficiently durable for clothing. To our knowledge there are no commercial finishing agents based on this approach currently available.

3.3 Treatments to develop soil-repellent and stain-resistant textiles

Soil repellency and stain resistance is usually achieved by the application of appropriate finishes through wet chemical treatment of textiles. The exact process employed in the finishing depends on the chemicals to be used, the fabric type and the available machinery and ranges from exhaustion in dyeing machines, to padding, spraying or foaming (Outdoor Industry, 2012). The most widely studied soil-repellent finishing agents for textiles are discussed further. Reference is also made to plasma treatment as an alternative, more environmentally friendly approach.

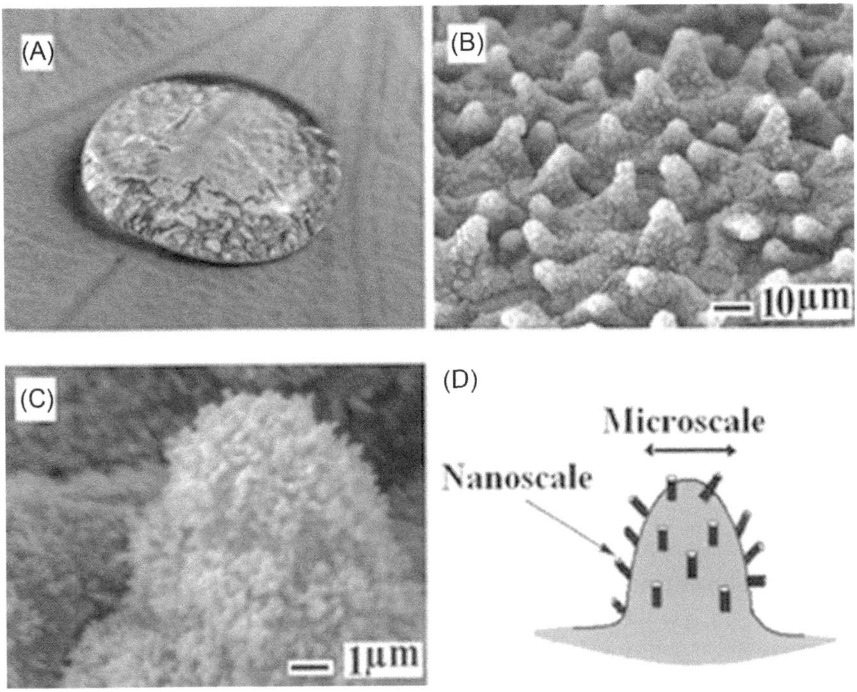

Fig. 3.2 Lotus effect. (A) Lotus leaf and beads; (B) SEM image of micropapillae present on the surface of lotus leaf; (C) Enlarged view of (B); (D) bi-modal roughness structure in the micro- and nanometre range (Mao et al., 2009).

3.3.1 Long chain and short chain fluorocarbon finishes

Fluorocarbons hold a unique place as repellent finishing agents, as they can provide the textile substrate with the lowest surface energies of any repellent finishes in the marketplace. They can be formulated to have a durable effect and can be water and oil repellent, but fluorocarbons and their breakdown products can result in health and environmental problems (Hastbacka, 2016).

From the chemical point of view, fluorochemicals or fluorinated compounds are organic chemicals containing at least one fluorine atom. Nevertheless, in the field of textile technology, fluorocarbons are defined as organic compounds with a high percentage of fluorine replacing the hydrogen bound to carbon. For typical industrial products fluorine contents of around 20% are reported. Typical fluorocarbons used nowadays for soil-repellent textile finishing are polymers with perfluorinated alkyl side chains (Mahltig, 2014).

In particular, fluorocarbon repellents are synthesized by incorporating perfluoro alkyl groups into acrylic or urethane monomers that can be polymerized to form fabric finishes. Originally, the perfluoro alkyl groups were produced by electrochemical fluorination, a production method which was later replaced by telomerization. The final polymer, when applied to a fibrous substrate, should form a structure that

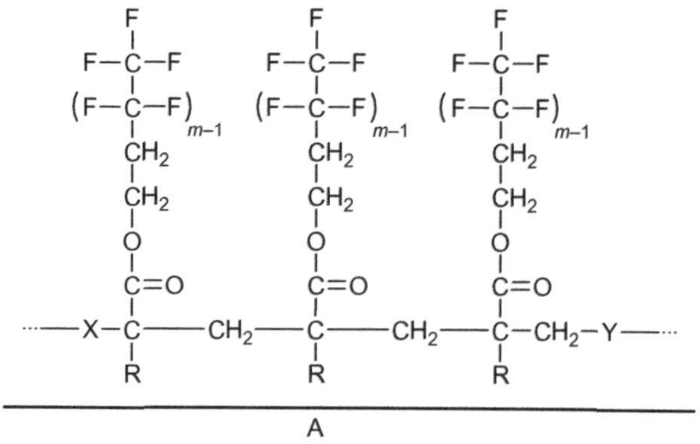

Fig. 3.3 Typical structure of fluorocarbon repellent on a textile surface. $m = 8–10$. X and Y are co-monomers, mainly stearylacrylates. R= H or CH_3. A is the textile surface.

presents a dense CF_3 outer surface for maximum repellency. A typical structure is shown in Fig. 3.3. The length of the perfluorinated side chains should be about 8–10 carbons. The small spacer group, mostly ethylene, can be modified to provide emulsification and solubility of the polymer. Co-monomers (X, Y; e.g. stearyl- or lauryl-methacrylate, butylacrylate, methylol- or epoxy- functional acrylates and block copolymers from α,ω-dihydroxydimethylpolysiloxane) affect the handle of the fabric, film formation capability and durability. In this way, and by adding appropriate emulsifiers, fluorocarbon products can be widely modified for many special performance profiles (Schindler and Hauser, 2004).

Most fluorocarbon repellents are padded, dried and cured. Heat treatment causes an orientation of the perfluoro side chains to almost crystalline structures, which is crucial for optimal repellency. Washing and dry cleaning disturb this orientation and reduce finish performance. The orientation must be regenerated by a new heat treatment (ironing, pressing or tumble drying) (Schindler and Hauser, 2004). Additional chemicals used during the application of fluorocarbons are also important in determining the effect achieved. Such chemicals could be cross-linkers or the so-called boosters that enhance the film-forming properties and film stability, and so improve the effectiveness of fluorocarbon finishing. The concept of booster technology is depicted in Fig. 3.4. Cross-linking agents can be based, for example, on isocyanate block copolymers and aziridine compounds. Other compositions contain butanetetracarboxylic acid. One key purpose of using cross-linkers is to improve the wash fastness of the finishing (Mahltig, 2014).

Following an alternative approach, Hanumansetty et al. (2012) formed fluoropolymer finishes on cotton knits by admicellar polymerization, a surface analogue of emulsion polymerization. This approach enabled the formation of durable finishes that exhibited high performance in stain resistance and stain repellency.

An important issue of fluorocarbon finishing agents is the ecology aspect, due to the fact that persistent fluorinated compounds can occur as by-products in the production of fluorocarbon polymers or fluorocarbon resins. The ecological disadvantages of fluorocarbon finishing are even discussed in daily newspapers, so the well-informed customer will be aware of these issues. The main concerns are related to the

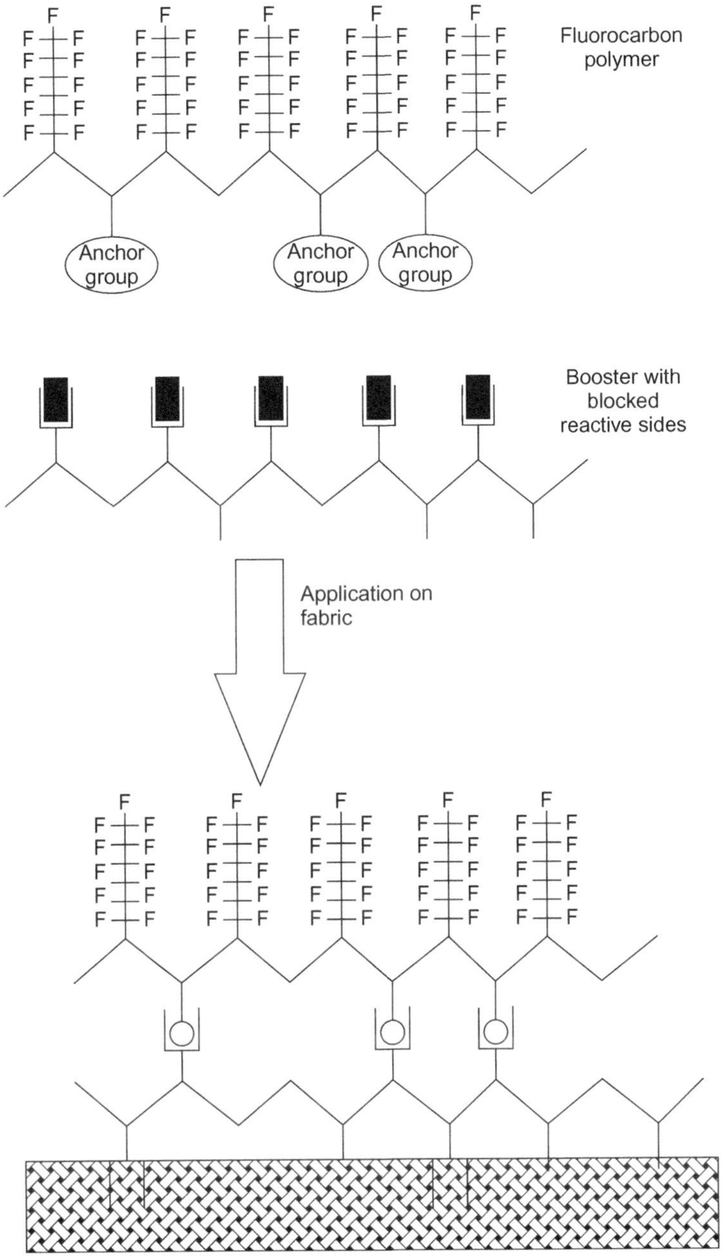

Fig. 3.4 Schematic drawing of booster technology for fixation and self-organization of fluorocarbon polymers on fibre surfaces (Mahltig, 2014).

persistence of fluorocarbons and their possible accumulation in air, food and drinking water (Mahltig, 2014). The first perfluorinated compound that gained attention for its toxicity and widespread occurrence in the environment was perfluorooctane (PFOS) and this substance is now restricted. The second substance to attract concern was perfluorooctanoic acid (PFOA). Both PFOS and PFOA are perfluoroalkyl acids (PFAAs) with a chain length of 8 carbons and derivatives of them, that are used industrially, will degrade or metabolize to the corresponding PFAA as an end-product (Hedegard, 2014).

As a result of their strong carbon-fluorine bonds, PFOA and PFOS do not readily break down in the environment. They have been shown to be persistent in the environment and have long elimination half-life in wildlife and humans. Numerous reports (Lin et al., 2014; Duong et al., 2015; Alder and van der Voet, 2015) have documented the presence of long-chain PFAAs in aquatic environments in Japan, United States, Germany and Italy, with PFOA and PFOS comprising the most detected chemicals (Hastbacka, 2016).

In light of the concerns associated with long-chain PFAAs, there is a shift towards repellent chemistries with shorter perfluoroalkyl chains (also termed 'C6' or 'C4' depending on the number of carbons in the perfluoroalkyl chain). Chemically, short-chain fluorinated chemistries are closely related to their long-chain homologues. Repellents containing short-chain fluorinated chemistries are produced using perfluoroalkyl raw materials such as fluorotelomer alcohols that are not expected to break down in the environment into PFOA and PFOS (Outdoor Industry, 2012).

Short-chain fluorinated repellent chemistries were promoted during the last decade by the chemical industry as having repellency and other performance attributes comparable to long-chain chemistries. The industry is, in fact, on a learning curve to match the performance levels of long-chain fluorinated chemistries. In general, short-chain fluorinated chemistries are not as effective as those with long-chains, particularly in repelling oil. For higher performance applications including 50 or more home laundering cycles, strong rain and aggressive stain resistance, there are reductions in performance levels achieved with short-chain fluorinated repellents. Although certain performance levels may eventually be achieved, it is understood that there are critical applications where the required performance levels may never be achieved by short-chain fluorinated chemistries. Furthermore, it has to be pointed out that substituting a long chain with a short-chain fluorinated repellent chemistry may require optimizing the application method, e.g. by increasing the amount of finishing applied on the textile (Outdoor Industry, 2012). Finally, although short-chain length fluorocarbons are considered to be less harmful than those of longer chain length, there are studies that indicate even these substances have toxic effects on the environment and human health (Hedegard, 2014). This may also be true for the other water repellent treatments.

The majority of the fluorocarbon-based chemicals are not regulated in any way; however, specific compounds are forbidden to sell, use or manufacture. PFOS is one of those which use is restricted by the EU in regulation number 850/2004. The phasing out of PFOS occurred voluntarily from the manufacturers due to this concern. In 2006 major chemical manufacturers also agreed to phase out the production of

PFOA by 95% by 2010 and a complete stop of production by 2015. Further, it is worth noting that environmental certification standards (e.g. Oeko-Tex and Bluesign) set very low threshold limits for the use of PFOS/PFOA, allowing only for trace amounts (Souma, 2012).

3.3.2 Dual-action stain-repellent and release fluorocarbon finishes

The introduction of soil repellency has been accepted by the apparel industry, but some consumers have experienced problems when soil repellency fails, allowing stains to become trapped or embedded in repellent-finished fabrics (Easter and Ankenman, 2005). This is a natural consequence of the fact that water repellency impedes the access of the washing liquid during laundering (Schindler and Hauser, 2004).

Therefore so-called dual-action fluorocarbon block copolymers have been developed, which combine repellency in the dry state and soil-release effects in an aqueous environment. Dual-action fluorocarbons enable a better removal of oily stains and dirt in domestic washing or laundering and may be better than existing ones in health/environmental issues. With conventional fluorocarbon products, the wash water is hindered from wetting and penetrating the fabric. Dual-action fluorocarbons are called hybrid fluorochemicals because they are block copolymers containing hydrophobic (like the usual fluorocarbons) and highly hydrophilic segments. In air, the perfluorinated side chains are, as usual, outwardly oriented and develop high repellency, but in water the double-face surface flips and the hydrophilic segments turn outwards to promote the wash effect as a soil-release finish. During heat drying or ironing the sandwich-like surface flips again to generate water, oil and soil repellency (Schindler and Hauser, 2004).

As an example, Pittman and Wasley (1972) disclosed the synthesis of polymers having the desirable combination of soil repellency and soil release. These are copolymerization products of at least two different monomers, one imparting oleophobic properties; the other hydrophilic properties. More specifically, the oleophobic monomer is a silane containing a terminal perfluoroalkyl group of 3 to 18 perfluorinated carbon atoms, such group being hereinafter designated as R_f. The hydrophilic monomer is a silane containing two or more groups of the structure —Alk—O— wherein Alk is an alkylene group containing 2–6 carbon atoms. The resulting siloxane copolymers have a silicon-to-oxygen backbone, plus pendant R_f groups that provide oleophobicity and groups of the structure —Alk—O— that provide hydrophilicity.

3.3.3 Dendrimers

Dendrimers are highly branched molecules, consisting of a number of branched organic units attached to a core molecule (Hedegard, 2014). They are known for their well-defined, regular structure, built up in several generations starting from a core and containing a surface with a high density of end groups (functional groups). The number of end groups increases exponentially with the number of generations building up the dendrimer. The costs of dendrimers also increase exponentially with increasing number of generations (Mahltig, 2014). This is because the synthesis of monodisperse polymers demands a high level of synthetic control, which can be achieved through step-by-step reaction, building the dendrimer up one monomer layer at a time (Outdoor Industry, 2012). In the future, with increased production or better technologies, the costs are expected to come down.

Thanks to their unique structure, dendrimers may be used in a wide variety of applications ranging from medicine to genetics, biology and chemistry, such as drug delivery, or as contrast agents for magnetic resonance imaging (Hedegard, 2014; Outdoor Industry, 2012). Examples of different dendrimer structures are illustrated in Fig. 3.5 (Vögtle et al., 2000).

Dendrimer-based repellent chemistry is a relatively new field of repellency. Depending on the chemical composition, dendrimers provide water and/or oil and soil repellency. In fact, two types of dendrimer applications have to be distinguished. First, the dendrimer fluorocarbons, which contain fluorinated alkyl groups and are supposed to lead to hydrophobic and oleophobic properties by themselves. Compared to other PFAS-based repellents, the fluorine content of the oleophobic dendrimer finishes is reduced. Second, non-fluorinated dendrimers applied in combination with conventional fluorocarbon finishing. In the latter case, dendrimers are applied to increase the oleophobic properties of the additionally applied fluorocarbons. The application of dendrimers together with fluorine-free polymers is reported to lead to excellent hydrophobic properties, however, without additional fluorinated compound no oleophobic properties are realized (Mahltig, 2014; The Danish Environmental Protection Agency, 2015).

Commonly, these finishes are applied as two-component systems consisting of an emulsion containing the dendrimers and a solution containing a cross-linking substance providing the fixation to the fibre. Fluorocarbon-free dendrimers are based on hydrocarbon or polyurethane chemistry. Cross-linking is commonly achieved by

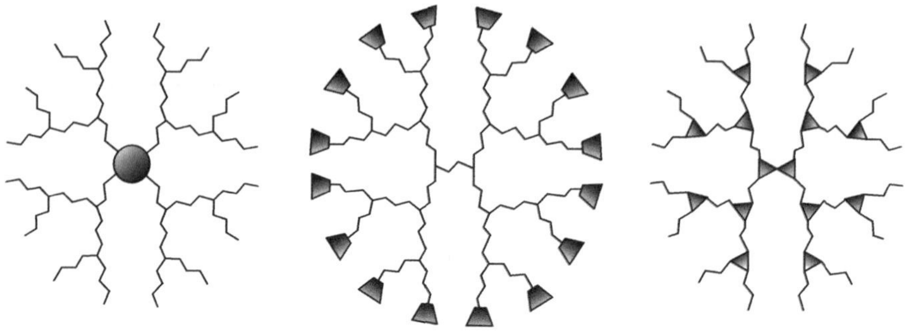

Fig. 3.5 Examples of dendrimers with different locations for the functional groups (Vögtle et al., 2000).

chemical binding of the dendrimers with isocyanates to the fibre. Glycols are added as solvents and cationic surfactants in small amounts act as emulsifiers (The Danish Environmental Protection Agency, 2015).

Although a number of dendrimer-based repellent finishes are already available on the market, e.g. RUCO-DRY ECO from Rudolf GmbH, relevant research efforts are still ongoing to optimize their formulation.

3.3.4 Nanotechnology and nanostructured surfaces

Cassie and Baxter (1944) were the first to observe that rough surfaces repel water due to the air enclosed between the gaps on the surface. This enlarges the water/air interface while the solid/water interface is minimized. In this situation, spreading does not occur and the water forms a spherical droplet. The self-cleaning propensity of plant leaves with a rough surface was investigated and reported by Neinhuis and Barthlott (1997). They observed that on water repellent surfaces, water contracted to form spherical droplets. It ran off the leaf very quickly, even at slight angles of inclination ($<5°$ angle), without leaving any residue. Particles of all kinds that adhered to the leaf surface were always removed entirely from water repellent leaves when subjected to natural or artificial rain, as long as the surface waxes were not destroyed (Gulrajani, 2013).

This phenomenon, known as the lotus effect, is the result of the combination of hydrophobic properties and combined micro- and nano-structures on the surface. To apply lotus-like behaviour to textiles, three features have to be introduced onto the textile surface: a microscopic structure, a nanoscopic structure and hydrophobic properties. However, most textiles already have a microscopic structure, because the diameters of fibres are on the micrometre scale. The challenge is to apply nanometre structures onto the fibres and combine these with hydrophobic properties (Mahltig, 2014).

The most common approach to achieve that involves the surface application of different types of nanoparticles in combination with hydrophobic components. In fact, the use of nanomaterials to impart water repellency and stain resistance effects is one of the most common ways nanotechnology is being used in the textile industry (The Danish Environmental Protection Agency, 2015). In this context, spherical nanoparticles, such as SiO_2 or TiO_2 particles, have been explored, as well as nanoparticles with different shapes, including nanowhiskers and nanotubes. In these cases, hydrophobic modification of the nanostructured surface is also required, through application of additional components such as butylacrylate, poly(glycidyl methacrylate), etc. (Mahltig, 2014).

It should be noted, however, in most cases the approach of using hydrophobic particles and film-forming agents to create super-hydrophobic surfaces has the drawback of poor durability on textile substrates. An alternative approach is to use a combination of both chemical and mechanical treatments to create super-hydrophobic nanostructures on the surface of textile materials. Mechanically roughened surfaces become an integral part of the product and are more durable. Mechanical roughening of the fabric can be carried out by treatments such as calendaring, embossing, etching, schreinering, sueding, sanding, abrading or emorizing. A subsequent treatment with cross-linkable fluorocarbon having nanoparticles of, for example, silica, colloidal silica, alumina, zirconia, titania, zinc oxide, precipitated calcium carbonate or PTFE of 10 to 50 nm size significantly improves the hydrophobicity (Gulrajani, 2013).

An even more innovative approach to address the issue of durable soil-resistant treatments was developed in a joint research project of the BMWi (Federal Ministry for Economic Affairs and Energy, Germany) carried out by the Hohenstein Institute and ITCF Denkendorf. By means of new fibre structuring, the soil-repellent effect has for the first time been realized through fibre immanent surface structuring. Particles that have been incorporated during manufacturing have shown only negligible effects on the behaviour of dyes and the dye absorbency of the polymer. More specifically, an additional nanostructured surface was generated with the help of super paramagnetic nanoparticles directly during the melt spinning process of man-made fibres. This so-called ferromagnetic fibre structuring with high-energy magnetic field coil takes place directly after the spinneret, when the spin-melt is still in a thermoplastic state, which ultimately allows for subsequent filament stretching as usual (Hohenstein Institute, 2017).

3.3.5 Hydrophobins

Hydrophobins are spherical proteins with diameters of few nanometres, having a hydrophobic half and a hydrophilic half. They naturally occur on the top of mushroom heads, where they prevent soiling of the mushroom and offer some water repellency. To copy this material from nature, a hydrophobin from aqueous solution is used as a textile finish. Depending on the properties of the textile, the hydrophobic or the hydrophilic part of the hydrophobin is placed on the fibre and the part with opposite properties is uppermost, leading to new textile properties. Therefore, hydrophilic cotton may become hydrophobic if a monolayer of hydrophobin is deposited on its surface. Where the hydrophobins are deposited in multilayers this effect is diminished. Although the application of hydrophobins on textiles has been demonstrated, the repellency achieved is not strong enough for commercial application and further research is required (Mahltig, 2014).

3.3.6 Plasma treatments

Conventional textile treatments, including dip-coating and padding with chemicals, require the utilization of large amounts of chemicals with significant environmental impact. This is especially true when it comes to application of hazardous substances such as fluorocarbons for stain resistance. Even more, conventional treatments may also affect the mechanical properties of the fabrics as, for instance, reducing the durability and the comfort to wear. Among the alternative techniques that have been investigated during the last 20 years to reduce the use of chemicals by the textile industry, plasma treatment attracts particular interest especially for its main peculiarities: the treatment modifies only the uppermost layers of the fabric surface without affecting the bulk properties, and it has low energy requirements and is environmental friendly as the use of chemicals is negligible (De Vietro et al., 2015).

Low-pressure plasma treatments have been applied to achieve different textile functionalities, including water repellency and stain resistance. Several research groups investigated the hydrophobicity enhancement of polymers and fabrics using plasmas fed with fluorocompounds, e.g. tetrafluoromethane (CF_4), sulphur hexafluoride (SF_6), hexafluoroethane (C_2F_6), hexafluoropropene (C_3F_6), etc. It was found that plasma fed with small molecules (e.g. CF_4, C_2F_6) did not result in treatments of good durability, probably for the formation of short polymer segments dangling on the treated surface. For example, a coating with F/C ratio of 1.04 deposited on silk and cotton in plasma fed with C_3F_6 shows a good hydrophobic character (water contact angle, WCA higher than 120° angle), but suffers a partial loss after water washing and alcohol-extraction (De Vietro et al., 2015).

Larger molecules like 2-(perfluorohexyl)ethyl acrylate can also be used in plasma polymerization processes. The advantage of using this monomer is that already a FC_6 fluorocarbon alkyl chain is present in the monomer and can be applied completely by plasma induced graft polymerization on the textile surface. Applications for this method are given in the literature for polyamide/cotton-blended yarn for military textiles. Another reason for plasma application is as a primer or pre-treatment used to enhance the fixation of fluorocarbons applied in a second step by a wet dipping procedure. Oxidative plasma is used to activate and equalize the textile surface by oxidation. In a second step, primer molecules are applied on the oxidized fibre surface, which is used as an anchor for grafting perfluorooctyl methacrylates (Mahltig, 2014).

While such low-pressure plasma treatments can be applied only on a batch scale, atmospheric plasma machines offer the possibility of treating textiles on a continuous scale. They can be used mostly for textile pretreatments and lack the versatility of low-pressure machines used for plasma polymerization. In this respect, new hybrid machines incorporating the characteristics of both low-pressure and atmospheric machines, as offered by GRINP srl Turin, Italy, are more aligned to the needs of the industry.

3.4 Assessment of textile soil repellency and stain resistance

The required performance level of repellent textiles directly depends on their intended use. There is not a single acceptable performance level for repellent finishes on textile products. In fact, the required performance levels are set by brands or retailers selling the finished textile products and vary considerably from one brand or retailer to the next and from one fabric or product to another (Outdoor Industry, 2012).

Accordingly there is no specific standard available from the International Standardization Organization or the American ASTM focusing wholly on stain resistance. This is also due to the fact that this property is closely interlinked with hydrophobicity and oleophobicity. ISO 23232, water/alcohol resistance test, can provide information but is not intended to measure penetration and stain resistance. ISO 6530, test method for resistance of materials to penetration of liquids, can be used to measure penetration, absorption and repellence. ISO 14419 oil repellency can provide a rough index of oil stain resistance but only generally, informing of the penetration levels (Textile Innovation Knowledge Platform, 2017).

Other standards of relevance are (Textile Innovation Knowledge Platform, 2017):

- The ASTM DM4265 standard for evaluating stain removal in home laundering.
- AATCC Test Method 130-2010 Soil Release: Oily Stain Release Method
- AATCC Test Method 175-2008 Stain Resistance: Pile Floor Coverings
- AATCC Test Method 193-2007Aqueous Liquid Repellency: Water/Alcohol Solution Resistance Test
- AATCC Test Method 42-2007 Water Resistance: Impact Penetration Test

3.5 Future trends and challenges

Demand for soil-repellent and stain-resistant finishing and relevant textile products is expected to continue into the future. In this direction, it is important to assess the potential health and environmental attributes of repellent chemistries, following a comprehensive approach to address their overall impacts. Lack of an association to PFOA and PFOS should not be the only criteria used to assess potential impacts. Instead, the overall risk associated with the repellent chemistries, not just hazard alone, should be taken into account. Chemistries not associated with PFOA and PFOS may be linked to other substances of concern. Byproducts of short-chain fluorinated chemistries are persistent in the environment and are subject to use restrictions in certain countries. Although these byproducts have been shown to be less toxic and bioaccumulative, their persistence is judged to be a characteristic of concern, particularly if they can be widely dispersed in the environment (Outdoor Industry, 2012).

Some market segments may possibly seek exemption to use C8 fluorocarbons where safety/survival of an individual is seen as vital. The main challenge in the research toward new repellent chemistries would be the development of fluorocarbon-free oleophobic finishing. Despite the research already implemented into new materials and processes (e.g. dendrimers, hydrophobins, plasma treatments) it is still not evident if a completely fluorine-free soil-repellent treatment is possible (Mahltig, 2014).

In any case, during the development of new repellent finishes, one should always ensure they permit the transfer of air and water vapour through fabrics (breathability) and are durable to repeated laundering and dry cleaning, as well as to abrasion. In addition to repellency, there are other important attributes considered in determining the performance of finishes. For example, repellent finishes applied on textiles can impact fabric colour, strength and handle (Outdoor Industry, 2012). In fact, for most fabrics finished with fluorochemicals the handle is not acceptable. Although selected conventional organic softeners can be used to overcome this issue, the softness they can provide has fallen short for many apparel applications. Thus the optimization of the finishing process along with the selection of the right softener becomes important to achieve premium hand and repellency (Vazquez, 2004).

3.6 Conclusion

Soil-repellent and stain-resistant textiles are developed through the application of hydrophobic and oleophobic treatments. The conventional finishing, based on fluorocarbons, has several health and environmental issues. So several alternative treatment processes are being developed, including nanostructured surfaces exhibiting the so-called lotus effect, plasma treatments and application of new materials such as dendrimers and hydrophobins. Notwithstanding these research efforts, it is still not clear whether a completely fluorine-free soil- and stain-resistant treatment could be achieved at commercial scale in the near future.

References

Alder, A.C., van der Voet, J., 2015. Occurrence and point source characterization of perfluoroalkyl acids in sewage sludge. Chemosphere 129, 62–73.

Baghaei, B., Mehmood, F., 2011. Alternatives to conventional fluorocarbon based soil resistant treatment for automotive interior textiles—an evaluative study. Master Thesis for the Master in Textile Technology, Swedish School of Textiles, University of Borås, Sweden.

Cassie, A.B.D., Baxter, S., 1944. Wettability of porous surfaces. Trans. Faraday Soc. 40, 546–551.

De Vietro, N., Bekforte, L., Lambertini, V., Placenza, B., Fracassi, F., 2015. Plasma treatment for preparing durable water repellent and anti-stain synthetic fabrics for automotive applications. J. Surf. Eng. Mater. Adv. Technol. 5, 103–109.

Duong, H.T., Kadokami, K., Shirasaka, H., Hidaka, R., Chau, H.T., Kong, L., Nguyen, T.Q., Nguyen, T.T., 2015. Occurrence of perfluoroalkyl acids in environmental waters in Vietnam. Chemosphere 122, 115–124.

Easter, E.P., Ankenman, B.E., 2005. A comparison of soil repellent vs dual action fluorocarbon finishes on cotton blends. AATCC Rev. 5(11).

Gulrajani, M.L., 2013. The use of nanotechnology in the finishing of technical textiles. In: Advances in the Dyeing and Finishing of Technical Textiles, Woodhead Publishing Series in Textiles, Cambridge.

Hanumansetty, S., Maity, J., Foster, R., O'Rear, E.A., 2012. Stain resistance of cotton fabrics before and after finishing with admicellar polymerization. Appl. Sci. 2, 192–205.

Hashim, A., 1986. Effects of soil repellent and smolder resistant finishes, types of soils and cleaning methods on soiling characteristics of upholstery fabrics. Pertanika 9 (3), 401–407.

Hastbacka, M., 2016. Non-fluorinated omniphobic coatings for stain resistant textiles. https://cfpub.epa.gov/ncer_abstracts/index.cfm/fuseaction/display.abstractDetail/abstract/10481/report/0 (accessed 07.02.17).

Hedegard, A., 2014. Durable oil and water repellent outdoor fabrics by atmospheric plasma treatment. Master of Science Thesis in Chemical and Biological Engineering. Chalmers University Of Technology Gothenburg, Sweden.

Hegemann, D., 2015. Stain repellent finishing on fabrics. Adv. Eng. Mater.

Hohenstein institute or an individual? A new approach optimises "lotus effect" of soil repellent textiles. https://www.hohenstein.de/en/inline/pressrelease_6980.xhtml (accessed 03.02.17).

Lin, A.Y., Panchangam, S.C., Tsai, Y.T., Yu, T.H., 2014. Occurrence of perfluorinated compounds in the aquatic environment as found in science park effluent, river water, rainwater, sediments, and biotissues. Environ. Monit. Assess. 186 (5), 3265–3275.

Mahltig, B., 2014. Hydrophobic and oleophobic finishes for textiles. In: Paul, R. (Ed.), Functional Finishes for Textiles. Woodhead Publishing Ltd., UK.

Mao, C., Liang, C., Luo, W., Bao, J., Shen, J., Hou, X., Zhao, W., 2009. Preparation of lotus-leaf-like polystyrene micro- and nanostructure films and its blood compatibility. J. Mater. Chem. 19, 9025–9029.

Neinhuis, C., Barthlott, W., 1997. Characterization and distribution of water-repellent, self-cleaning plant surfaces. Ann. Bot. (Oxford, UK) 79 (6), 667–677. https://doi.org/10.1006/anbo.1997.0400.

Outdoor Industry, 2012. Durable water and soil repellent chemistry in the textile industry—a research report. https://outdoorindustry.org/pdf/FINAL_ZDHC_P05_DWR%20Research_Nov2012.pdf (accessed 09.01.17).

Park, S., Kim, J., Park, C.H., 2016. Analysis of the wetting state of super-repellent fabrics with liquids of varying surface tension. RSC Adv. 6, 45884–45893.

Pittman, A.G., Wasley, W.L., 1972. Siloxane polymers for soil repellent and soil release textile finishes. US Patent No. 3,639,156.

Schindler, W.D., Hauser, P.J., 2004. Repellent finishes. In: Schindler, W.D., Hauser, P.J. (Eds.), Chemical Finishing of Textiles. Woodhead Publishing Ltd., UK.

Souma, D., 2012. Water and Stain repellent textiles, using new plasma technology. Master of Science Thesis in the Master Degree Programme Materials and Nanotechnology, Chalmers University Of Technology, Gothenburg, Sweden.

Textile Innovation Knowledge Platform, http://www.tikp.co.uk/knowledge/material-functionality/stain-resistance/ (accessed 10.01.17).

The Danish Environmental Protection Agency, 2015. Alternatives to perfluoroalkyl and polyfluoroalkyl substances (PFAS) in textiles. Survey of chemical substances in consumer products No. 137. The Danish Environmental Protection Agency.

Vazquez, F., 2004. Silicone Softeners for Stain Repellent and Stain Release Fabric Finishing. Dow Corning Corporation Greensboro, NC.

Vögtle, F., Gestermann, S., Hesse, R., Schwierz, H., Windisch, B., 2000. Functional dendrimers. Progr. Polym. Sci., 987–1041.

Toxicological and environmental issues associated with waterproofing and water repellent formulations

4

Margaret H. Whittaker, Lauren Heine*[†]
*ToxServices LLC, Washington, DC, United States, [†]Northwest Green Chemistry, Spokane, WA, United States

4.1 Introduction

Safer ingredient selection as part of the design of durable water repellents and waterproofing agents in textiles and clothing continues to challenge formulators due to the unique attributes of these formulations. Water repellents and waterproof formulations are expected to repel and/or retard moisture, and exhibit durability, breathability, and ideally, stain resistance, while at the same time being made of chemicals that pose no harm to humans or the environment, either during use or at end of life. Over the past 15 years, reports of bioaccumulation and long environmental persistence of certain chemicals in water repellents and waterproofing formulations have shone an unkind spotlight on specific chemistries (notably, long-chain fluorochemicals such as PFOA and PFOS, as well as cyclic siloxanes such as D4 and D5). The focus of most change in this industry is on the formulation of water repellents. A water repellent is also known as a durable water repellent (DWR), which is defined as a textile finish whose performance attributes (effects) may include water repellency, oil repellency, stain repellency, soil repellency, stain release, soil release and durability (ZDHC, 2012).

Regulatory, retailer and apparel/brand-level restrictions, and/or prohibitions on whole classes of chemicals in this industry have resulted in the relatively swift substitution of alternative water repellent chemistries and technologies, some of which have incomplete toxicity datasets, and some of which may prove to be regrettable substitutions. In 2011, members of the Zero Discharge of Hazardous Chemicals (ZDHC) coalition committed to establishing a timeline for the elimination of technologies that may contain or degrade into long-chain perfluoroalkyl and polyfluoroalkyl substances (PFAS) such as PFOA and PFOS (ZDHC, 2012), putting further pressure upon formulators to find safer substitutes. Similarly, the Outdoor Industry Association (OIA) applauded W.L. Gore for pledging to eliminate 100% of per- and polyfluorinated chemicals of environmental concern from its consumer laminate shipments by 2023, a move that will drive significant reduction in the use of these chemistries by the broader outdoor industry (OIA, 2017). Many of these changes

Waterproof and Water Repellent Textiles and Clothing. https://doi.org/10.1016/B978-0-08-101212-3.00004-6

are needed and are expected to have a positive effect on humans and the environment. However, it is critical that industry selects alternative chemistries carefully and with full awareness of data gaps, as well as demonstrate a commitment to completely and transparently assess the preferability of specific chemicals and chemical classes.

This chapter provides an overview of the primary toxicological and environmental issues associated with water repellents and waterproofing formulations, discusses initiatives intended to reduce and eliminate chemicals of concern in such formulations, and identifies sector-specific initiatives and assessment paradigms focused on the selection of safer and sustainable ingredients used in water repellents and waterproofing formulations.

4.2 Properties of chemicals used in water repellents and waterproofing formulations

As shown in Fig. 4.1, the ideal water repellent exhibits seven performance characteristics, including repelling and releasing water, oil and stains, as well as exhibiting durability. The successful commercialization of a non-hazardous, non-fluorinated water repellent displaying all seven of these performance attributes remains an elusive goal.

4.2.1 Historical usage of water repellents and waterproofing agents

Historically, few substances were available that could be applied to textiles or clothing to impart water repellency or waterproofing. Table 4.1 gives examples of a number of the historical classes of water repellent and waterproofing chemicals. In the 16th and 17th centuries, textiles, and clothing were coated with various fats and oils (such as boiled linseed oil) or waxes (such as paraffin wax) to resist wetting. Besides performing poorly, such treatments often imparted stiffness to the fabric (particularly

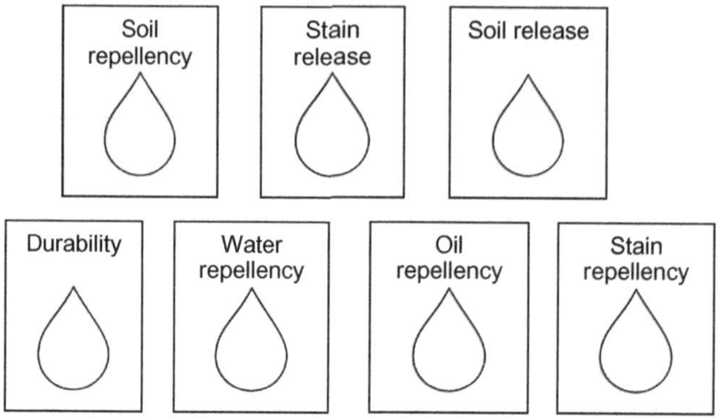

Fig. 4.1 Performance properties of water repellents and waterproofing agents.

Table 4.1 **Historical classes of water repellents and waterproofing chemicals**

Substance	Example of Product	Era of Use
Linseed oil	Oiled cloth	1600s/1700s
Natural rubber derivatives	Macintosh raincoat	1800s
Waxes and metal-wax emulsions (aluminium, chromium, zirconium, such as aluminium palmitate and aluminium stearate)	Tate electrolytic textile	Mid-1800s/ 1900s
Pyridinium compounds	ICI's Velan PF, DuPont's Zelan AP	1940s–1970s

at cold temperatures), and washed off when the garment underwent multiple washings. In the late 1700s and 1800s, patents began to be issued for waterproofing formulations based on natural rubber, which began to be exported from Central and South America. Two of the best known patents include one from 1791 that described waterproofing fabrics made from a solution of natural rubber and turpentine (Peal, 1791), as well as Scottish scientist Charles Macintosh's patent on a sandwiched rubber paste made by dissolving natural rubber in coal tar-derived naphtha to create a paste that when applied to a sandwiched fabric array, repelled water and in the process rendered, 'The texture of hemp, flax, wool, cotton, silk and also leather, paper and other substances impervious to water and air' (Macintosh, 1823; Commissioner of Patents, 1854; Tortora and Johnson, 2013).

The key ingredients in both patents (turpentine and coal-tar naphtha) are prime examples of ingredients in early formulations that posed multiple health hazards such as acute toxicity (vapours from terpene in turpentine are extremely irritating to eyes, skin and the respiratory tract), neurotoxicity (both turpentine and coal-tar naphtha contain chemicals that are central nervous system depressants) and carcinogenicity (industrial grade coal-tar naphtha contains benzene), along with physical safety hazards (both turpentine and coal-tar naphtha are mixtures of flammable hydrocarbons).

By the late 1880s/early 1900s, metal emulsions of waxes made with aluminium, chromium or zirconium began to be applied to textiles and clothing, often in an electrolytic current, and were considered a step forward as functional water repellents (Tate, 1917). Pyridinium-based water repellents were developed in the 1930s and sold until the 1970s, and were made by reacting chemicals such as pyridine hydrochloride, pyridine and paraformaldehyde (Kissa, 2001; Lichstein, 1977). Pyridinium-based formulations were curtailed over time because the pyridine liberated during the reaction had an unpleasant odour, and carcinogenic formaldehyde gas was liberated during the reaction process. Irrespective of these hazards, materials made with these treatments were at times foul smelling and impeded the diffusion of moisture vapours, leaving early users of water repellent and waterproofing formulations overheated, malodoured and uncomfortable.

4.2.2 Modern chemical classes used in water repellent and waterproofing agents

Currently, chemicals from five categories are used in water repellent and waterproofing formulations, as illustrated in Fig. 4.2. Fluorochemicals and siloxanes comprise the majority of commercial water repellents and waterproofing formulations. A number of comprehensive reviews on state-of-the-art water repellents and water proofing chemistries and technologies in textiles and clothing have been published (e.g. ZDHC, 2012; Holmquist et al., 2016; Swerea IVF, 2015). Table 4.2 identifies chemicals used in formulations in these five classes, along with their precursor molecules and breakdown products.

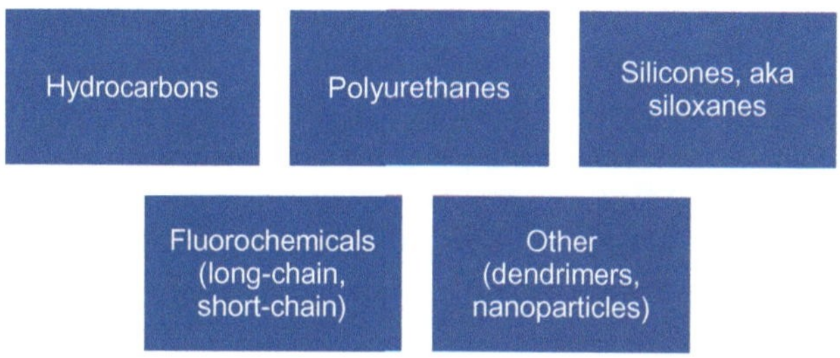

Fig. 4.2 Modern classes of water repellents and waterproofing agents.

Table 4.2 Chemicals used in water repellents

Substance	Abbreviation	CAS#	Occurrence
Long-chain fluorochemicals			
Perfluorooctanoic acid	PFOA	335-67-1	Stable degradation product
8:2-Fluorotelomer alcohol	8:2 FTOH	678-39-7	Precursor
Short-chain fluorochemicals			
Perfluorohexanoic acid	PFHxA	307-24-4	Stable degradation product
6:2 Fluorotelomer alcohol	6:2 FTOH	647-42-7	Precursor
Perfluorobutanesulfonic acid	PFBS	45187-15-3	Stable degradation product
Perfluorobutanoic sulfonamide alcohols	N/A	N/A	Precursor
Siloxanes			
Octamethyl cyclotetrasiloxane	D4	556-67-2	Precursor
Decamethyl cyclopentasiloxane	D5	541-02-6	Precursor

Table 4.2 **Continued**

Substance	Abbreviation	CAS#	Occurrence
Dodecamethyl cyclohexasiloxane	D6	540-97-6	Precursor
Hexamethyl disiloxane	MM (or HMDS)	107-46-0	Precursor
Octamethyl trisiloxane	MDM = L3	107-51-7	Precursor
Decamethyl tetrasiloxane	MD2M = L4	141-62-8	Precursor
Dodecamethyl pentasiloxane	MD3M = L5	141-63-9	Precursor
Hydrocarbons			
Paraffin wax	Unspecified	8002-74-2	Final product
Stearamidomethyl pyridine chloride	Unspecified	4261-72-7	Precursor
Modified melamine resins, blended with waxes	Unspecified	Unspecified	Final product
Polyurethanes			
Unspecified polyurethane	Unspecified	Unspecified	Final product
Other			
Dendrimers	Unspecified	Unspecified	Unspecified
Nanoparticles	Unspecified	Unspecified	Unspecified

Derived from SUPFES. Swerea IVF AB., 2015. Substitution in practice of prioritized fluorinated compounds for textile applications. Project report 24052.

4.3 Toxicological and ecotoxicological concerns associated with chemicals in waterproofing and water repellent agents

Over the past decade, a number of governmental and non-governmental organizations have assessed the toxicity of intentionally added chemicals, residuals and breakdown products of chemicals in waterproofing and water repellent formulations, including fluorinated and non-fluorinated chemicals (e.g. UNEP, 2009; ZDHC, 2012; Danish EPA, 2015; KEMI, 2015). Collectively, these publications indicate certain chemicals in water repellents are known to cause toxicity (notably, long-chain PFAS); however, there is much less certainty over the potential toxicity and long-term safety of all other classes of water repellent agents currently used in the marketplace. The Swedish Environmental Protection Agency (KEMI, 2015) reviewed alternatives to long-chain PFAS and emphasized the absence of transparency and significant data gaps among all of the alternatives to long-chain PFAS, including their potential degradation products. Despite these shortcomings, KEMI concluded that most of the alternatives (apart from those that are silicone-based) were preferable to long-chain PFAS, particularly from an environmental perspective.

To move away from the use of toxic chemicals in waterproofing and water repellent formulations, it is critical to recognize attributes that permit such chemicals to cause toxicity. Various definitions of toxicity exist in the published literature, and although they identify slightly different definitions, they all relate to the ability of a substance to cause an adverse change in an organism from normality, which may or may not be reversible (Wooley and Wooley, 2017). A key factor influencing the toxicity of any type of substance (whether chemical, biological, radiological or physical substance) is the level or degree of exposure. Additionally, toxic effects in an organism will not be manifest by a substance or its transformation product(s) unless that substance or its transformation product(s) reaches appropriate sites in an organism at a concentration and length of time sufficient to produce a toxic effect.

Because this chapter is focused on chemical agents, toxicity is explained in terms of chemicals, as opposed to other types of substances (e.g. microbiological, radioactivity, flammability, etc.). Whether a toxic response occurs is a function of the chemical and physical properties of that chemical, the exposure scenario, how the chemical is metabolized in an organism, the concentration of the active form of the chemical at a target site within the organism, and the overall susceptibility of the organism (Eaton and Gilbert, 2013).

Key terms pertaining to toxicity of water repellents and waterproofing agents are defined in Table 4.3.

Table 4.3 Key terms pertaining to toxicity

Term	Meaning	Reference
Toxicity	The degree to which a substance can harm humans, animals or plants. An adverse change in an organism from normality, which may or may not be reversible	Wooley and Wooley (2017)
Toxicant	Natural or synthetic substance with a potential to induce cellular and tissue damage under commonly encountered conditions of exposure, whether deliberate, accidental or unintentional	Burcham (2013)
Persistence	Substances that are persistent are highly stable and resist natural processes of degradation	Posner (2016a,b)
Bioaccumulation	Substances that bioaccumulate are readily absorbed in fatty tissue and can accumulate in the body fat of living organisms. These substances become more concentrated as they move up the food chain, especially into larger, longer-living organisms	Posner (2016a,b)

Table 4.3 **Continued**

Term	Meaning	Reference
POPs	Persistent organic pollutants: Organic chemical substances that possess a particular combination of physical and chemical properties such that, once released into the environment, they: • remain intact for exceptionally long periods of time • become widely distributed throughout the environment as a result of natural processes involving soil, water and, most notably, air • accumulate in the fatty tissue of living organisms including humans, and are found at higher concentrations at higher levels in the food chain and • are toxic to both humans and wildlife	Stockholm Convention (2017)
Hazard	An inherent property of a chemical such as carcinogenicity, flammability, aquatic toxicity. The potential to cause harm via a specific mode of action	Robinson and Thorn (2005), SSC (2000), Fjeld et al. (2007a)
Exposure	In its simplest form, contact with a chemical or physical agent and a target. Refers to ingestion, inhalation or dermal absorption of a chemical that increases the likelihood of risk when the chemical is hazardous	IPCS (2004), Fjeld et al. (2007b)
Risk	Risk is a function of hazard and exposure; the likelihood that a hazardous substance will cause harm	IPCS (2004), Fjeld et al. (2007b)

4.3.1 Chemical hazard assessment

Chemical hazard assessment is a systematic process of assessing and classifying hazards across a spectrum of endpoints and severity (Whittaker and Heine, 2013). Endpoints pertain to a specific adverse effect linked to mortality (death), morbidity (illness) or physical properties. Hazard endpoints are measured by changes in *in vivo*, *in vitro* or *in silico* test systems, and/or changes in physiochemical parameters. The Globally Harmonized System of Classification and Labelling of Chemicals (GHS) defines an endpoint as a physical, health and/or environmental hazard (GHS, 2015). Relevant to other life cycle impacts, the inherent hazard of a chemical cannot be changed, and although it may be possible to reduce or minimize exposure (and subsequently reduce risk), it is challenging to accurately, precisely or realistically assess exposure across all phases of a chemical's lifecycle.

To understand chemical hazard assessment, it is important to understand and classify hazards according to GHS. This is a system for harmonizing hazard classification criteria and chemical hazard communication elements worldwide. While GHS is voluntary, it is increasingly being adopted globally currently, in force in 72 countries (UNECE, 2017). GHS provides a framework with guidance for classifying and labelling chemical hazards for the purposes of harmonizing hazard communication and enhancing protection of human health and the environment. As shown in Table 4.4, GHS defines criteria for the classification of health, physical and

Table 4.4 GHS hazard classifications by endpoint

GHS health hazards	***High Hazard*** ↔ ***Low Hazard***
Acute toxicity	Categories 1, 2, 3, 4, 5
Skin corrosion/irritation	Categories 1 (1A, 1B, 1C), 2, 3
Serious eye damage/eye irritation	Categories 1, 2 (2A, 2B)
Respiratory or skin sensitization	Category 1 (1A, 1B)
Germ cell mutagenicity	Categories 1 (1A, 1B), 2
Carcinogenicity	Categories 1 (1A, 1B), 2
Reproductive toxicity	Categories 1 (1A, 1B), 2
Specific target organ toxicity single exposure	Categories 1, 2, 3
Specific target organ toxicity repeated exposure	Categories 1, 2
Aspiration hazard	Categories 1, 2
GHS physical hazards	***High Hazard*** ↔ ***Low Hazard***
Explosives	Divisions 1.1, 1.2, 1.3, 1.4, 1.5, 1.6
Flammable gases and chemically unstable gases	Categories 1, 2 & A, B
Aerosols	Categories 1, 2, 3
Oxidizing gases	Category 1
Gases under pressure	Compressed/liquefied/refrigerated liquefied/dissolved gas
Flammable liquids	Categories 1, 2, 3, 4
Flammable solids	Categories 1, 2
Self-reactive substances and mixtures	Types A, B, C and D, E and F, G
Pyrophoric liquids/solids	Category 1
Self-heating substances and mixtures	Categories 1, 2
Substances and mixtures which, in contact with water, emit flammable gases	Categories 1, 2, 3
Oxidizing liquids/solids	Categories 1, 2, 3
Organic peroxides	Types A, B, C, D, E, F, G
Corrosive to metals	Category 1
GHS environmental hazards	***High Hazard*** ↔ ***Low Hazard***
Acute aquatic hazard	Acute categories 1, 2, 3
Chronic aquatic hazard	Chronic categories 1, 2, (3), 4
Hazardous to the ozone layer	Category 1

environmental hazards. GHS Hazard Statements communicate potential dangers posed by a substance, and are categorized into 16 physical, 10 health and 3 environmental hazard classes. Such H statement classifications are communicated in GHS-compliant Safety Data Sheets (SDS) and are useful to characterize the hazards of potential ingredients in water repellent and waterproofing formulations.

Despite the broad range of hazard endpoints, GHS has limitations when applied to the classification of hazards posed by water repellent and waterproofing formulations. Hazards associated with long-chain PFAS are associated with persistence, bio-accumulation potential and toxicity. PFOS and its salts are listed as persistent organic pollutants (POPs) under the Stockholm Convention (UNEP, 2009). Yet, GHS does not classify chemicals that are persistent and/or bioaccumulative. Rather, GHS considers thresholds for persistence and bioaccumulation only in combination with aquatic toxicity to identify chemicals considered to be chronic aquatic toxicants. Although a chronic aquatic toxicant may or may not be a POP, not all POPs are chronic aquatic toxicants.

The Stockholm Convention Annex D defines information and screening criteria for defining POPs (Stockholm Convention Annex D). They include:

1. Chemical identity:
   - A. Names, including trade name or names, commercial name or names and synonyms, Chemical Abstracts Service (CAS) Registry number, International Union of Pure and Applied Chemistry (IUPAC) name; and
   - B. Structure, including specification of isomers, where applicable, and the structure of the chemical class;
2. Persistence:
   - A. Evidence the half-life of the chemical in water is greater than 2 months, or its half-life in soil is greater than 6 months, or its half-life in sediment is greater than 6 months; or
   - B. Evidence the chemical is otherwise sufficiently persistent to justify its consideration within the scope of this convention;
3. Bio-accumulation:
   - A. Evidence the bioconcentration factor (BCF) or bioaccumulation factor (BAF) in aquatic species for the chemical is greater than 5000 or, in the absence of such data, the log Kow (octanol/water partition coefficient) is greater than 5 (BCF, BAF and log Kow are indictors of a chemical's potential to accumulate);
   - B. Evidence a chemical presents other reasons for concern, such as high bioaccumulation in other species, high toxicity or ecotoxicity; or
   - C. Monitoring data in biota indicating the bioaccumulation potential of the chemical is sufficient to justify its consideration within the scope of this convention;
4. Potential for long-range environmental transport:
   - A. Measured levels of the chemical in locations distant from the sources of its release that are of potential concern;
   - B. Monitoring data showing long-range environmental transport of the chemical, with the potential for transfer to a receiving environment, may have occurred via air, water or migratory species; or,
   - C. Environmental fate properties and/or model results demonstrate the chemical has a potential for long-range environmental transport through air, water or migratory species, with the potential for transfer to a receiving environment in locations distant from the

sources of its release. For a chemical that migrates significantly through the air, its half-life in air should be greater than 2 days; and,

5. Adverse effects:
   A. Evidence of adverse effects to human health or to the environment that justifies consideration of the chemical within the scope of the convention; or
   B. Toxicity or ecotoxicity data that indicate the potential for damage to human health or to the environment.

Chemicals that are not classified as POPs are not necessarily safer alternatives. A chemical can be toxic and not fall within the classification of a POP. It is important to understand what is known and not known about the toxicity and fate of chemicals and also understand how chemicals are used in products. Chemical hazard assessment is a method that provides a framework and criteria for comparing chemical hazards, including persistence and bioaccumulation potential.

In addition to GHS, there are publicly accessible, transparent, non-regulatory, chemical hazard assessment schemes that build on GHS, but are designed to serve slightly different purposes. Best-known systems in the United States include US Environmental Protection Agency (EPA) Design for the Environment (DfE) Program Alternatives Assessment (AA) Criteria for Hazard Evaluation (US EPA, 2011) and Clean Production Action's GreenScreen for Safer Chemicals (CPA, 2016). The US EPA DfE AA Criteria were developed to support US EPA's Alternatives Assessment Partnerships. The hazard endpoints found in GHS, DfE and GreenScreen methodologies are summarized in Table 4.5. Hazard endpoints in italics are endpoints that are

Table 4.5 Example hazard endpoints used in chemical hazard assessment

Human health	Human health cont'd	Environmental toxicity and fate	Physical hazards/ climate hazards
Carcinogenicity	Acute toxicity (oral, dermal, inhalation)	Acute aquatic toxicity (fish, Daphnia, algae)	Reactivity
Mutagenicity and genotoxicity	Systemic toxicity and organ effects (oral, dermal, inhalation)	Chronic aquatic toxicity (fish, Daphnia, algae)	Flammability
Reproductive toxicity	Neurotoxicity	*Other ecotoxicity studies when available (earthworm, bird, bee)*	
Developmental toxicity	Skin sensitization Respiratory sensitization	*Persistence (air, water, sediment, soil)*	Ozone depletion potential
Endocrine activity	Skin irritation/ corrosivity Eye irritation/ corrosivity	*Bioaccumulation*	

not found as standalone endpoints in GHS. GreenScreen was developed to identify both chemicals of concern as well as safer alternatives, and goes a step beyond the DfE AA criteria by incorporating a numeric benchmarking system that facilitates an overall hazard comparison among chemicals and easy identification of chemicals to avoid.

Holmquist et al. (2016) used US EPA DfE AA criteria and GreenScreen criteria to evaluate and compare hazards associated with various classes of water repellents currently on the market. They assessed short-chain and long-chain fluorinated chemicals (PFOA, PFHxA, PFBS), silicones (including short-chain silanols, dimethylsilandiol; trimethylsilanol, D4 and D5), hydrocarbons (including paraffin wax and other chemistries), dendrimers and nanoparticles. The results of their hazard assessment are presented in Table 4.9, discussed later. As reflected in Table 4.9, Holmquist et al.'s comparative hazard assessment facilitates a hazard comparison of different chemical classes in addition to identifying data gaps among the assessed chemicals. Although the hazards of some chemicals such as PFOA are fairly well characterized, meaning that there are data associated with most key endpoints and few data gaps, other chemicals such as dendrimers and nanoparticles have little or no publicly available hazard data available, so it is difficult to conclude that dendrimers and nanoparticles are safer substitutes.

The US EPA DfE AA and Clean Production Action's GreenScreen methods employ a useful convention of indicating the level of confidence associated with each hazard classification. This convention was developed by the US EPA and adopted for GreenScreen. If data are estimated or come from limited or conflicting studies, the hazard classification level may be presented using an *italicized* letter to indicate lower confidence. If the classification is based on test data that are robust and multiple studies agree, then the hazard classification may be presented using bold text to indicate higher confidence. Where there are insufficient data to make a hazard classification, DG is indicated for data gap. This convention allows users to make decisions based on all existing data and also to have better understanding of the limitations of the available data.

GHS does not address persistence and bioaccumulation on their own as discrete hazard endpoints, which is a limitation in the methodology. In contrast, DfE AA and GreenScreen methods provide criteria for classifying biodegradation/persistence (P) and bioaccumulation potential (B), which is critical for assessing chemical classes that are known to persist and/or bioaccumulate. DfE AA and GreenScreen methods also require the identification and screening of hazards associated with environmental transformation products. Defining a chemical as persistent or not is an exercise in binary thinking that may miss chemicals that may be only slightly less persistent and bioaccumulative than those defined by POPs criteria while accumulating in the environment due to prevalence of use.

GreenScreen criteria also include evidence of long-range transport and presence in environmental and biomonitoring studies as part of the classification system which is also included as information sources under the Stockholm Convention. This is important for water repellents, particularly the fluorinated compounds and nanomaterials because they may exhibit different fate patterns than conventional chemicals for which standard/guideline tests like octanol/water coefficient determinations can be used, such as OECD Test Guidelines 107 or 117 (OECD, 1995, 2004). Standard bio-accumulation test methods such as measurement of log Kow or measurement of BCF in fish (OECD Test Guideline 305, OECD, 2012), which are designed to imitate fat solubility, may not capture perfluorinated substances that bioaccumulate in plants or bind to proteins (Abelkop et al., 2015; Inoue et al., 2011).

Criteria for persistence in the US EPA DfE and GreenScreen systems are presented in Table 4.6. Values are reported as half-life in days or as a specific test result.

Table 4.6 Classification criteria for persistence (US EPA, 2011; CPA, 2016)

Hazard level	EPA DfE criteria	Hazard level	GreenScreen criteria
Very high	Half-life >180 days or recalcitrant	Very high	Soil or sediment: >180 or recalcitrant Water: >60 or recalcitrant Air: >5 or recalcitrant
High	Half-life of 60–180 days	High	Soil or sediment: >60–180 Water: >40–60 Air: >2–5 Evidence of long range environmental transport
Moderate	Half-life <60 but ≥16 days	Moderate	Soil or sediment: 16–60 Water: 16–40 Air: NA Suggested evidence of long range environmental transport
Low	Half-life <16 days OR passes Ready Biodegradability test not including the 10-day window	Low	Soil or sediment: <16 OR GHS 'Rapid degradability' Water: <16 OR GHS 'Rapid degradability' Air: <2
Very low	Passes Ready Biodegradability test with 10-day window	Very low	Soil or sediment: Meets 10-day window in 'Ready Biodegradation Test' Water: Meets 10-day window in 'Ready Biodegradation Test' Air: NA
		Data gap	Insufficient data

Table 4.7 Classification criteria for bioaccumulation potential (US EPA, 2011; CPA, 2016)

Hazard level	EPA DfE criteria	Hazard level	GreenScreen criteria
Very high	BAF/BCF >5000; Log BAF/BCF >3.7	Very high	BAF >5000; BCF >5000; Log Kow >5.0
High	BAF/BCF 5000–1000; Log BAF/BCF 3.7–3	High	BAF >1000–5000; BCF >1000–5000; Log Kow >4.5–5.0; monitoring data: evidence
Moderate	BAF/BCF <1000–100; Log BAF/BCF <3–2	Moderate	BAF >500–1000; BCF >500–1000; Log Kow >4.0–4.5; monitoring data: suggestive evidence
Low	BAF/BCF <100; Log BAF/BCF <2	Low	BAF >100–500; BCF >100–500
		Very low	BAF ≤100; BCF ≤ 100; Log Kow ≤ 4
		Data gap	Insufficient data

Criteria for bioaccumulation in the US EPA and GreenScreen systems are presented in Table 4.7. Criteria are based on Bioaccumulation Factor (BAF), Bioconcentration Factor (BCF) and the Log of the octanol/water partition coefficient (log Kow).

4.3.2 Hazards associated with fluorochemicals

Both the performance and toxicity of fluorochemicals in water repellent and waterproofing formulations depend on the length of the fluorocarbon chain. Fluorochemicals used in water repellent and waterproofing formulations are polymers containing a perfluoroalkyl moiety ($C_nF_{2n+1}^-$) (Posner, 2016a,b). These compounds are referred to as perfluoroalkyl and polyfluoroalkyl substances (PFAS) in this chapter using the convention assigned by Buck et al. (2011). In the past, PFAS substances were often referred to as PFCs (per-and polyfluorinated chemicals).

Every PFAS substance contains a perfluorinated tail, meaning all hydrogens on the tail's carbons are replaced with fluorine (Posner, 2016a,b). The use of PFAS in textile production accounts for about 50% of global use of PFAS (Danish EPA, 2015).

The length of the perfluoroalkyl chain is at least four carbon atoms, with early fluorochemical formulations such as Scotchgard Fabric Protector and Zepel and Teflon Fabric Protectors being made of carbon chain lengths of eight carbons. Co-monomers used to modify physical properties of polymers include alkyl-, alkylamide- or polyether-substituted esters of acrylic or methacrylic acid (Howells, 2009).

Long-chain PFAS compounds are divided into two sub-categories (OECD, 2013):

- Long-chain perfluoroalkyl carboxylic acids (PFCAs) with seven or more perfluorinated carbons

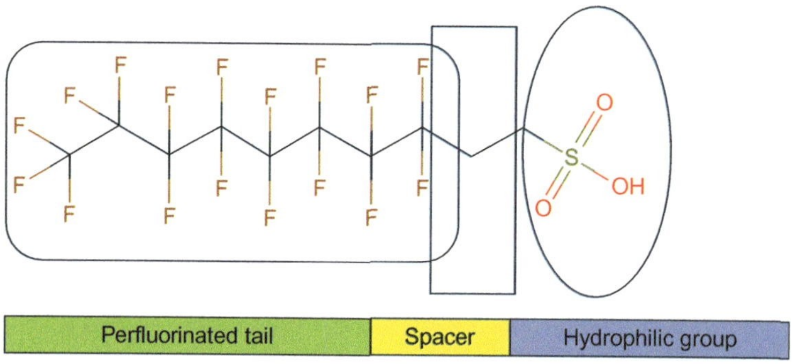

Fig. 4.3 Schematic PFAS molecule.

  - The best-known example is PFOA, which contains eight perfluorinated carbons
- Perfluoroalkane sulfonates (PFSAs) with five or more perfluorinated carbons, including:
  - Perfluorohexane sulfonic acid (PFHxS), which contains six perfluorinated carbons, and
  - Perfluorooctane sulfonic acid (PFOS), which contains eight perfluorinated carbons.

A simplified illustration of a PFAS molecule is shown in Fig. 4.3 (based upon general structures from Posner, 2015, and KEMI, 2015). The fluorinated tail of a PFAS molecule is hydrophobic (water repellent), a unique attribute which makes these chemicals extremely useful in water repellent applications.

PFCAs and PFSAs can reach the environment as impurities in other substances, as a result of degradation of precursor substances, as well as during their manufacture, use and disposal (Davies, 2014). Long-chain PFAS compounds were introduced commercially in the 1950s. Prior to 2000, almost half of all long-chain PFAS were used in the water repellent producing industry for textiles and apparel (UNEP, 2009).

3M's Scotchgard brand of stain-protection products was made with perfluoroalkyl sulfonamideethanols. These polymers are associated with PFOS ($C_8F_{17}SO_3H$), which is a sulfonamidoethanol degradation product. PFOA ($C_7F_{15}COOH$) in its ammonium salt form was used as a surfactant to solubilize fluorinated monomers during polymerization of fluorochemicals such as poly(tetrafluoroethylene). PFOA and related carboxylic acids are also degradation products of fluorotelomer alcohols, which are tetrafluoroethylene-based compounds used to attach fluoroalkyl groups to surfactants, polymers and material surfaces. PFOS and PFOA's armour-like properties are due to the eight carbons in their fluoroalkyl chain structure, which imparts rigidity and steric properties required to repel water and oil. Unfortunately, these structural attributes also contribute to the toxicity and long persistence of these compounds.

Public recognition of the pervasiveness and persistence of long-chain PFAS such as PFOS and PFOA is relatively recent, having only been publicly acknowledged over the past 20 years. The presence of PFOS in wildlife was first reported by Giesy and Kannan (2001), while Hansen et al. (2001) reported that PFOA and other PFAS were present in samples of human blood purchased from blood banks. Earlier publications reporting the presence of organofluorine substances in human blood samples and not in non-human blood samples led researchers to suspect an industrial or commercial source of exposure; however, the specific source was unspecified (e.g. Taves, 1971; Guy, et al., 1976; Ubel, et al., 1980; described in detail by Butenhoff and Rodricks, 2015).

Long-chain PFAS persist in the environment, bioaccumulate in wildlife and humans, and have shown to be reproductively, developmentally and systemically toxic in laboratory animals. The published literature on this topic is quite large, with thousands of journal articles published to date, in addition to industry, trade association and NGO-commissioned reports, often with conflicting and/or biased data summaries, adding to the complexity of impartially assessing PFAS substances. A number of detailed and comprehensive reports on the toxicity of PFAS compounds are available (e.g. DeWitt, 2015; OECD, 2013; Danish EPA, 2009, 2013, 2015). Long-chain perfluoroalkyl sulfonic acids (>C6) and perfluoroalkyl carboxylic acids (>C7) and their corresponding anions, have been shown to be more bioaccumulative than their short-chain analogues (Buck et al., 2011; Martin et al., 2003a,b; Conder et al., 2008; Olsen et al., 2009; US EPA, 2009; OECD, 2017).

In their public commitment to eliminate the use of hazardous per- and polyfluorinated chemicals by 2023, manufacturer W.L. Gore succinctly outlined the primary hazards traits of this class of chemicals (Gore, 2017) (Table 4.8). These endpoints represent a good 'stake in the earth' to identify attributes PFAS used in commerce must not exhibit. However, these attributes should not be considered the ultimate metrics to identify safer PFAS molecules. Instead, they should be considered a starting point.

Table 4.8 Attributes of chemicals of environmental concern

Trait	Definition	Reference
Highly fluorinated	Per- or polyfluorinated organic substances	Buck et al. (2011)
Small enough to be bioavailable	Capable of crossing a cellular membrane, molecular weight less than 3000 Da	De Mello (1987), OECD (2009), Alberts et al. (1994)
Persistent	Half-life >2 months (>60 days in water or soil)	Based on REACH persistence criteria for PBT and vPvP substances
Chemicals that degrade under normal conditions in the environment or during treatment at end of life	Fluorotelomer alcohols degrade in the environment to form degradants of concern Side-chain fluorinated polymers (fluorinated acrylate and methacrylate polymers) degrade in the environment over time to form degradants of concern	Joyce et al. (2004) Washington et al. (2009, 2015)

4.3.2.1 Transition from long-chain to short-chain fluorinated polymers

Short-chain PFASs are those containing either six or four fluorinated carbons (termed C6 or C4, respectively). With the phase-out of long-chain PFAS compounds, these now dominate the market. Although environmentally persistent, PFCAs with fewer than eight carbons, such as perfluorohexanoic acid (PFHxA), and PFSAs with fewer than six carbons, such as perfluorobutane sulfonic acid (PFBS), are thought to be generally less toxic and less bioaccumulative in wildlife and humans. Admittedly, there are numerous data gaps pertaining to long-term safety of short-chain PFAS, as well as uncertainties surrounding the human health and environmental toxicities of shorter-chain PFAS compounds (Ritter, 2015; UNEP, 2014a,b).

Short-chain PFAS are as persistent in the environment as long-chain PFAS, but they do not bioaccumulate to the same extent because their shorter, less rigid chains boost their water solubility (which in turn decreases bioaccumulation) (Danish EPA, 2015). However, a series of studies by Blaine et al. (2013, 2014a,b) indicate that PFAAs including the shorter chain lengths can be taken up from reclaimed water and biosolids in both the edible and non-edible portions of a variety of plants. The studies confirm PFAAs can enter and bioaccumulate in a range of food crops including lettuce, tomatoes, peas, radishes, strawberries and other foods. Bioaccumulation potential depends on analyte functional group and chain length, concentration in the reclaimed water or soil, and organic carbon content of the soil. Distribution was also seen to vary within the plant. For example, short-chain perfluorocarboxylates were the dominant fraction in the strawberry fruit and shoot compartments, whereas a more even distribution of all PFAAs appeared in the root compartment. While shorter-chain fluorinated polymers may be less bioaccumulating using traditional models of bioaccumulation, these studies demonstrating preferential uptake for shorter chain PFAAs deserve greater attention and portend caution.

Short-chain fluorinated PFAS cannot break down in the environment into PFOA and PFOS; however, degradation by-products of short-chain fluorinated compounds may turn out to be chemicals of concern. Potential by-products of short-chain fluorinated chemicals include perfluorohexanoic acid (PFHxA) and perfluorobutane sulfonic acid (PFBS). Both of these substances are persistent in the environment.

ENVIRON (2014, 2016) was contracted by the US FluoroCouncil to assess whether short-chain PFAS could be classified as POPs substances. ENVIRON assessed available toxicity data on five short-chain fluorinated chemicals linked to 6:2 fluorotelomer product chemistry. Specifically, they assessed a commercial short-chain fluorinated formulation, three intermediates, and one degradation product:

Commercial Product

- Short-chain polyfluoroalkyl acrylic polymer based on 6:2 fluorotelomer chemistry (Methacrylate Polymer; No CAS #)

Manufacturing Intermediates

- 6:2 Fluorotelomer alcohol (6:2 FTOH; CAS #647-42-7)
- 6:2 Fluorotelomer acrylate (6:2 FTAC; CAS #17527-29-6)
- 6:2 Fluorotelomer methacrylate (6:2 FTMAC; CAS #2144-53-8)

Degradation Product

- Perfluorohexanoic acid (PFHxA; CAS #307-24-4) and its anion perfluorohexanoate (PFHx)

ENVIRON classified available data for persistence, bioaccumulation, potential for long-term transport, and adverse effects. They concluded none of the five short-chain fluorinated chemicals was classified as a POP and each chemical met only one of four POPs criteria. ENVIRON's reports contain limited study details or insight into their hazard classifications, and are replete with numerous data gaps for multiple hazard endpoints. For example, ENVIRON cited studies confirming the presence of PFHxA in remote environments, but they argued the transport mechanism may not have been atmospheric, and therefore, they considered long range transport to be indeterminate. They also designated parent chemicals as unlikely to be persistent when known degradation products had been determined to be persistent, raising uncertainty about the conclusiveness or robustness of their assessment.

It is important to remember chemicals that are not POPs are not automatically classified as safer alternatives. It is vital to fully characterize hazards of a chemical across a full suite of human health and environmental endpoints, in addition to the endpoints of persistence and bioaccumulation that are evaluated during a POPs assessment.

Stockholm convention

The Stockholm Convention is a global treaty to protect human health and the environment from persistent organic pollutants (POPs). POPs are chemicals that remain intact in the environment for long periods, become widely distributed geographically, accumulate in the fatty tissue of living organisms, and are toxic to humans and wildlife. POPs circulate globally and can cause damage wherever they travel. In implementing the Stockholm Convention, governments will take measures to eliminate or reduce the release of POPs into the environment. Over 150 countries signed the Convention and it entered into force, on May 17, 2004, 90 days after the ratification by the fiftieth country (Stockholm Convention). In 2009, the Stockholm Convention added PFOS to its list of persistent organic pollutants (POPs).

Madrid statement

Blum et al. (2015) documents the scientific consensus of over 265 scientists regarding the persistence and potential for harm from poly- and perfluoroalkyl substances (PFASs). It builds on the Helsingør Statement by laying out a roadmap for gathering needed information and to prevent further harm. Recommended actions are outlined for scientists, governments, chemical manufacturers, product manufacturers and other professional users and purchasing organizations, retailers and individual consumers to limit the production of PFASs and to develop safer, non-fluorinated alternatives.

Helsingør statement

Scheringer et al. (2014) developed the Helsingør Statement to publicize concerns pertaining to potential impacts of fluorinated alternatives on human health and the environment to provide needed information for stakeholders, including the general public.

The authors are seven scientists who summarize key concerns including, 'The likelihood of fluorinated alternatives or their transformation products becoming ubiquitously present in the global environment; the need for more information on uses, properties and effects of fluorinated alternatives; the formation of persistent terminal transformation products including PFCAs and PFSAs; increasing environmental and human exposure and potential of adverse effects as a consequence of the high ultimate persistence and increasing usage of fluorinated alternatives; the high societal costs that would be caused if the uses, environmental fate, and adverse effects of fluorinated alternatives had to be investigated by publicly funded research; and the lack of consideration of non-persistent alternatives to long-chain PFAS.' The authors recommend that alternatives to long-chain PFAS be non-persistent and able to undergo complete degradation. Moreover, they recommend use of PFASs only in applications where they are shown to be indispensable.

US EPA's PFOA Stewardship Program

This program was a voluntary initiative begun in 2006 aimed at reducing uses of long-chain PFAAs. As part of the US EPA PFOA Stewardship Program, eight major manufacturers of PFOA committed to two goals: (1) to achieve, no later than 2010, a 95% reduction, measured from a year 2000 baseline, in both facility emissions to all media of PFOA, precursor chemicals that can break down to PFOA, and related higher homologue chemicals, and product content levels of these chemicals, and (2) to working towards the elimination of these chemicals from emissions and products by 2015. All participating companies met their commitments by December 31, 2015 (US EPA, 2017).

4.3.3 Hazards associated with silicones

Silicone polymers, more appropriately called siloxanes because of their repeating unit of $-[R_2Si-O]-$, provide good water repellency along with water-based stain resistance, particularly for cellulosic and synthetic fibres. Polydimethylsiloxane is the basic polymer used in silicone repellents. Cyclic siloxanes are used as precursors in the production of polydimethylsiloxanes. Continuous hydrolysis of chlorosilane monomers, such as dimethyldichlorosilane (Me_2SiCl_2), produces a mixture of cyclic and linear hydroxyl-terminated oligosiloxanes. The hydrolyzate is processed into siloxane polymers in one of two industrial processes: polymerization or condensation (Moisiewicz-Pieńkowska et al., 2016; Noll, 1968; Colas and Curtis, 2004; Nitzsche, et al., 1974).

- In the polymerization process, the hydrolyzate is first converted to a mixture of cyclic monomers, such as octamethylcyclotetrasiloxane (D4) and decamethyl-cyclopentasiloxane (D5), with subsequent ring opening. At the end of the reaction, the reactor contains a mixture of linear polysiloxanes and about 15% of cyclic oligomers.
- The polycondensation process also produces a mixture of linear siloxane polymers and cyclic oligomers, but with a low residual cyclic oligomer content of approximately 2%.

Dozens of reviews pertaining to the toxicity of siloxanes have been published, including TemaNord, 2005; SCCP, 2005; Danish EPA, 2015; Moisiewicz-

Pieńkowska et al., 2016. Human health and environmental hazards posed by siloxane polymers are generally attributed to the presence of residual cyclic monomers (cyclomethicones) and low molecular weight linear siloxanes. The low molecular weights of these compounds, in addition to their volatility contribute to their hazard potential and their ability to move through the environment (Moisiewicz-Pieńkowska et al., 2016; Genauldi et al., 2011; Kierkegaard and McLachlan, 2013; Wang, et al., 2013; Xu, et al., 2015). Cyclomethicone is the term generally used to describe residual cyclic dimethylpolysiloxane monomers, including octamethylcyclotetrasiloxane (D4, cyclotetrasiloxane), cyclotrisiloxane (D3), cyclopentasiloxane (D5), cyclohexasiloxane (D6) and cycloheptasiloxane (D7) (compounds with the general formula $(CH_3)_{2n}O_nSi_n$, where $n=3$–7).

Because siloxanes have been used for many decades, the collective literature on human health and environmental hazards is relatively robust. The Danish EPA (2015) presents a succinct overview of the toxicity of siloxanes when used as water repellents, and note that although this class of repellents is less toxic and less persistent than long-chain PFAS compounds, certain class members (notably, cyclic monomers) have the potential to be toxic. Lassen et al. (2005) published a detailed review of the hazards of siloxanes and specific cyclic monomers, including D4 and D5. PMDS is considered to pose a low hazard to humans and the environment. In contrast, toxicity data on D4 demonstrate that exposure impairs fertility in rats, in addition to targeting the liver, while the lung is the primary target organ for D5 exposure by inhalation. Despite negative *in vitro* and *in vivo* genotoxicity test data, studies in laboratory rats suggest that D5 may induce uterine tumours (US EPA, 2005).

In an encouraging development, Posner (2016a,b) reported the silicone industry has committed to reducing levels of residual cyclic siloxanes, which should improve the overall hazard profile of this class of chemicals in water repellent formulations.

4.3.4 Hazards associated with hydrocarbons

Hydrocarbon-based resins include paraffin/alkane formulations, as well as melamine resins blended with waxes. Paraffin repellents have been used for hundreds of years. Paraffins/alkanes are long-chain hydrocarbons with the general molecular formula C_nH_{2n+2}. Hydrocarbon-based water repellents also include modified melamine resins, blended with waxes. Stearic acid-melamine repellent chemistries are composed of compounds formed by a reaction between stearic acid, formaldehyde and melamine.

The primary ingredient in most products from this class is paraffin oil/wax (a mixture of linear aliphatic hydrocarbons), which are readily biodegradable, and do not bioconcentrate in organisms or food chains. Additionally, toxicity to aquatic and terrestrial organisms is low. Multiple assessments have concluded paraffin-based formulations pose relatively low hazards and are considered safer alternatives to long-chain PFAS (e.g. Danish EPA, 2015; ZDHC, 2012; Holmquist, et al., 2016; ToxServices, 2013). Note that not all hydrocarbon-based formulations are considered low hazard or low risk. For example, stearic acid-melamine resins can release formaldehyde, which is a known human carcinogen (ZDHC, 2012).

4.3.5 Hazards associated with polyurethanes

The Danish EPA (2015) and ZDHC (2012) note that polyurethane-based formulations represent a relatively new group of water repellents; however, details pertaining to specific chemistry and/or toxicity of this class of chemicals are lacking other than what is known about polyurethanes in general. Polyurethanes are polymers made by reacting a diisocyanate with a polyol. Diisocyanates are associated with a number of adverse health effects, and are known to be irritating to the eyes, nose, throat, lungs and skin (ACC, 2013). They also can cause allergic reactions of the skin and lungs (notably, allergic contact dermatitis and respiratory sensitization), underscoring the importance of minimizing exposure to workers and consumers when producing water repellents with this class of chemicals.

4.3.6 Hazards associated with dendrimers and nanoparticles

4.3.6.1 Hazards associated with dendrimers

Dendrimers are repetitively branched molecules leading to monodisperse, tree-like structures on the fabric surface (Danish EPA, 2015; Davies, 2014). The specific composition of chemicals in this class is unspecified, although Tang et al. (2010) describe the synthesis of hyperbranched hydrophobic dendrimeric polymers that consist of ester or polyurethane segments. Holmquist et al. (2016) reported that there was insufficient data in the available literature on this class of water repellents, and because of this, they were unable to conduct a chemical hazard assessment.

4.3.6.2 Hazards associated with nanoparticles

Holmquist et al. (2016) reported there was insufficient information in the available literature to characterize the chemical composition of nanoparticles in water repellents, and as such, they were unable to conduct a chemical hazard assessment. Although nanoparticles have unique properties, they too can undergo chemical hazard assessment. Sass et al. (2016) demonstrated that an existing method for chemical hazard assessment and communication can be used, with minor adaptations, to compare hazards across conventional and nano forms of a substance. The differences in data gaps and in hazard profiles support the argument that nano- and non-nano forms of a chemical should be considered as unique and subjected to hazard assessment to inform regulatory decisions and decisions about product design and development. A critical limitation of hazard assessments for nanoparticles is the lack of nano-specific hazard data, underscoring the need to invest in a testing/research program to fully assess human health and environmental hazards of nanoparticles. A robust evaluation of particle size attributes is vital, such as mean particle size, size range and agglomeration potential (Lin et al., 2014).

4.3.7 Chemical hazard assessment results for selected water repellents

Holmquist et al. (2016) assessed hazards of selected water repellents that reach the environment via diffuse emissions using the US EPA Design for the Environment Alternatives Assessment Criteria for Hazard Evaluation (US EPA, 2011), along with

Clean Production Action's GreenScreen for Safer Chemicals (CPA, 2016). Hazards for alternative water repellent chemicals were compared to PFOA, a known chemical of concern. Based on incomplete formulation disclosure and/or hazard data for dendrimer and inorganic nanoparticles, they assigned data gaps for each hazard endpoint for these two water repellent classes. Their assessment results, reproduced in Table 4.9, indicate that hydrocarbon-based water repellents are the most environmentally benign, followed by silicone and side-chain fluorinated polymer-based chemistries.

4.4 Green chemistry: Developing safer waterproofing and water repellent agents

Applying green chemistry and engineering principles, along with consistent adherence to a toxins reduction strategy can lead to a systematic transition to the design of waterproofing and water repellent formulations that are safer for humans and the environment. At a minimum, such formulations should not contain chemicals of concern or degrade in the environment to form a chemical of concern. Globally, a chemical of concern can be defined as a chemical that displays one of the following characteristics (US EPA, 2014):

- Carcinogenicity, mutagenicity, reproductive or developmental toxicity
- Potential concern for children's health (for example, because of potential adverse effects from endocrine disruption or known to cause asthma)
- Used in children's products or in products to which children may be highly exposed
- Neurotoxicity
- Persistent, bioaccumulative and toxic (PBT)
- Very persistent or very bioaccumulative in the environment (vPvB)
- Ozone depleting
- Detected in biomonitoring programs.

Simply banning or eliminating certain chemicals or classes of chemicals in water repellent and waterproofing formulations may be insufficient to permanently reduce risk or avoid regrettable substitutions. Simple bans can lead to substitution with the nearest replacement. This can result in essentially switching from a known chemical of concern (such as a long-chain PFAS) to a chemical with unknown or similar hazards. It is vital that substitutes are proven to be inherently safer to avoid unintended negative consequences.

4.4.1 Developing inherently safer products

To develop inherently safer water repellents and waterproofing agents, systematic changes will need to occur in the way products are developed and brought to market. Products should be developed based on the principles of green chemistry (Anastas and Warner, 1999). In this context, green chemistry can be practiced as green formulation or green synthetic chemistry, which is the design of new molecules that perform well, are cost effective, and pose inherently low hazards to humans and the environment.

Table 4.9 **Hazard assessment results for selected water repellent formulations**

Substance	Hazard classification per endpoint											
	Human health								Ecotoxicity		Fate	
	C	M	R	D	E	AT	ST	N	AA	CA	P	B
Benchmark chemical												
PFOA # *	**H**	**L**	*H*	**H**	PEA	**M**	**H**	DG	**L**	*L*	vH	**H**
Side-chain fluorinated polymers												
PFHxA # *	**L**	*L*	*L*	*M*	DG	**M**	**H**	DG	**M**	L	vH	**L**
PFBS # *	DG	**L**	*L*	**L**	PEA	**L**	*L*	DG	*L*	*L*	vH	**L**
Silicones												
Short-chain silanols #	DG	DG	DG	DG	DG	DG	DG	DG	DG	DG	DG	DG
Dimethylsilane diol #	DG	DG	DG	DG	DG	DG	*M*	DG	DG	DG	*vH*	*L**
Trimethyl silanol #	DG	*L*	DG	*L*	DG	**M**	*M*	DG	**L**	DG	DG	*L**
D4 *	NA	NA	**M**	NA	NA	NA	NA	NA	NA	vH	vH	vH
D5 *	*M*	**L**	**L**	*L*	DG	**L**	*M*	DG	**L**	**L**	vH	vH
Hydrocarbons												
Paraffin Wax #	**L**	**L**	*L*	*L*	DG	**L**	*M*	DG	*L*	*L*	**L**	*vL*
Other chemistries (dendrimers, inorganic nanoparticles)												
Unknown	DG	DG	DG	DG	DG	DG	DG	DG	DG	DG	DG	DG

Hazard classification abbreviations are: vL, very low; L, low; M, moderate; H, high; vH, very high; PEA, potentially endocrine active; DG, data gap. Classifications in italics are of low confidence and in bold of high confidence. Classifications based on estimated data are marked with an asterisk (*). The paraffin wax classification is based on a GreenScreen report (ToxServices LLC, 2016) (updated from the earlier ToxServices' GreenScreen assessment cited in Holmquist et al., 2016). Degradation products are denoted as # and impurities are denoted as *.

Typically, green formulation relates to the use of existing ingredients when formulating, while green synthetic chemistry refers to the development of new molecules as part of the product design process.

4.4.1.1 Steps to formulating inherently safer water repellent and waterproofing agents

Fig. 4.4 illustrates four steps to formulating water repellent and waterproofing agents that are inherently safer for human health and the environment. As a first step, there is need for greater transparency in the supply chain so information and data on chemical formulations, impurities or residuals, and transformation products can be communicated and the hazards associated with those chemicals can be assessed.

Secondly, each ingredient, residual chemical and potential breakdown product in a formulation should be screened against a comprehensive set of hazard endpoints. Formulations are typically proprietary and manufacturers are loathe to disclose confidential business information (CBI) to outside entities. Because complete formulation disclosure and robust hazard classification are critical to green formulation, there are ways to obtain CBI from suppliers to communicate hazard assessment results that protect CBI. A supplier can engage a third party in the assessment process or communicate results from a hazard assessment while masking the identity of specific chemicals.

The third step is to optimize formulations based on an understanding of hazards associated with specific ingredients, and eliminate chemicals of concern. Formulators can start by phasing out use of chemicals with known adverse human health hazards such as carcinogenicity, mutagenicity or genotoxicity, reproductive or developmental toxicity, or endocrine disrupting properties. Chemicals that are persistent, bioaccumulative and toxic to humans and the environment, or chemicals that are highly persistent and bioaccumulative, should also raise flags as likely chemicals of concern, whether currently regulated or not.

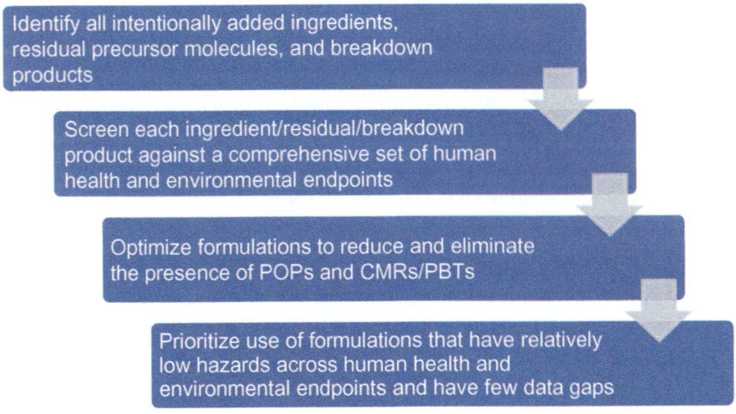

Fig. 4.4 Steps to formulate safer water repellents and waterproofing agents.

The final step in green formulation is to prioritize the use of formulations containing chemicals that are well characterized, with 100% of ingredients disclosed, in addition to identifying relevant and feasible transformation/breakdown products in such formulations. In addition, ingredients should be shown to perform well and pose low hazards across a wide range of human and environmental endpoints, especially for endpoints relevant to likely exposure scenarios.

Green formulation sounds like common sense, but it is challenging because often there is a paucity of reliable and comprehensive hazard assessment data on chemicals. The ideal chemical hazard assessment utilizes best available scientific data and applies expert judgment to evaluate a chemical across a broad suite of hazard endpoints. But this is only ideal. Often, data are incomplete, especially when assessments are based only on publicly accessible data.

4.4.1.2 Developing new classes of water repellent and waterproofing agents

New waterproofing or water repellent chemicals may have the disadvantage of being untested for human health, environmental toxicity and fate endpoints of interest. This disadvantage can be overcome by using a suite of rapid assessment tools to support internal decision making during new product development (Spencer, 2013). Cost-effective assessment of new chemicals can depend more on initial screening tests, including structure activity modeling, *in vitro* and *in silico* tests, and alternative assays based on cell cultures, rather than animal studies. Filling data gaps can be performed in a tiered way that allows ingredients to fail fast in order to succeed over the long term. Advances in predictive toxicology are moving towards the promise of increased speed and confidence, and reduced dependence on laboratory animals to fill needed data gaps.

4.4.1.3 Filling data gaps

Comprehensive chemical hazard assessments are limited by cost, time and technical resources. Standard toxicological testing is costly and animal-intensive. Historically, chemical hazards were investigated one chemical and one health effect endpoint at a time. Human health hazards comprise numerous endpoints, such as repeat dose toxicity, developmental toxicity, reproductive toxicity, neurotoxicity, carcinogenicity, sensitization potential, etc. Few chemicals in water repellent or water proofing formulations have been tested for all of these hazard endpoints. As a result, a chemical enters the supply chain with multiple data gaps pertaining to one or more hazard endpoints. Predictive toxicology is seen as the path forward. According to the National Research Council of the National Academies of Sciences:

> *Advances in molecular biology, biotechnology, and other fields are paving the way for major improvements in how scientists evaluate the health risks posed by potentially toxic chemicals found at low levels in the environment. Advances in predictive toxicology would make toxicity testing quicker, less expensive, and more directly*

> *relevant to human exposures. Predictive toxicology could also reduce the need for animal testing by substituting laboratory tests based on human cells.*
>
> *(National Research Council, 2007)*

Almost 15 years ago, the US National Toxicology Program described a vision for the 21st century to: 'Support the evolution of toxicology from a predominantly observational science at the level of disease-specific models to a predominantly predictive science focused upon a broad inclusion of target specific, mechanism-based, biological observations' (National Toxicology Program, 2004). These activities are collectively called *Tox21*.

Practically, this means adding or integrating new tools for predictive toxicology to standard methods and regulatory toxicological approaches. Methods are still emerging and in development. Toxicologists will need to learn new ways to integrate very different lines of evidence into their assessments. As predictive toxicology advances, assessing standard hazard endpoints may no longer be seen as the goal. Rather, an integrated approach to biological pathways, mechanisms and adverse outcome pathways will become the way to understand toxicity and fate of chemicals. When put into practice, it is hoped predictive toxicology will help make regrettable substitution a distant memory.

4.5 Conclusion

The history of water repellents and waterproofing agents demonstrates the principles of green chemistry have yet to be adopted in product design and development. Building chemical hazard assessment into the development of safer water repellents and waterproofing agents is critical and allows product developers to fail fast in order to succeed over the long term. The commercialization of water repellents and waterproofing chemicals that fail to undergo a thorough hazard screening is to risk bringing a product to market with undesirable hazard characteristics. This is a business risk that may result in greater investments in legal defence than product innovation. Enormous costs can be associated with bringing a product to the market based on limited hazard data, having it undergo success due to its performance attributes, and then having information emerge about its undesirable hazard properties leading to regulatory or market-based restrictions or de-selection. While it may be legal to bring products to the market with limited hazard data and data not available for public review, doing so can increase business risk.

The chemical industry is a vibrant and creative industry that continues to improve and bring new products to the marketplace. Chemical hazard assessment is a powerful tool for product development that can be internalized within a company because it allows for better understanding of the hazard properties of options prior to investing in product development and deployment. Additionally, greater disclosure about the use of specific chemicals in formulations along with their requisite hazards is needed to allow product manufacturers to improve products for their customers. This is true for both existing and emerging water repellent formulations. Pressure from downstream users and customers can drive improvements in the chemical supply chain, leading to safer formulations containing fewer toxic chemicals and impurities. Product manufacturers will need to trust but verify that long-chain fluorinated polymers are not continuing to show up in chemical formulations where there are agreements to remove such chemicals from the supply chain.

In the meantime, product manufacturers will need to incorporate consideration of human and environmental toxicity into their product formulation decisions. Preliminary hazard ranking suggests that hydrocarbon-based polymers are the most environmentally benign, followed by silicone-, and short-chain PFAS-based polymers. While short-chain fluoropolymers may provide better water and stain repellency, their use may not be worth the perceived and actual human health and environmental risks. Better understanding of the existing and emerging alternatives can only be resolved with greater transparency and with publicly available and peer-reviewed hazard profiles.

References

Abelkop, A.D., Graham, J.D., Royer, T.V., 2015. PBT determinations: science and standard procedures. In: Persistent, Bioaccumulative, and Toxic Chemicals: Technical Aspects, Policies, and Practices. CRC Press, Boca Raton, FL, pp. 29–64.

Alberts, B., Bray, D., Lewis, J., Raff, M., Roberts, K., Watson, J.D., 1994. Molecular Biology of the Cell, third ed. Garland Publishing, New York.

American Chemistry Council (ACC), 2013. Health Effects of Diisocyanates: Guidance for Medical Personnel. American Chemistry Council (ACC). https://polyurethane.americanchemistry.com/Resources-and-Document-Library/Health-Effects-of-Diisocyanates-Guidance-for-Medical-Personnel.pdf April 2013.

Anastas, P.T., Warner, J.C., 1999. Green Chemistry: Theory and Practice. Oxford University Press, Oxford, UK.

Blaine, A.C., Rich, C.D., Lakhwinder, S.H., Lau, C., Mills, M.A., Harris, K.M., Higgins, C.P., 2013. Update of Perfluoroalkyl acids into edible crops via land applied biosolids: field and greenhouse studies. Environ. Sci. Technol. 47, 14062–14069.

Blaine, A.C., Rich, C.D., Sedlacko, E.M., Lakhwinder, S.H., Kumar, K., Lau, C., Mills, M.A., Harris, K.M., Higgins, C.P., 2014a. Perfluoroalkyl acid distribution in various plant compartments of edible crops grown in biosolids-amended soils. Environ. Sci. Technol. 48, 7858–7865.

Blaine, A.C., Rich, C.D., Sedlacko, E.M., Hyland, K.C., Stushnoff, C., Dickenson, E.R.V., Higgins, C.P., 2014b. Perfluoroalkyl acid uptake in lettuce (Lactuca sativa) and strawberry (Fragaria ananassa) irrigated with reclaimed water. Environ. Sci. Technol. 48, 14361–14368.

Blum, A., Balan, S.A., Scheringer, M., Goldenman, G., Trier, X., Cousins, I., et al., 2015. The Madrid statement on poly- and perfluoroalkyl substances (PFASs). Environ. Health Perspect. 123 (5), A107–A111. https://www.ncbi.nlm.nih.gov/pmc/articles/PMC4421777/.

Buck, R.C., Franklin, J., Berger, U., Conder, J.M., Cousins, I.T., de Voogt, P., Jensen, A.A., Kannan, K., Mabury, S.A., van Leeuwen., S.P., 2011. Perfluoroalkyl and polyfluoroalkyl substances in the environment: terminology, classification, and origins. Integr. Environ. Assess. Manag. 7, 513–541. http://onlinelibrary.wiley.com/doi/10.1002/ieam.258/epdf.

Burcham, P.C., 2013. Core concepts in toxicology. In: An Introduction to Toxicology. Springer, New York, pp. 29–50 (Chapter 2).

Butenhoff, J.L., Rodricks, J.V., 2015. Human health risk assessment of perfluoroalkyl acids. In: Toxicological effects of Perfluoroalkyl and Polyfluoroalkyl Substances. Springer, Heidelberg, Germany, pp. 363–418 (Chapter 15).

Clean Production Action (CPA). 2016. GreenScreen® Version 1.3 Hazard Criteria. Dated: March 2016. Available at: http://www.greenscreenchemicals.org/.

Colas, A., Curtis, J., 2004. Silicone biomaterials: history and chemistry and medical applications of silicones. In: Ratner, B.D., Hoffman, A.S., Schoen, F.J., Lemons, J.E. (Eds.), Biomaterials Science: An Introduction to Materials in Medicine. Elsevier Academic Press, San Diego, CA.

Commissioner of Patents, 1854. Waterproofing and fireproofing. In: Subject-Matter Index (Made From Titles Only) of Patents of Inventions from March 2, 1617 to October 1, 1852. London Queen's Printing Office, London, UK, pp. 872–879.

Conder, J.M., Hoke, R.A., de Wolf, W., Russell, M.H., Buck, R.C., 2008. Are PFCAs bioaccumulative? A critical review and comparison with regulatory criteria and persistent lipophilic compounds. Environ. Sci. Technol. 42, 995–1003.

Danish Ministry of the Environment (Danish EPA), 2009. Short-chain polyfluoroalkyl substances (PFAS): a literature review of information on human health effects and environmental fate and effect aspects of short-chain PFAS Environmental Project No. 1707. Danish Ministry of the Environment (Danish EPA), Copenhagen, Denmark. http://www2.mst.dk/Udgiv/publications/2015/05/978-87-93352-15-5.pdf.

Danish Ministry of the Environment (Danish EPA)., 2013. Survey of PFOS, PFOA and other perfluoroalkyl and polyfluoroalkyl substances. Part of the LOUS-review. Environmental Project No. 1475. Danish Ministry of the Environment (Danish EPA), Copenhagen, Denmark.

Danish Ministry of the Environment (Danish EPA), 2015. Alternatives to perfluoroalkyl and polyfluoro-alkyl substances (PFAS) in textiles. Survey of chemical substances in consumer products. No. 137. Danish Ministry of the Environment (Danish EPA), Copenhagen, Denmark.

Davies, A., 2014. Durable Water Repellency-Study Phase I. An evaluation of the test methods used for assessing durable water repellent fabrics within the outdoor industry. http://www.europeanoutdoorgroup.com/files/DWR-Study_Alice_Davies__digital_.pdf.

De Mello, W.C., 1987. Cell-to-Cell Communication. Plenum Press, New York.

DeWitt, J.C., 2015. Toxicological Effects of Perfluoroalkyl and Polyfluoroalkyl Substances. Human Press, New York. pp. 1–495.

Eaton, D.L., Gilbert, S.G., 2013. Principles of toxicology. In: Casarett & Doull's Toxicology, eighth ed. McGraw Hill, New York, pp. 13–49 (Chapter 2).

ENVIRON (Ramboll ENVIRON), 2016. Assessment of POP Criteria for Specific Short-Chain Perfluorinated Alkyl Substances. (Prepared for FluoroCouncil, Washington, DC). ENVIRON, Arlington, VA. https://fluorocouncil.com/Resources/Research.

ENVIRON International Corporation, 2014. Assessment of POP Criteria for Specific Short-Chain Perfluorinated Alkyl Substances (Prepared for FluoroCouncil, Washington, DC). ENVIRON International Corporation, Arlington, VA. Project 0134304A. https://fluorocouncil.com/Resources/Research.

European Commission Scientific Steering Committee (SSC), 2000. First Report on the Harmonisation of Risk Assessment Procedures. Scientific Steering Committee, European Commission, Brussels, Belgium.

Fjeld, R.A., Eisenberg, N.A., Compton, K.L., 2007a. Introduction. In: Quantitative Environmental Risk Analysis for Human Health. Wiley, Hoboken, NJ.

Fjeld, R.A., Eisenberg, N.A., Compton, K.L., 2007b. Exposure assessment. In: Quantitative Environmental Risk Analysis for Human Health. Wiley, Hoboken, NJ, pp. 199–218.

Genualdi, S., Harner, T., Cheng, Y., Macleod, M., Hansen, K.M., van Egmond, R., Shoeib, M., Lee, S., 2011. Global distribution of linear and cyclic volatile methyl siloxanes in air. Environ. Sci. Technol. 8, 3349–3354.

Giesy, J.P., Kannan, K., 2001. Global distribution of perfluorooctane sulfonate in wildlife. Environ. Sci. Technol. 35, 1339–1342.

Gore, W.L., 2017. Gore Fabrics' Goal and Roadmap for Eliminating PFCs of Environmental Concern. https://www.gore-tex.com/pfcgoal. January 31, 2017.

Guy, W.S., Taves, D.R., Brey, W.S., 1976. Organic fluorocompounds in human plasma: prevalence and characterization. Am. Chem. Soc. 7, 117–134.

Hansen, K.J., Clemen, L.A., Ellefson, M.E., Johnson, H.O., 2001. Compound-specific, quantitative characterization of organic fluorochemicals in biological matrices. Environ. Sci. Technol. 35, 766–770.

Holmquist, H., Schellengerger, S., Van der Veen, I., Peters, G.M., Leonards, P.E.G., Cousins, I.T., 2016. Properties, performance and associated hazards of state-of-the-art durable water repellent (DWR) chemistry for textile finishing. Environ. Int. 91, 251–264.

Howells, R.D., 2009. Waterproofing and water and oil repellency. Kirk-Othmer Encyclopedia of Chemical Technology. John Wiley & Sons, Hoboken, NJ.

Inoue, Y., Hashizume, N., Yakata, N., Murakami, H., Suzuki, Y., Kikushima, E., Otsuka, M., 2011. Unique physicochemical properties of perfluorinated compounds and their bioconcentration in common carp Cyprinus carpio L. Arch. Environ. Contam. Toxicol. 62 (4), 672–680.

International Programme on Chemical Safety (IPCS), 2004. IPCS Risk Assessment Terminology: IPCS Harmonization Project Document No. 1: Part 1 IPCS/OECD Key Generic Terms and Part 2 IPCS Glossary of Key Exposure (Harmonization Project Documents). World Health Organization, Geneva. http://www.inchem.org/documents/harmproj/harmproj/harmproj1.pdf Accessed April 29, 2017.

Joyce, M.A., Dinglasan, Y.Y., Edwards, E.A., Mabury, S.A., 2004. Fluorotelomer Alcohol Biodegradation Yields Poly- and Perfluorinated Acids. Environ. Sci. Technol. 38 (10), 2857–2864.

KEMI, 2015. Occurrence and use of highly fluorinated substances and alternatives. Report from a government assignment, https://www.kemi.se/global/rapporter/2015/report-7-15-occurrence-and-use-of-highly-fluorinated-substances-and-alternatives.pdf.

Kierkegaard, A., McLachlan, M.S., 2013. Determination of linear and cyclic volatile methylsiloxanes in air at a regional background site in Sweden. Atmos. Environ. 80, 322–329.

Kissa, E., 2001. Fluorinated repellents. In: Fluorinated Surfactants and Repellents, second ed. Marcel Dekker, New York, pp. 516–556 (Chapter 12).

Lassen, C., Hansen, C.H., Mikkelsen, S.H., Maag, J., 2005. Siloxanes—Consumption, toxicity and alternatives. Environmental Project No. 103. Danish EPA, Copenhagen. www2.mst.dk/udgiv/publications/2005/87-7614-756-8/pdf/87-7614-757-6.pdf.

Lichstein, B.M., 1977. Stain and water repellency of textiles. In: Schick, M.J. (Ed.), Surface Characteristics of Fibers and Textile. Marcel Dekker, New York, pp. 495–524 (Chapter 13).

Lin, P.C., Lin, S., Wang, P.C., Rajagopalan, R., 2014. Techniques for characterization of nanomaterials. Biotechnol. Adv. 32 (4), 711–726. https://www.ncbi.nlm.nih.gov/pmc/articles/PMC4024087/.

Macintosh, C., 1823. Process and manufacture for rendering the texture of hemp, flax, wood, cotton, silk, also leather, paper and other substances impervious to water and air. Patent 4804. June 17, 1823.

Martin, J.W., Mabury, S.A., Solomon, K.R., Muir, D.C.G., 2003a. Dietary accumulation of perfluorinated acids in juvenile rainbow trout Oncorhynchus mykiss. Environ. Toxicol. Chem. 22, 189–195.

Martin, J.W., Mabury, S.A., Solomon, K.R., Muir, D.C.G., 2003b. Bioconcentration and tissue distribution of perfluorinated acids in rainbow trout (Oncorhynchus mykiss). Environ. Toxicol. Chem. 22, 196–204.

Moisiewicz-Pieńkowska, M., Jamrógiewicz, K., Szymkowska, K., Krenczkowska, D., 2016. Direct human contact with siloxanes (silicones)—Safety or risk part 1. Characteristics of siloxanes (silicones). Front. Pharmacol. https://www.ncbi.nlm.nih.gov/pmc/articles/PMC4884743/.

National Research Council, 2007. Toxicity Testing in the 21st Century: A Vision and a Strategy. https://www.nap.edu/catalog/11970/toxicity-testing-in-the-21st-century-a-vision-and-a.

National Toxicology Program (NTP), 2004. A National Toxicology Program for the 21st Century Roadmap to Achieve the NTP Vision. https://ntp.niehs.nih.gov/ntp/about_ntp/ntpvision/ntproadmap_508.pdf.

Nitzsche, S., Hechtl, W., Burghausen, E.W., 1974. Process for the condensation and/or equilibration of organosilicon compounds. U.S. Patent 3,839,388. U.S. Patent and Trademark Office, Washington, DC.

Noll, W., 1968. Chemistry and Technology of Silicones. Academic Press, New York.

Olsen, G.W., Chang, S.-C., Noker, P.E., Gorman, G.S., Ehresman, D.J., Lieder, P.H., Butenhoff, J.L., 2009. A comparison of the pharmacokinetics of perfluorobutanesulfonate (PFBS) in rats, monkeys, and humans. Toxicology 256, 65–74.

Organisation for Economic Cooperation and Development (OECD), 1995. Test No. 107: Partition Coefficient (n-octanol/water): Shake Flask Method. http://www.oecd-ilibrary.org/environment/test-no-107-partition-coefficient-n-octanol-water-shake-flask-method_9789264069626-en.

Organisation for Economic Cooperation and Development (OECD), 2004. Test No. 117: Partition Coefficient (n-octanol/water), HPLC Method. http://www.oecd-ilibrary.org/environment/test-no-117-partition-coefficient-n-octanol-water-hplc-method_9789264069824-en.

Organisation for Economic Cooperation and Development (OECD), 2009. Data Analysis of the Identification and Correlations Between Polymer Characteristics and Potential for Health or Ecotoxicological Concern. https://www.oecd.org/env/ehs/risk-assessment/42081261.pdf.

Organisation for Economic Co-operation and Development (OECD), 2013. OECD/UNEP Global PFC Group, Synthesis Paper on Per- and Polyfluorinated Chemicals (PFCs). Environment, Health and Safety, Environment Directorate, OECD. http://www.oecd.org/chemicalsafety/risk-management/synthesis-paper-on-per-and-polyfluorinated-chemicals.htm.

Organisation for Economic Co-operation and Development (OECD), 2017. OECD Portal on Perfluorinated Chemicals. Available from: https://www.oecd.org/ehs/pfc/.

Organisation for Economic Co-operation and Development (OECD). 2012. Test No. 305: Bioaccumulation in Fish: Aqueous and Dietary Exposure. http://www.oecd-ilibrary.org/environment/test-no-305-bioaccumulation-in-fish-aqueous-and-dietary-exposure_9789264185296-en.

Outdoor Industry Association (OIA). 2017. Is W.L. Gore's pledge to eliminate PFCs good for the Industry? Yes, and. https://outdoorindustry.org/article/w-l-gores-pledge-eliminate-pfcs-good-industry-yes/ (accessed 29.04.17.).

Peal, S., 1791. An improved method of making and rendering perfectly waterproof all kinds of leather, cotton, linnen, and woolen cloths, silks, stuffs, paper, wood and other maimfactures and substances, for the purpose of being worked up into shoes, boots, and other wearing apparel, and to be used on all occasions where dryness of power of repelling wet or moisture may be required. Patent 1801. May 2, 1791.

Posner, S., 2015. PFAS—Substances, Applications and Environmental Impacts. Umwelt Bundesamt. http://www.umweltbundesamt.de/dokument/pfas-substances-applications-environmental-impacts.

Posner, S., 2016a. EOG brands PFAS Q&A. European Outdoor Brands Backgrounder. https://outdoorindustry.org/resource/european-outdoor-group-eog-brands-pfas-backgrounder/.

Posner, S., 2016b. State of play for DWRs. Current and future (legal) challenges. Presentation at ISPO, January 26, 2016.

Ritter, S.K., 2015. The shrinking case for fluorochemicals. Chem. Eng. News 93 (28), 27–29. http://cen.acs.org/articles/93/i28/Shrinking-Case-Fluorochemicals.html.

Robinson, L., Thorn, I., 2005. Handling chemicals in the workplace. Toxicology and Ecotoxicology in Chemical Safety Assessment. CRC Press, Boca Raton, FL, pp. 100–114.

Sass, J., Heine, L.G., Hwang, N., 2016. Use of a modified GreenScreen tool to conduct a screening-level comparative hazard assessment of conventional silver and two forms of nanosilver. Environ. Health 15, 105. https://ehjournal.biomedcentral.com/articles/10.1186/s12940-016-0188-y.

Scheringer, M., Trier, X., Cousins, I.T., de Voogt, P., Fletcher, T., Wang, Z., Webster, T.F., 2014. Helsingør statement on poly- and perfluorinated alkyl substances (PFASs). Chemosphere 114, 337–339. http://www.sciencedirect.com/science/article/pii/S004565351400678X.

Scientific Committee on Consumer Safety (SCCS), 2010. Opinion on Cyclomethicone. Octamethylcyclotetrasiloxane (Cyclotetrasiloxane, D4) and Decamethylcyclopentasiloxane (Cyclopentasiloxane, D5). SCCS/1241/10. https://ec.europa.eu/health/scientific_committees/consumer_safety/docs/sccs_o_029.pdf.

Scientific Committee on Consumer Products (SCCP), 2005. Opinion on Octamethylcyclotetrasiloxane (D4). Cyclomethicone (INCI name) SCCP/0893/05. https://ec.europa.eu/health/ph_risk/committees/04_sccp/docs/sccp_o_035.pdf.

Spencer, P., 2013. Why good products fail: when chemistry and toxicology collide. Presented at Green Toxicology Day. Johns Hopkins Bloomberg School of Public Health.

Stockholm Convention, 2017. What are POPS? http://chm.pops.int/TheConvention/ThePOPs/tabid/673/Default.aspx.

Stockholm Convention on Persistent Organic Pollutants. Annex D. http://www.pops.int/documents/convtext/convtext_en.pdf (accessed 29.04.17.).

Swerea IVF AB, 2015. SUPFES—Substitution in practice of prioritized fluorinated compounds for textile applications. Project report 24052.

Tang, W., Huang, Y., Meng, W., Qing, F.-L., 2010. Synthesis of fluorinated hyperbranched polymers capable as highly hydrophobic and oleophobic coating materials. Eur. Polym. J. 46, 506–518.

Tate, A.O., 1917. Method of waterproofing fibrous materials. U.S. Patent 1,228,987, June 5, 1917.

Taves, D.R., 1971. Comparison of "organic" fluoride in human and nonhuman serums. J. Dent. Res. 50, 783.

TemaNord, 2005. Siloxanes in the Nordic Environment. Nordic Council of Ministers. http://nordicscreening.org/index.php?module=Pagesetter&func=viewpub&tid=5&pid=4.

Tortora, P.G., Johnson, I., 2013. Waterproof. The Fairchild Books Dictionary of Textiles. Fairchild Books, New York, NY.

ToxServices LLC, 2013. Paraffin Wax (Melting Point 45 to 48°C) (CAS #8002-74-2) GreenScreen™ Assessment. Washington, DC.

Ubel, F.A., Sorenson, S.D., Roach, D.E., 1980. Health status of plant workers exposure to fluorochemicals-a preliminary report. Am. Ind. Hyg. Assoc. J. 41, 584–589.

United National Economic Commission for Europe (UNECE), 2017. GHS Implementation. http://www.unece.org/trans/danger/publi/ghs/implementation_e.html.

United Nations, 2015. Globally Harmonized System of Classification and Labelling of Chemicals (GHS). Available at: http://www.unece.org/trans/danger/publi/ghs/ghs_welcome_e.html.

United Nations Environment Program (UNEP), 2009. Annotated outline for a guidance document on perfluorooctane sulfonate alternatives. In: Persistent Organic Pollutants Review Committee, Fifth meeting, Geneva, 12–16 October 2009. UNEP/POPS/POPRC.5/INF/10.

United Nations Environment Program (UNEP), 2014a. Report on the assessment of alternatives to perfluorooctane sulfonic acid, its salts and perfluorooctane sulfonyl fluoride. In: Persistent Organic Pollutants Review Committee, Tenth Meeting, Rome, 27–30 October 2014. UNEP/POPS/POPRC.10/INF/7/Rev.1*.

United Nations Environment Program (UNEP), 2014b. Factsheets on alternatives to perfluorooctane sulfonic acid, its salts and perfluorooctane sulfonyl fluoride. In: Persistent Organic Pollutants Review Committee, Tenth Meeting, Rome, 27–30 October 2014. UNEP/POPS/POPRC.10/INF/8/Rev.1.

United States Environmental Protection Agency (U.S. EPA), 2005. Siloxane D5 in Dry cleaning Applications Fact Sheet. Office of Pollution Prevention and Toxics. Available at: http://www.epa.gov/oppt/dfe/pubs/garment/d5fs3.pdf.

United States Environmental Protection Agency (U.S. EPA), 2009. Long-chain perfluorinated chemicals (PFCs) action plan. Available from: https://www.epa.gov/assessing-and-managing-chemicals-under-tsca/long-chain-perfluorinated-chemicals-pfcs-action-plan.

United States Environmental Protection Agency (U.S. EPA), 2011. U.S. EPA's Design for the Environment Program Alternatives Assessment Criteria for Hazard Evaluation. Version 2. https://www.epa.gov/sites/production/files/2014-01/documents/aa_criteria_v2.pdf.

United States Environmental Protection Agency (U.S. EPA), 2014. TSCA Work Plan for Chemical Assessments: 2014 Update. Office of Pollution Prevention and Toxics. https://www.epa.gov/sites/production/files/2015-01/documents/tsca_work_plan_chemicals_2014_update-final.pdf.

United States Environmental Protection Agency (U.S. EPA), 2017. PFOA Stewardship Plan. EPA's Non-CBI Summary Tables for 2015 Company Progress Reports (Final Progress Reports). February, 2017. https://www.epa.gov/assessing-and-managing-chemicals-under-tsca/and-polyfluoroalkyl-substances-pfass-under-tsca#tab-3.

Wang, G., Norwood, W., Alaee, M., Byer, J.D., Brimble, S., 2013. Review of recent advances in research on the toxicity, detection, occurrence and fate of cyclic volatile methylsiloxanes in the environment. Chemosphere 93, 711–725. https://doi.org/10.1016/j.chemosphere.2012.10.041.

Washington, J.W., Ellington, J.J., Jenkins, T.M., Evans, J.J., Yoo, J., Hafner, S.C., 2009. Degradability of an acrylate-linked, fluorotelomer polymer in soil. Environ. Sci. Technol. 43 (17), 6617–6623.

Washington, J.W., Jenkins, T.M., Rankin, K., Naile, J.E., 2015. Decades-Scale Degradation of Commercial, Side-Chain, Fluorotelomer-Based Polymers in Soils and Water. Environ. Sci. Technol. 49 (2), 915–923.

Whittaker, M.H., Heine, L., 2013. Chemicals alternatives assessment (CAA): tools for selecting less hazardous chemicals. In: Hester, R. (Ed.), Chemical Alternatives Assessments. Royal Society of Chemistry, Cambridge, pp. 1–43.

Wooley, D., Wooley, A., 2017. Introduction to toxicology: the necessity of measurement. In: Practical Toxicology: Evaluation, Prediction, and Risk, third ed. CRC Press, Boca Raton, FL (Chapter 1).

Xu, L., Shi, Y., Liu, N., Cai, Y., 2015. Methylsiloxanes in environmental matrices and human plasma/fat from both general industries and residential areas in China. Sci. Total Environ. 505, 454–463. https://doi.org/10.1016/j.scitotenv.2014.10.039.

Zero Discharge of Hazardous Chemicals (ZDHC). 2012. Durable Water and Soil Repellent Chemistry in the Textile Industry—A Research Report. P05Water Repellency Project. http://www.roadmaptozero.com/fileadmin/layout/media/downloads/en/DWR_Report.pdf.

Biomimetic principles for design of water repellent surfaces

5

Veronika Kapsali
University of the Arts London, London, United Kingdom

5.1 Introduction: Biomimetic design relevance to textile technology

Biomimetics is a compound word formed from the Greek *bio* (life) and *mimesis* (to imitate). The term was originally coined by polymath and biophysicist Otto Schmitt and used in his doctoral thesis to describe the notion of applying knowledge developed from the use of physics to describe/explain phenomena in biology into engineering. A parallel term, *bionics*, was created by Schmitt's contemporaries at the United States Air Force's Office of Scientific Research in 1960 to describe the 'science of systems which have some function copied from nature or which represent characteristics of natural systems or their analogues'. In the following decades, new terms such as *biomimicry*, *biognosis* and *bioinspiration* emerged in different parts of the world, but these all essentially mean the same thing (Vincent et al., 2006).

Since the conception of the term, biomimetics has been applied almost exclusively to innovation across the engineering fields including, electrical, mechanical and materials and constitutes an increasingly important multidisciplinary approach to STEM (Science, Technology, Engineering and Mathematics) disciplines. Activity in the academic sector has surged in the last twenty years, from 100 journal publications per year in the mid 1990s to over 3000 in 2013 (Lepora et al., 2013) However, terminology aside, observation-based inspiration from biology, be it emotional, aesthetic or functional, is human nature and was going on long before a word was invented for it.

Hidden in myth, many ancient cultures are woven with tales of nature-inspired invention, such as that of Talus (nephew and apprentice to Daedalus, a famous ancient Greek inventor). Although a young man, Talus demonstrated great inventive talent and was credited for innovations that gave him a reputation to rival his master's. The myth details one particular invention conceived during a walk on the beach, where Talus came across the intact spine of a dead fish. Following careful observation, he replicated the structure in metal, thus creating the first handsaw (Kapsali, 2016).

A very popular, but misleading legend of biomimetic innovation attributes the inspiration for Joseph Paxton's design of the Crystal Palace, the home of the 1851 Great Exhibition at London's Hyde Park to the structure of the leaf of the giant water lily (*Victoria amazonica*). Paxton was a horticulturist and brilliant engineer, and his drive to create large greenhouses led to the invention of a ridge-and-furrow system that enabled the creation of expansive, large-scale structures made entirely of glass and iron.

Waterproof and Water Repellent Textiles and Clothing. https://doi.org/10.1016/B978-0-08-101212-3.00005-8

Paxton designed a lily house, an enormous glass enclosure that successfully nurtured one of the first *Victoria amazonica* seedlings to reach England (Kapsali, 2016).

Paxton drew on his experience of the lily house construction to propose a highly ambitious building design for the 1851 Great Exhibition. The ideas were well received, and he was commissioned as the architect for the Crystal Palace, the world's largest structure built exclusively from iron and glass. After the exhibition, Paxton was invited to give a talk about his design at the Royal Society of Arts, during which he presented an illustration of the *Victoria amazonica* leaf. It is believed that an over-enthusiastic reporter from the audience misunderstood Paxton's talk and generated the legend (Vincent et al., 2006; Kapsali, 2016).

Biomimicry is inextricably linked to the modern textile industry. It could be said that the desire and subsequent efforts to synthesize artificial silk, which are believed to date back to ancient China in 3000 BC, resulted in a 5000-year journey that led to the birth of the man-made fibre industry as it is known today (Vincent et al., 2006; Kapsali, 2016).

More recently George de Mestral, a Swiss electrical engineer and keen mountaineer, wanted to find out why it was so difficult to remove burrs from his clothes and dog's fur after walks in the mountains. In 1941, a simple observation of burdock burrs (Fig. 5.1) under the microscope revealed that each spine ended with a curved tip, resembling a hook. Mestral spotted the potential for a reversible dry-adhesion system, to be used initially for clothing fasteners, so he worked with a French fabric mill to interpret his observations into a fibrous hook-and-loop textile system. The first

Fig. 5.1 Detail of prickly seed pod of Burdock, commonly known as burr. The pod is covered in spikes with a curved, hook like tip. This specialized design is central to the plant's seed dissemination mechanism; it enables the pod to attach itself to fibrous surfaces such as animal fur and human clothing during contact and be carried away. Burrs are notably difficult to remove, this property intrigued George de Mestral and lead to the invention of Velcro.

prototypes were made from cotton, but adhesion lasted only a few cycles. The newly invented nylon filament replaced cotton, increasing robustness and adhesion repeatability (US patent 2717437 A).

Mestral filed for a patent in 1951 and was granted in 1955. The first commercial applications were in space apparel this was followed by skiwear and active clothing systems. By 2016, Velcro's hook-and-loop system (Fig. 5.2) was used in a wide range of applications from children's shoes to protective clothing (Kapsali, 2016).

Biomimetics as an approach to innovation in textiles is a relatively unexplored field, although it holds significant potential for meaningful technological transfer. Biological materials such as skin, bone and wood demonstrate flexibility, strength and robustness, not because of the chemistry in their composition, but because of the way in which raw materials are organized into structures from nano- to macroscale. This hierarchical approach to design enables the optimization of resources critical to survival (Kapsali, 2016; Kapsali et al., 2013).

Molluscs such as the abalone and pearl oyster have shells whose primary function is to protect the fragile body of the animal from foreign objects and predators. The inner shell layer is commonly lined with an iridescent material known as mother of pearl or nacre. Nacre is composed mainly of the mineral calcium carbonate (95%), which makes it stiff and hard, and 5% protein, yet it demonstrates a level of toughness (resistance to cracking) three thousand times greater than the mineral from which it is made, due to the way the material is organized at the nano- to microscale. Nacre is a hierarchical structure (Fig. 5.3); the calcium carbonate mineral forms individual microscopic polygonal plates (5–15 μm wide and 1–5 μm thick) stacked like bricks, sometimes in columns like walls, and sometimes randomly like sheets. The calcium

Fig. 5.2 Detail of Velcro tape made from Nylon. To work as a mechanism Velcro required two components, one tape featuring a *prickly* surface representing the Burr hooks and a *fluffy* component representing animal fur or clothing.

Fig. 5.3 Coloured scanning electron micrograph (SEM) of a section through an abalone (Haliotis sp.) shell, an edible mollusc found in warm seas. The inner shell is composed of layers of overlapping platelets of calcium carbonate crystals that sit on thin layers of protein. This structure makes the shell significantly tougher than its individual components.

carbonate plates have the ability to slide along the surfaces of one another under tension. This system introduces the ductility responsible for nacre's highly amplified toughness and tolerance to damage (Sun and Bhushan, 2012; Meyers et al., 2008; Luz and Mano, 2009).

Just like biological structures, textiles are fibrous assemblies with a great degree of hierarchy. Fig. 5.4 compares the structure of the human bone with that of a textile. The properties of biological materials are mainly attributed to complex hierarchical subsystems interdependently shaping the structural properties across scales. This perspective is relatively unexplored in textiles (Fig. 5.5); textile technology presents a possible viable platform for a hierarchical approach to biomimetic innovation (Kapsali et al., 2013). This chapter is focused on opportunities for texture and surface morphology in the context of hydrophobic finishes for textiles.

5.2 Biomimetic principles in structural hydrophobicity

As a lotus flower is born in water, grows in water and rises out of water to stand above it unsoiled

Gautama Buddha (Horner, 1964)

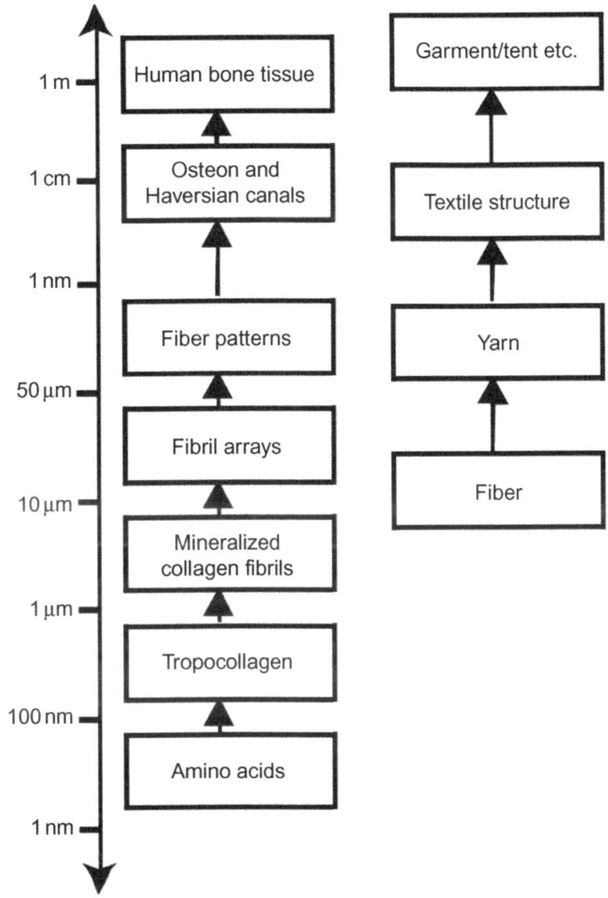

Fig. 5.4 Comparison of human bone structure with that of a conventional textile structure such as garments or tents. Just like biological structures, textiles are fibrous assemblies with a great degree of hierarchy. Fibril and fibre orientation are well studied factors in biomechanics yet are relatively unexplored in textiles; the basic building block in a textile designer's toolkit is the fibre, however advances in understanding and capability in material technologies at nano to microscale will enable a future where design begins at a much more elemental scale just like in nature.

The lotus plant, *Nelumbo nucifera*, holds a sacred place in many Eastern cultures. It is a symbol of purity, nonattachment and divine beauty, characteristics attributed to the plant because of its remarkable ability to remain clean despite its muddy habitat. Instead of getting wet when splashed, water droplets form on the surface of the leaf and simply roll off, removing dirt and contaminants in their wake. Although unique in its romance, the lotus leaf is not the only example of such counterintuitive, self-cleaning behaviour. Several types of plant, such as the wild and savoy cabbage, taro and prickly pear, as well as insect shells and wings, demonstrate similar properties (Ganesh et al., 2011; Fig. 5.6).

Fig. 5.5 View of current textile hierarchical system across scales from fibre to yarn, textile structure and garment.

Fig. 5.6 Detail of water droplets forming on the surface of a lotus leaf.

One of the primary functions of self-cleaning biological surfaces is to prevent adhesion of contaminants such as particles and microorganisms onto the plant surface and allow them to be washed away in the presence of water droplets. The mechanism behind this extraordinary behaviour remained a mystery until the early 1970s and the invention of the scanning electron microscope (SEM). The cuticle is a protective skin/epidermis that covers most parts of a plant (except the roots) which, due to its chemical composition, is usually hydrophobic. Wilhelm Barthlott and Neinhuis studied hundreds of leaf surface morphologies using SEM (Baker, 1982; Barthlott and Neinhuis, 1997; Barthlott, 1990, 1994; Cassie and Baxter, 1944; Jeffree, 1986). On analysis of a variety of leaf micro-surfaces, the team discovered that self-cleaning behaviour was observed in cases where the cuticle surface was rough. The lotus leaf,

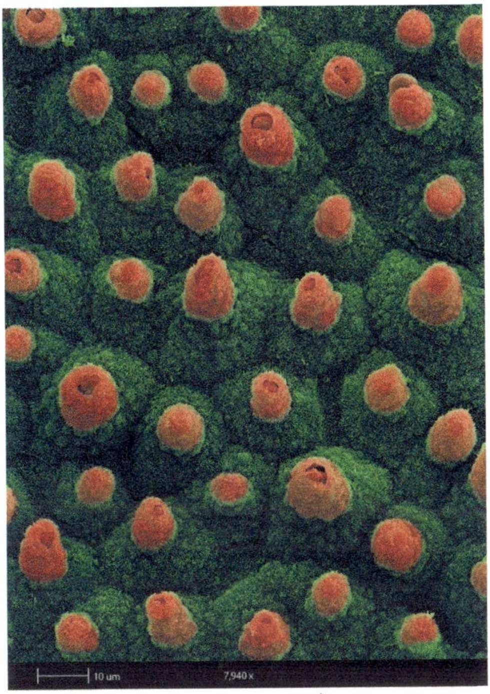

Fig. 5.7 SEM of the lotus leaf (*Nelumbo nucifera*) surface microstructure responsible for the super-hydrophobic properties of the leaf.

in particular, featured microscopic bumps about 10–20 μm high and 10–15 μm wide covering the cuticle surface, and topped with a waxy structure that resembles a cone with rounded tips (Fig. 5.7), ranging between 1 μm and 5 μm in height (Barthlott and Neinhuis, 1997; Barthlott, 1990, 1994; Cassie and Baxter, 1944; Jeffree, 1986).

Barthlott and Neinhuis (1997) studied the relationship between roughness and wettability of about 1000 types of leaf surfaces, combining SEM and contact angle measurements, and discovered that the specialized design of the texture combines with air to create a composite low energy surface which prevents water drops from spreading, by enlarging the water air surface and minimizing water solid contact area responsible for the formation of the spherical water droplets able to roll along the surface of the leaf. Contact angle (CA) measurements revealed that low CA <110 degrees resulted in wettable surfaces that allowed the spread of water, while 180 degrees $\leq$ CA corresponds to nonwetting.

Furthermore, the distance between the wax tips (15–25 μm) is smaller than the diameter of a dirt particle, pathogen or other contaminants, meaning that foreign particles sit on the tips of the wax crystalloids and away from the surface of the leaf, keeping contact area to a minimum. As a water droplet rolls along the surface of the leaf, it picks up the particles and carries them away thus cleaning the surface. Barthlott and Neinhuis coined the term *lotus effect* to describe this property. Their team identified significant opportunities for application of this functionality for self-cleaning surfaces and proceeded to explore, developing into surface coatings for external walls and textiles.

The first industrial application for the lotus effect was in the form of the self-cleaning masonry paint Lotusan. The paint creates a surface microstructure that resembles the wax tipped papillae of the lotus leaf in both form and scale. Façade renders containing Lotusan recreate the superhydrophobic properties of the lotus leaf, allowing passive self-cleaning during rainfall, thus requiring significantly less maintenance; according to the Lotusan information literature provided by STO (2017) (the company which currently produces and distributes Lotusan products) the effect preserves the cleanliness of the wall by preventing dirt and microbes from sticking onto the surface, thus requiring less-frequent repainting.

As scientific tools have enabled scholars to look more closely at biological systems and develop new knowledge and understanding of the way design is used to overcome limitations in materials and resources in general, there is much that can be learned from biology's lean operation. The present focus on micro-texture reveals a new world of possibilities afforded by understanding the functionality of texture.

Textile designers understand how to manipulate textures such as smooth, reflective, rough, soft, etc., to engineer aesthetic and ergonomic effects. However, nature shows there is more to bumps and lumps, and they could hold the key to advanced, sustainable, multifunctional textiles.

5.3 Brief overview of current industrial durable hydrophobic textile finishes

Water repellent coatings have been applied to textiles for hundreds of years. Originally, natural materials such as fish oils and grease were used to coat sails, but by the 1500s, these materials were beginning to be replaced by linseed oil. In the 1800s, paraffin wax coatings were applied to cotton fabrics and used to make outerwear garments for outdoor pursuits such as hunting and military tasks.

State-of-the-art water repellent systems can employ a variety of different polymers, depending on their functionality and end use, but are based on comparable structural elements: hydrophobic polymer sidechains, dendrimers or inorganic particles based on silicones, per- and polyfluroalkyl moieties or hydrocarbons linked to a polymer backbone. The repellent effect is created by the umbrella-like structure these sidechains create along the surface of the fibre (Holmquist et al., 2016). On an industrial scale, baths of aqueous solutions containing finishing materials are used to immerse textiles, which are then fed through a double roller system that squeezes out the water and concentrates the water repellent polymers on the textile fibres. Hydrophobic layers are formed as wet fabrics are cured in a heated oven, promoting crosslinking reactions and ultimately, creating a repellent finish. However, these processes are still very new, and not enough is known about their durability in the context of textiles in action, when exposed to abrasion and pressure, conditions to which the lotus and other superhydrophobic natural surfaces are not exposed.

Although these materials and techniques can simulate the superhydrophobic self-cleaning properties observed in natural systems (such as the lotus leaf) on a macroscale, the required processes rely on high-energy consumption and the use of toxic materials

(Slater, 2003). There is a need to develop novel technologies that can introduce advanced functionalities into textiles in a sustainable manner, meaning use of less materials (ideally nonhazardous to humans or the environment) under ambient conditions with low energy consumption. Progress in some, if not all, areas is an improvement.

5.4 Opportunities for novel biomimetic industrial approaches to engineered textile hydrophobic micro-textures

There has been significant advance in technology transfer of superhydrophobic functional textures from biology into man-made surfaces for a range of applications from biofouling (Bixler and Bhushan, 2012) to novel self-cleaning paints and finishes. Fabrication methods developed to this end are equally varied, from colloids containing nanoparticles, embossed plates derived directly from a biological surface (Nagaraja and Yao, 2007), to advanced micro-etching and embossing technologies.

Most commercial textiles are composed of fibres with diameters in the region of 11–22 μm if natural nonsilk (i.e. wool, cotton, flax) or 10–15 μm if silk. The diameter of synthetic fibres can be engineered with microfibres being the finest with thicknesses smaller that silk. A basic requirement for textiles is durability, thus fibres need to demonstrate specific tensile strength as well as flexibility. In terms of scale, the high aspect ratio of fibres positions them across micron (diameter) to macro (length ranges from cm to km) scale. Bundles of fibres are twisted together to form more robust systems of yarns (as compared to single fibres), which are then knitted or woven into the textile, resulting in very complex, flexible, fibrous 3D structures with very high surface areas. Also, relatively new electrospinning processes create finer webs with high surface areas.

Functional, superhydrophobic micro-textures in biology are located on the surface of leaves, wings and other biological structures, including fibrous/cellular systems, but at a submicron scale. The characteristic rough morphology of such surfaces is realized through microscopic 3D lumps, bumps and folds. In the case of the lotus leaf, the wax crystalloid tip is about 1–5 μm tall and sits on an epidermal bump (papillae) in the region of 10–15 μm wide and 10–20 μm high (Baker, 1982; Barthlott and Neinhuis, 1997). The dimensions of these surface features are comparable to the diameter of commercial textile fibres.

Technology transfer at a similar scale could be achieved through modification to existing technological processes. Rough fibre cross-section such as multilobal structures used in wicking technologies, often referred to as *Mickey Mouse ears* (Fig. 5.8), can be designed using novel fibre-extrusion nozzles and co-extrusion. Alternatively, the texture of the fibre surface could be modified or roughened using advanced etching methods such as enzymes currently used to clean and polish textiles. Further research is required to explore nonchemical, structural approaches to the functionalization of fibres by engineering microscale textures on the fibre surface through existing commercial and smaller, bespoke extrusion processes.

Fig. 5.8 Micrograph of cross-section of trilobal nylon fibres, also known as *mickymouse* ears.

Recent advances in plasma technologies (Brunon et al., 2017; Zanini et al., 2016; Gotoh et al., 2017) present a promising and viable alternative, with numerous studies conducted on superhydrophobic textile engineering. Plasma is a term coined in a seminal paper published by Tonks and Langmuir (1929) (as part of their studies of particle oscillations) to describe the electrically neutral medium of unbound positive and negative particles. Plasma treatment, today, is employed to introduce novel functionalities (such as superhydrophobicity) to surfaces including textiles. Unlike conventional finishing textile technologies, plasma is dry, clean, requires low or ambient pressures and uses minimal amounts of functional materials, and is thus clearly aligned with the sustainable ethical and environmental agenda of future manufacturing. However, most atmospheric treatments are currently used for niche applications on relatively small/individual items, rather than for fast throughput of a textile roll.

There are several methods for creating plasma in controlled conditions; the most relevant to textiles is cold plasma, in which gas temperature is ambient. In thermal plasma, the temperature of the gas is extreme, sometimes over 1000°C, and textiles cannot withstand such conditions. Cold plasma, or nonthermal plasma, treatments can be further categorized according to the required processing pressure, which can be low (vacuum) or atmospheric pressures. The former requires an expensive vacuum chamber for commercial upscale, but the latter does not. Several industrial atmospheric plasma systems for textile treatment have been developed; although superhydrophobic finishing can be engineered, the durability of the functionality is not yet optimized. The key function of current plasma systems in textiles is the preparation of fabric surfaces for optimized printing and dyeing.

Multiple Laser Surface Enhancement (MLSE) is a novel textile finishing process that combines atmospheric plasma treatment with a sol–gel delivery system, thus enabling a wider range and flexibility of surface engineering such as fire retardancy, with significant and measurable advantages in resource and energy consumption.

MTix Ltd., based at the Textile Centre of Excellence in Huddersfield (http://textilehouse.co.uk/innovations-and-projects/eco-innovation/mlse-technology/), has explored the commercial implications of the MLSE process as part of a project co-funded with the European Union. Specifically, the process can achieve a 90.0% reduction in greenhouse gases, and uses 94.8% less chemicals, 75.5% less water and 90.9% less energy than conventional textile finishing processes. This process also provides a breakthrough for atmospheric treatments in terms of output; current capability is in the region of 20 m/s, which is comparable to similar industrial finishing processes.

Interestingly, when compared to functional rough surfaces in biology, which are generally observed within the micron scale, structures created using plasma and sol–gels are much smaller, in the nanoscale (1000 nm in 1 μm). Usually, transfer of hierarchical structures from biology to engineering systems occurs within the same scale, or in some cases, examples of structures from microscale in biology can be recreated at macroscale (such as auxetics/cellular materials with negative Poisson's ratios). Plasma and MLSE technologies provide a technical platform that has the potential to recreate rough surfaces morphologies at microscale to introduce superhydrophobic properties to textiles. However, the durability of such technologies when exposed to rubbing, washing and other processes involved in garment wear and maintenance is not fully understood.

This chapter began to explore the role of texture in nature, and how the surface morphology of a system can introduce a host of passive properties essential to survival. Textural qualities such as smoothness, reflectivity, roughness, bumpiness, etc., play a significant role in the aesthetics and ergonomics of design. However, biology shows an entirely new perspective on texture as a platform for advanced products.

For example, shark skin-inspired wetsuits such as those originally developed by Speedo in 2000 featured specialized biomimetic textiles engineered to mimic the surface design of shark skin, specifically its drag resistance properties, and were claimed to improve the wearers' speed in the water while swimming by 3%, and were rapidly adopted by professional athletes. The actual functionality of these were the focus of significant debate, and they were subsequently banned from professional competitions in 2009. Shark skin, unlike fish skin, is covered not in scales, but in microscopic tooth-like structures (dentils) that feature a bumpy rib surface (Fig. 5.9). On flat or curved surfaces, these ribs divide the drag forces created as the structure moves through the water, reducing the overall effect significantly. The spacing between ribs varies among species (30–100 μm) (Lang et al., 2008) and exceeds the range observed for superhydrophobic textures in biology, yet creates a rough surface surrounding the shark that not only minimizes drag forces between the animal and contacting sheets of water, but also prevents the adhesion of parasitic microorganisms onto the skin.

This section demonstrated why microscopic bumps and lumps are favoured over flat, smooth surfaces in biology, and how the manipulation of scale and design of texture can be invaluable tools for resource optimization and introduction of multifunctional properties into man-made textile systems, and build upon the technical success of biomimetic textile innovations such as Velcro, Morphotex (Teijin Ltd) and Inotek (MMT Textiles Ltd).

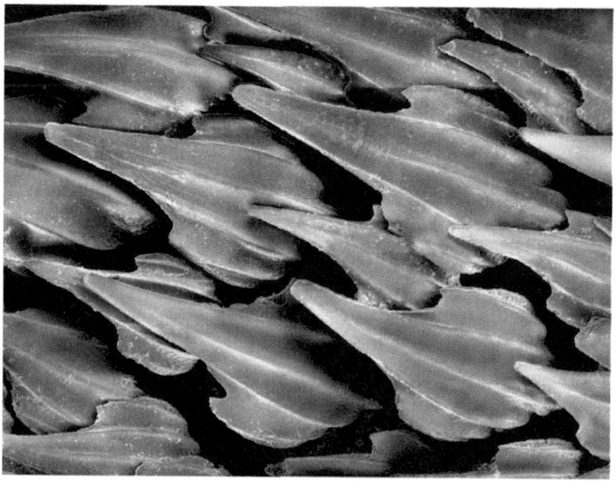

Fig. 5.9 Coloured scanning electron micrograph (SEM) of the skin of a shark. The surface is covered in ribbed dermal teeth also known as denticles.

5.5 Future applications

A research team set out to explore the fundamentally different problem-solving approaches utilized in technology and biology (Vincent et al., 2006). The team established a logical framework based on the mantra, *things* (substance and structure) *do things* (require energy and information) *somewhere* (space, time), and adapted a set of tools from TRIZ (translated from Russian, meaning, '*theory of inventive problem solving*', a system created in Russia to enable the translation of technical solutions from one field of engineering to another), to a biological context and conducted a study analysing thousands of examples from both biology and engineering. The outcomes of the study revealed that there is a significant difference between the way technical problems are solved in engineering and biology. These were roughly quantified and illustrated according to scale, from nano- to macroscale.

The findings revealed that at size levels up to 1 m (where most innovation occurs), manipulation of energy is the most significant variable in technical solutions; this is followed by use of material. Inversely, information and space are the prevailing variables in biology at the same scale.

Vincent contextualizes the study's findings by highlighting the sheer number of man-made polymers, currently exceeding 300 varieties, while biology mainly relies on only a few. The wide range of biological materials made from proteins (such as silk, muscle, enzymes) are made of similar monomers, yet the variety in emergent qualities and behaviour is unrelated to the energy required to synthesize the materials, but depends instead upon the way in which the basic building blocks (monomers) are arranged, which is essentially design. Take, for example, spider silk, which can have several forms including fibres and glue, all created at low temperature and pressure from the same monomers produced in specialized glands within the insect.

Analogical reasoning enables the transfer of properties, mechanisms and ideas from biology to technology. Shape, surface and structures are very important factors in both textile design and engineering. A biomimetic approach to textile design offers a new perspective on the relationship between materiality, form and function, leading to a new design space. The systematic approach to design and development of biomimetic textile systems is a novel and emergent space arising from the intersection of biomimetics, materials engineering and advances in textile technology, and is in need of further research.

This chapter focused on the superhydrophobic surface structure of the lotus leaf. However, there are many biological surfaces that draw on bumps, lumps, wrinkles and ridges to achieve advanced and complex functionalities such as antifouling and the manipulation of contact with air and water. Other potential surfaces and structures of interest for future research using novel technological processes such as plasma include (but are not limited to) shark skin, gecko feet, the Namib beetle, insect wings and thorny devil skin.

5.6 References, other sources of information

This section provides a comprehensive, but not exhaustive, list of resources for further information on biomimetic engineering and design.

References

Baker, E.A., 1982. Chemistry and morphology of plant epicuticular waxes. In: Linnean Society Symposium Series.

Barthlott, W., Neinhuis, C., 1997. Purity of the sacred lotus, or escape from contamination in biological surfaces. Planta 202 (1), 1–8.

Barthlott, W., 1990. Scanning electron microscopy of the epidermal surface in plants. Scanning electron microscopy in taxonomy and functional morphology. Syst. Assoc. Spec. Vol. 41, 69–94.

Barthlott, W., 1994. Epicuticular wax ultrastructure and systematics. Caryophyllales. Springer, Berlin, Heidelberg, pp. 75–86.

Bixler, G.D., Bhushan, B., 2012. Biofouling: lessons from nature. Phil. Trans. R. Soc. A 370, 2381–2417.

Brunon, C., Chadeau, E., Oulahal, N., Grossiord, C., Dubost, L., Simon, F., Bessueille, F., Degraeve, P., Leonard, D., 2017. Antimicrobial finishing of textiles intended for food processing industry by plasma enhanced chemical vapor deposition–physical vapor deposition of Ag-SiOCH composites coated with AlxOy or SiOCH encapsulation layers. Thin Solid Films 628, 132–141.

Cassie, A.B.D., Baxter, S., 1944. Wettability of porous surfaces. Trans. Faraday Soc. 40, 546–551.

Ganesh, V.A., Raut, H.K., Nair, A.S., Ramakrishna, S., 2011. A review on self-cleaning coatings. J. Mater. Chem. 21 (41), 16304–16322.

Gotoh, K., Shohbuke, E., Ryu, G., 2017. Application of atmospheric pressure plasma polymerization for soil guard finishing of textiles. Text. Res. J. https://doi.org/10.1177/0040517517698988.

Holmquist, H., Schellenberger, S., van der Veen, I., Peters, G.M., Leonards, P.E.G., Cousins, I.T., 2016. Properties, performance and associated hazards of state-of-the-art durable water repellent (DWR) chemistry for textile finishing. Environ. Int. 91, 251–264.
Horner, I.B., 1964. Milinda's Questions [Vol. 1]. Pali Text Society.
Jeffree, C.E., 1986. The cuticle, epicuticular waxes and trichomes of plants, with reference to their structure, functions and evolution. Insects and the Plant Surface. Edward Arnold, London, pp. 23–64.
Kapsali, V., 2016. Biomimetics for Designers. Thames & Hudson, London.
Kapsali, V., Toomey, A., Tandler, L., 2013. Biomimetic spatial and temporal (4D) design and fabrication. In: Conference on Biomimetic and Biohybrid Systems. Springer, Berlin, Heidelberg, pp. 387–389.
Lang, A.W., Motta, P., Hidalgo, P., Westcott, M., 2008. Bristled shark skin: a microgeometry for boundary layer control. Bioinspir. Biomim. 3 (4), 046005.
Lepora, N.F., Verschure, P.F., Prescott, T.J., 2013. Towards a roadmap for living machines. In: Conference on Biomimetic and Biohybrid Systems, July. Springer, Berlin, Heidelberg, pp. 396–398.
Luz, G.M., Mano, J.F., 2009. Biomimetic design of materials and biomaterials inspired by the structure of nacre. Phil. Trans. R. Soc. Lond. A: Math. Phys. Eng. Sci. 367 (1893), 1587–1605.
Meyers, M.A., Lin, A.Y.M., Chen, P.Y., Muyco, J., 2008. Mechanical strength of abalone nacre: role of the soft organic layer. J. Mech. Behav. Biomed. Mater. 1 (1), 76–85.
Nagaraja, P., Yao, D., 2007. Rapid pattern transfer of biomimetic surface structures onto thermoplastic polymers. Mater. Sci. Eng. C 27 (4), 794–797.
Slater, K., 2003. Environmental Impact of Textiles: Production, Processes and Protection. Vol. 27. Woodhead Publishing, Cambridge.
STO, 2017. http://www.sto.co.uk/en/products/facade-finishes/facade-renders/stolotusan.html.
Sun, J., Bhushan, B., 2012. Hierarchical structure and mechanical properties of nacre: a review. RSC Adv. 2 (20), 7617–7632.
Tonks, L., Langmuir, I., 1929. A general theory of the plasma of an arc. Phys. Rev. 34 (6), 876.
Vincent, J.F., Bogatyreva, O.A., Bogatyrev, N.R., Bowyer, A., Pahl, A.K., 2006. Biomimetics: its practice and theory. J. R. Soc. Interface 3 (9), 471–482.
Zanini, S., Freti, S., Citterio, A., Riccardi, C., 2016. Characterization of hydro-and oleo-repellent pure cashmere and wool/nylon textiles obtained by atmospheric pressure plasma pre-treatment and coating with a fluorocarbon resin. Surf. Coat. Technol. 292, 155–160.

Further reading

Bar-Cohen, Y., 2005. Biomimetics: Biologically Inspired Technologies. CRC Press, Boca Raton, FL.
Benyus, J.M., 1997. Biomimicry. William Morrow, New York, NY, p. 1.
Gordon, J.E., Gordon, J.E., 1991. The New Science of Strong Materials: Or Why You Don't Fall Through the Floor. Penguin, London.
Kapsali, V., 2008. Biomimetic principles in the design of clothing. In: Abbott, A., Ellison, M. (Eds.), Biologically Inspired Textiles. Woodhead Publishing Limited, Cambridge.
Kapsali, V., 2009. Biomimetics and its impact on outdoor clothing. In: Williams, J.T. (Ed.), Textiles for Cold Weather Protection. Woodhead Publishing Limited, Cambridge, pp. 113–139.

Kapsali, V., 2012a. Biomimetic approaches to the design of smart textiles for protection. In: Chapman, R. (Ed.), Smart Textiles for Protection. Woodhead Publishing Limited, Cambridge.

Kapsali, V., 2012b. Impact of biomimetic principles on sustainability in the fashion industry. In: Black, S. (Ed.), The Sustainable Fashion Handbook. Thames and Hudson, London.

Kapsali, V., 2015. Bionic sportswear. In: Shishoo, R. (Ed.), Textiles for Sportswear. Woodhead Publishing Limited, Cambridge.

Mazzoleni, I., 2013. Architecture Follows Nature-Biomimetic Principles for Innovative Design. vol. 2. CRC Press, Boca Raton, FL.

McDonough, W., Braungart, M., 2010. Cradle to Cradle: Remaking the Way We Make Things. MacMillan, New York, NY.

Vincent, J.F., 2012. Structural Biomaterials. Princeton University Press, Princeton, NJ.

Vogel, S., 1988. Life's Devices: The Physical World of Animals and Plants. Princeton University Press, Princeton, NJ.

Vogel, S., 2003. Prime Mover: A Natural History of Muscle. WW Norton & Company, New York, NY.

Vogel, S., 2009. Glimpses of Creatures in Their Physical Worlds. Princeton University Press, Princeton, NJ.

Vogel, S., 2012. The Life of a Leaf. University of Chicago Press, Chicago, IL.

Vogel, S., 2013. Comparative Biomechanics: Life's Physical World. Princeton University Press, Princeton, NJ.

Vogel, S., Berg, H.C., 1998. Cats' Paws and Catapults: Mechanical Worlds of Nature and People. WW Norton & Company, New York, NY.

Online Resources

Kapsali, V., 2017. http://www.bioinspireddesign.com/.

MMT Textiles Limited, 2017. http://mmttextiles.com.

Ruano, D.S., 2017. https://biomimeticdesign.wordpress.com/.

The Biomimicry Institute, 2017. https://biomimicry.org/.

Part Two

Types of water repellent textile finishes

Finishing of textiles with fluorocarbons

6

Usha Sayed, Prince Dabhi
Institute of Chemical Technology, Matunga, India

6.1 Introduction

Value addition is achieved by introducing some novel finishing. Finishing is the heart of the textile and by applying some novel finishes on apparel many desired properties like the fabric absorbency, stains to be easily removed from fabric and apparel faster, rise in the wicking action for better comfort in wear, fabric are easily dry-cleanable, good laundry durability and maintains brightness after respective repeated laundering can be easily achieved. Hydrophobicity is one of the most widely spread textile finishes today. Finishes that repel water, oil and dry dirt are important in all parts of the textile market—for clothing, home and technical textiles (Malik and Parmar, n.d.). The oldest water repellant finishes for fabrics were simply coatings of paraffin or wax—and they are generally washed out eventually (Corpart and Dessaint, 1997a; Audenaert et al., 1999a; Lammermann, 1991a; Nuyttens, 1995a). Perfluorochemicals (PFC's) are the only chemicals capable of repelling water, oil and other liquids that cause stains. Fabrics finished with PFCs have nonstick properties; this family of chemicals is used in almost all the stain repellant finishes in the market today (Corpart and Dessaint, 1997a; Audenaert et al., 1999a; Lammermann, 1991a; Nuyttens, 1995a). Other materials can be made to perform some of these functions but when subjected to oil and other liquids are considerably less durable (Prasad, 2007; Roach et al., 2008).

The purpose of this finish is self-evident. Drops of water should not spread on the surface of the textile and should not wet the fabric. The drops should stay on the surface and easily drip off. Similarly, oil repellent finishes should prevent oily fluids from wetting treated textiles (Schindler and Hauser, 2004). In a similar manner, soil-repellent finishes should protect textiles from both dry and wet soils. In all cases, the air permeability of the finished fabric should not be significantly reduced. In addition to the desired repellency effects, other undesirable fabric properties are often found with repellent finishes (Schindler and Hauser, 2004). These include problems with static electricity, poor soil removal in aqueous laundering, stiffer fabric hand, greying (soil re-deposition) during aqueous laundering and increased flammability (Prasad, 2007; Roach et al., 2008). Some fabric properties that are often improved by repellent finishes include better durable press properties, more rapid drying and increased resistance to acids, bases and other chemicals (Schindler and Hauser, 2004).

Waterproof and Water Repellent Textiles and Clothing. https://doi.org/10.1016/B978-0-08-101212-3.00006-X
Mahendra Publications/2014

6.2 Fluorochemical finishes

Fluorochemical finishes (also termed fluorocarbon finishes) were first applied to textiles in the 1960s and their growth, particularly during the 1990s, has been greatly stimulated by the consumer demands for easy-care properties such as water and oil-repellency, stain repellency and soil and stain-release properties. Fluorochemicals (fluorocarbons) are a class of synthetically produced organic chemicals that contain a perfluoroalkyl residue in which all the hydrogen atoms have been replaced by fluorine (Audenaert et al., 1999b; Higgins and Moilliet, 1963; Lewin et al., 1984; Corpart and Dessaint, 1997b; Lammermann, 1991b; Nuyttens, 1995b; Grottenmüller, 1998a). Fluoro chemicals exhibit outstanding chemical and thermal stability, low reactivity through their incompatibility with water and oil, and considerable reduction in surface tension (Lewin et al., 1984; Corpart and Dessaint, 1997b; Lammermann, 1991b; Nuyttens, 1995b; Grottenmüller, 1998a; Lewis, 1992). It is this latter property which is of particular importance in the context of water- and oil-repellency, while their chemical and thermal stability contribute towards the durability of the surface finish to fabric care treatments such as laundering, dry-cleaning and tumble-drying.

Fluorine, first isolated by Moissan in 1886, is the most strongly electronegative element in nature and is extremely chemically reactive (Nuyttens, 1995b). In the fluorine atom the electrons are held close to the atomic nucleus. The interaction of fluorine atoms with carbon atoms to form strong carbon-fluorine bonds leads to the carbon atoms in fluorocarbons being closely surrounded by fluorine atoms (Lammermann, 1991b; Nuyttens, 1995b). Thus the basis for the low reactivity of fluorochemical finishes is the high degree of protection provided by the fluorine atoms to the shielded carbon scaffold. In addition, the extremely low critical surface tension and high-surface activity associated with fluorochemicals are also a consequence of this unique molecular topography (Audenaert et al., 1999b; Lewin et al., 1984; Lammermann, 1991b; Nuyttens, 1995b; Grottenmüller, 1998a; Holme, 1993). Fluorochemicals are used in many fields, for example, as durable lubricants, corrosion protection coatings for metals, flame-retardant polymers, fluorine elastomers used in the rubber industry and heat transfer fluids in refrigeration technology (Nuyttens, 1995b). They are widely used as protective agents against water and oil, stains and soiling in the textile, paper and leather industries. Some fluorochemicals are used for specialist wetting applications in the textile field; others are used in firefighting, and in the electroplating and electronics industries (Lammermann, 1991b).

6.3 Techniques for the formation of fluorochemicals

The current techniques available for the formation of fluoro chemicals depend upon (Audenaert et al., 1999b; Corpart and Dessaint, 1997b; Lammermann, 1991b; Nuyttens, 1995b; Grottenmüller, 1998a):

telomerization
electrochemical fluorination
oligomerization

However, the first two techniques have been the most widely used to manufacture fluorochemicals for repellent finishes (Audenaert et al., 1999b), but finishes based on electrochemical fluorination were discontinued by 3M (Minnesota Mining and Manufacturing Company) in 2001 on eco-toxicological grounds concerning the type of fluorochemicals produced (Risk and Policy Analysts Limited, 2004; 3M Company, 2001). In telomerization, a fluorinated ethylene derivative (taxogen) is reacted with an initiator for chain growth polymerization (telogen), producing the desired FC.

Telomerization was used by Du Point. This process involves a free radical polymerization process of tetra fluoroethylene as described in Fig. 6.1 (Banks et al., 1994).

In electro fluorination, a hydrocarbon product is dissolved in hydrogen fluoride. Under an applied electrical current, the hydrogen atoms on the hydrocarbon are replaced by fluorine atoms (Banks et al., 1994). The oligomerization process is similar to telomerization, but results in high-molecular weight fluoropolymers, and thus is not generally used for production of FC finishes.

6.3.1 Fluorochemical finish formulations

Fluorochemical finishes are based upon fluorocarbon dispersions and are mainly acrylate polymers containing a perfluorinated acrylate as the major component (Audenaert et al., 1999b; Corpart and Dessaint, 1997b; Lammermann, 1991b; Nuyttens, 1995b; Grottenmüller, 1998a). However, practically none of the commercial fluorochemicals consist of pure perfluoroacrylate homopolymer and normally there are various comonomers present. The main components that may be present in typical fluorochemical dispersion are illustrated in Fig. 6.2 (Grottenmüller, 1998a). Oil-repellency and water repellency depend crucially upon the perfluoroacrylate component but the presence of long-chain fatty alcohol acrylates, such as lauryl and stearyl acrylate, has been shown to produce synergistic improvement in water repellency with the perfluoroacrylate without adverse impairment of the oil-repellency (Grottenmüller, 1998a). Vinyl chloride, vinylidene chloride, methyl methacrylate and acrylonitrile are also frequently incorporated as co-monomers where special performance characteristics are required, such as soil-repellency or resistance to organic solvents and white spirit (Fig. 6.3).

$$F_3C{-}CF_2{-}CF_2{-}CF_2{-}CF_2{-}CF_2{-}CF_2{-}CF_2{-}CH_2{-}CH_2{-}OH$$

Fig. 6.1 Perfluorooctane telomere alcohol.

$$\underset{\text{Cell feed}}{C_8H_{17}SO_2F + 17\ HF} \xrightarrow{e^-} \underset{\text{Cell output}}{C_8H_{17}SO_2F + 17H_2}$$

Fig. 6.2 Electrochemical fluorination.

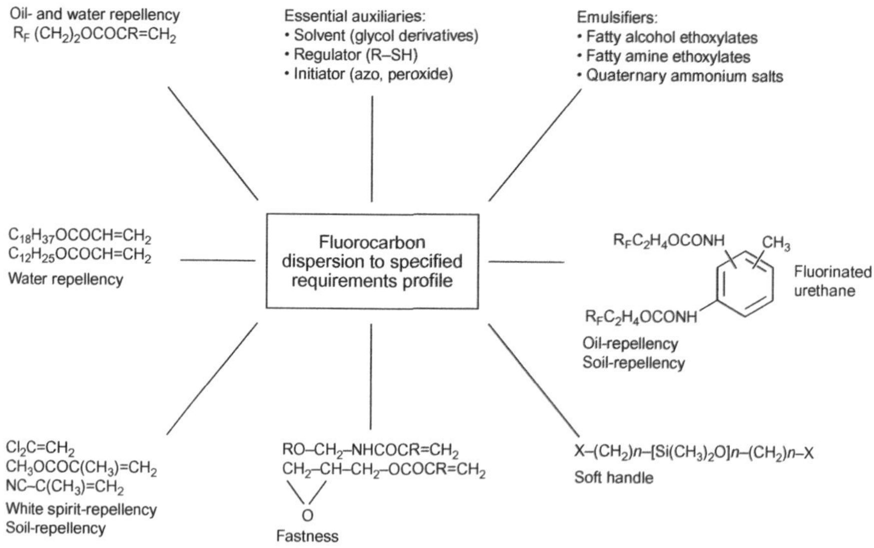

Fig. 6.3 Major components of fluorocarbon dispersion.

6.3.1.1 Mechanisms of repellency

Repellent finishes achieve their properties by reducing the free energy at fibre surfaces. If the adhesive interactions between a fibre and a drop of liquid placed on the fibre are greater than the internal cohesive interactions within the liquid, internal cohesive interactions within the liquid, the drop will spread (Holmes et al., 2000). If the adhesive interactions between the fibre and the liquid are less than the internal cohesive interactions within the liquid, the drop will not spread. Surfaces that exhibit low interactions with liquids are referred to as low energy surfaces (Kissa et al., 1984). Their critical surface energy or surface tension γ_C must be lower than the surface tension of the liquid γ_L (the internal cohesive interaction) that is repelled. γ_L of water, at 73 mN m^{-1}, is two to three times greater than γ_L of oils (20–35 mN m^{-1}). Therefore, oil repellency finishes with fluorocarbons (γ_C=10–20 mN m^{-1}) always achieve water repellency, but fluorine-free products, for example, silicones (γ_C=24–30 mN m^{-1}) will not repel oil. Low energy surfaces also provides measure of dry soil repellency by preventing soil particles from strongly adhering to fibre surfaces (Corpart and Dessaint, 1997a; Grottenmüller, 1998a; Lewis, 1992; Weast, 1988–89). This low interaction allows the soil particles to be easily dislodged and removed by mechanical action (Fig. 6.4).

6.3.1.2 Application methods for fluorochemical finishes

The method of application of the fluorochemical finish to the textile depends upon the form of the material—for example, flat woven, weft knitted fabric, velvet or pile fabric or carpet—and whether the fluorochemical finish is required to penetrate the

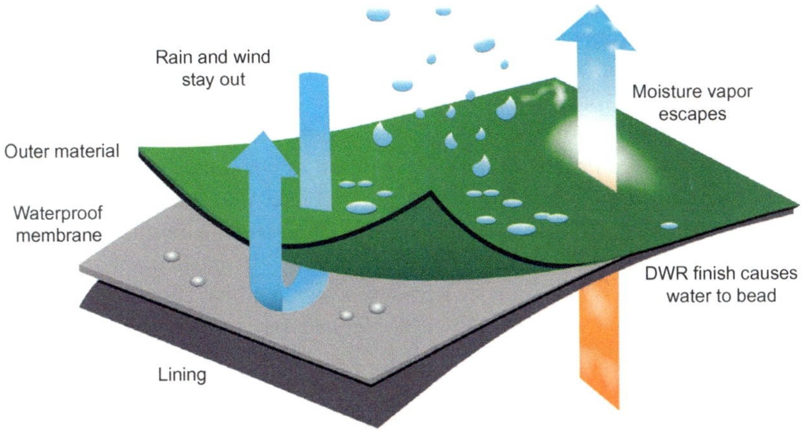

Fig. 6.4 Mechanism of repellency.

structure and provide a film covering all the fibres uniformly, or whether a topical (or surface) application is all that is required (Audenaert et al., 1999b; Corpart and Dessaint, 1997b; Lammermann, 1991b; Nuyttens, 1995b; Grottenmüller, 1998a; Sahin, 1996; Wilhelm, 1993). Light and medium weight fabrics (linings, apparel and home furnishing fabrics) may be saturated by immersion in a pad bath containing the fluorochemical and any appropriate auxiliaries followed by squeezing in a two-bowl or three-bowl nip (Corpart and Dessaint, 1997b; Sahin, 1996; Wilhelm, 1993). The fibre content and fabric construction normally dictate the wet pick-up range that is possible, and the wet pick-up is controlled by the nip bowl pressure. The chemical add-on of fluorochemical deposited on the fabric is typically around 0.15%–0.3% on the weight of the fabric. The chemical add-on is controlled principally by regulating the fluorochemical finish concentration in the pad bath, rather than by regulation of the wet pick-up. As an alternative to padding, other methods of minimum application may be used such as lick-roll or nip padding or vacuum extraction. In the latter case, excess removal may be recycled back to the pad bath.

The vast majority of the fluorochemical finishes used in the treatment of textiles are cationic in nature and the recommended pH of the finishing bath should therefore preferably be around pH 5, which can be attained usually by addition of a small amount of acetic or formic acid (Audenaert et al., 1999b; Corpart and Dessaint, 1997b; Lammermann, 1991b; Nuyttens, 1995b; Grottenmüller, 1998a; Spencer, 1993). This ensures that any residual alkalinity in the fabric is neutralized and also that the finish bath is stabilized because the cationic emulsifiers used are much more effective at this pH level.

After impregnation, the fabric is dried at temperatures of about 110–130°C followed by heat curing at a temperature normally in the range 150–170°C. This may be carried out in separate stages in circulating hot air ovens or more usually in a single-pass treatment on a stenter. The higher the curing temperature, the shorter is the curing time to avoid yellowing of the textile material. Thus, heat-curing treatments at 150°C for 3–4 min, or 160–180°C for 30–45 s, are typical of the conditions to be encountered in practice (Spencer, 1993). However, carpet products may be heat-cured at a minimum temperature of 110°C while other products may be treated for 2 min at 170°C, or for as little as 30 s at 190°C. The heat-curing conditions used must provide optimal film formation and are dependent upon the chemical nature of the fluorochemical, the presence of extenders used to boost the fluorochemical performance and the durability performance required. Fibres that are heat-sensitive, such as wool or polypropylene, are cured at lower temperatures (Sahin, 1996). Where multifunctional finishes are used, for example, fluorochemical combined with an easy-care finish—the heat-curing conditions selected must ensure that both the fluorochemical and the easy-care finish are satisfactorily cross linked to achieve the optimum fabric performance (Fig. 6.5).

6.3.1.3 Other products

Chemical additives, such as wetting agent, thermosetting resin, extenders, softeners, antistatic agents, builders, flame-retardants, or fungicidal or antibacterial agents, may also be incorporated into the pad bath to obtain multifunctional finishes (Audenaert et al., 1999b; Lewin et al., 1984; Corpart and Dessaint, 1997b; Sahin, 1996; Needles, 1996). Their effects upon the pad bath stability, or problems due to chemical incompatibility that may impair the final finished fabric performance, must be evaluated to optimize the process and product performance (Audenaert et al., 1999b; Lewin et al., 1984; Corpart and Dessaint, 1997b; Sahin, 1996; Needles, 1996).

The final polymer, when applied to a fibre, should form a structure that presents a dense —CF_3 outer surface for maximum repellency. The length of the perfluorinated side chains should be about 8–10 carbons (Audenaert et al., 1999b; Lewin et al., 1984; Corpart and Dessaint, 1997b; Sahin, 1996; Needles, 1996). The small spacer group, mostly ethylene, can be modified to improve emulsification and solubility of the polymer. Co-monomers (X, Y, for example, stearyl- or lauryl methacrylate, butyl acrylate, methylol- or epoxy functional acrylates and block copolymers from α,ω-dihydroxy dimethylpolysiloxane) affect fabric hand, film formation and durability (Audenaert et al., 1999b; Lewin et al., 1984; Corpart and Dessaint, 1997b; Sahin, 1996; Needles, 1996).

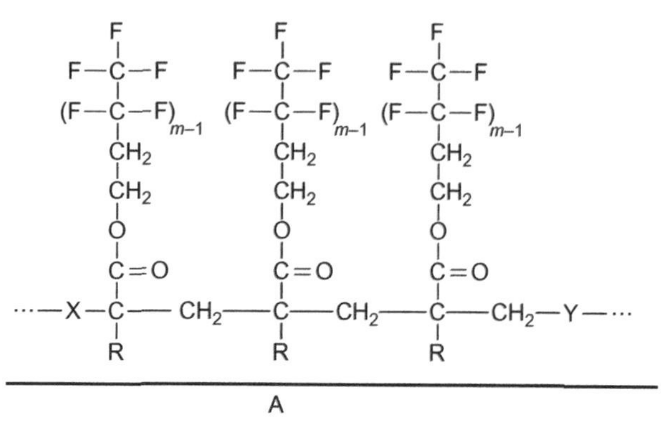

Fig. 6.5 Fluorocarbon repellent on fibre surface. $m=8$–10. X and Y are comonomers, mainly stearylacrylates. R=H or CH_3 (polyacrylic or polymethacrylic acid esters). A is the fibre surface.

6.4 Effect of fluorochemical chain length

The effect of the fluorochemical chain length upon the critical surface tension (γ_C) is illustrated in Fig. 6.6 (Audenaert et al., 1999b). It can be clearly seen that the critical surface tension decreases fairly rapidly as the chain length (n) increases from 1 to 8, after which little further decrease in critical surface tension is noted (Audenaert et al., 1999b; Corpart and Dessaint, 1997b). As the critical surface tension of the fluorochemical film on the fibre surface is decreased, the water- and oil-repellency rise, reaching their maximum protective value around a chain length of $n=9$. This dramatic improvement in repellency performance is illustrated in below table in which the oil-repellency (as measured according to AATCC Test Method 118) rises from 0 with a $—CF_3$ terminal group to 8 with a $—(CF_2)_8—CF_3$ side chain. In addition, the water repellency (as determined by the spray test ISO 4920) also rises from 50 up to 80. Thus an increase in the perfluorinated chain length from CF_3 to C_9F_{19} (i.e., $—(CF_2)_8—CF_3$) gradually enhances the oil repellency and, to a lesser extent, the water repellency.

Oil- and water repellency of fabrics treated with acrylic polymers

Perfluorinated groups	Measurement of oil-repellency (AATCC 118)	Spray test (ISO 4920)
$—CF_3$	0	50
$—CF_2—CF_3$	3–4	70
$—(CF_2)_2—CF_3$	6–7	70
$—(CF_2)_4—CF_3$	7–8	70
$—(CF_2)_6—CF_3$	7–8	70
$—(CF_2)_8—CF_3$	8	80

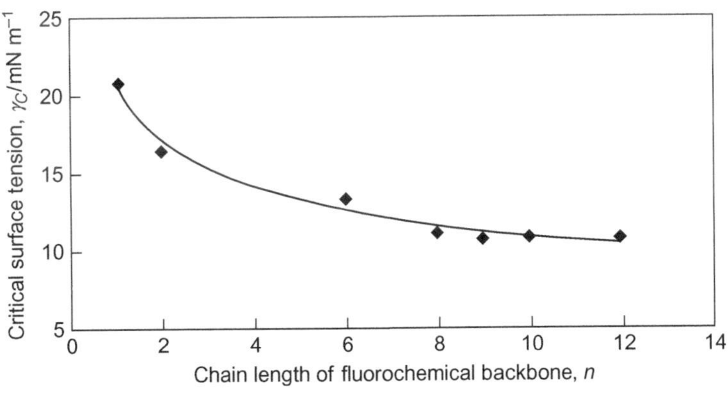

Fig. 6.6 Effect of the fluorochemical chain length on surface tension.

6.5 Evaluation

Test methods for fabric with repellent finish (Schindler and Hauser, 2004)

Test method	Procedure summary
AATCC TM 22—Water repellency: spray test	Treated fabric is stretched taut, held at a 45 degrees angle and sprayed with 250 mL of water from above. The resulting wetting pattern, if any, is rated using photographic standards. This is a simple, rapid method suitable for the plant floor
AATCC TM 42—Water resistance: impact penetration test	Similar to AATCC TM 22, but a weighed piece of blotter paper is placed under the fabric. The weight gain of the paper after 500 mL of water has been sprayed on the fabric is recorded
AATCC TM 35—Water resistance: rain test	The treated fabric, backed by a weighed piece of blotter paper, is sprayed with water under constant hydrostatic pressure for 5 min. The weight gain of the paper after the test is recorded. This test requires a special apparatus
AATCC TM 118—Oil repellency: hydrocarbon resistance test, equivalent to ISO 14 419	Drops of eight standard fluids, consisting of a selected series of eight hydrocarbons with decreasing surface tensions, are placed on the treated fabric and observed for wetting, wicking and contact angle. The highest numbered fluid that does not wet the fabric is recorded as the oil repellency rating with grade numbers from 0 to 8. This method is a simple, rapid test suitable for the plant floor but it is a static test involving no mechanical stress that is often common in real life
3M dry soil resistance test	Samples of treated fabric are shaken in a sealed Samples of treated fabric are shaken in a sealed container with a standard dry soil. Loose soil is removed with compressed air and the appearance of the fabrics is compared to photographic standards

6.5.1 Toxicology and ecological considerations concerning PFCS

Traditional commercial FC products consist of 15%–30% fluorocarbon polymer, 1%–3% emulsifier, 8%–25% organic solvents and water. The exhaust air of the drying and curing processes of fluorocarbon finishes therefore often contains high amounts of volatile organic compounds (VOC), like glycols and other organic solvents and a lower content of residual monomers (Grottenmüller, 1998b; Thumm, 2000). Cellulosic fibres retain most of the glycols (Grottenmüller, 1998b; Thumm, 2000). The exhaust air problem is greater with FC finishes on synthetic fibres. VOCs can be reduced by more than 90% by the addition of corresponding boosters, providing equal effects with lower quantities of finish products. PFCs are a group of chemicals used to make fluoropolymer coatings and products that resist heat, oil, stains, grease and water. During Electrofluorination and Telomerization a small amount of *PFOS* (perfluorooctane sulfonate) and *PFOA* (perfluorooctanoic acid) is produced (PFOA Puzzle, n.d.; Bjorhus, 2003).

PFOA persists in the environment and does not breakdown. Scientists noticed that PFOS (the C_8 fluorocarbon) began showing up everywhere: in polar bears, dolphins, baby eagles, tap water and human blood. They kill laboratory rats at higher doses, and there are potential links to tissue problems, developmental delays and some forms of cancer. Below are results which the US Environmental Protection Agency released from data collected by 3M and DuPont; some humans have more PFOA in their blood than the estimated levels in animals in this study (PFOA Puzzle, n.d.; Bjorhus, 2003) (Figs. 6.7 and 6.8).

PFOA and PFOS, according to the US EPA:

- Are very persistent in the environment.
- Are found at very low levels both in the environment and in the blood of the US population.

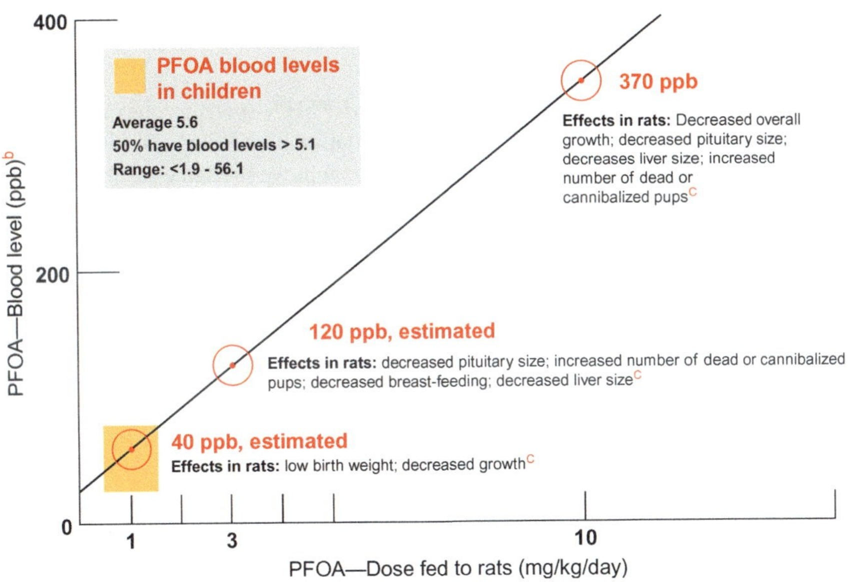

Fig. 6.7 PFOA levels in human blood overlap with level in laboratory studies that cause harmful effect.

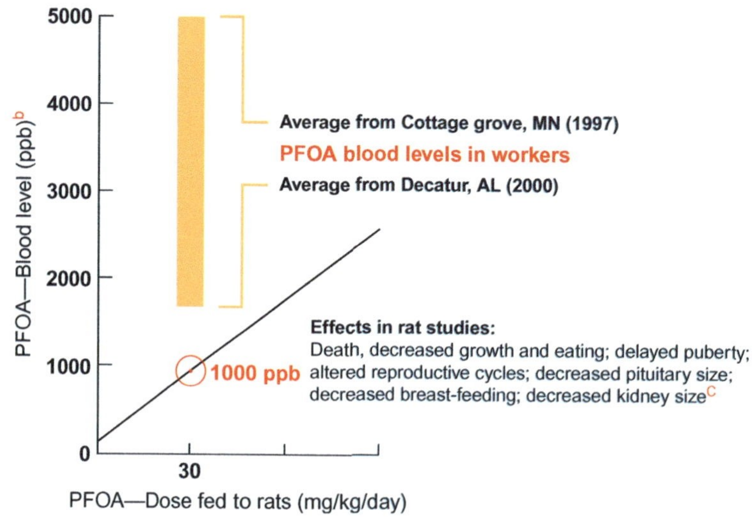

Fig. 6.8 PFOA levels in human blood overlap with level in laboratory studies that cause harmful effect.

- Remain in people for a very long time.

PFOA and PFOS cause developmental and other adverse effects in laboratory animals (Peter and Rouette, 1989; Benjamin et al., 2007).

6.5.2 C_6 and C_4 chemistry

C_6—this chemistry produces a by-product called PFHA (perfluorohexanoic acid), which is supposed to be 40 times less bio accumulative than PFOA. But it's also less effective, so more of the chemical has to be used to achieve the same result (Peter and Rouette, 1989; Benjamin et al., 2007).

Manufacturers are trying to find smaller and smaller perfluorocarbon segments in their products, and even C_4 has been used. The smaller the fluorocarbon, the more rapidly it breaks down in the environment. Unfortunately, the desired textile performance goes down as the size of the perfluor carbon goes down. 'C_6 is closest chemically to C_8, but it contains no PFOA. It breaks down in the environment—a positive trait—but it doesn't stick as well to outerwear and it doesn't repel water and oil as well as C_8, which means it falls short of meeting a vague industry standard, as well as individual company standards for durability and repellency' (Peter and Rouette, 1989; Benjamin et al., 2007).

3M settled on perfluorobutane sulfonate, or *PFBS*, a four-carbon cousin of the old chemicals. This new C4-based chemistry is completely safe. According to 3M, the results show that under federal EPA guidelines, *PFBS* isn't toxic and doesn't accumulate the way the old chemical did. It does persist in the environment, but 3M concluded that isn't a problem if it isn't accumulating or toxic. *PFBS* can enter the bloodstream of people and animals but 'it's eliminated very quickly' and does no harm at typical very low levels.

6.6 Recent development

A number of novel approaches to repellent finishing, several of which deserve mention, have been examined in recent years (Rios et al., 2007; Badyal et al., 2001a). Each of the novel approaches has shown significant benefits and could be a potentially important finishing option in the near future.

Nature is now being copied by bio-engineering textile materials using nano coatings and other approaches that generate a low surface energy finish with a surface roughness that creates an ultra hydrophobic and self cleaning effect. One approach is that marketed by Nano Tex LLC (Greensboro, NC, United States) which produces self cleaning cotton or cotton-blended fabric finish termed Nano-Care (Forbes, 2005). The Nano-Care finish utilizes whisker shaped 10 nm molecules of a fluorinated monomer copolymerized with a carboxylic acid oligomer and converted to the anhydride to react with the fibre. Nano Pel treatments are similar to Nano Care and are designed for use on wool and multiple layer fabrics but Nano-Tex technology has also been said to simulate the surface of a peach which has very fine, short hairs which are durably bonded to the fibres in the fabric. Nano-Tex technology is then licensed to other companies (Wadia and Nimkar, 2006; Cafaggi, 2004) and such fabrics are used in work wear, casual wear and sportswear (Cafaggi, 2004).

Schoeller Textiles AG (Sevelen, Switzerland) has introduced its Nano Sphere technology which imparts water repellency, soil-repellent, antiadhesive and self cleaning properties (Overington and Croskell, 2001). Their technology leads to the formation of micro-rough three-dimensional surface structures from which water; dirt and oil simply roll off. Using their Nano Sphere finish dirt particles on the surface of the finished textile cling to water droplets and when the water droplets run off the surface the dirt particles are simply washed away. Nano Sphere technology is said to use a so-called guest-host system of substances that allow spatial self-organization and result in the formation of a micro-rough surface. In parallel with this structure formation, gel-forming additives lead to the development of the porous system of a membrane.

Other approaches rely upon treatments that modify the fibre surface topography in combination with chemical finishes. photonic surface treatment using a pulsed UV-laser (excimer laser) to yield a regular micron-scaled structure over a wide range of pulse energies (Holme, 2002).

Irradiation in the presence of perfluoro-4-methylpent-2-ene can then lead to radically bound hydrophobic end groups. Pulsed plasma polymerization of monomers containing long perfluoroalkyl chains linked to a polymerizable carbon-carbon double bond can provide ultrahydrophobic effects which can be used not only for water-, oil- and stain-repellency (Badyal, 2001), but can also be used for chemical protection, e.g., in nuclear, biological, chemical protection suits, for protection against nerve agents

(Badyal, 2001; Holme, 2005, 2007; Badyal et al., 2001b; Hutton et al., 2000; Hegemann, 2006).

There are very clearly many opportunities for liquid proof and microorganism-impermeable medical textiles, e.g., for reusable surgical gowns made from polyester microfilament fabrics, suitable for repeated washing and sterilizing. Treatment of such fabrics with plasma treated/C_2F_6, plasma treated/$H_2+C_2F_6$ (1:2) and plasma treated/$C_2H_4+C_2F_6$ (1:2) provides contact angles of around 140°C (Islam et al., 2004).

A recent development has been the introduction of a special fluorocarbon product Anthydrin NK (Zschimmer & Schwarz Mohsdorf GmbH & Co., Burgstädt, Germany) which is a low temperature cure product (Rasch et al., 2005). Use of a special cross linking system allows curing at around 110°C compared with conventional cross linking at around 160°C. A booster product Polappret VIB is co-applied if the finish must be fast to washing.

6.7 Conclusion

Fluorochemical based chemicals shown in the recent review are viable water repellent, water proofing, durable and high-performing finish. Fluorocarbon based water repellent has found wide application in a variety of consumer goods including non-stick cookware, paints and coatings and stain release treatment for carpets. The water repellency is achieved by lowering the surface energy of the fabric, so that water will bead on the surface and doesn't wet out the garments. Although similar finish can be achieved with many types of finishes including waxes, oils and silicones, they are not repellent, oils and lotions and hence fluorocarbon is most effective in repelling both oil and water and hence it is commonly used in water repellence finishes or outwear clothing.

References

3M Company, 2001. 8#HQ-998-374. Submitted to US EPA.

Audenaert, F., Lens, H., Rolly, D., Van, P., 1999. Fluoro-chemical textile repellents—synthesis, and applications: a 3M perspective. J. Text. Inst. 90 (3), 76.

Badyal, J., 2001. Beyond the surface-cold plasmas are streamlining the surface coating. Chem. Br. 37 (1), 45.

Badyal, J., Coulson, S., Willis, C., Brewer, S., 2001a. Defence Research Agency, UK Patent 2, 864 B. 341.

Badyal, J., Coulson, S., Willis, C., Brewer, S., 2001b. GB2341864B (Defense Research Agency).

Banks, E., Smart, E., Tatlow, C., 1994. Organofluorine Chemistry, Principles and Commercial Applications. Plenum Press, New York. NY 321.

Benjamin, J., Frank, R., Julie, B., Antonia, M., Rolf, U., Larry, L., Lynn, R., 2007. Cord serum concentrations of perfluorooctane sulfonate (PFOS) and perfluorooc-tanoate (PFOA) in relation to weight and size at birth. Environ. Health Perspect. 115, 1670–1676.

Bjorhus, J., 2003. Scotchgard is Attractive Again. Pioneer Press, St. Paul.

Cafaggi, G., 2004. Quality finish on worsted fabric. Int. Dyer 189 (4), 4.

Corpart, J., Dessaint, A., 1997. Fluorine-base textile finishing. Melliand Textilber. 78, 625 E135.

Forbes, P., 2005. The Gecko's Foot, Bio-Inspiration: Engineered From Nature. Fourth Estate, London 272.

Grottenmüller, R., 1998a. Fluorocarbons—an innovation aid to the finishing of textiles. Melliand Textilber. 79 (10), 743 E195.

Grottenmüller, R., 1998b. Fluorocarbons—an innovative auxiliary for the finish of textile surfaces. Melliand Textilber. 79 (10), 743–746 E 195.

Hegemann, D., 2006. Plasma polymerization and its application in textiles. Indian J. Fibre Text. Res. 31, 99–115.

Higgins, E., Moilliet, J., 1963. Waterproofing and Water-Repellency. Elsevier, Amsterdam 188.

Holme, I., 1993. New developments in the chemical finishing of textiles. J. Text. Inst. 84, 520.

Holme, I., 2002. REACH: the cost of compliance. Int. Dyer 187 (10), 11.

Holme, I., 2005. Finishes for protective and military textiles. Int. Dyer 190 (4), 9.

Holme, I., 2007. Innovative technologies for high performance textiles. Color. Technol. 123, 59–73.

Holmes, D., Horrocks, A., Anand, S. (Eds.), 2000. Handbook of Technical Textiles. Woodhead Publishing Ltd, Cambridge p. 461.

Hutton, S., Crowther, J., Badyal, J.P.S., 2000. Complexation of fluorosurfactants to functionalized solid surfaces: smart behavior. Chem. Mater. 12, 2282.

Islam, N., Lappan, U., Voit, B., 2004. Barrier fabrics for medical application. Int. Dyer 12–17.

Kissa, E., Lewin, M., Sello, S., 1984. Handbook of Fiber Science and Technology. Chemical Processing of Fibers and Fabrics, Part B Functional Finishes, vol. 2. Marcel Dekker, New York, NY 143.

Lammermann, D., 1991. Fluorocarbons in textile finishing. Melliand Textilber. 72 (11), 949 E380.

Lewin, M., Sello, S., Kissam, E., 1984. Handbook of Fiber Science and Technology. Chemical Processing of Fibers and Fabrics, Part B Functional Finishes, vol. 2. Marcel Dekker, New York 143.

Lewis, D., 1992. Review of progress in coloration and related topics. J. Soc. Dye. Colour. 22 (1), 1–13.

Malik, T., Parmar, S., Special Finishes to Garment—An Overview, www.fiber2fashion.com.

Needles, H., 1996. Textile Fibers, Dyes, Finishes, and Processes. NOYES Publication, Park Ridge, NJ pp. 193–220.

Nuyttens, R., 1995. Fluorochemicals: high performance finishes for textile protection. Text. Tech. Int. 167–169.

Overington, Y., Croskell, R., 2001. Standards for breathable fabrics. Int. Dyer 186 (5), 23.

Peter, M., Rouette, H., 1989. Grundlagen Der Textile Veredlung. Deutscher Fachverlag, Frankfurt.

PFOA Puzzle—Textile Insights http:/www.textileinsight.com/articles.php?id=37.

Prasad, A., 2007. Novel effects in garment processing and value added finishes. J. Text. Assoc. 68, 39–42.

Rasch, M., Pahlke, E., Knoll, I., Paulini, T., 2005. Fluoro-carbon oil and water resist treatment requires no curing. Int. Dyer 190 (1), 37.

Rios, P., Dodiuk, H., Kenig, S., McCarthy, S., Dotan, A., 2007. Transparent ultra-hydrophobic surfaces. J. Adhes. Sci. Technol. 21 (5–6), 399–408.

Risk & Policy Analysts Limited, 2004. Perfluorooctane Sulphonate Risk Reduction Strategy and Analysis of Advantage and Drawbacks. Department for Environment, Foods and Rural Affairs and the Environment Agency for England and Wales.

Roach, P., Shirtcliffe, N., Newton, M., 2008. Progress in superhydrophobic surface development. Soft Matter 4, 224–240.

Sahin, B., 1996. Fluorochemicals in textile finishing. Int. Text. Bull. 42 (3), 26 (Dyeing/Printing/Finishing).

Schindler, W., Hauser, P., 2004. Chemical Finishing of Textiles. vol. 4. Woodhead Publishing Ltd, Boca Raton, FL pp. 74–91.

Spencer, D., 1993. Knitting Technology. Woodhead Publishing Ltd, Lancaster, PA.

Thumm, S., 2000. LAD-fluorocarbon technology for high-tech sports-wear. Int. Text. Bull. 46 (1), 56–61.

Wadia, M., Nimkar, U., 2006. Nanotechnology—a boon to textile auxiliary manufactures. Colourage 53 (1), 51.

Weast, R., 1988–89. Handbook of Chemistry and Physics, 69th ed. CRC Press, Boca Raton, FL, F-34.

Wilhelm, M., 1993. Textile finishing. Int. Text. Bull. 3, 57 (Dyeing/Printing/Finishing).

Further reading

Ferrero, F., Periolatto, M., Udrescu, C., 2012. Water and oil-repellent coatings of perfluoro-poly-acrylate resins on cotton fibers: UV curing in comparison with thermal polymerization. Fibers Polym. 13 (2), 191–198.

Silicone-based water repellents

7

Hikmet Ziya Özek
University of Namik Kemal, Tekirdağ, Turkey

7.1 Introduction

Silicones are inspiring materials with many unique potentials to improve various materials which are used in daily life. Silicone compounds came from a combination of patient thought and brilliant laboratory technique. In general, silicone refers to materials in which silicon is bonded to oxygen. Although silicates have been used in ceramic arts and pottery for the past 5000 years, silicon was not known as an element until Berzelius (1824) first reported its isolation (Newton, 2010). Further research on silicon compounds was carried out by Wöhler (1856), Friedel and Crafts (1865) who had prepared the first known organosilicon compound, tetraethylsilane. Kipping pioneered the work with silanols to form a range of commercially useful silicones by condensation polymerization (Kipping, 1904) and is considered today to be the father of classical organosilicon chemistry (Seyferth, 2001; Thomas, 2010). In textiles industry, silicone compounds have gained wide acceptance and are used commercially either as coatings or as additives because of their unique surface properties. With proven silicone technology, textile producers worldwide can create fabrics, clothing, sporting goods, and technical textiles which are ideal for today's active lifestyles and industrial usages.

The first example of an organosilicon compound possessing at least one Si—C bond was reported in 1863 (Liles, 2012). Eventually, the first commercial heat- and chemical-resistant silicone rubber was developed by scientists at General Electric, based on earlier works initiated by the Corning Company of New York. The new material was found to be inert to many chemical agents. It could not be wet by water, resisted the attack of oils and greases, and could also be made so pure as to seem biologically inert when used as medical implants inside the body (Libretex, 2013). Since their introduction in 1943, silicones have been applied in an exceptional variety of industries. The initial target of the material was the electrical industry, but today silicone is widely used in many industries such as electronics, aerospace, construction, food processing, medical products and textiles. Silicone materials exhibit unique properties based on their chemical structure.

All silicone chemistry finds its roots in silicon, which is a chemical element with atomic number 14, and is the most elemental raw material. Silicon was discovered by Berzelius through an isolation process by reducing potassium fluorosilicate with potassium. It is a tetravalent metalloid and exists in two allotropic forms. One allotrope consists of a hard and brittle crystalline solid with a greyish black metallic lustre. The other allotrope has no crystal structure and usually occurs as a brown powder. Silicon is known as the most-used semiconductor material for discrete devices. It is a relatively inactive element at room temperature and does not combine with oxygen

Waterproof and Water Repellent Textiles and Clothing. https://doi.org/10.1016/B978-0-08-101212-3.00026-5

or most other elements. Water, steam and most acids have very little effect on the element. At higher temperatures, however, silicon becomes much more reactive and has great chemical affinity for oxygen. In addition to linking with oxygen to form a polymeric chain, the silicon atoms also bond with organic groups, typically methyl groups (Colas and Curtis, 2005). Silicon never occurs as a pure element in nature and always occurs as a compound with oxygen, magnesium, calcium, phosphorus or other elements. It is the eighth most common element in the universe by mass. In its combined form, it accounts for about 27.6% of the earth's crust, and ranks second only to oxygen. It is most widely distributed in dusts, sands, planetoids and planets as various forms of silicon dioxide (silica) or silicates (Newton, 2010).

Silicones are industrially produced compounds derived from the element silicon. In many cases, silicone compounds have gained acceptance and are used commercially either as coatings or as additives because of their unique surface properties (Owen, 2005; Seyferth, 2001). Some of the general properties and typical characteristics of silicone materials are as follows below.

Good wetting and spreading: Silicones flow easily and evenly over surfaces that would cause other materials to bead. Silicones also flow over themselves, giving them a 'self-healing' characteristic.

Long term elasticity, flexibility and pliability: Silicones resist stretching, twisting and compressing. As soon as the external force is released, they will quickly return to their original shape.

Stability under challenging conditions: Silicones are highly resistant to damage and degradation from

- thermal shocks
- extreme high and low temperatures
- chemicals
- oxidation

Water repellency: This is one of the most distinctive characteristics of silicones.

Weather resistance: Silicones are known for their long-term endurance against rain, snow, sleet, ozone, acid rain and the effects of UV rays.

Resistance to ageing: Silicones do not harden, crack, crumble, dry out, peel, rot or become brittle with age, unlike organic (carbon-based) materials.

Inertness: Silicones do not react with most materials unless they are designed to do so.

Strong adhesion properties: Silicones have a strong ability to make things stick

Excellent dielectric properties: Silicones are a good insulator of electricity.

With proven silicone technology, textile producers worldwide can create fabrics, clothing, sporting goods and technical textiles which are ideal for today's active lifestyles and industrial usage. In this context, they are known to have revolutionized the textiles industry. Silicones are, simply, capable of making fabrics and clothing more stylish, comfortable and easier to care for. It can be said that silicones excel in both fashion and industrial textiles applications. In modern textile manufacturing, they are used in all stages of the process, on the fibre during production, on the fabric and directly on the finished goods. Silicones are applied from different delivery systems to provide various benefits like lubrication, softening, foam control or hydrophobic coatings. The use of silicones can be very cost-effective because they are easy to process and effective at low-use levels. Textile applications with silicones cover three main areas (Lenoble, 2007; ACC Silicone, n.d.):

- Fluids and emulsions for thread & fabric lubrication.
- Antifoams as processing aids.
- Emulsions for use as functional textile modifiers.

Lubrication: Silicones improve gliding properties as a lubricant during yarn and thread production so that the thread breakage is minimized. Sewing properties are also improved, and the lustre of the sewing threads enhanced. The use of lubricant also improves resistance to high temperatures that might occur during the process.

Processing aids: Silicones are well known as antifoam agents used during the sizing and dyeing processes. They control the amount of foam generated optimizing productivity, continuing to perform in difficult conditions such as high temperatures and high pH values. They can be added directly to wet processes to cut down or prevent foam and used as formulation ingredients to achieve low-foaming textile solutions. Modified silicones are used as wetting agents to reduce the surface tension of aqueous systems.

Functional textile modifiers: Silicone fabric finishes bring a wide range of functional benefits to textiles. They add softness, reduce shrinkage and improve wrinkle recovery, tear strength, abrasion resistance and loft recovery. Silicones are very often used as softeners in the aqueous phase of conventional finishing processes. Silicone finishes are employed to increase the softness of fabrics so as imparting a particular hand and enhancing their aesthetic feel. They can also improve water repellency or water absorbency. The water repellent properties of the fabric can be enhanced without interfering with the permeability of the material because the pores of fabric are not occluded. With this treatment, the textile becomes more resistant to abrasion and wet stains. They can provide either water absorption or water repellency and make fabrics more comfortable and more desirable to touch, purchase and wear. Other functional benefits of silicones are the bettering of various physical properties, such as tear strength, abrasion, wrinkle resistance, stretch recovery and shrinkage reduction.

7.2 Characteristics of silicone water repellents

In fabrication work as diverse as textiles, building materials, home care products, paints and protective coatings, excellent water repellence may be achieved thanks to the convergence of three key features of hydrophobic silicone formulations. First is the inherent presence of a hydrophobic functional group in the silicone molecule. Second is excellent wetting and high spreading capability, which allows the silicone to penetrate the substrate and distribute the hydrophobic functional group deeply in or through the material. Third is the stability of silicones and their resistance to chemical, heat, ultraviolet or oxidative degeneration. Combining these features with the fact that silicones are highly gas permeable, and the result is a durable, yet breathable water repellent end product. Silicone is, therefore, widely utilized to take advantage of this chemistry as a highly

Fig. 7.1 Wetting of nontreated fabric and typical water repellency effect on silicone-treated fabric.

satisfactory water repellent substance. Typical water repellency behaviour of silicone-treated fabric and a fabric with no treatment is shown in Fig. 7.1.

The understanding of the functionality of silicone compounds is directly related to three components of its structure: construction, functionalization and derivatization. Each component is critical to functionality. Construction relates to the constitution of the silicone backbone, while functionalization is relevant to the groups that are on the silicone backbone. Derivatization refers to the chemistry conducted on the groups added by functionalization (The Silicone Conundrum, 2008).

The configuration of silicones has water-resistant hydrocarbon groups arrayed along the outside of the chain of Si—O atoms. Silicones act like hydrocarbon polymers, as in the case of polyethylene. It can be said that water is kept 'unaware' that the hydrophilic Si—O chain is even there. The hydrophobicity of silicones was discovered by Patnode at General Electric laboratories in 1940 (Patnode, 1942). The discovery that methylchlorosilanes in the vapour phase will react with many types of surfaces to produce water repellent films led to one of the most important developments in the organosilicon field. The effect was an entirely unexpected outcome of the preparation and handling of large volumes of methylchlorosilanes by the direct synthesis method developed by Rochow (1947). It was observed that paper processed with chloromethylsilanes ($(CH_3)_2SiCl_2$) became water repellent when exposed to moist air (Kissa, 1984; Libretex, 2013). Treated articles like paper and cloth would not absorbed water, which just beaded up on the surface. Today, dozens of silane materials are used for protection from water. Chloromethylsilanes are hydrolyzed by water to silanols, which condense spontaneously to siloxanes.

Silicones provide very hydrophobic finishes on various fabrics. This treatment involves full fabric impregnation from silicone-in-water emulsions, mostly by means of a padding process. The silicone phase of such emulsions contains SiH functional polymers because of their reactivity towards the fabric, but also because these polymers can cross-link with each other into a hydrophobic and durable fabric treatment, particularly if formulated with a suitable catalyst. The basic repeating unit is known as 'siloxane'.

The name 'silicone' was given in 1901 by Kipping, as analogous to ketones, to describe new compounds of the complete formula R_2SiO (Colas, 2007). These were rapidly identified as being polymeric and corresponding to polydialkylsiloxanes, with the formulation shown as siloxane in Fig. 7.2.

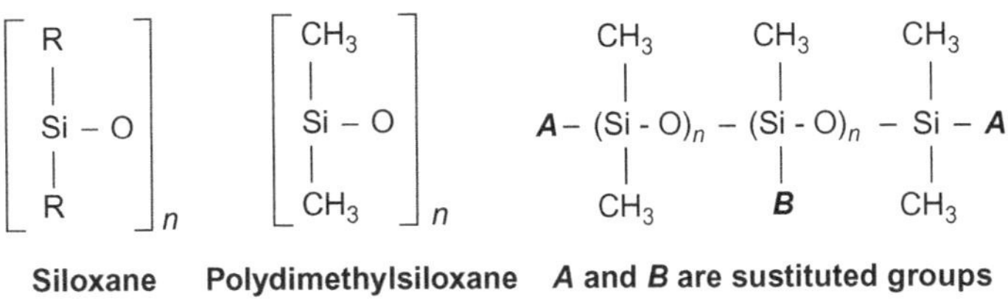

Fig. 7.2 Siloxane, polydimethylsiloxane chemical structure and penetrating water repellent arrangement.

The generic term 'silicone' was adopted by the industry, and it refers to a kind of manmade polymer based on a framework of alternating silicon and oxygen (siloxane) bonds, with organic substituents attached to the silicon. In common use, this term is assumed to refer to particular linear polymers where R is replaced by methyl groups. This is called as 'polydimethylsiloxane' which is the most common silicone and is generally abbreviated as PDMS, shown in Fig. 7.2.

Silicone water repellents or waterproofing agents generally come in two forms (Dow Corning, n.d.):

- Elastomeric polydimethylsiloxanes—elastomeric coatings that adhere to the substrate and cure to form a flexible, protective membrane.
- Penetrating water repellent chemicals—reactive silanes and siloxane resins with cross-linkable side chains. These materials have smaller molecular structures, which enable them to penetrate deeply into the substrate, where they chemically bond with it.

Either of these materials can be delivered via solvent or aqueous emulsion. The list of functional groups for penetrating water repellents is given in Table 7.1.

The methyl groups along the chain can be substituted by many other groups such as phenyl, vinyl or trifluoropropyl. The simultaneous presence of 'organic' groups attached to an 'inorganic' backbone gives silicones a combination of unique properties, making it possible to use them as fluids, emulsions, compounds, resins and elastomers in numerous applications as different as aerospace (low and high temperature performance), electronics (electrical insulation), health care (excellent biocompatibility), construction (resistance to weathering) and in the textile industries (repellency to water) (Colas, 2007).

Table 7.1 Functional groups for penetrating water repellents

Group	Position	Reactive	Function
Alkyl	A or B	No	Water repellency
Amino	A or B	No	Catalytic
Alkoxy	A or B	Yes	Crosslinking
Hydroxyl	A	Yes	Crosslinking
Hydrogen	B	Yes	Crosslinking

A: terminal position and *B*: side group position, as shown in Fig. 7.2.

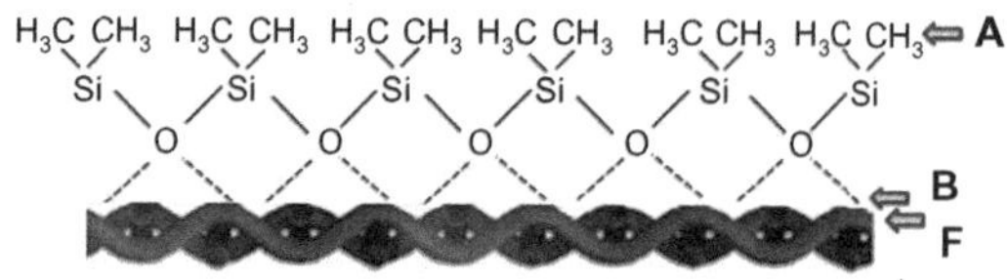

Fig. 7.3 Mechanism of water repellency by polydimethylsiloxane. *A*, hydrophobic outer surface; *B*, polar hydrogen bonds; *F*, fabric.

Silicones are capable of spreading and soaking easily into a substrate's pores because of their low surface tension. Their highly flexible and mobile siloxane backbone enables the water-repelling methyl groups to orient themselves towards the surface, creating a waterproof 'shield'. Thanks to this characteristic, silicones can also be used as a waterproof substance. Its ability to breathe makes this 'shield' very unique. Since silicones have a very open molecular structure, water vapour trapped in the substrate can easily slip through. Water molecules, however, are too large to enter. This is the mechanism of waterproof behaviour.

The unique structure of the polydimethyl-siloxanes (PDMS) provides an ability to form hydrogen bonds with fibres as well as exhibiting a hydrophobic outer surface, as seen in Fig. 7.3. The PDMS molecules are attached to the fibre surface (F) covering the top of the fabric through polar hydrogen bonds (B). On the outer side of PDMS, there occurs a hydrophobic surface (A) combining hydrocarbon groups which acts as a water repellent film. The hydrophobic methyl groups on the siloxane backbone tend to orient away from the fibre surface fabric. Meanwhile fibre bonding occurs between the hydrophilic O—Si—O chain and polar surface groups. Due to the difference in electronegativity of silicon and oxygen atoms, hydrogen bonding occurs between surface groups that carry electropositive hydrogen atoms and the O—Si—O chain. On the other hand, the presence of relatively weak intermolecular interaction enables several reactive groups to be introduced in the siloxane chain.

Hydrophobic agents are expected to reduce water adsorption by more than 70%, and vapour permeability by no more than 10% (GE Bayer Silicones, 2003). Silicones are the ideal material for achieving this due to their excellent hydrophobic properties. When silicone is applied as a water repellent agent, the liquid water is prevented from passing through the substrate by the methyl barrier at the outer surface. This formation is represented in Fig. 7.4A. On the other hand, gaseous water is able to pass through the small pores in silicone film crosslinked on the fabric surface. The physical structure of a polydimethylsiloxane molecule is shown in Fig. 7.4B. The angles, lengths and energy level of chemical bonds of this structure is given below (Kuo, 1999; Mark, 2004; Lenoble, 2007).

Bond angles[a]	Bond lengths	Bond energy
C—Si—C ~106 degrees Si—O—Si ~143 degrees and O—Si—O ~109 degrees	1.92 Å 1.65 Å	Si—C 326 kJ mol^{-1} Si—O 443 kJ mol^{-1}

[a]Measured at $[(CH_3)_2SiO—]_4$ at −50°C.

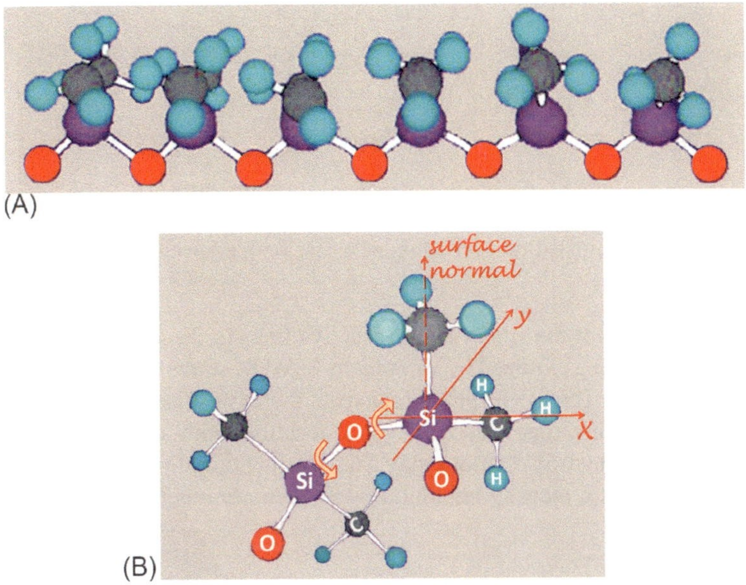

Fig. 7.4 Three-dimensional illustration of polydimethylsiloxane molecular arrangement. (A) Orientation of polymer chain; (B) Molecular structure and the coordinate system.

The siloxane system is known to have a comparatively open structure with the flattest angle and the longest bond length (Mark, 2004; Lenoble, 2007). Freedom of rotation around Si—O and Si—C bonds results in low energy barrier. Typical physical properties of silicones are:

- Lower intermolecular forces.
- Low glass transition point (GTP) temperature ($<$ -127°C/146 K).
- Higher intermolecular distances.
  - High gas permeability.
  - High compressibility.
  - Small changes of physical properties with temperature changes.

Silicones designed as water repellent agents are expected to consist of three components to maintain some durability. These components are a silanol, a silane and a catalyst such as tin octoate, as shown in Fig. 7.5. The catalyst enables not only moderate condensation conditions, but also promotes orientation of the silicone film on the fibre surface, whereas the outward-oriented methyl groups generate the function of water repellency.

During the drying step after padding application, the silanol and silane components can react to form a three-dimensional crosslinked layer around the fibre (Schindler and Hauser, 2004; Hill, 2006). This reaction is generally expected to be completed after storage of about one day; in sequence, the function of full repellency appears.

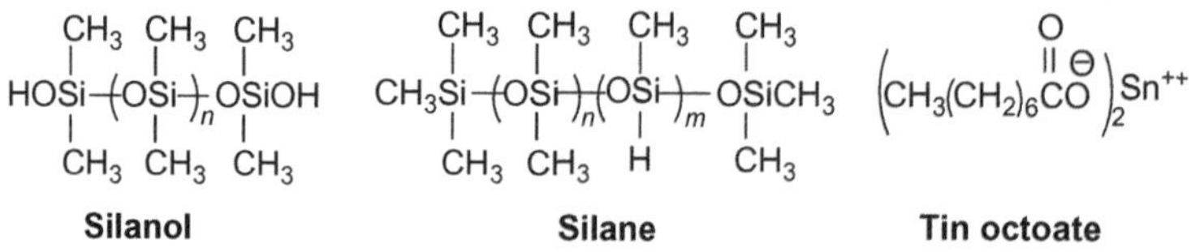

Fig. 7.5 Components of typical silicone material polydimethylsiloxane.
Data from Schindler, W.D., Hauser, P.J., 2004. Chemical Finishing of Textiles, CRC Press and Woodhead Publishing Limited (in association with The Textile Institute), Cambridge.

The Si—H groups of the silane are the reactive links in the silicone chain, generating crosslinks or being oxidized by air or hydrolyzed by water to hydroxyl groups as illustrated in Fig. 7.6. The Si—H groups can react thereby either directly or as hydrolyzed hydroxyl groups that can undergo a reaction with crosslinkers such as diisocyanates. These hydroxyl groups may cause further crosslinking, but if too many of them stay unreacted, their hydrophilicity will decrease repellency (Schindler and Hauser, 2004).

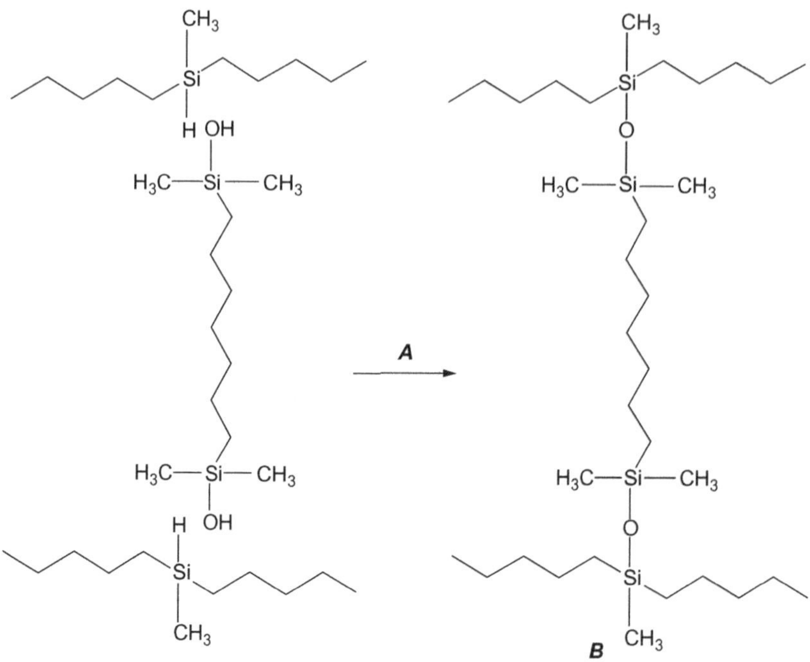

Fig. 7.6 Silanol-silane reaction where A refers to the removal of H_2 and B shows Si—O—Si crosslinked polymer, while tin octoate is the catalyst.
Modified from Schindler, W.D., Hauser, P.J., 2004. Chemical Finishing of Textiles. CRC Press and Woodhead Publishing Limited (in association with The Textile Institute), Cambridge.

The other additional features of silicone which are beneficial for water repellency and waterproofing can be summarized as follows:

- High spreading and wetting capabilities—the ability to thoroughly cover a surface or penetrate a porous substrate.
- Permeability to gas and water vapour.
- Durability, flexibility and mechanical and chemical adhesion.
- Resistance to ultraviolet light (UV), heat and oxidative degeneration.

7.3 Methods of silicone applications for developing water repellent fabrics

Silicone finishes can provide good water repellency at relatively low concentrations. Most commonly, silicone finishes are applied as an aqueous polysiloxane emulsion by padding, followed by drying and curing. Application by an exhaustion process is also possible for some silicone finishes. In both of these processes, finishing substances of silicone compounds are transferred from aqueous solution phase to the fabric surface and then cured through the application of heat, leading to the desired surface finish. Full repellency is expected to manifest after ageing for 24 hours. While the finishing compounds are transferred to the fabric through physical entrapment in the padding process, the actual exhaustion of the compound from the solution phase onto the fabric is accomplished through specific adsorption in the exhaust process (Purohit et al., 2006). For some commercial products, application by spraying is also possible.

Water repellency of a treated fabric is the ability to withstand penetration by liquid under static conditions involving no other forces except for capillary forces and the weight of the drop (Kissa, 1984, 1996). It is defined as the ability of a surface to resist wetting by applied liquid water. Repellency results in limited wettability, while wetting a textile material involves the processes of immersion, capillary sorption, adhesion and spreading. A water repellent for textiles is anything that lowers the surface energy of the textile so that water will not wet it. Repellent finishes achieve this particular effect by reducing the free energy at fibre surfaces. The interactions between the drop of liquid and textile material surface, which is the fibre surface, may vary depending on the cohesive forces. If the adhesive interactions between a fibre and a splash of liquid placed on the fibre are greater than the internal cohesive interactions within the liquid, the splash will spread as seen in Fig. 7.7. The splash of liquid will not spread when the adhesive interactions between the fibre and the liquid are less than the internal cohesive interactions within the liquid, and the contact angle between the surface and the liquid exceeds 90 degrees.

Surfaces exhibiting low interactions with liquids are referred to as low-energy surfaces. Their critical surface energy or surface tension (γ_c) must be lower than the surface tension of the liquid (γ_L), which is defined as the internal cohesive interaction necessary for that particular liquid to be repelled. In the case of silicones, surface energy varies (γ_c) about 24–30 mN/m. Since it is two to three times lower than that of water ($\gamma_W = 73$ mN/m), water repellency is always obtained. However, the

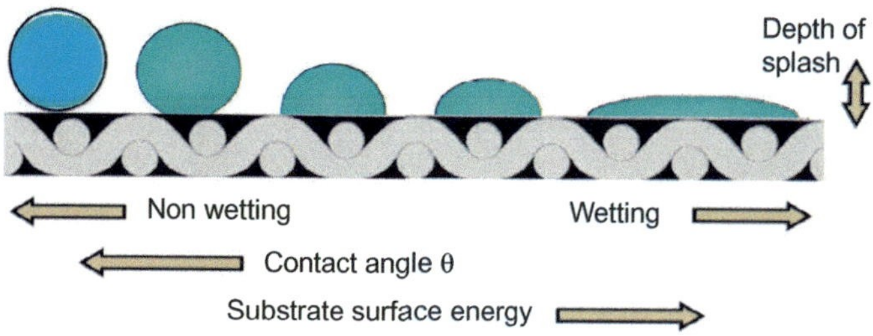

Fig. 7.7 The effect of substrate surface energy and the variation of contact angle on wetting.

efficiency of water repellency will not be as strong as with fluorocarbon finishes, because their surface energy is even lower: 10–20 mN/m. It should also be noted that silicones will not repel oil, unlike fluorocarbon finishes, because of their higher surface energy. It means that silicone-finished textiles are likely to be dirtied by oil or oil-bearing substances.

Silicones for textiles may be broadly classified into the following three categories:

1. Nonreactive silicone polymers—These are based on polydimethyl siloxanes, which are available as an oily material and insoluble in water. The emulsions can be made using cationic, anionic and nonionic emulsifier. For the sake of compatibility with other textile finishing auxiliaries, nonionic emulsifiers are normally preferred.
2. Reactive silicone polymers—Hydrogen or silanol functional groups are added to the PDMS for producing such polymers. These compounds do not bind to the fabric surface through any known chemical reaction.
3. Chemically modified silicones—Silicone is initially formed as dimethylpolysiloxane only; the organic groups attached to the silicone atom via silicone—carbon bonds define the class of silicone. Depending on the characteristics of the attached group, different products and functionality can be achieved.

Silicone repellents are methylhydrogen and dimethylsiloxane copolymers which react in the presence of a catalyst to form water repellent crosslinked films. Additional melamine prepolymers or reaction resins are preferred when used with cellulose fibres. Typical catalysts are tin, zirconium and aluminium salts, for example, zinc octoate, tin oleate, tin octoate, copper naphthenate or metal-free compounds such as epoxy amines or amides (Lacasse and Baumann, 2004). The repellents are usually applied using a pad and dry cure process in the form of emulsified oils. Application as foam is also possible.

Silicone finishes are applied to fabrics either within an organic solvent or with water as an emulsion. Some commercial products synthesized with organic solvent can be employed by padding and foulard techniques but this may cause the acquired film layer to lose elastomeric character. When cationic emulsifiers are used to make an emulsion, the finish may be applied by exhaustion because the negative fibre surface charges tend to attract positively charged particles. It should also be noted that silicone water repellents are applied together with a durable press finish. Durable press resins enhance the durability of the water repellent finish. Chlorinated solvent solutions may be sprayed onto upholstery fabrics by the retailer as a customer option

(Tomasino, 1992; Lacasse and Baumann, 2004). Most complications involved in using emulsions for delivering silicone to substrates arise from the fact that the silicone is delivered from a micelle. When silicone is delivered from a micelle, the energetics of delivery to the substrate are complicated by the presence and type of emulsifiers and the particle size of the silicone. All three factors must be optimized for the best performance of the emulsion in the formulation (O'Lenick and Siltech, 1999).

When surface active agents are added to water, the first observable effect is that the surface tension at the air–water interface drops. With additional surfactant, the critical micelle concentration is reached, and eventually, micelles of silicon particles are formed. Equilibrium is achieved between surfactant in the micelle and surfactant the interface, reducing the surface tension. Additionally, emulsions have some inherent shear instability, and may demonstrate freeze-thaw instability. Finally, there are limitations as to the type of additional surfactant that can be added to an emulsion-containing system.

Surface treatment with silicone tends to alter surface properties. The morphology of fibres was examined using AFM in order to determine the manner of alteration (Purohit et al., 2006). It was observed that the treated fibre is more uniform, well stacked and relaxed compared to the untreated fibre. It also appears that a single fibre structure is divided into many fibres after silicone treatment. This indicates that the silicone treatment is responsible for changes in the microproperties of the fibres, and can be used to modify microstructural properties of fibres to induce smoothness, bounciness and other such desirable properties in the fabric. The individual properties imparted to the fabric can be correlated with the molecular architecture of the silicone chains. It can be said that the bounciness of the fabric substrates may be due to the proposed helical conformations of silicone chains.

Silicone-based durable water and oil repellents (DWOR) have been identified as one of the most environmentally friendly alternatives to fluorocarbon-based substances for achieving water repellency of fabrics (Holmquist et al., 2016; MIDWOR, 2016). Compared to their fluorocarbon-based analogues, silicon-based DWORs' manufacture and application result in far fewer environmental issues.

7.4 Durable water repellents and silicone

In the case of water repellency, it is almost impossible to achieve a permanent effect by means of finishes. The category of durable water repellent (DWR) for textiles (ZDHC Programme, 2012) is specified as a textile finish whose performance attributes may include water, oil, stain and soil repellency, stain and soil release, and durability (e.g. to laundering, dry cleaning, abrasion, light exposure, rain, etc.).

In general, finishing materials can be classified in several ways depending upon their functions, performance and nature. On the basis of degree of performance, they are grouped into four main categories (Vidyanagar, 2015):

1. Temporary finishes are not durable and run off after first washing or dry cleaning.
2. Semidurable finishes stay on the fabric surface for several washings.

3. Durable finishes last throughout the life of a fabric or a garment, but may lose their effectiveness after many washes.
4. Permanent finishes are usually achieved through chemical treatment. This modifies the fibre structure and remains for the entire life of a fabric.

DWR finishes are frequently used within the outdoor apparel industry to provide fabrics with water- and oil-repellent properties. DWRs help to ensure protection against changing weather conditions when engaging in outdoor activities. For many years, most textile manufacturers relied upon fluorocarbon-based DWR chemistries. Unfortunately, these chemicals have recently been revealed to be toxic and persistent in the environment.

The Green Peace Detox Campaign (2011) starting in 2011 targeted the elimination of hazardous chemicals including PFCs, which continue to suffer increasing scrutiny. Since then, Greenpeace has been calling on the clothing industry to eliminate all hazardous chemicals from its supply chain. Certain legislative and voluntary restrictions were consequently implemented across the industry. Hence a group of textile associations and some brands made a commitment in 2011 to reach the goal of zero discharge of hazardous chemicals (ZDHC) in the supply chain by 2020. The Joint Roadmap report of this collaboration was published in 2012 (ZDHC Programme, 2012). One of the roadmap projects has been focused on actively researching durable water repellent chemistries that will afford high performance with less environmental impact.

A Green Peace Study (2015) found traces of PFCs in the waters of high-altitude lakes around the world, including the Lago di Pilato in the Apennine mountain range of central Italy. A follow-up investigation (Green Peace, 2016) analysed a range of outdoor gear for hazardous per- and polyfluorinated chemicals (PFCs). This survey was carried out in November 2015 as part of a joint effort between 19 countries. It involved 40 popular products from 11 well-known international brands, selected by consumers and purchased at local stores. The study revealed that not only outdoor clothing and footwear but also camping and hiking equipment such as backpacks, tents and sleeping bags contain chemicals that are hazardous to the environment and to human health. PFCs were detected in all product categories, aside from gloves. Concentrations of volatile PFCs for jackets in the study varied between 0 and 630 $\mu g/m^2$, while the concentration of total ionic PFCs also varied within a similar range of 0–684 $\mu g/m^2$. The highest values are well over the current limit of 1 $\mu g/m^2$ for all materials with a water- and oil-repellent finish or coating specified by the Standard 100 Oeko-Tex (Oeko-Tex, 2017). The level of PFCs concentration found was similar for other product ranges. Unfortunately, it was observed that 36 of the 40 tested products contained different forms of PFCs, despite most brands having already pledged to cease or phase out the use of such toxins.

In light of scientific findings that PFC-based DWRs lead to residues of perfluorinated chemicals (PFCs) around the world, the outdoor industry has set off on a quest for new, environmentally friendly, durable, water repellent solutions. The Association of the German Sporting Goods Industry, the Outdoor Industry

Association and the European Outdoor Group are supporting outdoor companies on this path and have initiated the 'Durable Water Repellent (DWR) Project'. As part of this project, a survey about the expectations of end users and brands was conducted in cooperation with De Montfort University in the United Kingdom (Burman, 2014). Key findings of the survey indicated that the end user intends to use their jacket for a total of 4–5 years, which is half the lifetime recommended by the brands. The results suggested that the majority of end users do not provide the required home treatment such as washing, re-impregnating, tumble drying and ironing to ensure the water repellent finish is re-energized and to improve the performance and longevity of the garment. Furthermore, the end user is not aware of the type of factory-applied finish on their current garment, so they are unable to provide the correct home treatment. It was apparent that the end user is not conscious of the environmental impact of fluorine-based finishes or the changes in technology. It was suggested that outdoor brands need to adjust their marketing behaviour to further educate the end user on the new technology and on how they can provide the correct home treatment for their water repellent garment.

This report also compiled a summary of commercially available alternative DWR technologies and chemistries and described the steps involved in moving from long-chain to short-chain and nonfluorinated technologies and chemistries. The brands owners were also asked if they are working on any technological alternatives to fluorine-based finishes. The results suggested that 21 out of 53 brands said that they were either already successfully using green chemistry or are still working on finding alternatives. Promising alternatives include acrylic- and urethane-based chemicals together with paraffin, silicone and stearic acid-melamine.

Silicone repellents offer a high degree of water repellency at relatively low (0.5%–1%) on weight of fabric concentrations (Hill, 2006). This is a typical positive feature of silicone water repellents. On the other hand, their repellency can be reduced if excess amounts are applied. Regrettably, silicone repellents have only moderate durability to laundering and dry cleaning, and almost none to oil and soil repellency. Hence, they are likely to require frequent laundering. This fact raises the question of how durable the finish is to laundering and possible degradation.

Most of the DWR-finished garments and other products undergo a reduction in repellency during lifetime usage (Arunyadej et al., 1998; Leonas, 1998; Holmquist et al., 2016). These particular loss mechanisms of DWR-based substances from a DWR-treated fabric occur due to leaching to water, evaporation to air and possible release of DWR-related substances by wear and tear. There are many relevant studies in the literature identifying and evaluating these matters with reference to the DWR chemistries. Although the studies are focused on the use phase of a DWR-finished fabric, many of the mechanisms are dependent on the substance used and are also applicable for diffusive releases after the end of use life phase of the fabric or relevant product. These phenomena causing possible loss of chemicals from a garment are schematically illustrated in Fig. 7.8. The change of this macroscopic material property is a clear argument for the loss of chemicals from the surface of DWR-treated fabrics, and several mechanisms are conceivable. Loss processes can occur from the surface of

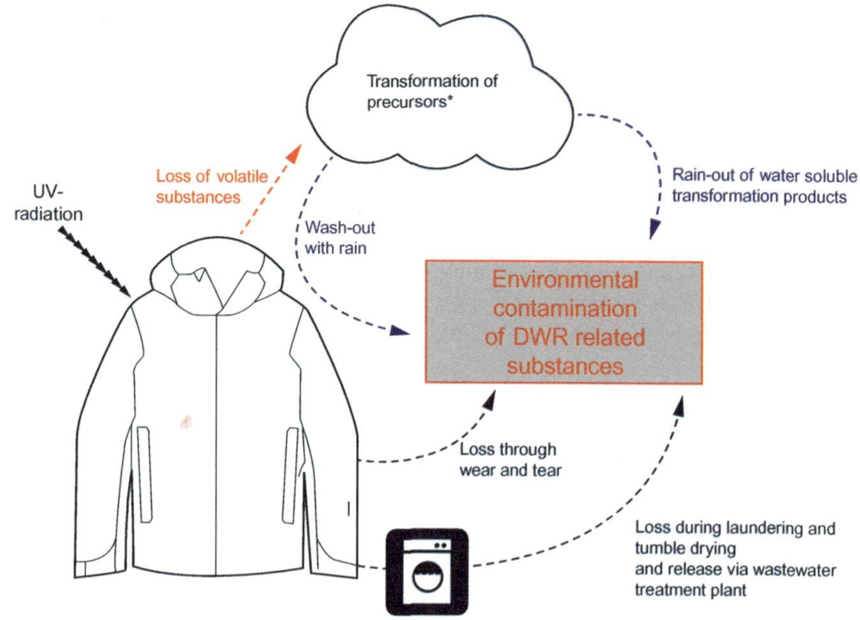

Fig. 7.8 Schematic view of loss of chemicals from DWR-finished fabrics during usage. *The transformation of precursors might also occur in other environmental compartments. Reproduced with permission from Holmquist, H., Schellenberger, S., Van der Veen, I., Peters, G.M., Leonards, P.E.G., Cousins, I.T., 2016. Properties, performance and associated hazards of state-of-the-art durable water repellent (DWR) chemistry for textile finishing. Environ. Int. 91, 251–264.

the fabric or from fragments like fibres or particles into the environmental compartments of atmosphere, hydrosphere, lithosphere and biosphere (Ozdil et al., 2012).

An important requirement for hydrolysis which takes place during laundering is the accessibility of positions on the polymers for degradation (e.g. ester bonds) under the conditions in which the textiles are used. Since hydrophobic moieties like CH_3 and CF_3 groups build a hydrophobic shield around the DWR backbone, the point of degradation could be inaccessible (Holmquist et al., 2016). The chemical nature of degradation and the range of degradation naturally depends on the chemical class of the DWR agent.

The increased surface area of these material fragments is likely to lead to faster loss rates (Hu, 2008; Ozdil et al., 2012).

It was also proven that rough surfaces such as DWR-finished cotton fabrics might have a higher tendency towards abrasion and fragment formation, in comparison with the smooth surfaces of synthetic fibres (Beyler and Hirschler, 2002). Additionally, the type of linkage between the DWR compound and fibre surface greatly influences the rate of losses. For example, chemicals without covalent bonds, such as residuals from production or nonreacted monomers, can be released during laundering or washed out by rain penetrating the fibre surface (Holmquist et al., 2016). They may also be lost to the air, depending on their volatility. Polymer degradation, thermal decomposition and the relevant breakage of covalent bonds could be another mechanism for the loss of chemicals from garments with DWR finishes, and DWR polymers may also be subject to UV degradation (Beyler and Hirschler, 2002) because of exposure to sunlight, depending on the radiation intensity.

Another additional factor for possible chemical loss is reimpregnation treatment of garments to refresh the initial performance of textiles (Casper and Kelemen, 2014). Specially designed fluorinated or nonfluorinated reimpregnation agents (3M, 2013; Rudolf Chemie, 2015) are added in the washing process, or used as aerosol spray. Depending on the chemical structure, they bind to the fabric either physically (nonpermanent) or chemically (permanent). This process is often used in work wear and protective clothing. Nonpermanent reimpregnation agents, in particular, are prone to chemical release.

7.5 Performance of silicone-treated water repellent fabrics

In addition to the hydrophobic character of the PDMS chain, fabrics treated with silicone-based DWR polymers result in finishes with a soft hand. This special characteristic is due to the structure of the polysiloxane backbone, which is the most flexible chain known in polymer chemistry (Schindler and Hauser, 2004). The combination of flexibility and long molecular bonds permits an easy change in polymer conformation, and CH3— groups can freely rotate around the Si—O—Si backbone.

Despite some of the advantages such as soft feel and shape retention, the majority of these fabrics have only a moderate durability to laundering, and silicone DWRs can deliver no oil repellency unless modified.

The formation of a sheath of finish around the fibre gives rise to durability. Therefore, when the sheath cracks, durability is lost. Adsorption of hydrophilic substances found in dry cleaning and laundry products also impairs water repellency. The list of advantages and disadvantages of silicone water repellents is given below.

Advantages of silicone water repellents:

- A high degree of water repellency at relatively low (0.5%–1%) on weight of fabric concentrations.
- Very soft fabric hand, resistance to water-borne stains.
- Resistance to ultraviolet light (UV), heat and oxidative degeneration.
- Permeability to gas and water vapour, no significant hindrance of breathability of fabric.
- Improved sewability.
- Improved shape retention.
- Improved appearance and feel of pile fabrics.
- High spreading and wetting capabilities (the ability to thoroughly cover a surface or penetrate a porous substrate).
- Degradable in environment.
- No residues of PFCs.

- Some modified silicone repellents can be exhaust applied (to pressure-sensitive fabrics).
- Water repellent treatment ability for leather.

Disadvantages of silicone water repellents:

- Moderate durability to laundering (through hydrolysis of siloxane and rupture of the film by strong cellulose fibre swelling).
- Moderate durability to dry cleaning (adsorption of surfactants).
- Increased pilling and seam slippage.
- Reduced repellency if excessive amounts are applied (for example silicone double layer with polar outside only).
- No oil and soil repellency (silicone finish may enhance the attraction of hydrophobic dirt).
- Toxic waste water, especially the residual baths, from these finish application processes might be toxic to fish.

Specific examples of commercially available silicone-based agents may be given to justify these characteristics. Dow Corning DWR-7000 Soft Hydro Guard (Dow Corning, 2012) and ShinEtsu (2016) KS-7002 and POLON-MF-33A solvent based water repellents are claimed to provide very soft hand. Momentive Magnasoft NFR, in addition to soft hand, also maintains repellency to water, even after multiple home laundry cycles (Momentive Perf. Inc., 2011). WACKER Finish WS 60 by Wacker Chemie (Wacker Chemie, n.d.) is known to produce a soft supple feel while retaining full breathability by allowing permeability to water vapour. The prescreening tests of Protex Dryol SOL (which is a silicone-based water and oil repellent) as part of the MIDWOR project have demonstrated that treated cotton and polyester blend trousers fabric show no oleoproof effect before and after washing (MIDWOR, 2016), whereas water repellency effect is reported as no wetting before washing and slight random wetting after 5 washes.

Silicone-based DWRs have been identified as one of the most environmentally friendly alternatives for achieving water repellency of fabrics (Holmquist et al., 2016). It is reported that degradation of siloxane compounds in the environment occurs quickly, taking just weeks in the atmosphere and days in soil (Mota, 2016).

Siloxane surfactants are a class of novel surfactants consisting of a permethylated siloxane backbone with one or more hydrophilic groups attached. Most siloxane surfactants are intermediate molecular weight copolymers (Hill, 1999). These surfactants are highly surface active in both aqueous and nonaqueous media, resulting in a surface tension as low as 20 mN/m. Generally, it remains difficult to correlate specific features of molecular architecture with performance.

The required performance level of the repellent finishes depends on their expected usage as apparel products and other important factors such as their durability to laundering and dry cleaning, resistance to abrasion and fabric breathability. Silicone-based water repellent finishes are more durable than wax and melamine's repellents, but less durable than fluorochemical finishes (Davies, 2014). Silicone repellents on cellulosic fibres provide good dry cleaning fastness but poor washing fastness. On the other hand, good fastness is achieved with synthetics. In recent years, organo-modified silicones with various reactive functional groups have also become available, and they are demonstrating greater potential durability. However, their use as a water repellent is not common and they are more frequently used as softeners. From the cost point of view, silicones are more expensive than wax repellents and less expensive than fluorochemical repellents. Treatment with silicones, however, requires a relatively small amount of additive, so less chemical is used. Modification of the silicone backbone corrects the lack of resistance to hydrolysis shown by typical super wetters. This results in fast drying and minimizes process time and costs of drying (Wacker Report, n.d.).

In a project called TEX-SHIELD (Graves, 2016) (within the EU 7th Framework Programme), silicone-based water repellents were synthesized to develop novel fluorine-free coating solutions based on multiscaled nanoparticles with high repellent functionality bonded directly to the textile surface. The project aimed to offer alternatives to replace the incumbent PFC chemistry while maintaining the same level of repellency performance. The materials that are being developed use a combination of multiscale structure and surface energy to control repellency performance. The views of cotton fibre after treatment taken by a scanning electron microscope microscopy are given in Fig. 7.9. The nano-roughening effect which could be created by the silica nanoparticle deposition is clearly observed, which has further increased the ultimate hydrophobicity and partial oleophobicity. Nanoparticles deposition is known to enhance the water repellency of silicone-based compounds (Manatunga et al., 2016).

7.6 Environmental and health aspects of silicone water repellent

Silicone-based water repellents have been identified as one of the most environmentally friendly alternatives for attaining water repellency in fabrics (MIDWOR, 2016;

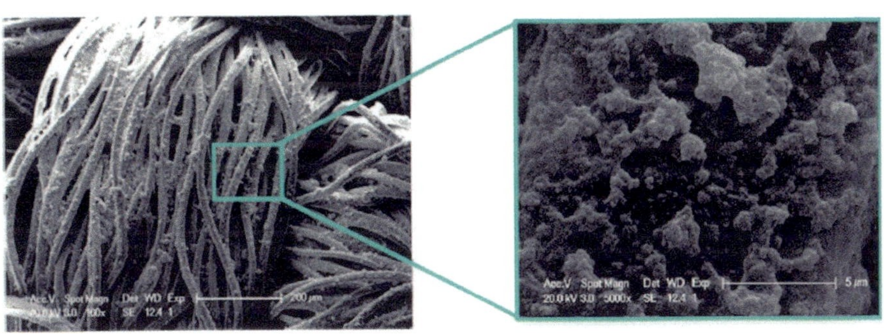

Fig. 7.9 Silicone and multiscale silica (80/20 wt.%) water repellent-treated cotton fabric. Magnification ×100 on the yarn and ×5000 on the fibre.
Reproduced with permission from Graves, C., 2016. D6.4 Technical Publication "TEX-SHIELD Environmental friendly and Durable Oil and Water Repellence Finish on Technical Textiles" Collaborative Projects, SME-2012-2. Available from: http:/www.texshield-project.eu/publications/Deliverable%20report%206.4%20Technical%20publication.pdf (accessed January 2017).

Mota, 2016; Holmquist et al., 2016). Silicones are well-known, useful materials varying in structure, reactivity and chemical and physical properties, but they all contain a covalent bond between the silicon atom and an organic group. The most common of these polymers are based on PDMS. These polymers are manmade, with an organosilicon linkage not found in nature, and therefore, natural degradation in the environment was thought somehow to be nonexistent. Work by Graiver et al. (2003), however, showed that degradation takes place by hydrolysis of PDMS. Various siloxanes, especially the cyclic siloxanes known as D4, D5 and D6, and the linear siloxanes HMDSO, MDM, MD2M and MD3M are used as intermediates for synthesis of silicone polymers, which, in turn, is used for several treatments in textile manufacturing.

The presence of residual cyclic siloxanes, and the presence of solvents (if any) in its composition might give rise to environmental and human health hazards. For some siloxanes used as intermediates in the production of silicon polymers, some health and environmental information is given, but it is unclear to what degree these siloxanes are ingredients or impurities in commercial products or remain in textile products. Nevertheless, it is known that the levels of chemicals in textiles and garments are very low, because residual silicone compounds evaporate during the curing stage of textile production.

Residual cyclic methyl siloxanes are considered the main environmental hazard of PFC-free silicone-based DWORs (Holmquist et al., 2016). Various silicone DWORs are reckoned to contain residual levels of cyclic volatile methyl siloxanes, such as octamethylcyclotetrasiloxane (D4) or decamethylcyclopentasiloxane (D5), in concentrations varying within the range of 400–3360 and 570–3110 ppm, respectively, as a consequence of the formation of DWORs of high molecular weight based on silicones (Brooke et al., 2009a,b). On the other hand, the hazard of these compounds is not clear, and they can also be found in personal care products (such as cosmetic products, and skin and hair care products) and household/industrial cleaning products. However residual levels of D4 or D5 due to the application of silicon-based DWORs are expected to evaporate during the curing stage in fibre treatment, thus their presence in the finished garment should be lower (Holmquist et al., 2016).

Studies on PDMS polymers showed a relatively fast depolymerization process which leads to biodegradable compounds. According to the risk evaluation carried out by the UK Environment Agency, no risks were identified from production and uses of D5 for the air, water, soil or humans exposed via the environment. The study on PDMS degradation conducted by Graiver et al. (2003) concluded that most of these cyclic compounds end up in the soil as result of applying sludge from waste water treatment plants as fertilizer. In the soil, PDMS undergoes depolymerization through cleavage of the siloxane bonds to yield low molecular weight silanols. These silanols are sufficiently volatile to be transported into the atmosphere, where they are degraded by hydroxyl radicals to yield silica, water and carbonyl compounds. The rate of degradation of these compounds in the upper atmosphere is 10–30 days, while in dry soil, they degrade much faster, in about 4–7 days. The presence of other chemicals in the water repellent formulation may have also a relevant impact, depending on the type of compound and its composition.

With regard to health assessment due to the use of siloxanes, there have been several studies (Lassen et al., 2005, 2015; Greve et al., 2008; ECETOC, 2011). Humans may be exposed to PDMS via oral ingestion and dermal contact, but it appears to have a low potential for absorption via these routes. In humans, PDMS has no effect on the immune system. Low-molecular-weight poly(dimethylsiloxanes) have been studied by siloxane manufacturers, and it was concluded that all the poly(dimethylsiloxanes) which were studied possessed a very low toxicity potential.

In a study for the Danish EPA, the toxicological information which primarily existed for D4, D5 and HMDSO was reviewed (Lassen et al., 2005). The three siloxanes demonstrated a relatively low order of acute toxicity by oral, dermal and inhalation routes, and do not require EU classification for this effect. They are not shown to be irritating to skin or eyes, and are also not found to be sensitizing by skin contact. Other studies of siloxanes, however, indicate that they appear to be harmful when inhaled, and that exposure may induce serious damage to eyes. A risk assessment of DWOR in textile finishing process as part of the MIDWOR Project (Mota, 2016) has been scheduled but no data is disclosed, yet. In short it can be said that knowledge of the toxicity of siloxanes is still incomplete (Lassen et al., 2015). Among the eco-friendly chemicals in processing, silicones appear to be capable of offering one the best available solutions to meet the multidimensional demands of customers with respect to environmental concern (Manickam, 2009). The harmonized classification according to the CLP Regulation (Regulation (EC) No 1272/2008) and the classification as notified to the C&L inventory is indicated in Table 7.2.

7.7 Review of commercial products and future trends

New discovery and development in silicone technology enables textile manufacturers to create various functional fabrics and intelligent textiles which can satisfy the demands of consumers. Since General Electric and Corning launched commercial silicones in the 1940s, the market has continued to flourish, not only in terms of overall silicone consumption, but also in terms of the diverse uses that silicones serve, the number of products developed, and the number of enterprises who produce and distribute these fascinating materials. The size of the global silicone market, on the basis of 2013 turnover, is estimated to be $11 billion US dollars, with 2,122,000 metric tons of formulated silicone products (Global Silicone Council, 2016).

According to a recent report (PRNewswire, 2016) published by Allied Market Research, titled 'World Silicones Market', the silicones market is expected to reach $20.7 million US dollars by 2022, from the value of $15.3 billion US dollars (2015), growing at a CAGR of 4.9% from 2016 to 2022. The silicone elastomers segment dominated with more than one-third share of the market revenue and volume in 2015, among other forms of fluid, gel and resin. On the basis of the end user segment, building and construction has the largest share, while the share of the textile industry is expected to be around 5%–6% in volume.

Table 7.2 Harmonized classification and labelling and packaging information for siloxanes compounds notified by manufacturers and importers (C&L 'Classification & Labelling' inventory)

Substance	CAS No.	Harmonized[a] C&L list (ECHA; 2014D)			
		Hazard class and category code(s)	Hazard statement codes	Number of notifiers	Labelling hazard stat. codes
Octamethylcyclo-tetrasiloxane (D4)	556-67-2	Repr. 2	H361f	Harmonized classification*	H361f
		Aquatic chronic 4	H413		H413
Polydimethylsiloxanes (PDMS)	63148-62-9	Total		1405	
		Not classified		356	
		Aquatic chronic 4	H413	131	H413
		Eye Irrit. 2	H319	51	H319
		Aquatic chronic 2	H411	13	H411
		Repr. 2	H361	25	H361
		Flam. liq. 3	H226	13	H226
		Skin corr. 1A	H314	13	H314
		Eye dam. 1	H318	1	H318
		Asp. tox. 1	H304	1	H304
		Acute tox. 1	H300	1	H300
		Acute tox. 2	H300		H300
Decamethylcyclo-pentasiloxane (D5)	541-02-6	Total		1479	
		Not classified		1313	
		Aquatic chronic 4	H413	65	H413
		Acute tox. 3	H331	46	H331
		Skin irrit. 2	H315	27	H315
		Eye irrit. 2	H319	31	H319
		STOT SE 3	H335	2	H335
Dodecamethylcyclo-hexasiloxane (D6)	540-97-6	Total		266	
		Not classified		230	
		Eye Irrit. 2	H319	19	H319
		Aquatic chronic 4	H413	16	H413

Octamethyltrisiloxane (MDM)	107-51-7	Total		224	
		Not classified		83	
		Flam. liq. 3	H226	141	H226
		Aquatic chronic	H413	3	H413
Hexamethyldisiloxane (HMDSO)	107-46-0	Total		924	
		Not classified		99	
		Flam. liq. 2	H225	793	H225
		Aquatic acute 1	H400	596	H400
		Aquatic chronic 1	H410	408	H410
		Carc. 2	H351	63	H351
		Aquatic chronic 2	H411	73	H411
		Acute tox. 3	H301	18	H301
		Acute tox. 4	H332	18	H332
		Flam sol. 1	H228	1	H228
		Water react. 1	H260	1	H260
		Eye irrit. 2	H319	1	H319
		Flam liq. 3	H226	1	H226
		Asp. tox. 1	H304	1	H304
		Skin irrit. 1	H315	1	H315

[a]According to Annex VI of Regulation (EC) No 1272/2008 on Classification, Labelling and Packaging of substances and mixtures (CLP REGULATION).
Modified from Lassen, C., Jensen, A.A., Warming, M., 2015. Alternatives to Perfluoroalkyl and Polyfluoroalkyl Substances (PFAS) in Textiles. The Danish Environmental Protection Agency, Copenhagen. ISBN No: 978-87-93352-16-2.

Water repellents based on silicone chemistry have been used for many years, and many commercial products are available in the market worldwide There are various manufacturers of silicone-based water repellent products. Dow Corning and GE Silicone, which continues as Momentive Performance Materials, Inc., following the takeover in 2006, are the pioneering companies in this field. Bluestar Silicones, Wacker Chemie, Huntsman Textile Effects, Protex International, Resil Chemicals, Rudolf GmbH and Shin-Etsu are other well-known chemical manufacturers offering such products. Typical examples of marketed commercial products and details regarding classification and claimed performance properties are listed in Table 7.3. These products are generally offered in emulsion form, either water- or solvent-based. One or two component types are available, and softness of the fabric appears to be most outstanding characteristic. The most common application technique is padding, preceding the drying and curing process. Most of the silicone-based products are curable at room temperature. It is also common to combine with PFA chains to impart oil and soil repellency, as well as to enhance water repellency performance.

Phobotex products by Huntsman are included in the ZDHC Manufacturing Restricted Substance List. The Phobotex range is also reported to entail a clever stain release effect after washing, together with quick drying properties, wicking away perspiration and enhancing the comfort level for the wearer (Huntsman, 2015). The WACKER HC 303 silicone emulsion by Wacker Chemie AG is claimed to be a breakthrough in water-based impregnation systems. Until this development, it was by no means certain that good impregnation results could be achieved with water-based silicone emulsions. The reason for this is that the stabilizing emulsifiers form water-permeable microchannels in the impregnation film, which severely compromise the impregnation effect (Wacker, 2014). This finely dispersed water-based silicone emulsion is based on aminofunctional PDMS particles typically 25 nm in diameter. The repellent finish can be applied by impregnation or by spraying, followed by drying and curing. During the course of drying, the colourless, low viscosity, oil in water micro emulsion forms a water repellent silicone film which remains unaffected by emulsifiers or other inclusions. When used in textiles or leather, the emulsion envelops individual fibres in a thin and uniform layer without sticking them together. It enables the impregnated surface to breathe and stay dry, while water and other aqueous liquids immediately bead off. Repellent finishes are usually composed of a number of active substances and the exact nature of the formulations are not usually disclosed. For example, Wacker HC 303 contains; Poly [3-((2- aminoethyl)amino) propyl] methl (dimethyl) siloxane, hydroxy-terminated; Diethyleneglycol monobutylether; Ethylene glycol monohexyl ether and Octamethylcyclotetrasiloxane in addition to the main ingredient PDMS (TEX-SHIELD Project, 2012).

DWR-7000 Soft Hydro Guard is another patent-pending silicone introduced by Dow Corning in 2012. It was claimed to be an eco-friendly solution offering softer hand, comfort and performance benefits with durable water repellency and waterborne stain repellency. This concept is challenging conventional fluorocarbon-based technologies in the area of DWR textile finishes, especially those applied to cellulosic and blended fabrics (Dow Corning, 2012). It has been confirmed through testing to meet or exceed the durability of conventional fluorocarbon finishes, withstanding

Table 7.3 **Specific properties of commercially available silicone based water repellents**

Product name	Chemical properties	Application	Performance
Manufacturer: Dow Corning (United States) (Dow Corning, 2012; TEX-SHIELD Project, 2012)			
DWR-7000 Soft Hydro Guard finish	Silicone-based emulsion, no solvent or APEOs	Padding	Durable water repellency (approx. 30 washes), waterborne stain repellency and much softer hand
Dow Corning 8060	Water-based effective catalyst and crosslinker		Enhances water repellent performance and durability
Dow Corning C2-0563 XIAMETER FBL-0563	Silicone-based emulsion, improved dilution stability, with wide range of diluents to use (aliphatic, aromatic and chlorinated solvents) air-drying, air-curing silicone	Padding	Resistant to waterborne stains, shed water easily and dry rapidly; advantageous for wash- and wear clothing, sports clothing and children's clothing
Manufacturer: Momentive (United States) (Momentive, Inc., 2011)			
Magnasoft NFR-A	Two-component specialty silicone emulsion, Contains no PFOS, PFOA and APEO, weakly cationic	Water dispersible; easy-to-use in any type of finishing process	Excellent water repellency, durable effect on cotton fabrics even after multiple home laundry cycles, superior smoothness and softness on all types of fabric
Magnasoft NFR-B	Catalyst for Magnasoft NFR-A	Water-dispersible	Enhances durability of water repellency and softness
RPS-116	Emulsion based on a curable silicone fluid and it cures at room temperature by forming a film with excellent water repellency. It can be diluted to any level with an appropriate solvent to control water repellency.	Curable silicone fluid, ideal for aerosol products	Usages for textiles, fibre and thread lubricants, nonwoven textile treatments

Continued

Table 7.3 **Continued**

Product name	Chemical properties	Application	Performance
Manufacturer: Huntsman (United States) (Huntsman Textile Effects, 2015)			
Phobotex SSR	Dispersion containing hydrophilic polysiloxane and hydrophilic polyester	Padding	Very high durability to washing, particularly on polyester and aramid fibres
Phobotex Catalyst BC	Emulsion of polydimethylsiloxane catalyst, a self-crosslinking condensation product, Cationic	Padding only	Improves stability to weathering, including UV light
Phobotex WS Conc	Polysiloxane-based emulsion and contains cocoalkyl bis (2-hydroxyethyl) ethoxylat-ed methyl ammonium chloride	Padding only	Good durability to laundering, improved fabric resilience and applicable to pile, velvet fabrics and casual outerwear
Manufacturer: Bluestar Silicones (China) (Bluestar Silicones, 2008; Bluestar Silicones, n.d.)			
Rhodorsil TCS 7001 Bluesil TCS 7001	Alkyl polysiloxane solution in solvent (aliphatic hydrocarbon) phase	Spray or impregnation	High level of wash resistance, excellent abrasion resistance in humid environments and long-lasting waterproof effect
Bluesil EMUL 2400 Bluesil EMUL 2600	Nonionic oil in water silicone emulsion of a methyl-hydrogen polysiloxane. BLUESIL EMUL 2600 is a siloxane macro emulsion	Padding or spraying	Good water repellency with good tear strength and abrasion resistance, suitable for synthetic or blended woven, knitted, nonwoven textile fabrics
Manufacturer: Protex (S. Korea) (Protex International, n.d.; MIDWOR Project, 2016)			
DRYOLS	Silicone-based white emulsion, excellent emulsion stability and nonionic	Padding and impregnation	Good softening effect to the cotton and its blends, nonyellowing, especially good for nylon/cotton and nylon/ polyester blends

Table 7.3 **Continued**

Product name	Chemical properties	Application	Performance
DRYOL S 600	Silicone-based white emulsion, excellent emulsion stability and nonionic	Padding and impregnation	Excellent water repellency and good durability
Manufacturer: Resil Chemicals (India) (Resil, n.d.)			
Ultrafab WR A	Two-component silicone-based emulsion, free of PFOS, PFOA and APEO	Water dispersible; any type of finishing process possible	Excellent repellency with softness on all kinds of substrate and high degree of smoothness
Ultrafab WR B	Catalyst for Ultrafab WR A	Water-dispersible	Enhances durability
Manufacturer: Rudolf GmbH (Germany) (Rudolf, n.d.)			
RUCO 1410	Solvent-based silicone (polysiloxanes), free of reactive amino-silicones and metal-organic catalysts, nonionic	Suitable for aerosol spray can applications	Resistant to aqueous soil, for all fibre types and leather
RUCO 1420	Solvent-based silicone (polysiloxanes), free of reactive amino-silicones and metal-organic catalysts	Dry cleaning	Confers a soft fabric handle, excellent abrasion resistance, for all fibre types and leather
RUCO 1450	Solvent-based silicone (polysiloxanes), Dilutable with hydrocarbons and organic polar solvents, nonionic	Dry cleaning	Full performance after drying at room temperature, soft fabric handle and excellent abrasion resistance, for all fibre types and leather
RUCO 1460	Aqueous emulsion of modified polysiloxanes, cationic	Padding or exhaust process	No curing, for all fibre types and leather
Manufacturer: Shin-Etsu Silicone (Japan) (ShinEtsu, 2016)			
KS-7002	One-component solvent (isopropyl alcohol) based silicone emulsion	Padding	Efficient at room temperature, water repellent with a soft and silky wet texture. It is also applicable to leathers

Continued

Table 7.3 Continued

Product name	Chemical properties	Application	Performance
POLON-MF-33A	High-viscosity solvent (toluene) based silicone emulsion with good stability	Padding	Improved weatherability, durability and softness, reported to have good compatibility with phenolic resins
Manufacturer: Texchem (United Kingdom) (Texchem, n.d.)			
Texfin WRSi	Aqueous emulsion of modified silicone compounds	Padding	Suitable for all the major fibres, provides a durable hydrophobic barrier, exceptionally soft handle and very low friction surface, good performance for outdoor textiles
Texfin-HTF and Texfin-SWR-A	Aqueous emulsion of modified silicone compounds	Padding	Suitable for all the major fibres, Good performance for outdoor textiles
Manufacturer: Ultrakim (Turkey) (Ultrakim, n.d.)			
NEVOSIL TWR 45	Methyl hydrogen polysiloxane-based emulsion	When used with catalyst C 7125, it enables all types of textile materials to be waterproof	Water repellent or waterproof properties with no adverse effects on the breathing ability of fabrics. Resistance to washing (40°C) and dry cleaning together with soft and slippery touch
Manufacturer: Wacker Chemie AG (Germany) (Wacker Chemie, n.d.; Wacker, 2014)			
WACKER FINISH WS 60 E	Aqueous macro emulsion containing hydrogen siloxane, active substance organic tin compound, nonionic	Impregnation	Good water repellency with better tear strength and abrasion resistance
Wacker HC 303	Finely dispersed water based silicone emulsion, contains aminofunctional PDMS particles typically 25 nm in diameter	Impregnation or spraying	Good water repellency with better tear strength and abrasion resistance

30 and more home laundry cycles. It is free of fluorine, solvents and APEOs, and is reported to exhibit no adverse toxicological effects or evidence of bioaccumulation, and also to have no effects on soil living species or on crop growth (Dow Corning, 2012). The performance comparison of this new product with fluorochemical C6 type water repellent is given in Fig. 7.10. The durability of this product is almost similar to that of the fluorochemical C6, while the softness appears to be much better. The company suggests the use of Dow Corning 8060 water-based effective catalyst and crosslinker with DWR-7000 Soft Hydro Guard systems to maximize and enhance water repellent performance and durability, particularly when used for finishing of cellulosic textiles and blends (TEX-SHIELD Project, 2012).

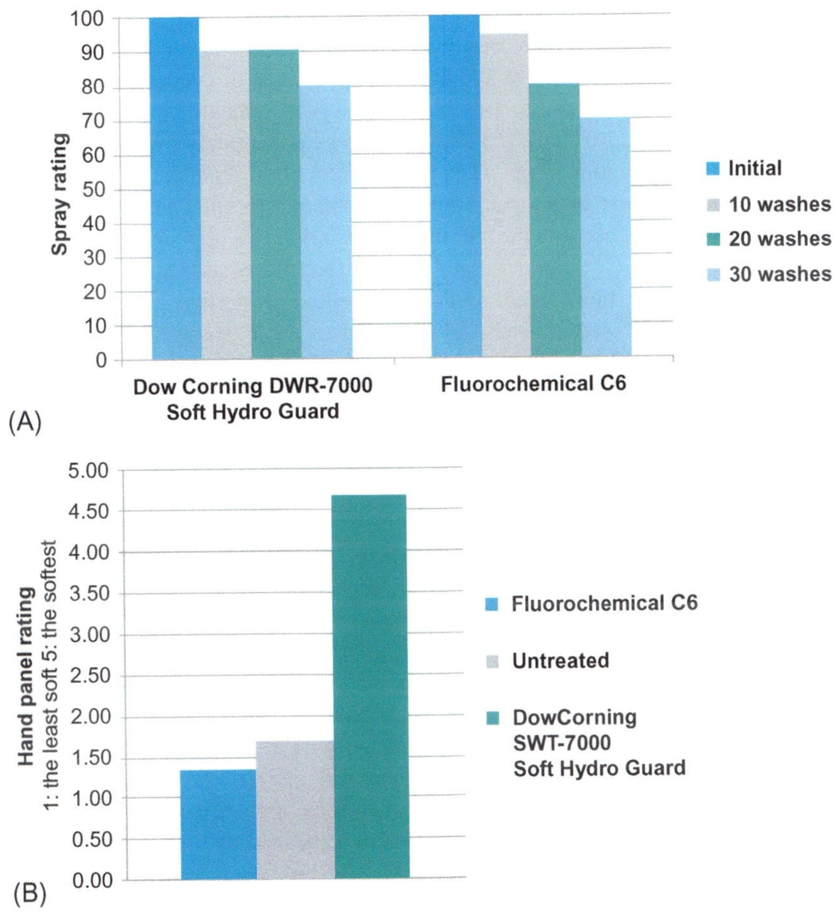

Fig. 7.10 Comparative water repellency and hand rating of khaki twill treated with Dow Corning DWR-7000 silicon-based water repellent.
Data from Dow Corning, 2012. Durable Water Repellency Plus Soft Hand. Text. World, April 17.

Inadequacy of conventional silicone DWRs for delivering oil repellency can be improved with chemical modification. DWRs based on PDMS can additionally be modified with per- or polyfluoroalkyl side chains (Holmquist et al., 2016). These modified types are called hybrid systems have similar properties to PDMS based DWRs, but the addition of per- and polyfluoroalkyl chains can result in an additional oil and soil repellency of the finished fabric.

UNIDYNE of Daikin is an example of this generation of DWOR (Daikin, n.d.), based on a fluoro-silicone hybrid technology, developed in cooperation with Dow Corning Corporation, with a good soft touch of the treated textiles. The broader functionality of Unidyne is said to allow use in a wider array of end-use applications and is used for various textiles, in particular, for apparel like casual shirts, jeans, rainwear and skiwear. The new product improves stain release during laundering. Thanks to the presence of fluorine within a hybrid structure, UNIDYNE repels oil as well as water. Silicones, by contrast, are only effective at repelling water.

Fluorosilicone hybrid coatings combine fluorine and silicone chemistry to create a safe, durable material that exhibits the best characteristics of each technology. Silicones, with their natural polymer flexibility and UV resistance, and fluorine, with durable and oleophobic surfaces, are combined to produce many different polymers and resins for coating applications. Despite excellent stain repellency, conventional fluorochemicals are quite harsh to the touch, in contrast to silicone. Hence, amino-modified silicones have become the natural choice to obtain a high degree of softness. However, when both are combined together, it is found that softness is nevertheless obtained at the cost of stain release and stain-repellent properties. Organo-modified silicones are therefore recommended in cases where both properties are required. These are silicone copolymers containing hydrophilic groups, such as polyalkylene oxide polymers, and some additional amine, amide or epoxide groups, which offer reactivity, so that durable softness is imparted (Vazquez, 2011; Paul, 2015).

Admicellar polymerization of two silicon compounds, methacryloxy methyltrimethylsilane (MSi) and methacryloxy propylpentamethyldisiloxane (MDSi) for obtaining water repellent cotton, has been investigated using two silicone surfactants, a nonionic and a cationic (Tragoonwichian et al., 2011). Adsorption of the cationic surfactant on cotton was found to be higher than that of the nonionic surfactant. The results of characterization analyses showed that a thin polymer coating was successfully formed. The treated fabrics had good water repellency as determined by contact angle measurement, and the MSi-treated cotton showed better water repellency than the MDSi-treated cotton in terms of wetting time.

Silicone polymers have dual characteristics, because of which they can either be used as emulsifiers or act as the continuous/dispersed phase of the emulsion. Various types of silicone emulsions and their applications are discussed in a review (Somasundaran et al., 2006). A silicone emulsion stabilized by hydrocarbon emulsifiers shows drastic changes in their electrokinetic and optical properties under external perturbations, for example, pH change. An analytic study by atomic force microscopy (AFM) illustrated that a coating of silicone emulsion causes a solid substrate-like fabric to become smooth. Another study also confirmed that the silicone-treated fabric tends to be smoother and more relaxed as compared to untreated fabric (Purohit et al., 2006). Surface and bulk characteristics of fabrics such as softness (surface or inner), bounciness, dry feel, wet feel, hydrophobicity, hydrophilicity and several other fabric properties can be enhanced significantly by modifying silicones for desired transport and interfacial properties.

Mohamed et al. (2013) showed that Confocal Raman microscopy (CRM) has been successfully used to monitor the distribution of different molecular weights of linear PDMS as a model of molecules interpenetrate into cotton fibres, as a function of depth in fibres treated in aqueous or hexane medium. The cotton fibres were treated with linear PDMS compound (0.5%, w/w) with various molecular weights (1.5, 9.6 and 69 kDa) in two different media, namely aqueous solution (oil/water emulsion) and hexane, over 30 minutes at different temperatures (25, 50 and 60°C) and drying at 25°C. Scanning electron microscopy images of the treated fibres at various temperatures are given in Fig. 7.11. The improvement on the smoothness of the fibre surface is visible in these images. The CRM results which were also compared with the FT-IR analysis, proved the deposition and the infiltration of PDMS compound with low molecular weight on the surface and in the bulk of cotton fibres in aqueous solution at 50°C, while hexane media support the infiltration only at 60°C. It was reported that treatment at room temperature had no effect in either infiltration or deposition of PDMS molecules in either treatment medium. The use of aqueous medium appears to provide a good infiltration of low and medium molecular weight PDMS into cotton fibres at 50°C and 60°C reaction temperature. PDMS compound with three different molecular weights can infiltrate into cotton fibres by chosen the suitable treatment medium.

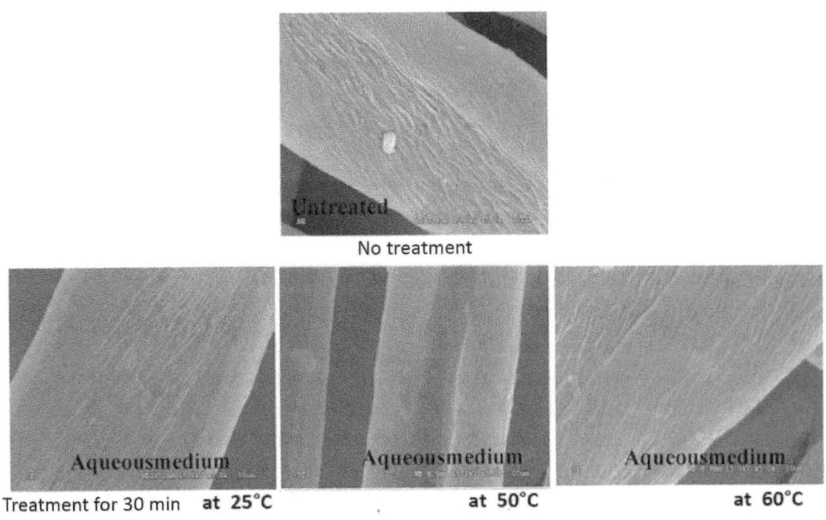

Fig. 7.11 SEM micrographs of untreated and PDMS treated cotton fibre in aqueous media at different temperature.

Reproduced with permission from Mohamed, A.L., Er-Rafik, M., Moller, M., 2013. Suitability of confocal Raman microscopy for monitoring the penetration of PDMS compounds into cotton fibres. Carbohydr. Polym. 96, 305–313.

The interactions of functionalized silicones with various fabric substrates has been studied by Jahagirdar and Tiwari (2004a), where modified amino silicones (PDMS-based amino silicone) were emulsified using nonionic surfactants (lauryl and nonylphenol ethoxylates) to form oil-in-water emulsions. The emulsion acted as a delivery agent for the silicone droplets, which imparted the desired physical properties to the fabric substrates. In order to identify the reason for the improvement in the fabric properties, the fibre structure was monitored using atomic force microscopy. It was shown that the treated fibres were far smoother, more relaxed, and more uniform as compared to the untreated fibres. Thus, the morphology of the fabric was modified in a specific way by treatment with specialty silicones.

The effect of dichlorodimethylsilane (DCDMS) on plasma-treated cotton fabric has also been observed by Jahagirdar and Tiwari (2004b). The physical properties of cotton fabrics, treated with DCDMS in the presence and absence of air plasma, proved that cotton fabric can be made water repellent by treating it with DCDMS solution. On the other hand, the fabric loses its original strength after treatment. Prior exposure of cotton fabric to plasma before treatment with DCDMS solution modifies the fabric, leading to deposition of more and more silane groups, making it water repellent as compared to control cotton fabric.

Silicone coatings were used by Di et al. (2001) to study the variation of the surface energy of coated fabrics in various experimental conditions, with different thicknesses of coating. The method adopted was liquid drop analysis on the coated fabrics, and the contact angle between a liquid drop and the fabric surface was measured, with the surface energy calculated by means of the Owens method. Kinetic measurement was adopted in this study, and it was found that the contact angle of liquids on the fabric coated with silicone decreased over time. A compound solution, DX, has been found which makes it so that the contact angle of liquids on DX-washed fabric becomes constant, and the surface energy of the fabric can be reduced to as low as 15 mN/m (15 mJ/m^2).

Silicones are also used in most leather applications, namely footwear, home and automotive upholstery, garments and other leather goods. The feel, lubricating effect, low surface tension, hydrophobicity and durability of silicones make them of particular benefit to the leather industry (Mallen and Godefroid, 2004).

Since silicone compounds are predominantly used in aqueous textile processes, they are likely to enter the environment via waste water, either from scouring operations or through softener baths. Silicone compounds have a long history of safe use in the textiles industry. Early studies (Chandra, 1995; Rajadhyaksha and Keskar, 2002) report that they were found to make up a minor component of the wastewater stream, becoming attached to particulate matter in wastewater treatment sites. Tests indicated no adverse effects from PDMS on either the sludge treatment processes or the microorganisms necessary for effective sewage degradation. An added benefit of these essentially nontoxic products is their ability to reduce unwanted foam. Methyl silicones have undergone intensive eco-toxicological testing and have been found to be inert without significant potential for negative environmental impact. Silicones have been extensively tested by Rajadhyaksha and Keskar (2002) in both fresh and marine water, and it was reported to show no harmful effect on the environment

Thus, it was claimed that the use of silicones in the textile industry will not cause any environmental harm.

These findings are confirmed by Manickam (2009). Nonvolatile PDMS fluid, being insoluble in water, become a minor component of the sludge in the treatment plant. When the sludge is incinerated, the silicone content converts to amorphous silica, water and carbon dioxide. Silicone materials are highly resistant to biodegradation by microorganisms, but they undergo very effective degradation via natural chemical processes such as catalysed hydrolysis and oxidation. PDMS breaks down into siloxanols and silanols. PDMS is ecologically inert and has been found to have no effect on aerobic or anaerobic bacteria. It does not inhibit the biological process by which waste water is treated.

The silicones most used in textile impregnation agents are based on PDMSs. These siloxanes are registered in REACH, they are inert, and they generally have no adverse effects from the health point of view (Lassen et al., 2015; Davies, 2014). To what extent cyclic and small linear siloxanes may be present at trace levels in the agents is unclear yet. Specifically, D4 is suspected of damaging fertility, and D5 is a potential carcinogen. From the environmental point of view, not enough data for a detailed evaluation is available yet. Which siloxanes may be present at trace levels in the agents is unclear. Some cyclic siloxanes are PBTs. Low molecular weight PDMS have low toxicity and are not considered PBT substances.

A general comparison of common water repellents is shown in Fig. 7.12. It is shown that silicone, wax and polyurethane repellents behave in similar manners, while

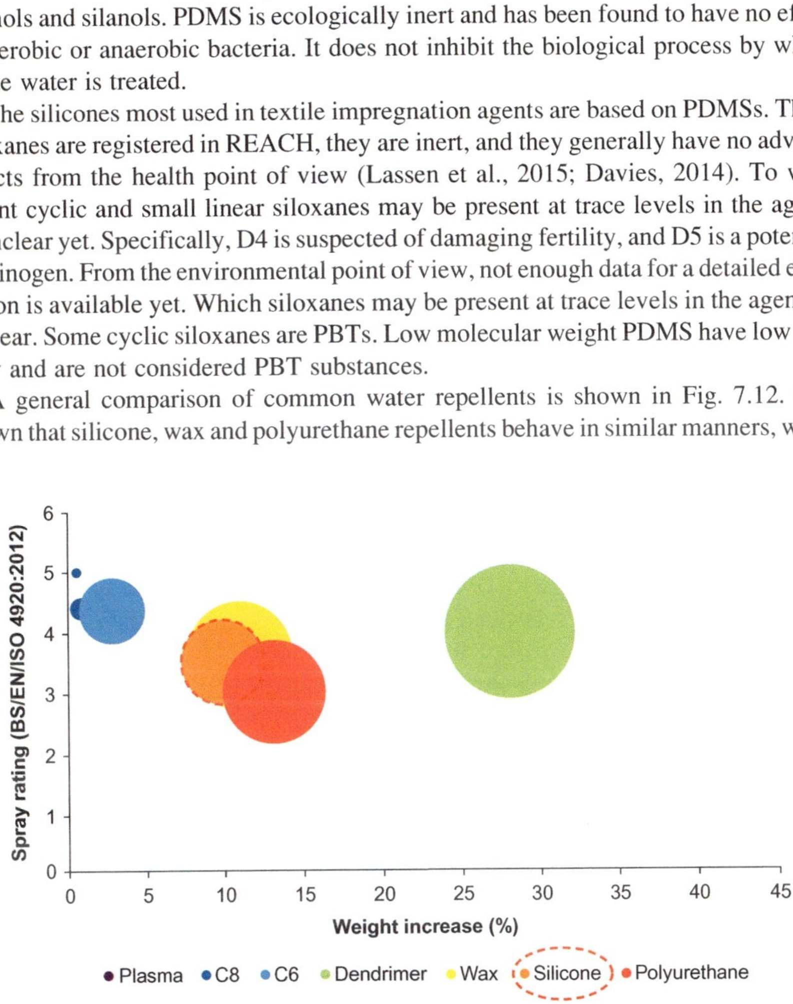

Fig. 7.12 Comparison of water repellency performances of common chemicals (chemicals applied to new fabrics).
Adapted by Davies, A., 2014. DWR Project—Study Phase I, an Evaluation of the Test Methods Used for Assessing Durable Water Repellent Fabrics Within the Outdoor Industry. Available from: http:/www.europeanoutdoorgroup.com/files/DWR-Study_Alice_Davies__digital_.pdf

C8 and C6 groups have the best ratings. This proves that silicon compounds do really have potential as water repellents.

7.8 Conclusions

In the last decade, the demand for and awareness of fluorocarbon-free water repellents have been increasing. Thanks to the positive outcomes of the ZDHC movement, this demand is backed by various apparel brands and some responsible manufacturers and consumers. Consumer demand has driven the need for brands to find alternatives and spurred the rapid development of PFC-free technologies by suppliers. Progress towards technologies free from hazardous PFCs is not yet complete. The search continues for a new technology or substances which would provide both a sustainable environmental solution and durable water repellency. This is a good example of how consumer pressure encourages responsive brands and suppliers to develop and implement better solutions. Other industries that use PFCs, as well as those using other hazardous chemicals, could learn from this example and take on such a challenge to transform their own sector.

The available sources and commercial products show that DWR alternatives can be divided into four broad groups reflecting their basic chemistry: side-chain fluorinated polymers, silicones, hydrocarbons representing hydrophobic moieties and other chemistries based hydrophobic moieties that are bound to organic or inorganic particles, including dendrimer and inorganic nanoparticles. The current studies show that there are large differences in performance between the alternative DWRs, most importantly the lack of oil repellency in nonfluorinated alternatives. The fact that the impurities and/or degradation products of the DWR chemistries are diffusively emitted to the environment is another important concern for all alternatives. According to existing studies, hydrocarbon-based DWRs are the most environmentally favourable, followed by silicones and side-chain fluorinated, polymer-based DWR chemistries. Industrial commitments to reduce the levels of impurities in silicone-based and side-chain fluorinated polymer-based DWR formulations will lower the actual risks.

Silicone-based water repellents had been in the market long before the PFC-based products. Their moderate repellency effect discarding oil and soil repellency, has led the discovery of PFC-based water repellents. None of the alternatives, including silicone compounds, can challenge the overall durability and oleophobicity of PFC-based products. Despite many advantages from durability, cost, health and environmental points of view, silicone-based products' level of durability and poor repellency to oil and stain are major concerns remaining to be tackled.

Currently, specialty organo-modified silicones are used in stain release and stain repellency textile finishing, imparting various degrees of softness without significantly degrading the properties provided by fluorochemicals. These materials are generally silicone copolymers or terpolymers and have chemical structures in which hydrophilic groups, such as polyalkylene oxide polymers, are arranged in different ways onto the main siloxane backbone. These organo-modified silicone polymers may also contain additional reactive organic groups, such as amines, amides and epoxides, which normally contribute to the softness and/or durability.

In addition to modified versions of the above, new silicone chemistries are also being introduced which offer improved softness coupled with the benefit of not impacting stain repellency and/or stain release properties. More interestingly, some of these silicones have even shown synergistic effects. While manufacturers around the world continue to seek environmentally friendly alternatives for DWOR chemicals in the textile industry, the inherent eco-safety of silicones, together with low usage levels, indicate the likelihood of their sustained presence in textile manufacturing. Silicones may not yet be the first choice for the top performers, but they are certainly an unmatchable alternative for applications where moderate durability is sufficient. Silicone manufacturers will continue their search and work towards offering novel ways to incorporate silicone's properties, such as low surface tension, nonionic/nonpolar nature, unique solubility, hydrophobicity, thermal stability, oxidation resistance, shear stability, low flammability, relative inertness, nontoxicity and environmental safety.

References

3 M, 2013. MSDS Scotchgard TM Outdoor Water Shield; Scotchgard TM (Cat. No. 5020, 5019) (US).

ACC Silicone, n.d. Silicone in Textiles. Available from: https://www.acc-silicones.com/applications/textiles.ashx (accessed August 2016).

Arunyadej, S., Mitchell, R., Walton, J., Carr, C.M., 1998. An investigation into the effect of laundering on the repellency behaviour of a fluorochemical-treated cotton fabric. J. Text. Inst. 89, 696–702.

Berzelius, J., 1824. Zersetzung der ussspatsauren Kieselerde durch Kalium, Ann. Phys. Chem. 2 (1), 169–208.

Beyler, C., Hirschler, M.M., 2002. Thermal decomposition of polymers. In: SFPE Handbook of Fire Protection Engineering 2, third ed., National Fire Protection Association, Quincy, MA, pp. 111–131 (Section 1, Chapter 7).

Bluestar Silicones, 2008. Technical Data Sheet SIL 08 Silicones for Technical Fabrics and Leathers 232 3—May 2008. Available from: http://www.siliconiitalia.it/public/schede/486/TCS-7001-INGLESE.pdf (accessed November 2016).

Bluestar Silicones, n.d. Available from: https://www.bluestarsilicones.com/EN/our_offer/Product/90054184/_/BLUESIL-EMULSION-2400 (accessed November 2016).

Brooke, D.N., Crookes, M.J., Gray, D., Robertson, S., 2009a. Environmental Risk Assessment Report: Decamethylcyclopentasiloxane. Environment Agency, Bristol. Available from: http://www.cyclosiloxanes.org/uploads/Modules/Links/16.-uk-ea-environmental-risk-assessment-report-dodecamethylcyclohexasiloxane.pdf.

Brooke, D.N., Crookes, M.J., Gray, D., Robertson, S., 2009b. Environmental Risk Assessment Report: Octamethylcyclotetrasiloxane. Environment Agency, Bristol, UK.

Burman, G., 2014. DWR Project "Expectations of Durable Water Repellent Fabric Finishes", Durable Water Repellency—Study Phase I. Available from: http://www.europeanoutdoorgroup.com/files/DWR-Study_Georgina_Burman__digital_.pdf.

Casper, P., Kelemen, M.C., 2014. Renewable textile finishing in the context of textile reprocessing. In: 8 Aachen-Dresden International Textile Conference (Dresden) November 2014.

Chandra, G., 1995. Review of the environmental fate and effects of silicone materials in textile applications. Text. Chem. Color. 27 (4), 21–24.

Colas, A., 2007. Silicones: preparation, properties and performance. In: de Jaeger, R., Gleria, M. (Eds.), "Silicones in Industrial Applications" of Inorganic Polymers. Nova Science Publishers, Hauppauge, NY. Available from: https://www.dowcorning.com/content/publishedlit/silicones-in-industrial-applications.pdf.

Colas, A., Curtis, J., 2005. Silicone Biomaterials: History and Chemistry & Medical Applications of Silicones. Dow Corning Corporation (Reprinted from Chapter 2.3, Biomaterials Science: An Introduction to Materials in Medicine, second ed. Elsevier Academic Press).

Daikin, n.d. Available from: https://www.daikin.com/chm/pro/kasei/unidyne_multi/feature/multi.html (accessed January 2017).

Davies, A., 2014. DWR Project—Study Phase I, an Evaluation of the Test Methods Used for Assessing Durable Water Repellent Fabrics Within the Outdoor Industry. Available from: http://www.europeanoutdoorgroup.com/files/DWR-Study_Alice_Davies__digital_.pdf.

Di, J., Perwuelz, A., Gueguen, V., 2001. The surface energy of PET fabrics coated with silicone. J. Text. Inst. 92 (2), 184–192.

Dow Corning, 2012. Durable water repellency plus soft hand. Text. World. April 17.

Dow Corning, n.d. How Silicone Water Repellents Work. Available from: http://www.dowcorning.com/content/discover/discoverchem/si-repellents.aspx (accessed May 2016).

ECETOC, 2011. Linear Polydimethylsiloxanes Cas no. 63148-62-9. Second Edition. JACC No: 55, Brussels, Available from: http://www.ecetoc.org/wp-content/uploads/2014/08/JACC-055-Linear-Polydimethylsiloxanes-CAS-No.-63148-62-9-Second-Edition.pdf (accessed January 2017).

Friedel, C., Crafts, J.M., 1865. Ueber das Siliciummethyl und die Kieselsäure-Methylathe. Liebigs Ann. Chem. 136, 203–212.

GE Bayer Silicones, 2003. Silicones for Textiles. Available from: www.korsil.ru/content/files/catalog1/peno5.pdf (accessed May 2016).

Global Silicone Council, 2016. Socio-Economic Evaluation of the Global Silicone Industry. March, Amec Foster Wheeler Env. & Inf. UK Ltd, London, UK.

Graiver, D., Farminer, K.W., Narayan, R., 2003. Review of the fate and effects of silicones in the environment. J. Polym. Environ. 11 (4), 129–136.

Graves, C., 2016. D6.4 Technical Publication "TEX-SHIELD Environmental friendly and Durable Oil and Water Repellence Finish on Technical Textiles" Collaborative Projects, SME-2012-2. Available from: http://www.texshield-project.eu/publications/Deliverable%20report%206.4%20Technical%20publication.pdf (accessed January 2017).

Green Peace, 2011. The Detox Campaign. Available from: http://www.greenpeace.org/international/en/campaigns/detox/water/detox/intro/ (accessed August 2016).

Green Peace, 2015. Footprints in the snow: Hazardous PFCs in remote locations around the globe, a report. Available from: http://detox-outdoor.org/assets/uploads/Report%20RAE/RAE_report_08_2015_english_final.pdf (accessed August 2016).

Green Peace, 2016. Leaving traces: the hidden hazardous chemicals in outdoor gear, Greenpeace product test 2016. Available from: http://www.greenpeace.org/international/Global/international/publications/detox/2016/Leaving-Traces.pdf (accessed August 2016).

Greve, K., Nielsen, E., Ladefoged, O., 2008. Toxic effects of siloxanes: group evaluation of D3, D4, D5, D6 and HMDS in order to set a health-based quality criterion in ambient air. Toxicol. Lett. 180, S67.

Hill, R.M., 2006. Silicone surfactants. In: Farn, R.J. (Ed.), Chemistry and Technology of Surfactants. Blackwell Publishing Ltd, Oxford (Chapter 6.3).

Holmquist, H., Schellenberger, S., Van der Veen, I., Peters, G.M., Leonards, P.E.G., Cousins, I.T., 2016. Properties, performance and associated hazards of state-of-the-art durable water repellent (DWR) chemistry for textile finishing. Environ. Int. 91, 251–264.

Hu, J. (Ed.), 2008. Fabric Testing, first ed. Cambridge, Woodhead Publishing (Elsevier).

Huntsman Textile Effects, 2015. PHOBOTEX® Advanced fluorine-free rain and stain management. Available from: http://www.huntsman.com/textile_effects/Media%20Library/global/files/phobotex_brochure_LR.pdf (accessed January 2017).
Jahagirdar, C.J., Tiwari, L.B., 2004a. Study of plasma polymerization of dichloromethane on cotton and polyester fabrics. J. Appl. Polym. Sci. 94 (5), 2014–2021.
Jahagirdar, C.J., Tiwari, L.B., 2004b. Effect of dichlorodimethylsilane on plasma-treated cotton fabric. Pramana J. Phys. 62, 1099.
Kipping, F.S., 1904. Organic derivative of silicon. Preparation of alkylsilicon chlorides. Proc. Chem. Soc. 20, 15–16.
Kissa, E., 1984. Repellent finishes. In: Lewin, M., Sello, S.B. (Eds.), Handbook of Fiber Science and Technology Volume 2: Chemical Processing of Fibers of Fabrics Functional Finished Part B. Marcel Dekker, New York, NY (Chapter 2).
Kissa, E., 1996. Wetting and wicking. Text. Res. J. 66 (10), 660–668.
Kuo, A.C.M., 1999. Poly(dimethylsiloxane). In: Mark, J.E. (Ed.), Polymer Data Handbook. Oxford University Press, New York, NY.
Lacasse, K., Baumann, W., 2004. Textile Chemicals, Environmental Data and Facts. Springer-Verlag, Berlin.
Lassen, C., Hansen, C.H., Mikkelsen, A.H., Maag, J., 2005. Siloxanes—Consumption, toxicity and alternatives. Environmental Project No. 103, Danish EPA, Copenhagen.
Lassen, C., Jensen, A.A., Warming, M., 2015. Alternatives to Perfluoroalkyl and Polyfluoroalkyl Substances (PFAS) in Textiles. The Danish Environmental Protection Agency, Copenhagen, K, Denmark. ISBN 978-87-93352-16-2.
Lenoble, B., 2007. Silicones in the textile industries. In: de Jaeger, R., Gleria, M. (Eds.), "Silicones in Industrial Applications" of Inorganic Polymers. Nova Science Publishers, Hauppauge, NY, pp. 61–161. (Chapter 2). Available from: www.dowcorning.com/content/publishedlit/Chapter5.pdf.
Leonas, K.K., 1998. Effect of laundering on the barrier properties of reusable surgical gown fabrics. Am. J. Infect. Control 26, 495–501.
Libretex Chemistry, 2013. Silicones 13. Thousands of Useful Products. Available from: http://chem.libretexts.org/Exemplars_and_Case_Studies/ChemCases/Heat_and_Chemical_Resistant_Silicone_Rubber/Silicones_x13._Thousands_of_Useful_Products (accessed November 2016).
Liles, D.T., 2012. The fascinating world of silicone. J. Coating Technol. 9 (5), 58–66. Available from: http://www.dowcorning.com/content/publishedlit/The_Fascinating_World_of_SiliconesPart1.pdf (accessed December 2016).
Mallen, L., Godefroid, L., 2004. The benefits of silicones in leather finishing. World Leather 2004, 31–34.
Manatunga, D.C., De Silva, R.M., De Silva, K.M.N., 2016. Double layer approach to create durable superhydrophobicity oncotton fabric using nano silica and auxiliary non fluorinated materials. Appl. Surf. Sci. 360, 777–788.
Manickam, M.M., 2009. Silicone chemistry for fabric care. Colourage 56 (11), 86.
Mark, J.E., 2004. Some interesting things about polysiloxanes. Acc. Chem. Res. 37 (12), 946–953.
MIDWOR Project, 2016. LIFE14 ENV/ES/000670, MIDWOR Deliverable A2.1 Report on the state of the art technical and environmental data of the selected conventional and alternative DWOR. Available from: http://www.midwor-life.eu/wp-content/uploads/2016/07/Deliverable-A2-1-public_version-PROTECTED.pdf (accessed January 2017).
Mohamed, A.L., Er-Rafik, M., Moller, M., 2013. Suitability of Confocal Raman microscopy for monitoring the penetration of PDMS compounds into cotton fibres. Carbohydr. Polym. 96, 305–313.

Momentive Performance Mat. Inc, 2011. Magnasoft* NFR-A & NFR-B Textile Water Repellent Finish Marketing Report. Available from: https://www.momentive.com/products/show-technical-datasheet.aspx?id=22415 (accessed November 2016).

Mota, J., 2016. MIDWOR-LIFE project LIFE14ENV/ES/000670, Workshop: A review of the work done in Actions A1, A2 and B1. Available from: http://www.pointex.eu/uploads/Workshop_Pointex_Midwor_2016.11.29_Jordi_MOTA.pdf (accessed January 2017).

Newton, D.E., 2010. In: Edgar, K.J. (Ed.), Chemical Elements, second ed. Gale, Cengage Learning, Detroit.

O'Lenick, A. J., Siltech, Jr., 1999. Basic silicone chemistry—a review. Silicone Spectator. Available from: http://www.scientificspectator.com/documents/silicone%20spectator/Silicone_Spectator_January_2009.pdf (accessed August 2016).

Oeko-Tex, 2017. Standard 100 by OEKO-TEX, Appendix 4, Limit values and fastness, part 1, Edition 02.2017.

Owen, M.J., 2005. Why Silicones Behave Funny, Dow Corning Corporation [original article : Chimie Nouvelle, 85, 27 (2004)].

Ozdil, N., Kayseri, G.O., Mengüç, G.S., 2012. Analysis of abrasion characteristics in textiles. In: Adamiak, M. (Ed.), Abrasion Resistance of Materials. In: Vol. 119. Intech Open Science, Rijeka, Croatia (Chapter 7).

Patnode, W.I. 1942. Method of rendering materials water repellent. US Patent 2 036 222 A, Issued December 22.

Paul, R. (Ed.), 2015. Functional Finishes for Textiles: Improving Comfort, Performance and Protection. Woodhead Publishing Limited (in association with The Textile Institute), Elsevier, Cambridge, England.

PRNewswire, 2016. Silicones Market is Expected to Reach $20,784 Million, by 2022" Available from: The Telegraph India. https://www.telegraphindia.com/pressrelease/prnw/15059/silicones-market-is-expected-to-reach-20-784-million-by-2022.html (accessed February 2017).

Protex International, n.d. Available from: http://www.protexkorea.co.kr/textile-finishing-agent/Water-Repellent-agent.asp (accessed January 2017).

Purohit, P., Somasundaran, P., Kulkarni, R., 2006. Study of properties of modified silicones at solid–liquid interface: fabric–silicone interactions. J. Colloid Interface Sci. 298 (2), 987–990.

Rajadhyaksha, P.B., Keskar, V.H., 2002. Silicone based chemicals in textile application. Text. Assoc. 63, 195–197.

Resil Ultrafab TDS, n.d. Available from: http://resil.com/functional-finishes/silicone-water-repellents/ (accessed November 2016).

Rochow, E.G., 1947. An Introduction to the Chemistry of the Silicones, Second Printing John Wiley & Sons, New York, NY.

Rudolf Chemie, 2015. RUCO 1514: Intelligent self-organisation for perfect water repellent features—fluorine-free. Available from: http://www.rudolf-duraner.com.tr/en/products/product-brochures-archive/details/brochure/ruco-1514-11-2013.html (accessed November 2016).

Rudolf Duraner, n.d. Available from: http://www.rudolfduraner.com.tr/en/products/coproducerb2b/50-silcones.html (accessed November 2016).

Schindler, W.D., Hauser, P.J., 2004. Chemical Finishing of Textiles. CRC Press and Woodhead Publishing Limited (in association with The Textile Institute), Cambridge.

Seyferth, D., 2001. Dimethyldichlorosilane and the direct synthesis of methylchlorosilanes. The key to the silicones industry. Organometallics 20, 4978–4992.

ShinEtsu, 2016. Textile Treatments—2011-2016. Available from: http://www.shinetsusilicone-global.com/catalog/pdf/TextileTreatments_e.pdf (accessed November 2016).

Hill, R.M. (Ed.), 1999. Silicone Surfactants. Marcel Dekker, New York, NY (Chapter 1).
Somasundaran, P., Mehta, S.C., Purohit, P., 2006. Silicone emulsions. Adv. Colloid Interface Sci. 128–130, 103–109.
Texchem, U.K. n.d. Available from: http://www.texchem.co.uk/TexfinHTF.html (accessed November 2016).
TEX-SHIELD Project, 2012. Environmental friendly and Durable Oil and water repellence finish on Technical Textiles. Collaborative Projects, SME-2012-2. Available from: http://www.unitex.be/index.php/texshield.html?file=tl_files/users/TexShield+no2.pdf (accessed November 2016).
The Silicone Conundrum, 2008. A review. Silicone Spectator. Available from: http://www.scientificspectator.com/documents/silicone%20spectator/September_2008.pdf (accessed August 2016).
Thomas, N.R., 2010. Frederic Stanley Kipping—pioneer in silicon chemistry: his life & legacy. Silicon 2, 87–193.
Tomasino, C., 1992. Chemistry & Technology of Fabric Preparation & Finishing. Department of Textile Engineering, Chemistry & Science College of Textiles, North Carolina State University.
Tragoonwichian, S., Kothary, P., Siriviriyanun, A., O'Rear, E.A., Yanumet, N., 2011. Silicon compound coating for preparation of water repellent cotton fabric by admicellar polymerization. Colloids Surf. A: Physicochem. Eng. Aspects 384, 381–387.
Ultrakim Kimya San. ve Tic. AS, n.d., Ultrasil TWR 45 TDS. Available from: http://www.ultrakim.com/info/17032014180025ultrasil_twr_45_tds_tr.pdf (accessed November 2016).
Vazquez, F., 2011. Silicone Softeners For Stain Repellent and Stain Release Fabric Finishing, Dow Corning Corp. Technical Doc. Available from: http://www4.dowcorning.com/content/publishedlit/26-1277-01.pdf (accessed November 2016).
Vidyanagar, V., 2015. Application of Silicone Finishes To Enhance Performance Of Textiles, Ph.D Thesis, Sardar Patel University, India.
Wacker, 2014. Wacker presents silicone impregnant for household and textile applications. Text. India Progr. 40–41.
Wacker Chemie, Available from: https://www.wacker.com/cms/en/products/product/product.jsp?product=9271 (accessed November 2016).
Wacker Report, n.d. Silicone Products for Household Fabric Care. Available from: https://www.wacker.com/cms/media/documents/markets_brands/consumer_care_1/household_science/Household_Science_FabricCare.pdf (accessed January 2017).
Wöhler, F., 1856. Ueber das Silicium. Ann. Chem. Pharm. 97, 266.
ZDHC Programme, 2012. Durable Water and Soil Repellent Chemistry in the Textile Industry—A Research Report, P05 Water Repellency Project, November 2012. Available from: https://outdoorindustry.org/pdf/FINAL_ZDHC_P05_DWR%20Research_Nov2012.pdf (accessed August 2016).

Further reading

Abdullah, I., Blackburn, R.S., Russell, S.J., Taylor, J., 2006. Abrasion phenomena in twill tencel fabric. J. Appl. Polym. Sci. 102, 1391–1398.
Sung, L.P., Vicini, S., Ho, D.L., Hedhli, L., Olmstead, C., Wood, K.A., 2004. Effect of microstructure of fluorinated acrylic coatings on UV degradation testing. Polymer 45, 6639–6646.

Dendritic molecules and their use in water repellency treatments of textile materials

8

Riza Atav
University of Namık Kemal, Tekirdağ, Turkey

8.1 Introduction

Polymer chemistry and technology have traditionally focused on commonly used linear polymers (Klajnert and Bryszewska, 2001). Most synthetic and naturally occurring macromolecules have a simple linear structure. They consist of a covalently linked chain of repeating monomeric building blocks, with an end group at each end of the chain (Froehling, 2001). Linear macromolecules only occasionally contain some smaller or longer branches (Klajnert and Bryszewska, 2001). Over the last 20 years, polymer chemistry has created a series of nonlinear variations on this simple polymer structure. These variations are commonly denoted as 'macromolecular architectures' (Froehling, 2001). The evolution of three major macromolecular architectures, namely: linear (class I), cross-linked (bridged; class II) and branched types (class III). are recognized as traditional synthetic polymers (Esfand and Tomalia, 2001). In all these classes, structures or architectures are produced by largely statistical polymerization processes, rather than exact distribution processes (Doshi, 2011). Increasing scientific efforts have gone into the design and synthesis of dendrimers (Sashiwa and Aiba, 2004). As early as 1979, the first synthetic strategies to produce monodispersed, structure-controlled, dendritic macromolecules in ordinary laboratory glassware were initiated (Esfand and Tomalia, 2001). Dendritic polymers are considered the fourth largest class of macromolecular architecture due to their unique repertoire of new properties. Dendritic architecture consists of the four subsets that are related to degree of structural control and are listed in ascending order: namely random hyperbranched polymers, dendrigraft polymers, dendrons and dendrimers (Akbarı, 2013). Four major classes of macromolecular architecture are shown in Fig. 8.1.

Dendrimer chemistry was developed as a 'spin-off' from conventional polymer chemistry and dendrimers are gaining importance due to their versatility (Burkinshaw et al., 2000). However, it is difficult to place dendritic polymers, especially dendrons and dendrimers, definitely in any branch of chemistry. Sometimes they are defined in terms of supramolecular chemistry or polymer chemistry, although in some cases they are not actually supramolecules or polymers, but are treated as macromolecules (Patel and Patel, 2013).

Waterproof and Water Repellent Textiles and Clothing. https://doi.org/10.1016/B978-0-08-101212-3.00007-1

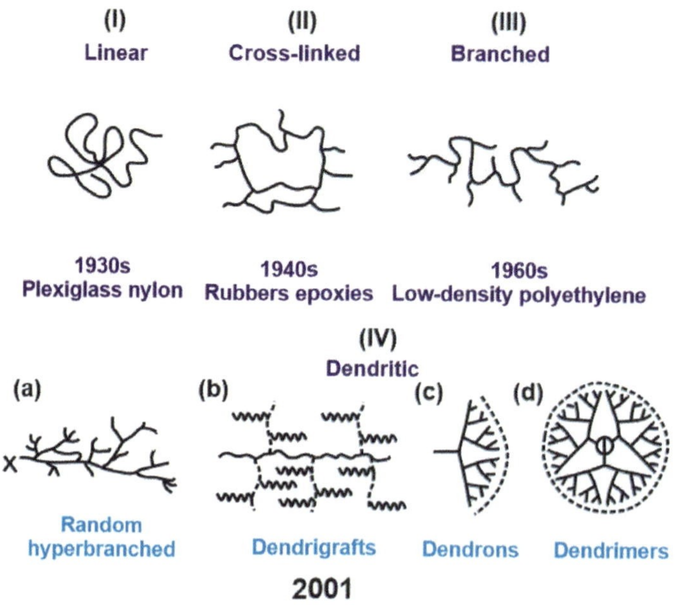

Fig. 8.1 Four major classes of macromolecular architecture. Traditional synthetic polymers: (I) linear, (II) cross-linked (bridged) and (III) branched. Structure controlled polymers (IV) dendritic (Esfand and Tomalia, 2001).

Dendrimers are nanosized, nearly monodisperse macromolecules that contain symmetric branching units built around a small molecule or a linear polymer core (Abbasi et al., 2014; Agrawal and Kulkarni, 2015). They are like tree-shape artificial macromolecules that are extremely branched, globular, multivalent, monodisperse and three-dimensional molecules having a high number of functional groups on their surface (Gupta and Nayak, 2015; Baig et al., 2015; Munir et al., 2016). The structure of these materials has a great impact on their physical and chemical properties (Agrawal and Kulkarni, 2015).

Dendritic architecture is one of the most pervasive topologies observed in nature at dimensional length scales measured in metres (trees) to nanometres (amylopectin) (Akbarı, 2013). They appear in many diverse prototypes including those that can be measured in metres, in circulatory topologies in the human anatomy that are found in mm and cm, or in cerebral neurons in μm (Doshi, 2011). The reasons for such extensive mimicry at all dimensional length scales are not entirely clear. However, it could be speculated that these architectures have evolved to provide maximum interfaces for optimum energy extraction and distribution, nutrient extraction and distribution, or information storage and retrieval (Esfand and Tomalia, 2001). For example, the branching of a tree serves to display its leaves to the outside world to harness the power of the sun. The leaves are held at the surface of the tree to maximize the amount of light harvested. On a smaller scale, nature uses dendritic structures in the body. Neurons (brain cells) have dendritic structures to maximize their connections with each other, thereby forming efficient pathways in the brain. At the molecular level, dendritic structures can be found in natural polysaccharides like glycogen, amylopectin and dextran. Nature uses these branched polysaccharides for energy storage; the presence of many chain ends means large quantities of glucose may be released very quickly when needed (Dykes, 2001).

The versatility of dendrimers as a class of regularly branched macromolecules with unique structural and topological features makes them valuable tools for exploitation by scientists and technologists (Toit et al., 2010). There have been more than 1000 published patents relating to dendrimers since the first dendrimer patent was issued to D.A. Tomalia (Majoros and Baker, 2008). Their treelike structure and a high molecular density of surface functionality makes them very useful building blocks and carrier molecules at the nanometre scale (Sarkar et al., 2005). Due to unique chemical and physical properties, dendrimers and hyperbranched polymers have received much attention over the past two decades (Zhang et al., 2008a,b).

Due to these unique features this class of polymeric nanomaterials have various applications in different fields including drug delivery, dendrimer-based nanomedicine, gene delivery, light harvesting, dendritic nanomaterials, electrode design, solubility enhancers and for various biotech applications (Gupta and Nayak, 2015). In recent years, several excellent reviews (e.g. Klajnert and Bryszewska, 2001; Dykes, 2001; Sadler and Tam, 2002; Gao and Yan, 2004; Atav and Yavaş, 2010; Froehling, 2001; Doshi, 2011; Namırtı and Atav, 2011; Akbarı, 2013; Patel and Patel, 2013; Abbasi et al., 2014; Agrawal and Kulkarni, 2015; Baig et al., 2015; Gupta and Nayak, 2015; Munir et al., 2016) and books (e.g. Newkome et al., 2001; Boas et al., 2006; Vögtle et al., 2009; Tomalia et al., 2012) on dendrimers and hyperbranched polymers have been published. The preparation, properties and end uses of dendrimers have been extensively explained. As Akbarı (2013) previously stated, an area of application that has remained almost unconsidered in scientific discussions is the field of textile engineering. Atav and Yavaş (2010), had given information about applications of dendrimers in textile finishing, and then Akbarı (2013) published the review article in which detailed information about the use of dendrimers in the textile field was presented. However, there is still lack of knowledge in the literature about potential applications of dendrimers in water- and oil-repellency treatments which have become important in several textile application areas in recent years (Schindler and Hauser, 2004). As known, perfluorinated compounds (PFCs) have been widely used as durable water and oil-repellent treatments in outdoor clothing for more than 50 years. However, there are health and environment al concerns around these useful compounds including reproduction toxicity, cytotoxicity and ecotoxicity (Akerblom and Göranzon, 2013). Due to increased attention to the adverse effects of C8 substances on human health and the environment, the application of polymers with polyfluoroalkylated side chains based on short-chain fluorine chemistry (C4–C6 chemistry) has been growing in recent years (Lassen et al., 2015). But there are studies that indicate even the short-chain-length fluorocarbons have toxic effects on the environment and human health, although these substances are considered to be less harmful than ones of longer chain length (Hedegard, 2014). For this reason, there is a need for more environmental and human-health-friendly, hence greener, water repellent treatments (Akerblom and Göranzon, 2013). Within this context, it can be said that the latest approaches in water repellency are based on dendrimer technology. To better understand the role of dendrimers in water repellency, it is believed that providing some knowledge about this technology would be beneficial to readers. Therefore in this review, after giving information about the historical background, classes, structure, synthesis, types and properties of dendrimers, their applications in water repellency treatments of textile material are discussed.

8.2 Historical background of dendritic molecules

The first example of an iterative synthetic procedure towards well-defined branched structures was reported by Vögtle in 1978, who named this procedure a 'cascade synthesis' (Holister et al., 2003). Cascade molecules are accepted to be the first tangible representatives of compounds exhibiting potentially perpetual branching (Vögtle et al., 2009).

In the early 1980s, Denkewalter patented the synthesis of L-lysine-based dendrimers. However, no detailed characteristics of the materials were given (Holister et al., 2003; Baig et al., 2015). On the other hand, the term dendrimer was first proposed by Tomalia in 1985 (Toit et al., 2010). Dendrimer is derived from the Greek word 'dendron' which is used for tree and the Greek suffix 'mer' (segment) which describes the synthetic, three-dimensional molecules having branching parts (Munir et al., 2016). These might also be called cascade molecules, but this term is not as much established as dendrimers (Agrawal and Kulkarni, 2015).

At the same time, Newkome's group independently reported synthesis of similar macromolecules. They called them arborols from the Latin word 'arbor' also meaning a tree. The term cascade molecule is also used, but dendrimer is the best established one (Klajnert and Bryszewska, 2001).

8.3 Classification of dendritic molecules

As the fourth major class of polymer architecture, coming after traditional types which include linear, cross-linked and branched architectures, dendritic polymers may be further subdivided into four subclasses: (a) random hyperbranched polymers; (b) dendrigrafts; (c) dendrons; (d) dendrimers as shown in Fig. 8.1. The order of this subset, from IVa to IVd, reflects the relative degree of structural control present in each of these dendritic architectures (Tomalia, 2005).

A class of branched polymers is hyperbranched polymers. In contrast to dendrimers, hyperbranched polymers are neither defect-free nor perfectly monodisperse. Hyperbranched polymers, however, are typically obtained in a one-pot reaction and as a result can be easily prepared in larger quantities (Scholl et al., 2009).

The other class of dendritic polymers is the dendrigraft systems, introduced simultaneously in 1991 as comb-burst polymers by Tomalia et al. and as arborescent polymers by Gauthier and Möller. Dendrigraft polymers may be viewed as semi-controlled branched polymer architectures intermediate in terms of structure control between dendrimers and hyperbranched polymers (Akbarı, 2013). Dendrigrafts are a class of dendritic polymers like dendrimers that can be constructed with a well-defined molecular structure, i.e. being monodisperse. They are centred on a linear polymer chain, to which branches consisting of copolymer chains are attached. These copolymer chains are further modified with other copolymer chains, giving a hyperbranched motif built up by a finite number of combined polymers (Doshi, 2011). Dendrigraft polymers are generally larger structures than dendrimers that grow faster and amplify surface groups more dramatically as a function of generational development (Tomalia et al., 2012).

Dendrons and dendrimers are the most intensely investigated subsets of dendritic polymers (Tomalia, 2005). Dendrons is the term used to describe a dendritic wedge without a core, and the dendrimer can be prepared from assembling two or more dendrons. Dendrons are useful tools in the synthesis of dendrimers by the segment coupling strategy. These dendrons have been used in the creation of numerous dendrimers having different structures and functions (Doshi, 2011).

Dendrimers are a novel class of regularly branched 3D, man-made molecules produced by an unusual synthetic route incorporating repetitive branching sequences to create a unique novel architecture by an unusual and high level of uniformity of control the size (Dastjerdi and Montazer, 2010). Ideally dendrimers are spherical, defect-free and perfectly monodisperse compounds. Dendrimer synthesis however, is usually based on tedious multistep protocols (Scholl et al., 2009). The multistep synthesis and slow molecular mass growth of dendrimers is often considered to be a major drawback hindering the use of these materials on a large scale (Akbarı, 2013).

8.4 Structure of dendrimers

Dendrimers are built from a starting atom, such as nitrogen, to which carbon and other elements are added by a repeating series of chemical reactions that produce a spherical branching structure. As the process repeats, successive layers are added and the sphere can be expanded to the desired size by the investigator (Agrawal and Kulkarni, 2015). It is possible to control their shape, dimension, molecular size, flexibility, density and solubility by choosing surface functional groups and different building/branching units. Moreover, they can also possess polymers and organic molecules as a part of their structure, and thus acquire special chemical and physical properties (Munir et al., 2016).

Dendrimers possess three distinguished architectural components, namely an initiator core, interior layers (generations) composed of repeating units radically attached to the central core, and exterior (terminal functionality) attached to the outermost interior generations (Doshi, 2011). An initiator core determines the size and shape of the dendrimer (Agrawal and Kulkarni, 2015). As an example, the core may be synthesized from ammonia or ethylenediamine (Toit et al., 2010). Depending on the structure of the core molecule, dendrimers may commonly have 3 or 4 branches (Munir et al., 2016). The building blocks with several interior layers are composed of repeating units (e.g. polyamidoamine (PAMAM), polyaminoacids, polyphenols, polyporphyrins and polyethers), which created a flexible space within the voids for the possible encapsulation of various small guest molecules (Toit et al., 2010). The repeated units or branches covalently attached to the central core are organized in

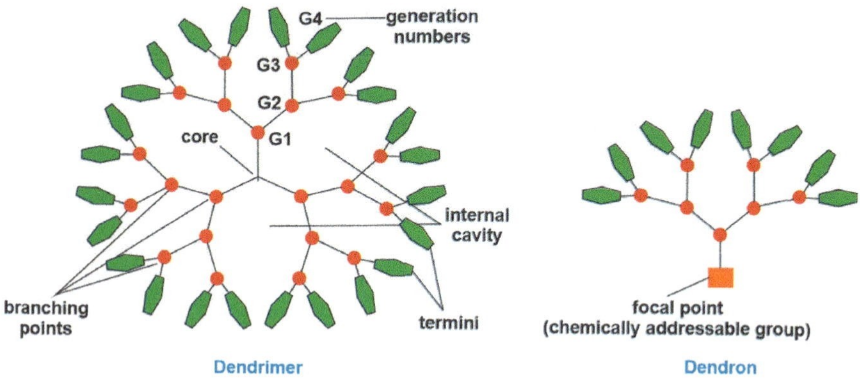

Fig. 8.2 Representation of fourth-generation dendrimers (Agrawal and Kulkarni, 2015).

a series of radially homocentric layers called 'generations' (G) (Akbarı, 2013). The number of focal points when going from the core to the dendrimer surface is the generation number, in other words, a dendrimer with four focal points when going from the centre to the periphery is designated as the fourth generation dendrimer. The core part of the dendrimer is sometimes denoted generation 'zero', or in the terminology presented as 'G0' (Fig. 8.2). The core structure thus presents no focal points, as hydrogen substituents are not considered focal points (Agrawal and Kulkarni, 2015).

Dendrimer's diameter increases with the rise in generation number; similarly, the surface group number also increases geometrically with the rise in generation number as shown in Fig. 8.3 (Munir et al., 2016).

For the first three generations, dendrimers resemble ordinary organic molecules. They have no consistent or specific three-dimensional structure, but instead appear

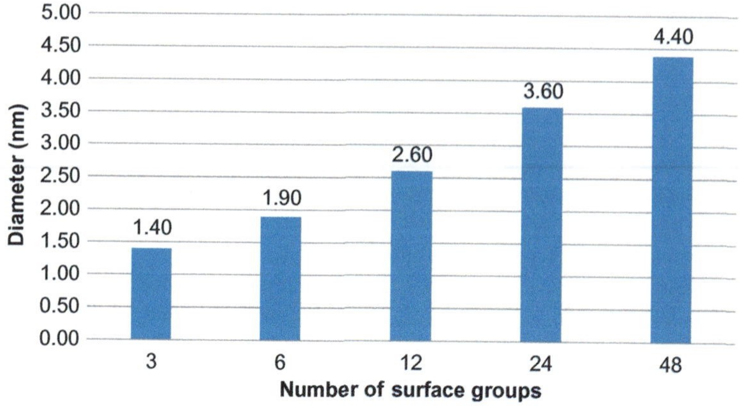

Fig. 8.3 Change of number of surface groups and diameter of dendrimer with the increase in generation number (Munir et al., 2016).

small and floppy-shaped. At generation 4 (G4), they appear to be spherical in shape. After G5 they show consistent and specific three-dimensional structures and become highly spherical in structure (Munir et al., 2016).

End-group is also generally referred to as the 'terminal group' or the 'surface group' of the dendrimer. The term surface group is slightly more inaccurate, in the sense that the dendrimer branches can sometimes fold into the interior of the dendrimer (Doshi, 2011). Reactive end-groups may be used for continuation of dendritic growth or for modification of reactivity of the dendrimer composition. On the other hand, passive groups may be used to modify physical properties of dendrimer molecules and the type of their interactions with each other or with the surrounding environment (Sarkar et al., 2005). Functional groups present at the surface of dendrimers determine their physical properties in aqueous solutions or in solid state. Molecular weight, number of functional groups and dendrimer size are related with the generation of dendrimer and are controlled during the dendrimer synthesis (Munir et al., 2016). The increase of the number of dendrimer end groups is consistent with the geometric progression (Klajnert and Bryszewska, 2001):

$$Z = n_c \cdot n_m^G \tag{8.1}$$

where Z is the number of terminal groups, n_c is the core multiplicity, n_m is the branch-juncture multiplicity and G is the generation number.

As the chains growing from the core molecule become longer and more branched (in 4 and higher generations), dendrimers take on a globular structure. Dendrimers become densely packed as they extend out to the periphery, which forms a closed membrane-like structure. When a critical branched state is reached, dendrimers cannot grow because of a lack of space. This is called the 'starburst effect' (Klajnert and Bryszewska, 2001). By altering the nature of the core and repeating units, the number of layers and the composition of the surface layer, it is possible to synthesize a single polymeric dendrimer of defined three-dimensional structure and size with predictable physicochemical properties, thereby defining the macroscopic properties of the dendrimer (Toit et al., 2010).

8.5 Synthesis of dendrimers

Dendrimers are just in between molecular chemistry and polymer chemistry. They relate to the molecular chemistry world by virtue of their step-by-step controlled synthesis, and they relate to the polymer world because of their repetitive structure made of monomers (Abbasi et al., 2014). However, dendrimer synthesis differs from conventional polymerization reactions, which are either based on repeated addition of monomers to a growing chain with an activated end group (e.g. radical addition polymerization of vinyl monomers) or on the stepwise random combination of monomers and oligomers in polycondensation reactions (Froehling, 2001).

The idea of repetitive growth with branching was first reported by Vögtle. This was followed by independent development of the divergent, macromolecular synthesis of 'true dendrimers' by Tomalia. They were the first who coined the term dendrimer for these macromolecules and described the synthesis of poly(amidoamine) (PAMAM) dendrimer in detailed manner. Then a convergent synthetic method of dendrimer synthesis was introduced by Frechet. Afterwards, there was explosion of scientific interest in dendrimers as a new class of polymeric nanomaterial because of its various cutting edge structural advantages over simple polymers (Gupta and Nayak, 2015).

Initial applications for the compounds were restrained by the lack of reliable and practical methods for dendrimer preparation. The later introduction of so-called convergent and divergent methods of preparation allowed the materials to become available on an industrial scale (Burkinshaw et al., 2000). Divergent growth is based on the stepwise addition of low molecular mass building blocks starting from a multifunctional core molecule and results in a radial growth of the dendrimers. Convergent dendrimer synthesis involves the coupling of preformed dendrons onto a central core molecule (Scholl et al., 2009). In both processes it is possible to distinguish between generations, which are the intramolecular layers of branch cells from the lowest molecular weight derivative (generation 0) to the geometrically progressive higher homologues (Sarkar et al., 2005). The synthesis used for dendrimer preparation permit almost entire control over the critical molecular design parameters such as size, shape, surface/interior chemistry, flexibility and topology (Doshi, 2011). Their step-by-step synthesis also allows for control in the quantity of functional groups present on a dendrimer, depending on its generation (Goddard and Hotchkiss, 2007). General knowledge about the two main dendrimer synthesis methods, namely divergent and convergent methods, and also more recent approaches on dendrimer synthesis, called accelerated approaches, are given below.

Divergent method: In this method (Fig. 8.4A), the growth of dendrimers originates from a core site. The core is reacted with two or more units (in mole) of reagent containing at least two protecting branching sites, followed by removal of the protecting groups, which lead to the first generation dendrimers. This process is repeated until

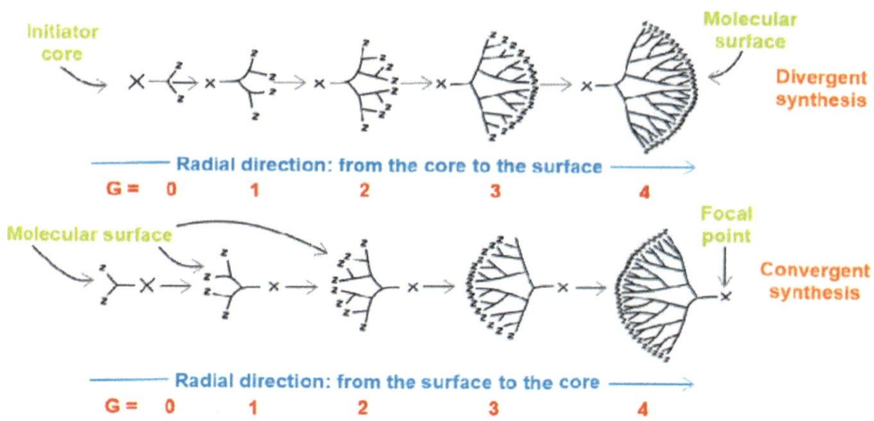

Fig. 8.4 Two principle synthetic methods for constructing dendritic macromolecules (Esfand and Tomalia, 2001).

the dendrimer of the desired size is obtained (Agrawal and Kulkarni, 2015). In each generation, the number of arms doubles from the previous generation (Gupta and Nayak, 2015). Each step of the reaction must be determined to full completion to prevent mistakes in the dendrimer. The major disadvantage of this approach is that the incomplete growth and the side reactions lead to imperfect dendrimers (Baig et al., 2015). To prevent side reactions and force reactions to completion, a large excess of reagent is required (Akbarı, 2013). This divergent route is advantageous to get higher yield of dendrimer with lower purity or purity is compromised for getting higher yield (Gupta and Nayak, 2015). For this reason, this approach is currently the preferred commercial route used by worldwide producers including Dendrimax (Ann Arbor, MI, USA), DSM Fine Chemicals (Geleen, the Netherlands) and The Perstorp Group (Perstorp, Sweden) (Doshi, 2011).

Convergent method: The convergent methods were developed as a response to the weaknesses of divergent synthesis (Akbarı, 2013). This approach of dendrimer synthesis overcomes the purity and structural defect issues of divergent synthesis. By this approach, more uniform and symmetric dendrimers can be synthesized but with lower overall yield. For this reason, it is generally used for laboratory scale dendrimer synthesis (Gupta and Nayak, 2015). In the convergent approach (Fig. 8.4B), dendrimers are constructed stepwise, starting from the end groups and progressing inwards. When the growing branched polymeric arms, called dendrons, are large enough, they are attached to a multifunctional core molecule (Akbarı, 2013). This method makes it much easier to eliminate impurities and shorter branches along the way, so that the final dendrimer is more monodisperse (Baig et al., 2015). In this method, purification is done after each step, whereas in the divergent method, because the reactant and product remain the same, it is difficult to purify using the chromatographic technique (Agrawal and Kulkarni, 2015). Furthermore, this approach has the potential to attract dendrons of different types to one dendrimer (Munir et al., 2016). However, the convergent approach does not allow the formation of high generations because steric problems occur in the reactions of the dendrons and the core molecules (Akbarı, 2013). This size limitation is not seen with the divergent approach of dendrimer synthesis (Gupta and Nayak, 2015).

The divergent and convergent methods are commonly used in synthesis of various dendrimers successfully, but both suffer from the common problems of multistep and laborious procedures. These are time consuming processes, and overall low yield causes macromolecules to become more expensive. For this reason, recently accelerated approaches are being used for the synthesis of dendrimers. These are double exponential and hyper core techniques. The common feature of both is to reduce the number of steps for the synthesis of dendrimers drastically. This greatly accelerates the synthetic process and thereby decreases the work load and ultimately the cost of the polymer (Doshi, 2011).

(a) *Double exponential and mixed growth*: Double exponential growth, similar to a rapid growth technique for linear polymers, involves an AB2 monomer with orthogonal protecting groups for the A and B functionalities (Doshi, 2011). In this approach, two products (monomers for both convergent and divergent growth) are reacted together to give an orthogonally protected trimer, which may be used to repeat the growth process. Strength of double exponential growth is more subtle than the ability to build large dendrimers in relatively few steps (Agrawal and Kulkarni, 2015).

(b) *Hyper core and branched monomers growth*: This method involves the preassembly of oligomeric species, which can be linked together to give dendrimers in fewer steps or higher yields in a radial, branch-upon-branch configuration. The core is reacted with two or more moles of reagent containing at least two protecting branching sites, followed by removal of the protecting groups. The subsequent liberated reactive sites lead to the first generation dendrimers (Agrawal and Kulkarni, 2015). This approach is more like a convergent approach with fewer steps required for the generation of dendrimers (Doshi, 2011).

8.6 Types of dendrimers

Dendrimers can be classified on the basis of their shape, structure, branching, solubility, chirality and attachment (Baig et al., 2015). In this section, dendrimers will be classified according to the chirality.

8.6.1 Achiral dendrimers

8.6.1.1 PAMAM

PAMAM or poly-amidoamine were the first dendrimers commercialized and synthesized by the divergent method (Munir et al., 2016). These dendrimers represent the most extensively characterized and best-understood series at this time (Akbarı, 2013). They are also known as starburst dendrimers (Klajnert and Bryszewska, 2001). Starburst dendrimers is applied as a trademark name for a subclass of PAMAM dendrimers based on a trisaminoethylene-imine core. The name refers to the star-like pattern observed when looking at the structure of the high generation dendrimers of this type in two-dimensions (Agrawal and Kulkarni, 2015). Dendritech (A.B.D.) produces PAMAM dendrimers (Klajnert and Bryszewska, 2001). Of the several varieties of dendrimers, Starburst poly(amidoamine) (PAMAM) dendrimers have been commercialized by Dendritech (Midland, MI, USA), leading to its widespread investigation (Majoros and Baker, 2008).

These dendrimers are synthesized by the divergent approach (Fig. 8.5). Typically, ethylenediamine or ammonia are used as cores and allowed to undergo reiterative two-step reaction sequences involving: (a) exhaustive alkylation of primary amines by Michael addition with methyl acrylate, and (b) amidation of amplified ester groups with a large excess of ethylenediamine to produce primary amine terminal groups (Agrawal and Kulkarni, 2015). After repetition of the amidation reaction and Michael addition with a step-by-step manner, dendrimers of high-generation can be achieved. The products which have amine in the terminal group as a functional group are referred to as full-generation dendrimers denoted by G2, G3 and G4. After coming in contact with other groups like ester or carboxylate groups, they can be considered half-generation dendrimers denoted by G2.5, G3.5 and G4.5 (Munir et al., 2016). The half-generation PAMAM dendrimers possess anionic surfaces of carboxylic groups (Akbarı, 2013).

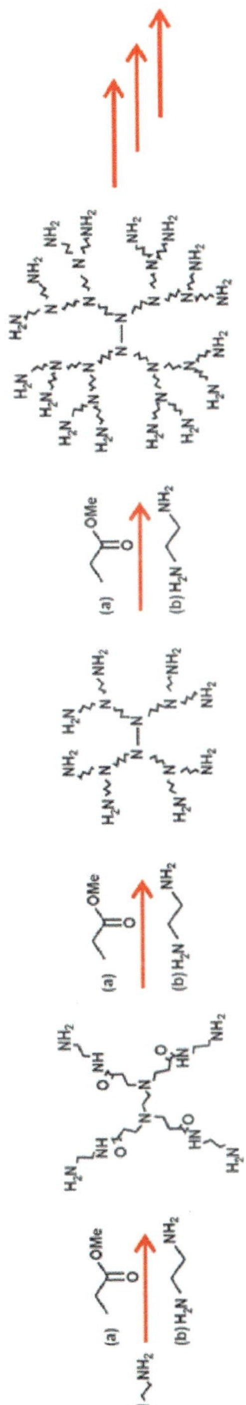

Fig. 8.5 Synthesis of tetra-functional poly(amidoamine) (PAMAM) dendrimers: exhaustive Michael addition of amino groups with methyl acrylate, followed by amidation of the resulting esters with ethylenediamine (Esfand and Tomalia, 2001).

Modification of terminal groups with different functionalities, such as acetamide, hydroxyl, carboxyl, or quaternary ammonium, further leads to the versatile applicability of these materials. For example, quarternized PAMAM dendrimers have a potent antimicrobial activity. In another aspect, dendrimer-like hyperbranched polyamidoamine can be incorporated into silica or chitosan powder surface, resulting in hybrid materials with versatile functionalities (Klaykruayat et al., 2010). In recent years, many articles have appeared on the use of dendrimers in antimicrobial finishing of textiles (Zhang et al., 2008b, 2009; Ghosh et al., 2010; Abkenar and Malek, 2012).

The high solubility and reactivity of these are due to presence of a number of functional end groups and empty internal cavities (Agrawal and Kulkarni, 2015). Products up to generation 10 have been obtained (Holister et al., 2003). Size range of PAMAM dendrimers lies between 1.1 and 12.4 nm as generations pass from 1 to 10, respectively (Munir et al., 2016). The dimension of PAMAM from 3rd to 7th generation number is in the range of nanodimension. Furthermore, PAMAM dendrimer size and structure is very close to natural proteins such as insulin, cytochrome *c*, haemoglobin, prealbumin and hemerythrin (Akbarı, 2013).

8.6.1.2 POPAM

PPI or poly(propylene imine)/poly(propylene amine) is one of the oldest known dendrimers, developed initially by Vögtle (Agrawal and Kulkarni, 2015). DSM (Netherlands) has developed production of poly(propylene imine) dendrimers. Butylenediamine (BDA) is used as the core molecule. The repetitive reaction sequence involves Michael addition of acrylonitrile to a primary amino group, followed by hydrogenation of nitrile groups to primary amino groups (Klajnert and Bryszewska, 2001). Its core structure is based on diamino butane with primary amines as end groups and tertiary propylene amines as interior. Thus, these dendrimers are also sometimes referred to as DAB-dendrimers, where DAB refers to the core structure, which is usually based on diaminobutane. PPI dendrimers are commercially available up to G5, and have widespread applications in the fields of material science and biology (Agrawal and Kulkarni, 2015).

8.6.1.3 POMAM

Majoros et al. (2008) synthesized new POMAM hybrid dendrimers constructed from poly(propylenimine) core and poly(amidoamine) shells and carried out their characterization by HPLC, GPC, NMR and noncontact atomic force microscopy (AFM). POMAM hybrid dendrimers are monodisperse with narrow distribution. For the higher generations, bimodal narrow distribution was observed. Vögtle et al. (2009) developed POPAM/PAMAM hybrid dendrimers up to the third generation.

8.6.2 Polylysine dendrimers

Like PAMAM dendrimers, polylysine dendrimers are largely characterized by amide bonds. Synthesis of these dendrimers is performed by condensation of the amino acid lysine, whose amino functions have previously been protected with *tert*-butyloxycarbonyl groups (Boc), onto an (activated) L-lysine *p*-nitrophenyl ester.

The resulting coupling product is then deprotected with trifluoroacetic acid and thus activated for renewed reaction. Iteration of the assembly and activation step ultimately lead to a polylysine dendrimer with 1024 terminal butyloxycarbonyl groups (Vögtle et al., 2009).

8.6.3 Dendritic hydrocarbons

Hardly any purely aliphatic hydrocarbon dendrimers have hitherto been prepared, probably owing to fundamental synthetic problems in the formation of unsymmetrical C–C bonds. The dendritic hydrocarbons described so far are nearly all made up of condensed or coupled arene or/and multiply bonded components, which is easier to achieve synthetically (Vögtle et al., 2009).

8.6.4 Carbon/oxygen-based (and Fréchet) dendrimers

A more recent type of dendrimer developed by Hawker and Frechet is based on a polybenzyl ether hyper branched skeleton. These dendrimers usually have carboxylic acid groups as surface groups, serving as a good anchoring point for further surface fuctionalization, and as polar surface groups to increase the solubility of this hydrophobic dendrimer type in polar solvents or aqueous media (Agrawal and Kulkarni, 2015).

8.6.5 Porphyrin-based dendrimers

Dendrimers built from porphyrin subunits are quite interesting from many viewpoints. Dendrimers containing porphyrins as cores and different branching units have been studied as synthetic analogous of biological systems, for example, to explore and clarify the redox properties of heme species, as well as to investigate long-distance electron transfer at electrodes (Loiseau et al., 2005). It was succeeded in synthesizing and characterizing a defect-free highly symmetrical, para-phenylene-spaced porphyrin-heneicosameric nickel complex. The molecule contains 21 porphyrin units. The synthesis utilizes the condensation of formyl-substituted phenyl end groups with pyrrole units, with these reaction steps giving almost quantitative yields (Vögtle et al., 2009).

8.6.6 Ionic dendrimers

Ionic dendrimers can bear positive or negative electric charges in the interior of the molecule (core, branching units) or on terminal (functional) groups. Some polyanionic dendrimers act as micellar molecules. Newkome et al. first synthesized water-soluble dendrimers, whose hydrophobic molecular niches can accommodate guests. The periphery of such dendrimers was furnished with carboxylate functions. On the other hand, polycationic dendritic molecules can generally be prepared in the following ways:

- Introduction of positively charged centres in the course of synthesis
- Transformation of neutral groups of the dendrimer into cationic ones
- Incorporation of transition metals into the dendrimer (Vögtle et al., 2009).

8.6.7 Silicone-based dendrimers

Radially layered poly(amidoamine-organosilicon) (PAMAMOS) dendrimers are inverted unimolecular micelles that consist of hydrophilic, nucleophilic polyamidoamine (PAMAM) interiors and hydrophobic organosilicon (OS) exteriors. This unique, new family of dendrimers was discovered by Dvornic et al in the 1990s. It includes many compositional and functional variants, among which are those with alkoxysilyl endgroups (Dvornic et al., 2000). These dendrimers are exceptionally useful precursors for the preparation of honeycomb-like networks with nanoscopic PAMAM and OS domains (Agrawal and Kulkarni, 2015).

8.6.8 Phosphorus-based dendrimers

Among all types of dendrimers, phosphorus-containing dendrimers play an important role, with applications ranging from catalysis, materials and even biology/nanomedicine (Caminade et al., 2012). The first neutral phosphorous-containing dendrimers (phosphodendrimers) were described in 1994. The phosphorous was bonded to other heteroatoms such as N and O and also to carbon. However, dendrimers are also known in which phosphorus atoms are located in the core unit or in the periphery, and also both in the core and at the branching points as well as the periphery (Vögtle et al., 2009).

8.6.9 Metallodendrimers (and Newkome dendrimers)

Dendrimers attached with the metal ion to form the complexation either in the interior or on the peripheral, which may be regarded as metallodendrimers (Baig et al., 2015). The introduction of metal centres on the surface or within dendrimers has been a recent trend reflecting a shift from the initial synthetic directions to a more applied emphasis. In general, surface modification is the easier mode of attachment as compared to the incorporation of metal centres within the interior of the dendrimers or hyperbranched polymers. The first synthesis of metallodendrimers possessing homo- and hetero-metallic branching centres (ruthenium and osmium) was reported by Balzani et al. The synthetic strategy developed by Balzani's group allows the preparation of dendrimers possessing two or more different metals and/or ligands (Newkome et al., 2001).

8.6.10 Chiral dendrimers

In chiral dendrimers the chirality is based on the building of four constitutionally assorted but chemically alike branches to the achiral core, e.g. chiral dendrimers obtained from pentaerythritol (Baig et al., 2015). Their potential use as chiral hosts for enantiomeric resolutions and as chiral catalysts for asymmetric synthesis has been recognized (Agrawal and Kulkarni, 2015). In the classification of chiral dendrimers, the way in which chirality was introduced into the molecule, or the site at which chiral modules are implanted, can be employed as distinguishing criterion. The chiral dendrimers can accordingly be assigned to one of the following groups:

(A) Dendrimers with chiral core and achiral branching scaffold
(B) Dendrimers with chiral end groups
(C) Dendrimers with chiral spacers or chiral branching units
(D) Dendrimers with achiral core unit and at least three constitutionally different dendrons
(E) Dendrimers with both chiral core as well as chiral branching units and chiral end groups (Vögtle et al., 2009).

8.7 Properties of dendrimers

Dendrimers are characterized by a highly branched structure, a compact shape and a large number of (reactive) end groups (Burkinshaw et al., 2002). Due to their molecular architecture, dendrimers show some significantly improved physical and chemical properties when compared to traditional linear polymers (Klajnert and Bryszewska, 2001) like no entanglement, high surface functionalities, low viscosity, high reactivity, higher solubility, interior cavity, etc. (Doshi, 2011). Many of the dendrimer's chemical and physical properties, such as state of aggregation, reactivity, stability and solubility are closely related to the nature of the end groups. Thus it is possible to tailor the dendrimer's properties by appropriate chemical modification of the end groups (Froehling, 2001). The peculiar physical properties of dendrimers constitute one of the reasons explaining the current interest for this type of molecule. Different theoretical models and simulation methods have been devoted to understand these properties. Thus Monte Carlo, Brownian dynamics and molecular dynamics methods have been applied to different ideal and atomistic representations of the dendrimer molecules (Rodriguez et al., 2007).

8.7.1 Solubility

Dendrimers have high solubility, miscibility and reactivity due to the presence of many chain ends (Agrawal and Kulkarni, 2015). Dendrimers' solubility is strongly influenced by the natüre of surface groups (Nair, 2008). Because of having hydrophilic groups, solubility of dendrimers is greater in polar solvents, whereas those dendrimers which have hydrophobic groups show greater solubility in nonpolar solvents (Munir et al., 2016). In a solubility test with tetrahydrofuran (THF) as the solvent, the solubility of dendritic polyester was found to be remarkably higher than that of analogous linear polyester. A marked difference was also observed in chemical reactivity (Klajnert and Bryszewska, 2001).

8.7.2 Host–guest interaction

Host–guest chemistry involves the binding of a guest molecule in a receptor molecule as a host one. The design and construction of hosts that are capable of selectively binding guest molecules requires precise control over geometrical features and interacting complementarily generation (Akbarı, 2013). In the field of dendrimer host–guest chemistry, PAMAM dendrimers have been extensively exploited. In the correct solvent conditions, hydrophobic guest molecules can be induced to reside in the hydrophobic interior of the dendrimer. This process is reversible on changing to a hydrophobic solvent. The core–shell molecular behaviour that dendrimers possess was initially put forward by Maciejewski in 1982, although it was not until later that experimental evidence was reported. Goddard and Tomalia were among the first to apply dendrimers in the field of host–guest chemistry. They proposed that guest molecules might not necessarily be fully encapsulated within the interior but could alternatively be partially incorporated at the dendrimer surface. The micelle-like properties of such host–guest interactions were identified by Newkome et al. and reported in the synthesis of a 'Unimolecular Micelle' from a nonpolar dendrimer with charged peripheral groups (Gittins and Twyman, 2003). The potential of dendrimers as hosts for small molecules was strikingly demonstrated in 1994 by Bert Meijer, chemistry professor at Eindhoven University of Technology in the Netherlands (Holister et al., 2003). Meijer and coworkers trapped p-nitrobenzoic acid/rose bengal into the 'dendritic box' of poly propylene imine dendrimers possessing about 64 branches on its corners (Munir et al., 2016). It is generally possible for the dendrimer to take up larger species as well. The nature of the host–guest interaction is varied. Ionic (acid/base), donor–acceptor, van der Waals and hydrogen-bonding interactions are known (Froehling, 2001). Atomic/molecular dispersion of the guest in a dendrimer host has created unique physical as well as chemical properties (Dastjerdi and Montazer, 2010). The nature of the host–guest interaction is varied by increasing generation numbers. PAMAM dendrimers reveals the periodic properties as a function of generation ($G=0$–10). It demonstrates flexible scaffolding for generations beyond 4, container properties for generations between 4 and 6 and rigid surface scaffolding for generations after 7. By increasing generation, the shape of dendrimer is changed from flat, floppy conformations to robust spheroids. As a result, entrapment of guest molecules is difficult in earlier generations due to the low cavity, whereas the steric hindrance has limited penetration of the guest into the higher generation (Akbarı, 2013).

8.7.3 Molecular weight

The classical polymerization process which results in linear polymers is usually random in nature and produces molecules of different sizes, whereas size and molecular mass of dendrimers can be specifically controlled during synthesis (Klajnert and Bryszewska, 2001). The first and second generations belong to the low-molecular-weight region, whereas the fifth and following generations are high-molecular-weight compounds (Pavlov et al., 2002). The molar mass (M) of the dendrimer can be predicted mathematically:

$$M = M_c + n_c \cdot \left[M_m \cdot \left(\frac{n_m^{\mathrm{G}} - 1}{n_m - 1} \right) + M_t \cdot n_m^{\mathrm{G}} \right] \tag{8.2}$$

where M_c is the molar mass of the core, M_m is the molar mass of the branched monomer, M_t is the molar mass of the terminal groups, n_c is the core multiplicity, n_m is the branch-juncture multiplicity and G is the generation number (Klajnert and Bryszewska, 2001).

8.7.4 Viscosity

The monodispersity and spherical steric expansion of dendrimers leads to a variety of interesting properties. In solution, linear chains exist as flexible coils; in contrast, dendrimers form a tightly packed ball. This has a great impact on their rheological properties. Dendrimer solutions have significantly lower viscosity than linear polymers (Nair, 2008). It is reported that the intrinsic viscosity of dendrimer molecules is small and has a very low dependency on the molecular weight, or that this dependency has a small maximum depending on the number of generations (Pavlov et al., 2002). As the molecular mass increases in a dendrimer, intrinsic viscosity also approaches its highest level at generation 4 and then starts to decrease. This type of behaviour differs from the linear polymers. In the case of classical polymers, increase in molecular mass will increase the intrinsic viscosity continuously (Mourey et al., 1992).

8.7.5 Toxicity

Biological properties of dendrimers are crucial because of the growing interest in using them in biomedical applications. 'Cationic' dendrimers (e.g. amine terminated PAMAM and poly(propylene imine) dendrimers that form cationic groups at low pH) are generally haemolytic and cytotoxic. Their toxicity is generation-dependent and increases with the number of surface groups. PAMAM dendrimers (generations 2, 3 and 4) interact with erythrocyte membrane proteins causing changes in protein conformation. These changes increase with generation number and the concentration of dendrimers. The interactions between proteins and half-generation PAMAM dendrimers (2.5 and 3.5) are weaker. Anionic dendrimers, bearing a carboxylate surface, are not cytotoxic over a broad concentration range (Klajnert and Bryszewska, 2001).

8.8 The use of dendrimers in water repellency treatments of textile materials

The increase in demand for comfortable, aesthetic, durable, functional and reliable textile products forces the development of new production techniques. The chemical finishing processes are commonly used for improving the properties of natural fibres, synthetic fibres and fabrics. For this purpose textile materials are subjected to various functional finishing processes such as water repellency, oil repellency, flame retardancy, antistatic and antimicrobial activity (Cerne and Simoncic, 2004).

Water and oil repellency processes have become more important in several textile application areas. Repellent finishes achieve their properties by reducing the free energy at fibre surfaces. If the adhesive interactions between a fibre and a drop of liquid placed on the fibre are greater than the internal cohesive interactions within the liquid, the drop will spread or vice versa. Surfaces that exhibit low interactions with liquids are referred to as low-energy surfaces (Schindler and Hauser, 2004). The energy accumulated in these surface molecules is expressed as the surface tension and is measured in millinewtons per metre ($mN\ m^{-1}$). When water and/or oil repellency is required, different chemicals can be impregnated or coated on a textile substrate to decrease the surface energy of the textile (Akerblom and Göranzon, 2013). Their critical surface energy or surface tension γ_c must be lower than the surface tension of the liquid γ_L (the internal cohesive interaction) that is repelled. γ_L of water, at 73 $mN\ m^{-1}$, is two to three times greater than γ_L of oils (20–35 $mN\ m^{-1}$). Therefore, oil repellency finishes with fluorocarbons ($\gamma_c = 10$–$20\ mN\ m^{-1}$) always achieve water repellency, but fluorine-free products, for example, silicones ($\gamma_c = 24$–$30\ mN\ m^{-1}$), will not repel oil (Schindler and Hauser, 2004).

The constitution of the fluorocarbon molecule is a carbon backbone where all hydrogen has been replaced with fluorine, and this carbon–fluorine bond is one of the strongest known in organic chemistry (Akerblom and Göranzon, 2013). The efficiency of PFC compounds is due to the structure of the bond between the F and C atoms. While the length of a C–F bond is 1.35 Å, that of a C–C bond is 1.54 Å. Because a C–F bond is shorter, the movement of fluorinated alkyl groups is lower, the F atom being strongly bonded with the C atom. This causes fluorocarbon compounds to have a very low boundary surface tension (Çoban, 1999). PFCs are usable in a wide range of commercial products including water and oil repellency agents for textiles (Akerblom and Göranzon, 2013). In the 1950s, silicone-based materials were introduced as water repellents. One common silicone-based water repellent is polydimethylsiloxane. It consists of a siloxane skeleton with methyl groups and creates a closely packed hydrocarbon surface (Hedegard, 2014). Newer environmentally friendly approaches in water repellency are based on dendrimer technology. The performance and environmental and health effects of dendrimer-based water repellency agents are considered here as fluoro-free alternatives.

Characteristics of dendritic polymers (such as highly branched structures, large numbers of reactive end groups and room between the branches for taking up guest molecules) create these fourth classes of polymer a suitable candidate for use in the textile engineering processes such as water repellency (Akbarı, 2013). The water repellent mechanism of dendrimers derives from the macromolecules' ability to build-up crystal structures in the nano range with the benefit of not impairing the textile material's air and vapour permeability. When combined with fluoro polymers, dendrimers force them to cocrystallize, leading to a self-organization of the whole system and an enrichment of the fluoro polymers on the outermost layer of the textile. Functionality and properties of dendrimers can be changed by filling their cavities or modifying the core and chain ends (Colleoni et al., 2011). The only limiting factor in the widespread applications is the price of these compounds. If dendritic polymers can be produced in high volumes at lower costs, a wide variety of uses can be considered in textiles engineering (Akbarı, 2013).

Depending on the chemical composition, dendrimers can provide water and/or oil repellency. There are currently some commercial products available that offer different textile finishing chemicals based on dendrimer technology. The oil- and soil-repellent dendrimer finishes include a fluorocarbon resin in their structure, while the fluorocarbon-free dendrimers only provide water repellency. Compared to other PFAS-based repellents, the fluorine content of the oleophobic dendrimer finishes is reduced. The fluorocarbon-free dendrimers provide water repellency through closely packed surface groups, i.e. methyl groups (Lassen et al., 2015).

FC polymers are applied together with dendrimers, causing self-organization where the fluorocarbon chains are enriched on the surface and cocrystallize with the dendrimers. Dendrimers are highly branched oligomers with nonpolar chains forming a starbush structure. They force the polar parts of FC polymers to form the surface structure. It is claimed that the sandwich structures obtained are highly ordered, resulting in equal or better effects with significantly lower fluorocarbon quantities compared to a conventional FC-finish (Schindler and Hauser, 2004).

Some of the scientific studies related to the use of dendrimer-based commercial products in water, oil and stain repellency finishing processes are summarized below.

Sancaktaroğlu (2008) investigated the changes in the values of water and oil repellency and also the colour through treating cotton and cotton/polyester fabrics with products containing dendrimer, fluorocarbon and combination of dendrimer/fluorocarbon. Once the results were compared as the dendrimers oriented the placement of fluorocarbons, it was determined that the combination of dendrimer/fluorocarbon showed better oil and water repellency effect.

Cotton fabrics have been treated with chemicals such as polymeric dendrimer containing fluorocarbon, nanosized fluorocarbon polymer, nano-silica acid, conventional fluorocarbon and paraffin emulsion containing zirconium salt by Namlıgöz et al. (2009). When the results are compared, it was understood that cotton fabrics treated with fluorocarbon/dendrimer combination had better water-, oil,- and stain-repellency properties.

Colleoni et al. (2011) treated the dyed cotton/polyester fabrics with dendrimer, fluorocarbon/dendrimer combination and fluorocarbon compounds provided by Rudolf Chemie. They focused on the assessment of colour alterations occurring on the dyed cotton/polyester blended fabrics due to various compounds used for repellency treatment. In the combination of dendrimers with fluorocarbon polymers, a high level of water- and oil-repellency was obtained, and the reason for this was grounded on orientation of fluorocarbon chains with the effect of dendrimers.

Akerblom and Göranzon (2013) compared the reference sample impregnated with fluorocarbons with aliphatic polyurethane, organic silicone and acid and hydrocarbon (dendrimer). Samples were subjected to abrasion, UV-radiation and washing. After each destructive treatment, oil- and water repellency tests were conducted. The hydrocarbons were found to be inferior to the fluorocarbons that showed greater durability to UV-radiation and washing. On the contrary, the hydrocarbon showed better abrasion resistance than the fluorocarbon that was more degraded. Furthermore, the hydrocarbon dendrimer did not pass any of the oil liquids. They stated that when only water repellency is needed, replacing fluoro chemicals with alternative treatments is possible. But for areas where also oil repellency is required, neither the chemicals tested in their study, nor any other alternatives to fluorochemicals, seem to be good enough.

Davies (2014) stated that dendrimer-based repellents exhibited little difference in terms of spray rating; however, the weight increase measured was significantly higher than not only long-chain PFCs but all other repellent types. That said, these weight increase results were not consistent between samples of this repellent type, and this could be due to any number of reasons including differences in processing conditions used or the specific fabric constructions.

Atav et al. (2014) searched the effects of treatments with products containing dendrimer, fluorocarbon and mixture of dendrimer/fluorocarbon provided by Rudolf Chemie on the colour of reactive dyed knitted cotton fabrics. Water- and oil-repellency efficiencies of these commercial products were also compared. They found that effects of all three products on colour change were acceptable, and the most effective was found to be the fluorocarbon/dendrimer mixture-based product in terms of water- and oil-repellency.

A study by Atav and Barış (2016) was conducted to determine whether dendrimer-based products replaced classic fluorocarbon-based products in bringing water repellency to bleached undyed woven cotton fabrics or not. Furthermore, it was examined whether or not water- and oil-repellency effect and its permanence increased in the case of adding dendrimer into fluorocarbon. For that purpose, fluorocarbon, dendrimer and fluorocarbon/dendrimer mixture-based commercial products provided by Rudolf Chemie were applied to cotton fabrics using the pad-dry-cure method. The water- and oil-repellency performance and washing resistance of these chemicals were investigated. According to the experimental results, it was found that the same water repellency values could be obtained in lower concentrations or in more temperate curing conditions, in case of fluorocarbon/dendrimer mixture-based product, compared to fluorocarbon-based product.

In the view of the literature, it can be said that dendrimers could be a good choice for water repellency treatments, but in cases where oil repellency is desired, it should be noted that dendrimers do not have oil-repellency property. On the other hand, better water- and oil-repellency values can be obtained when fluorocarbon/dendrimer mixture-based product is used. Within this context, it can be said that dendrimers have a good potential for obtaining repellent textiles.

Besides the water- and oil-repellency finishes, according to a US patent (No: 20080282480 A1), dendrimers can also be used for multifunctional finishes in which all water and oil repellency, water proof, antistatic, UV protective, antibacterial and other properties can be obtained in a single step. A finishing liquid which contains at least one type of dendrimer molecules with terminal groups of $-NH_2$, –OH, –COOH, –COOR, –R, –Si–O–R, PAMAM-OS, –R-halogen, etc., or with groups which can be modified to achieve these groups was invented. It is claimed that by employing the sol–gel or solution method, the finished textiles gain some of the following multi-functions: breathability, waterproofness, oil repellency, soil repellency, antistatic property, anti-UV irradiation, antibacterial, dyeing acceleration and controlled release of drug and perfume. Inorganic functional chemicals (inorganic precursors such as TiO_2 and SiO_2) through the sol–gel method can react with dendrimer's terminal groups. After the finishing process, a special cubic net comprising a nano dendrimer and functional inorganic chemicals are formed on the textile substrate. The dendrimers with –COOH, $-NH_2$, etc. terminal groups can bring finished textiles with hydrophilic, antibacterial, antistatic, etc., properties simultaneously. The dendrimers with alkyl terminal groups provide textile material with waterproof, soft handle and soil-repellency properties. The amphiphilic dendrimers and dendrimers having siloxane groups bring textiles with soft feeling and play a role as carrier and phase transfer material enhancing the dyeability of hydrophobic fibres. Besides, because of internal voids present in their structures, dendrimers offer controlled drug or perfume release properties for textiles (Hu et al., 2008).

8.9 Conclusion

In recent years consumer demands in the textile field have grown, and to meet these demands, many finishing chemicals and processes are being developed. One of the most important areas in which new chemicals are developed is repellency treatment including water, oil and soil repellencies, especially as the very effective C8 fluorocarbon finishes are being withdrawn due to environmental/toxicity concerns. One of the latest approaches in this field is based on dendrimer technology. Dendrimers are characterized by a highly branched structure of great regularity, a compact shape, a large number of reactive end groups and gaps between the branches, which makes them able to include guest molecules. These unique features make them attractive for use in many different application fields. The number of patents and research papers on dendrimer technology is increasing every year, and for this reason there is increasing progress on their applications in both academic research and industrial applications. Dendrimers have various applications in many fields including textile engineering, which is one of the most recent fields for these macromolecular architectures. Dendrimer-based products for water repellency treatments are commercially available today. They provide water repellency through closely packed surface groups, i.e. methyl groups. In addition, when they are combined with fluorocarbon polymers, high level of water repellency together with oil repellency can be obtained. The limiting factor of dendrimer technology is the long and costly synthesis procedure even though there is some progress with this issue. When new synthesis procedures allowing faster and mass-cycle synthesis to an industrially acceptable level are developed, it is believed the use of dendrimer technology in textile finishing processes will become widespread. As it is one of the newest fields of polymer technology, collaborative research where various branches of science take part is essential for dendrimer technology and its applications to be developed.

References

Abbasi, E., Aval, S.F., Akbarzadeh, A., Milani, M., Nasrabadi, H.T., Joo, S.W., Hanifehpour, Y., Nejati-Koshki, K., Pashaei-Asl, R., 2014. Dendrimers: synthesis, applications and properties. Nanoscale Res. Lett. 9, 247.

Abkenar, S.S., Malek, R.M.A., 2012. Preparation, characterization, and antimicrobial property of cotton cellulose fabric grafted with poly (propylene imine) dendrimer. Cellulose 19 (5), 1701–1714.

Agrawal, A., Kulkarni, S., 2015. Dendrimers: a new generation carrier. Int. J. Res. Dev. Pharm. L. Sci. 4 (5), 1700–1712.

Akbarı, S., 2013. The application of dendritic material in textile engineering. Sci. Bull. 7, 11–26.

Akerblom, D., Göranzon, E., 2013. Greener Water Repellency? Feasible Alternatives to Fluoro Chemicals for DWOR Treatments on Textiles. Bachelor in Textile Engineering, Swedish School of Textiles.

Atav, R., Barış, B., 2016. Dendrimer technology for water and oil repellent cotton textiles. AATCC J. Res. 3 (2), 16–24.

Atav, R., Yavaş, A., 2010. Dendrimers and Their Applications in Textile Finishing. Ulusal Nanobilim ve Nanoteknoloji Konferansı, Çeşme-İzmir/Türkiye.

Atav, R., Bılgıc, E., Yuksel, M.F., 2014. Obtaining water and oil repellent reactive dyed cotton functional textiles with the aid of dendrimer technology. In: 14th AUTEX World Textile Conference, Book of Abstracts, 72–73, Bursa/Turkey.

Baig, T., Nayak, J., Dwivedi, V., Singh, A., Srivastava, A., Tripathi, P.K., 2015. A review about dendrimers: synthesis, types, characterization and applications. J. Adv. Pharm. Biol. Chem. 4 (1), 44–59.

Boas, U., Christensen, J.B., Heegaard, P.M.H., 2006. Dendrimers in Medicine and Biotechnology: New Molecular Tools. The Royal Society of Chemistry, Cambridge, UK.

Burkinshaw, S.M., Mignanelli, M., Froehling, P.E., Bide, M.J., 2000. The use of dendrimers to modify the dyeing behaviour of reactive dyes on cotton. Dyes Pigments 47 (3), 259–267.

Burkinshaw, S.M., Froehling, P.E., Mignanelli, M., 2002. The effect of hyperbranched polymers on the dyeing of polypropylene fibre. Dyes Pigments 53, 229–235.

Caminade, A.M., Laurent, R., Zablocka, M., Majoral, J.-P., 2012. Organophosphorus chemistry for the synthesis of dendrimers. Molecules 17, 13605–13621.

Cerne, L., Simoncic, B., 2004. Influence of repellent finishing on the free surface energy of cellulosic textile substrates. Text. Res. J. 74 (5), 426–432.

Çoban, S., 1999. Genel Tekstil Terbiyesi ve Bitim İşlemleri. Ege Üniversitesi Tekstil ve Konfeksiyon Araştırma-Uygulama Merkezi Yayını, İzmir/Türkiye.

Colleoni, C., Massafra, M.R., Migani, V., Rosace, G., 2011. Dendrimer finishing influence on CO/PES blended fabrics color assessment. J. Appl. Polym. Sci. 120, 2122–2129.

Dastjerdi, R., Montazer, M., 2010. A review on the application of inorganic nano-structured materials in the modification of textiles: focus on anti-microbial properties. Colloids Surf. B: Biointerfaces 79, 5–18.

Davies, A., 2014. Durable Water Repellency – Study Phase I, An Evaluation of the Test Methods Used for Assessing Durable Water Repellent Fabrics Within the Outdoor Industry. De Montfort University, United Kingdom.

Doshi, M., 2011. Dendrimer and its application. Int. J. Pharm. Sci. Rev. Res. 7 (2), 104–111.

Dvornic, P.R., de Leuze-Jallouli, A.M., Owen, M.J., Perz, S.V., 2000. Radially layered poly (amidoamine-organosilicon) dendrimers. Macromolecules 33 (15), 5366–5378.

Dykes, G.M., 2001. Review dendrimers: a review of their appeal and applications. J. Chem. Technol. Biotechnol. 76, 903–918.

Esfand, R., Tomalia, D.A., 2001. Poly(amidoamine) (PAMAM) dendrimers: from biomimicry to drug delivery and biomedical applications. Drug Discov. Today 6 (8), 427–436.

Froehling, P.E., 2001. Dendrimers and dyes. Dyes Pigments 48 (3), 187–195.

Gao, C., Yan, D., 2004. Hyperbranched polymers: from synthesis to applications. Prog. Polym. Sci. 29 (3), 183–275.

Ghosh, S., Yadav, S., Vasanthan, N., Sekosan, G., 2010. A study of antimicrobial property of textile fabric treated with modified dendrimers. J. Appl. Polym. Sci. 115 (2), 716–722.

Gittins, P.J., Twyman, L.J., 2003. Dendrimers and supramolecular chemistry. Supramol. Chem. 15 (1), 523.

Goddard, J.M., Hotchkiss, J.H., 2007. Polymer surface modification for the attachment of bioactive compounds. Prog. Polym. Sci. 32 (7), 698–725.

Gupta, V., Nayak, S.K., 2015. Dendrimers: a review on synthetic approaches. J. Appl. Pharmacol. Sci. 5 (03), 117–122.

Hedegard, A., 2014. Durable Oil and Water Repellent Outdoor Fabrics by Atmospheric Plasma Treatment Reducing the Use of Perfluorinated Compounds (M.Sc. thesis). Chemical and Biological Engineering, Chalmers University of Technology, Gothenburg, Sweden.

Holister, P., Vas, C.R., Harper, T., 2003. Dendrimers. Technology White Papers, No: 6. Available from: http://www.romnet.net/ro/e-newsletter/e-newsletter13/FreeNanotechnology-download/dendrimers_WP.pdf (Accessed 19 February 2017).

Hu, J., Nie, K., Meng, Q., Zheng, G., 2008. Multifunction Finishing Liquids Containing Dendrimers and the Application of the Liquids in Textile Finishing. US Patent No: 20080282480 A1.

Klajnert, B., Bryszewska, M., 2001. Dendrimers: properties and applications. Acta Biochim. Pol. 48 (1), 199–208.

Klaykruayat, B., Siralertmukul, K., Srikulkit, K., 2010. Chemical modification of chitosan with cationic hyperbranched dendritic polyamidoamine and its antimicrobial activity on cotton fabric. Carbohydr. Polym. 80 (1), 197–207.

Lassen, C., Jensen, A.A., Warming, M., 2015. Alternatives to Perfluoroalkyl and Polyfluoroalkyl Substances (PFAS) in Textiles. Survey of Chemical Substances in Consumer Products No. 137. The Danish Environmental Protection Agency, Copenhagen, Denmark.

Loiseau, F., Campagna, S., Hameurlaine, A., Dehaen, W., 2005. Dendrimers made of porphyrin cores and carbazole chromophores as peripheral units. absorption spectra, luminescence properties, and oxidation behavior. J. Am. Chem. Soc. 127 (32), 11352–11363.

Majoros, I.J., Baker, J.R., 2008. Dendrimer-Based Nanomedicine. Pan Stanford Publishing Pte Ltd, Singapore.

Majoros, I.J., Williams, C.R., Tomalia, D.A., Baker, J.R., 2008. New dendrimers: synthesis and characterization of POPAM–PAMAM hybrid dendrimers. Macromolecules 41 (22), 8372–8379.

Mourey, T., Turner, S., Rubinstein, M., Fréchet, J., Hawker, C., Wooley, K., 1992. Unique behavior of dendritic macromolecules: intrinsic viscosity of polyether dendrimers. Macromolecules 25 (9), 2401–2406.

Munir, M., Hanif, M., Ranjha, N.M., 2016. Dendrimers and their applications. J. Pharm. Res. 2 (1), 55–66.

Nair, J.R., 2008. Dendritic polymers: design tools for property control in coatings. Paintindia 77–96 (March).

Namırtı, O., Atav, R., 2011. Tekstilde Yeni Bir Konsept Olan Dendrimerlerin Tarihçesi, Sınıflandırılması, Molekül Yapısı ve Özellikleri. Pamukkale Üniversitesi Mühendislik Fakültesi Mühendislik Bilimleri Dergisi 17 (2), 109–115.

Namlıgöz, E.S., Bahtiyari, M.İ., Hoşaf, E., Çoban, S., 2009. Performance comparison of new (dendrimer, nanoproduct) and conventional water, oil and stain repellents. Fibres Text. East. Eur. 17 (5), 76–81.

Newkome, G.R., Moorefield, C.N., Vögtle, F., 2001. Dendrimers and Dendrons, Concepts, Syntheses Applications. WILEY-VCH Verlag GmbH, Weinheim.

Patel, H.N., Patel, P.M., 2013. Dendrimer applications – a review. Int. J. Pharm. Biol. Sci. 4 (2), 454–463.

Pavlov, G.M., Korneeva, E.V., Meijer, E.W., 2002. Molecular characteristics of poly (propylene imine) dendrimers as studied with translational diffusion and viscometry. Colloid Polym. Sci. 280 (5), 416–423.

Rodriguez, E., Freire, J.J., Echenique, G.R., Hernandez, C.J.G., Torre, J.G., 2007. Improved simulation method for the calculation of the intrinsic viscosity of some dendrimer molecules. Polymer 48 (4), 1155–1163.

Sadler, K., Tam, J.P., 2002. Peptide dendrimers: applications and synthesis. Rev. Mol. Biotechnol. 90 (34), 195–229.

Sancaktaroğlu, E., 2008. Bitim İşlemlerinde Pamuk, Pamuk/Poliester Karışımlı Kumaşlarda Dendrimerlerin Renk Üzerine Etkisi. Yüksek Lisans Tezi, Fen Bilimleri Enstitüsü, Uludağ Üniversitesi, Bursa-Türkiye.

Sarkar, A., Kaganove, S.N., Dvornic, P.R., Satoh, P.S., 2005. Colorimetric biosensors based on polydiacetylene (PDA) and polyamidoamine (PAMAM) dendrimers. Polym. News 30 (12), 370–377.

Sashiwa, H., Aiba, S., 2004. Chemically modified chitin and chitosan as biomaterials. Prog. Polym. Sci. 29 (9), 887–908.

Schindler, W.D., Hauser, P.J., 2004. Chemical Finishing of Textiles. Wood Head Publishing Limited, Cambridge, England.

Scholl, M., Kadlecova, Z., Klok, H.A., 2009. Dendritic and hyperbranched polyamides. Prog. Polym. Sci. 34 (1), 24–61.

Toit, L.C.D., Pillay, V., Choonara, Y.E., 2010. Nano-microbicides: challenges in drug delivery, patient ethics and intellectual property in the war against HIV/AIDS. Adv. Drug Deliv. Rev. 62, 532–546.

Tomalia, D.A., 2005. The dendritic state. Mater. Today 8 (3), 34–46.

Tomalia, D.A., Christensen, J.B., Boas, U., 2012. Dendrimers, Dendrons, and Dendritic Polymers Discovery, Applications, and the Future. Cambridge University Press, Cambridge, ISBN: 9780521515801.

Vögtle, F., Richardt, G., Werner, N., 2009. Dendrimer Chemistry. Strauss GmbH, Mörlenbach. ISBN: 978-3-527-32066-0.

Zhang, F., Chen, Y., Lin, H., Wang, H., Zhao, B., 2008a. HBP-NH_2 grafted cotton fiber: preparation and salt-free dyeing properties. Carbohydr. Polym. 74 (2), 250–256.

Zhang, F., Chen, Y.Y., Lin, H., Zhang, D.S., 2008b. Performance of cotton fabric treated with an amino-terminated hyperbranched polymer. Fibers Polym. 9 (5), 515–520.

Zhang, F., Zhang, D., Chen, Y., Lin, H., 2009. The antimicrobial activity of the cotton fabric grafted with an amino-terminated hyperbranched polymer. Cellulose 16 (2), 281–288.

Plasma-based treatments of textiles for water repellency

9

Nicholas W.M. Edwards, Parikshit Goswami
University of Leeds, Leeds, United Kingdom

9.1 Plasma treatments

First described by Langmuir in 1928, plasma, considered as the fourth aggregation state of matter, can be described as,

> *a mixture of partially ionised gases where the constituents are achieved by external energy addition*
>
> *Blackburn (2009).*

Plasmas are generally classified as hot/thermal and cold/nonthermal depending on the temperature of the plasma zone. A concise account of different technical plasma processes used for material processing can be found in Fig. 9.1.

9.1.1 Thermal plasma

In thermal plasma the gas density is sufficiently high and so the frequency of collisions between electrons, ions and neutral species composing the plasma is such that an energy exchange is possible. Thermal plasmas are characterized by thermal equilibrium between all the different species contained in the gas. Temperatures in the order of thousands of degrees Celsius are reached in thermal plasmas, so they are not suitable for surface modification of heat-sensitive polymeric and textile materials (Jelil, 2015).

9.1.2 Nonthermal plasma

Cold plasmas can be produced at room temperature by applying an electrical field over two electrodes with a gas in-between, or by inducing radiofrequency (RF) resonant current in a coil (Jelil, 2015). The electrons acquire higher energies than ions and molecules, and, due to low density of gas, collisions are rare and thermal equilibrium is not reached (Shishoo, 2007). These electrons make up >1/1,000,000 of the total mass, their influence is negligible and the plasma has a low overall temperature, making this treatment type suitable for use in surface modification of textile materials. The main advantage of nonthermal plasma treatments include:

Waterproof and Water Repellent Textiles and Clothing. https://doi.org/10.1016/B978-0-08-101212-3.00008-3

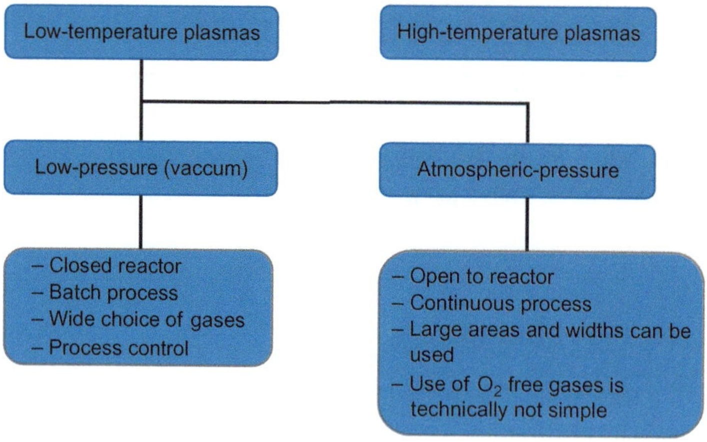

Fig. 9.1 Definitions and characteristics of technical plasma processes used for material processing.

- The electrons are able to cleave covalent bonds at the surface of the treated material, producing physical and chemical modification at the surface (on a micron scale) without changing the bulk properties of the material.
- Compared to conventional finishing processes, plasma treatment consumes minimal chemicals and no costly drying process is required.
- The plasma processes has high environmental compatibility.
- Low temperature plasma can be applied to most kinds of fibre, textile or polymeric material.

9.2 Plasma treatment for materials

Regarding large-scale plasma treatment lines, the 'Plasmatreat' pilot workshop includes a conveyor system with integrated plasma for treating widths up to 300 mm and processing speeds of up to 400 m min^{-1} for treatment of textiles and nonwovens. Some atmospheric plasma treatment systems can handle widths of 2 m at speeds of 20 m min^{-1} (Sparavigna, 2008). Subatmospheric and low-pressure plasma systems tend to be batch processes (Buyle, 2009).

9.2.1 *Atmospheric plasma treatment*

Atmospheric plasma treatment is, as the name suggests, carried out under ambient conditions (see Fig. 9.2). This is advantageous as material can be processed continuously, so this system can be integrated into an existing textile processing or treatment line. Atmospheric plasma treatment can be split into corona treatment, dielectric barrier discharge and glow discharge (Kale and Desai, 2011).

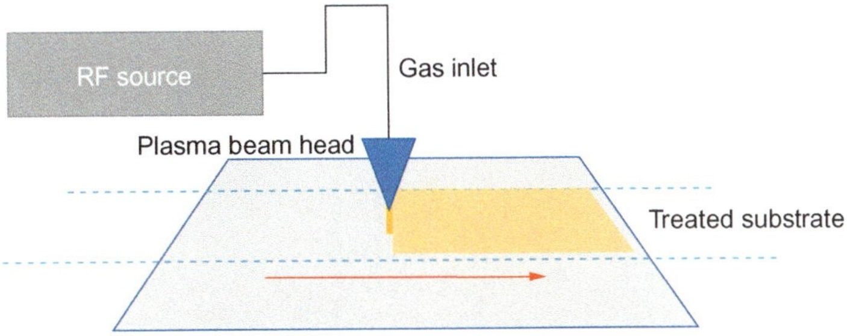

Fig. 9.2 Simple schematic of atmospheric plasma treatment.

9.2.1.1 Corona plasma treatment

Corona plasma treatment uses low temperature corona discharge plasma to impart changes in the properties of a textile surface. A corona discharge is an electrical discharge resulting from the ionization of a fluid around an electrically charged conductor. Corona treatment of cotton, for instance, can improve spinnability and the strength and abrasion resistance of the fabric (Thorsen, 1971). Corona treatment can increase the hydrophilicity of a textile material, especially when used in combination with a pretreatment (Carneiro et al., 2001); or if the treatment is repeated multiple times (Ryu et al., 1991).

9.2.1.2 Dielectric barrier discharge plasma

Dielectric barrier discharge plasma is an electrical discharge between two electrodes, separated by an electrical insulator that can be polarized by an applied electric field (dielectric). It is used to clean or modify substrates, including textiles (Kogelschatz et al., 1997). This usually involves RF to microwave frequency, high voltage AC current (Borra, 2008).

9.2.2 Low pressure plasma

Low-pressure plasma is a cost-effective, environmental friendly technique used to modify the surface of polymeric materials and textiles on a micro- to nano-level (see Fig. 9.3). The material to be treated is placed in a vacuum chamber to facilitate the low pressure (typically below 500 mT). The advantages of this type of treatment include no heat, minimal surface ablation, a uniform 3-D treatment, repeatable results and little to no environmental or health concerns (Schutze et al., 1998).

9.2.2.1 Glow discharge plasma

Glow-discharge plasma is created by the passage of electric current through a low-pressure gas. This type of treatment can modify the wettability of polyester (Morent et al., 2008) or cotton (Temmerman and Leys, 2005), with an effect lasting for several days.

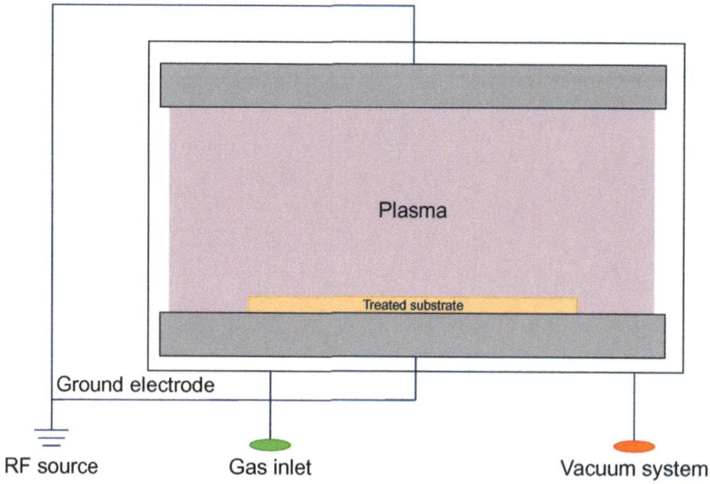

Fig. 9.3 Simple schematic of low-pressure plasma.

9.3 Surface modification with plasma

Plasma treatment can be used to modify materials either chemically, physically, or a mixture of both.

9.3.1 Physical

9.3.1.1 Etching

Plasma etching is material removal from a surface via a plasma process. This involves a sample being treated with an appropriate plasma gas mixture being pulsed at a sample. The plasma source, known as etch species, can be either charged (ions) or neutral (atoms and radicals). Applications include the microstructuring of surfaces, removal of oxide layers and etching of semiconductors (Donnelly and Kornblit, 2013).

9.3.2 Chemical

9.3.2.1 Plasma enhanced vapour deposition

Plasma enhanced vapour deposition (PEVD) is used to deposit thin solid films on a substrate, from the gas state. Plasma is created from this gas by either DC, or more commonly radiofrequency (RF; AC) discharge in the reaction chamber. PEVD is different from plasma grafting; PEVD is a one-step in situ method, while plasma grafting has separate radical-forming and graft-polymerization steps (Morent et al., 2008). Plasma polymer films are often deposited by PEVD (Khelifa et al., 2016). Plasma deposition is limited to very thin layers (up to 200 nm) on the individual fibres.

9.3.2.2 Plasma cleaning

Plasma treatment can be used to remove contaminants from a fibre surface, for example, desizing cotton. Plasma cleaning is not relevant to this discussion.

9.3.2.3 Grafting

Grafting of copolymers involves fixing polymeric chains to a structurally different polymeric substrate, to change surface functionality whilst preserving bulk mechanical properties. Grafting copolymers to a textile surface can be facilitated by using atmospheric or pressure-dependent plasma processing, allowing tailoring of the textile to a specific end, including hydrophobicity (De Vietro et al., 2015).

9.4 Hydrophobic and hydrophilic materials

The wettability of a given surface is an important property of a material. It is controlled by the chemical composition and the geometry of the said surface (Daoud et al., 2004). The wetting behaviour of a surface is determined by the relation of the interfacial energies between the solid substrate and the liquid (Y_{sl}), between the solid and gaseous atmosphere (Y_{sv}) and between the liquid and the atmosphere (Y_{lv}). This is particularly important for the assessment of the success of any plasma treatment of textiles to confer hydrophobicity.

9.4.1 Contact angle

Contact angle is described by Young's equation (White, 1977; Sing and Williams, 2012):

$$\gamma_{lv}\cos\gamma_{\theta} = \gamma_{sv} - \gamma_{sl} \tag{9.1}$$

where $\gamma\theta$ indicates Young's contact angle, γ_{lv} indicates liquid-vapour interface tension; γ_{sv} indicates solid-vapour interface tension; γ_{sl} indicates solid-liquid interface tension.

Lower contact angle values ($\theta < 90$ degree, hydrophilic) indicates surface wetting is favourable, and the fluid will spread over a large area on that surface; higher contact angle values ($\theta > 90$ degree, hydrophobic) indicates wetting of the surface is unfavourable so the fluid will minimize its contact with said surface and form a compact liquid droplet (Kalin and Polajnar, 2014; Yuan and Lee, 2013) (Fig. 9.4). A material is conventionally defined as super-hydrophobic if the static water contact angle is >150 degree (Lafuma and Quere, 2003).

In sessile drop contact angle measurement, the angle formed between the liquid/solid interface and the liquid/gas interface is the contact angle (measured through the liquid phase) (Bachmann et al., 2000; Skinner et al., 1989). Direct optical contact angle measurement is the most widely used method of contact angle measurement (Hunter, 2001). The small size of the liquid and substrate means there is a higher risk/impact of impurities (Yuan and Lee, 2013).

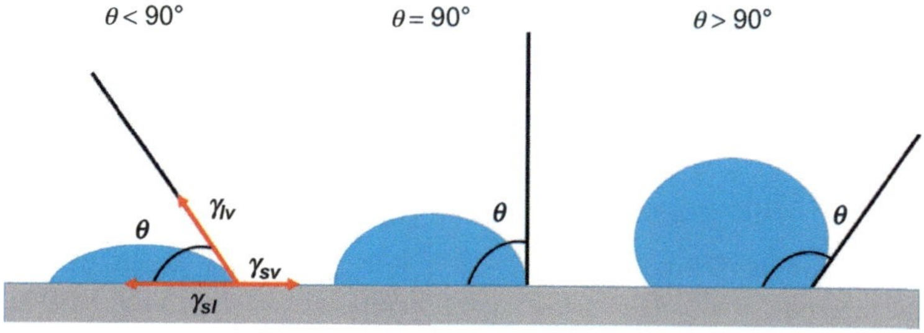

Fig. 9.4 Illustration of contact angles formed by sessile drops on smooth, homogenous solid surfaces. θ indicates contact angle, γ_{lv} indicates liquid-vapour interface tension; γ_{sv} indicates solid-vapour interface tension; γ_{sl} indicates solid-liquid interface tension.

From Young's equation, it can be concluded that, in general, hydrophobicity of a material could be increased by either lowering the surface energy of the fabric (decreasing the adhesion of fabric) or by increasing the surface tension of the liquid (increasing the cohesion of liquid). As it is not possible to alter the surface tension of the liquid in the majority of applications, the only way of achieving a hydrophobic/super-hydrophobic surface is by lowering the surface energy of the fabric. However, it has long been recognized that a super-hydrophobic surface requires a unique combination of two fundamental properties, e.g. surface roughness and low-surface energy (Balu et al., 2008).

9.4.2 Contact angle and hysteresis

Surface wetting is not a static state (Yuan and Lee, 2013). Contact angle hysteresis is the difference between the advancing (maximal) and receding (minimal) contact angles. This is generally thought to arise due to the roughness and heterogeneity of a surface (Sajadinia and Sharif, 2010; Schwartz and Garoff, 1985), as Young's equation does not take surface topography into account (Gao and McCarthy, 2006). This 'equilibrium' contact angle reflects the relative strengths of the molecular interactions between solid, liquid and vapour interfaces (Mognetti and Yeomans, 2010). Therefore, the advancing and receding contact angles of the surface samples are measured to obtain a more accurate contact angle; typically via goniometry (Li et al., 2017).

9.4.3 Textiles and surface characterization

Textile samples significantly differ from the more or less planar surfaces typically used in the literature (see Fig. 9.4 and Young's equation). The characteristics and often complex 3D geometry of a textile substrate has notable effects on the initial wetting behaviour of a droplet, and the dynamics of wetting. While many researchers employ contact angle measurements for the characterization of the wetting properties of textiles, the absolute value obtained by this method is questionable. The apparent contact angle on textiles will always be determined by the combined effect of microscopic surface properties (i.e. fibres), macroscopic surface geometry (i.e. fabric) and capillary phenomena. Moreover, it is difficult to determine the base line on uneven textile substrates, largely true for nonwovens (Russell, 2004). Surface roughness is known to influence wetting properties (Miller et al., 1996; Spori et al., 2008). Therefore, the apparent contact angle will always differ from the true contact angle (Bahners et al., 2008); nevertheless, the contact angle measurements are used extensively for textiles to qualitatively analyse the effect of different surface treatments including plasma treatments.

9.5 Water and oil repellency

Water repellency, or hydrophobicity, is the physical property of materials that makes them not easily penetrated by water, especially as a result of being treated for such a purpose with a surface coating such as DWR (durable water repellent). Water repellent fabrics are coated with a finish (such as DWR) that is resistant but not impervious to penetration by water; whereas waterproof, breathable fabrics help regulate heat and release moisture, and are impervious to water. Oil repellency, or oleophobicity, is the physical property of a molecule that is repelled from oil. In addition to being hydrophobic, fluorocarbons are oleophobic (Kovalchuk et al., 2014).

9.5.1 *Traditional hydrophobic treatments for textiles*

Textile materials are flexible, lightweight, strong and have a large surface-to-volume ratio; all intrinsic properties that make them valuable commodities (Buyle, 2009). To impart additional functionality would increase their value. Typical examples of such functionalities are hydrophobicity, oleophobicity, or antibacterial activity.

Very few textiles are inherently water repellent and none are oil repellent, so additional processing must take place to confer these properties. Traditionally, water repellency is achieved by jig or pad application of functional chemicals, followed by careful drying on a stenter or in a festoon. The functional chemicals used to achieve water repellent finishing vary significantly in their chemistry however, their role is to reduce the surface energy of the fabric by adding lower surface energy chemical groups to the surface (Shahidi et al., 2013).

The different chemicals traditionally used to achieve water repellence may be divided into a few groups (Heywood, 2003):

- Metal salts
- Soap/metal salt
- Wax
- Pyridinium-based finishes
- Organo-metallic complexes

- *N*-Methylol derivatives
- Silicone finishes and
- Fluorochemical finishes

Among these, only fluorocarbon finishes can repel both oil and water (Yeh et al., 2007), while the others only repel water. In general, the traditional wet-processing route requires huge amount of water and associated problem of water pollution, and high energy consumption in removing the water from the fabric (Kalliala and Talvenmaa, 2000). Plasma treatment does not require these large volumes of water or wet chemicals, or the large amount of energy required to dry the fabric.

9.5.2 Outline of problems associated with C_8 fluorochemistry

Briefly, fluorochemicals are synthetically produced agents that repel water and oil. They are also inert due to the strong C–F bond (Fromme et al., 2009), and so are persistent towards the physical and chemical conditions in industrial processing. Fluorocarbons are organofluorine compounds (Banks et al., 1994) with the formula C_xF_y. Many compounds used to functionalize textiles against water and oil are based on C_8 telomeres (contains 8 carbon atoms and 17 fluorine atoms). Most factory applied DWR treatments are fluoropolymer based (Alderson, 2015). In manufacturing C_8-based repellent materials, trace amounts of PFCs (perfluorinated compounds) can be generated as an unintended byproduct (Frisbee et al., 2009).

PFCs break down very slowly in the environment, and so can accumulate in human and animal life, with a $t_{1/2}$ of several years. PFCs and their precursors have been linked with (ATSDR, 2016):

- Affecting the developing foetus and child, (changes in growth, learning and behaviour)
- Decreased fertility
- Increasing cholesterol
- Affecting the immune system and
- Increasing the risk of cancer

With respect to textiles, PFCs came into the public consciousness when the Greenpeace report, ‘Chemistry for any weather’ (Greenpeace, 2012) called upon the outdoor clothing industry to:

> *...ban hazardous substances such as PFCs from its production processes and immediately switch to safe functional alternatives wherever possible. The industry must immediately take steps to further develop safer alternatives and use them in production. Greenpeace calls on all manufacturers of outdoor clothing to immediately address this issue and begin phasing out the entire group of PFC compounds...*

Plasma treatment is one viable replacement; however the finish durability of chemically treated samples is thought to be much higher than that of plasma treated samples (Morent et al., 2008), as plasma treatment only affects the surface of the treated substrate.

9.6 Plasma treatment of textiles to confer hydrophobicity

Buyle suggests that for fabric, fibres and other textile materials, plasma treatment can be used to:

- Impart hydrophilic properties
- Increase adhesion
- Influence printability and dyeability
- Change the electrical conductivity
- Impart hydrophobic and oleophobic properties
- Apply antibacterial or fire retarding agents
- Antishrink treat wool
- Sterilize materials
- Desize cotton (Buyle, 2009).

Hydrophobicity can be conferred on a textile substrate by plasma treatment. Polymer deposition onto a textile via plasma treatment with fluorocarbons or novel silicones is commonly used to this end (Zille et al., 2015).

Cellulosic fibres are a common natural fibre used throughout the textile industry (Hearle and Morton, 2008). Plasmas treatment with fluorocompounds such as tetrafluoromethane (CF_4) (Sigurdsson and Shishoo, 1997), sulphur hexafluoride (SF_6) (Hodak et al., 2008), hexafluoroethane (C_2F_6) (Sun and Stylios, 2006), or hexafluoropropene (C_3F_6) (Li and Jinjin, 2007) have been investigated for hydrophobicity enhancement of polymers and fabrics. In cellulosic fibres, the hydroxyl groups are the only functional groups generally available for reaction, limiting the reactions to esterification and etherification. By use of cold plasma, cellulose can be oxidized, reduced and/or substituted in new unique ways. Cellulose can also be grafted with monomers by using plasma as free radical initiator at the cellulose surface (Shishoo, 2007).

A major factor in the plasma process design is the choice of process gas (see Table 9.1 for general examples). As discussed earlier, fluorocarbon finishes can repel both oil and water (Laguardia et al., 2007). Treatment of cotton with different kinds of plasma gas is a well-studied subject; the literature suggests that after hexafluoroethane (C_2F_6) plasma, material made up with cotton become highly hydrophobic, without influencing the water vapour transmission (Sun and Stylios, 2006). This kind of fluorocarbon finishing in conjugation with increased specific surface area (resulting from,

Table 9.1 Examples of plasma process gas and properties conferred

Treatment gas (substrate)	Properties conferred
N_2 (wool); O_2 (cotton)	Enhance mechanical properties
O_2/air/NH_3 (PA; PE; PP; OET; PTFE)	Hydrophilicity
Perfluorocarbon/siloxan (cotton; PET)	Hydrophobicity
O_2 (wool; cotton); $SiCl_4$ (PET); Ar (PA)	Capillarity improvement

Adapted from Sparavigna, A., 2008. Plasma treatment advantages for textiles. Los Alamos Natl. Lab. Prepr. Arch. Phys. 1–16, arXiv:0801.3727v1 [physics.pop-ph].

Fig. 9.5 Superhydrophobicity resulting from surface roughening of an already hydrophobic surface.

for example, pretreatment with oxygen plasma) results in an effect generally known as Lotus effect (super-hydrophobic effect outlined in Fig. 9.5).

In general, a hydrophobic treatment introduces certain functional groups via a coating or a graft copolymer, removes hydrophilic functional groups or changes hydrophilic groups into nonhydrophilic once (Morent et al., 2008). Plasma polymerization using fluorocarbons is a well-studied subject. Fluorocarbon gas plasmas can change surface properties via either surface treatment or polymerization and deposition of a thin film (Sardella et al., 2009). Fluorocarbon gases can have a saturated structure (CF_4, C_2F_6), an unsaturated structure (C_3F_6), or a cyclic structure (C_4F_8). Plasma based fluorination, using saturated fluorine compounds such as CF_4, CHF_3 and C_2F_6, has been used by various research groups to improve the water repellency and barrier properties of polymers and textiles.

Sahun et al. (2002) have shown that CF_4 radiofrequency (RF) plasma environments can induce intense surface fluorination of a paper (i.e. cellulosic) surface. Sigurdsson and Shishoo (1997) used CF_4 microwave-frequency plasma to treat various materials (PET, PP and cellophane) and reported extensive incorporation of fluorine containing groups onto the surface of the treated materials (Sigurdsson and Shishoo, 1997). They have also shown the formation of CHF, CF_2 and CF_3 groups in the surface of cellophane. They attributed the decrease in surface energy of cellophane after plasma treatment to the incorporation of nonpolar CF_2 and CF_3 groups, as CF_3 and CF_2 make the main contribution to the hydrophobic property of modified fibres. It is well known that production of polymers from saturated fluorocarbons is negligible, and that addition of methane or hydrogen enhances plasma polymerization of saturated fluorocarbons (Masuoka and Yasuda, 1982; Wang et al., 1993).

Using saturated, unsaturated and cyclic fluorocarbons, Wang et al. (1993) reported adding small amounts of methane to pure fluorocarbons causes significant increase in plasma polymerization. However, addition of excess CH_4 decreases polymer deposition. From wettability studies, they concluded that pure fluorocarbon plasmas and fluorocarbon-dominated plasmas can achieve a hydrophobic surface. The F/C ratio is a prime parameter (Bahners et al., 2008) in determining the hydrophobicity. As the concentration of CH_4 increases, the F/C ratio decreases, accompanied by a corresponding decrease in hydrophobicity. With the same molar gas feed composition, the deposition of polymer depends on molecular structure of the fluorocarbon and the following ranking was observed: $C_3F_6 > C_4F_8 > C_2F_6 > CF_4$ (Takahashi and Tachibana, 2001).

Similar results were reported by McCord et al. (2003). In this work they treated cotton fabric by radio-frequency plasma with CF_4 and C_3F_6 gases and concluded that changes in surface energy were much greater for the samples treated in C_3F_6 gas plasma than for those treated in CF_4 gas plasma. They concluded from this work that $-CF_3$ group/F_{1s} ratio increased more with the C_3F_6 plasma treatment than with CF_4 plasma, resulting in higher contact angles of samples treated with C_3F_6 plasma. Wettability measurements indicated the CF_3 group was more efficient in increasing surface hydrophobicity than either the CF of CF_2 groups. Work by Inagaki et al. (1991) also speculated that surface fluorocarbon chemical composition is more important to determine hydrophobicity than the overall surface atomic fluorine content. Li and Jinjin (2007) used C_3F_6 plasma treatment and reported remarkable improvement of the hydrophobicity of silk and cotton. They also reported that the hydrophobicity of both the fibres does not change with time and that the tensile strength increases after plasma treatment.

The durability of plasma treatments is an issue, as thin functional layers are deposited onto soft textile substrates (Morent et al., 2008). Iriyama and Yasuda found that CF_4 and C_2F_6 plasma treated nylon fabrics did not have sufficient resilience to washing (Iriyama et al., 1990), though the saturated fluorocarbon treatments were found to have better durability than unsaturated treatments.

These treatments still use fluorine gas however, and are able to circumvent any ecological concerns as in theory, they do not contain C_8 chemistry, having shorter perfluoroalkyl chains. The longer the perfluorinated chain in a compound, the more it is persistent, bioaccumulative and potentially toxic.

Textiles plasma-functionalized to be hydrophobic may be water repellent however, because their innate fibrous structure allows water vapour to diffuse from one side of the textile to the other (Morent et al., 2008). This can improve garment comfort properties, and be applied in sportswear and protective garments.

Other plasma treatments that can be used to confer water repellency to textiles use reduced add-on weights or volume of precursor. Examples of these include the combination of wet chemical and plasma treatment that can be employed to generate an oleophobic finish. Hegemann (2005) used RF plasma with a mixture of argon and O_2 to functionalize polyethylene terephthalate (PET) fabric, followed by a wet-chemical fluorocarbon impregnation to this end.

De Vietro et al. developed a novel nonequilibrium low-pressure plasma treatment process to deposit a durable, water repellent film onto synthetic textiles. Using RF plasma with 1H, 1H, 2H-perfluoro-1-decene vapour, 5 mm thickness PU foams that were treated became wear resistant, particularly if a high plasma power was used during the said treatment (De Vietro et al., 2015). This has a potential application in the automotive market, where water repellency and the resistance to motor oil stains is particularly desirable.

9.7 Nanoparticle deposition via plasma treatment

Hydrophobic functionalization through nanoscale particle deposition; altering the nano-structure of the textile substrate; or deposition of a sol–gel (nanoparticles dispersed in a liquid, agglomerated together to form a continuous 3D network;

Brinker and Scherer, 1990) can all be carried out via plasma treatment. Cotton coated with a film of hydrophobic nanoparticles using fluorocarbon AC plasma had softness, water retention, colour retention, abrasion, friction and permeability superior to those of ScotchGard-sprayed control samples. The nanoparticle-coated-cotton also had a super hydrophobic contact angle of 164 degree (Zhang et al., 2003).

9.8 Plasma treatment and fibre surface nano-roughness

Plasma treatment of lyocell results in modification of fibre surface morphology via physical etching, introducing nano-roughness, as well as chemical modification. Both can lead to a change in surface wetting and therefore, adhesion behaviour—particularly with respect to bacteria. It was found that Gram-negative *E. coli* were able to overcome changes in nanoroughness to adhere to nonwoven materials of 'ideal' surface energy, whereas Gram-positive *S. Aureus* and *E. faecalis* were not (Edwards et al., 2017).

Surface roughness of plasma treated fibres can be measured via atomic force microscopy (AFM) (Fig. 9.4). R_a is the arithmetic average of the roughness profile, and calculated at 3 μm resolution. For lyocell, untreated fibres (Fig. 9.6A; R_a 0.2 nm) show a smooth, flat surface with no discernible nano-irregularities, while plasma treatment increases fibre surface roughness (Fig. 9.6B; R_a 2.3 nm) (Edwards et al., 2017).

Polyethylene terephthalate (PET) fabrics treated with $SiCl_4$ plasma saw significant increases in fibre roughness over 60 second and 120 second treatments (Poletti et al., 2003). Woven carbon/polyetheretherketone fibres treated with O_2 plasma had increased surface roughness after 180 seconds of treatment (Jang and Kim, 1997). This is required to increase the surface reactivity of the carbon fibres. Acrylic fibres modified with low-pressure nitrogen glow plasma showed surface roughness that increased with treatment time length (Liu et al., 2006).

In low-pressure plasma treatment, surface etching occurs as fibres are bombarded with charged ions and electrons, subjecting their surfaces to a physical sputtering effect, alongside any chemical effects (Shishoo, 2007). The sputtering can lead to micro- or nano-roughness on the fibre surfaces, exposing the underlying structure by removing surface material as a result of the energy input during the plasma treatment process (Smiley and Delgass, 1993). Low temperature plasma techniques are surface selective in this regard (Kou et al., 2003). Fibre etching is more commonly observed with nonpolymerizing gases such as O_2 than with depositing gases such as C_2F_6 (Kale and Desai, 2011). This has been observed by Kale and Desai and is attributed to plasma etching.

As low-pressure plasma is not viable for large industrial scale production due to the depressurization steps required, atmospheric plasma can be used. Commercially available atmospheric plasma treatments include AcXys ULS Plasma Spot and Plasma Curtain models or the Diener PICO low temperature low-pressure plasma coaters can etch textile fibres in the aforementioned manner.

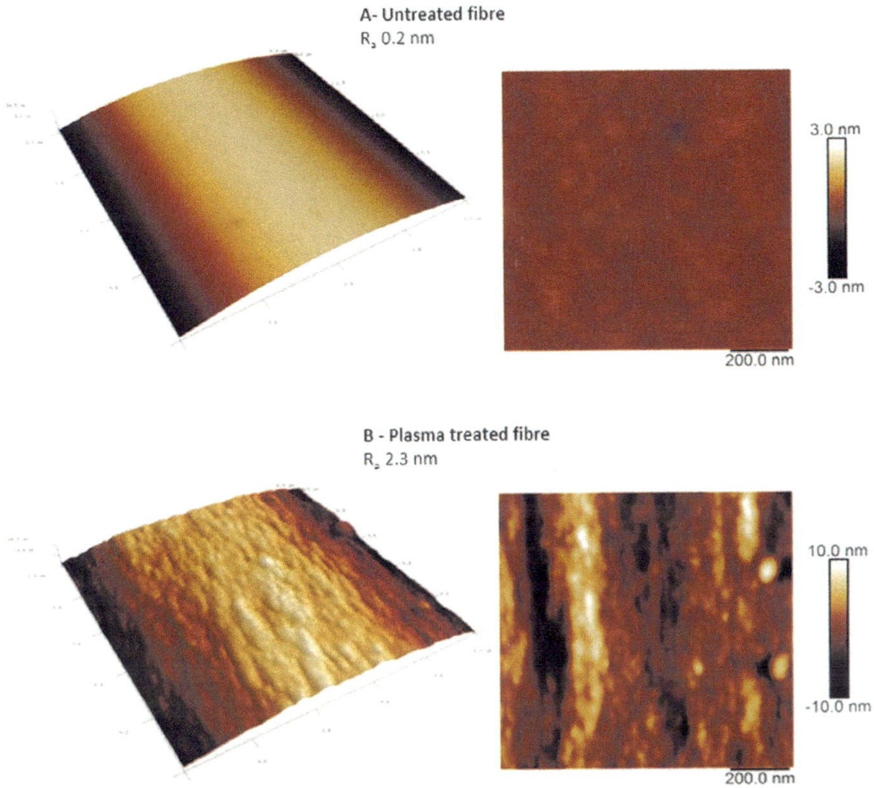

Fig. 9.6 Typical AFM images of cellulosic fibre surfaces before (A) and after (B) plasma treatment (Edwards et al., 2017).

9.9 Silicon chemistry in plasma

Cold plasma treatment can be used to deposit organosilicon films onto textiles, resulting in an ultra-hydrophobic coating. This is achieved through use of the organosilicon precursor hexamethyldisiloxane (HMDSO). HMDSO plasma-polymerized films have CH_3 groups within a Si–O network resulting in a hydrophobic effect, as well as reduced friction, when used to treat polyester fibres (Keller et al., 2007).

HMDSO has also been used in combination with atmospheric plasma treatment by Ji et al. PET fibres were treated to give super-hydrophobic and water repellent functionality. Under testing, functionalized PET gave a water repellency rating of 90 (slight random sticking or wetting of upper surface) versus the untreated PET's rating of 0 (complete wetting of whole upper and lower surfaces) according to the AATCC 58-248 spray test standards (Ji et al., 2008).

Polyester and polyamide have been functionalized with hydrophobic siloxane plasma coatings by Bertaux et al. (2009). They found that the water repellence of the siloxane-coated fabrics reduced the friction coefficient in wet conditions, with positive impact on clothing comfort; one of the major current concerns for textile manufacturers

Silicon plasma treatments have limited durability and will also mean the fabric soils, requiring frequent washing. Information regarding number of washes after which repellency is lost is currently lacking; and reproofing of said garment would likely require retreatment with the silicone plasma.

9.10 Multifunctional plasma treatments

There is much demand for multifunctional finishes on textiles ranging from repellency to flame retardancy, to antibacterial activity. It is possible to utilize multifunctional plasma treatments to confer multiple properties to a textile item or garment, besides hydrophobicity. These include functionalization of cotton by plasma-induced vapour phase graft polymerization of acrylic acid and subsequent silver nanoparticle synthesis. This confers antibacterial activity, self-cleaning, thermal stability and wash fastness to the cotton fabric (Wang et al., 2017).

Davis et al. used atmospheric plasma treatment to confer DWR and antimicrobial functionality on a cotton/polyester blend. The fabric surface was activated with plasma, depositing 1,1,2,2-tetrahydroperfluorodecyl acrylate and 1,1,2,2-tetrahydroperfluorododecyl acrylate monomers; then graft polymerizing the monomers with a second plasma exposure (Davis et al., 2011). Levalois-Grützmacher et al. (2012) used argon plasma-induced graft-polymerization of diethyl(acryloyloxyethyl)phosphoramidate to produce a wash-resistant, flame-retardant, water repellent multifunctional coating on natural fabrics like cotton and silk.

9.11 Summary

Plasma modification of textiles and fibres is an environmentally friendly method for textile surface functionalization, without altering the bulk properties of the material. Plasma treatment doesn't require the use of water and doesn't create C_8 by-products that persist in the environment, so is a process that is both more economical and ecologically sound.

As textile manufacturers and modifiers move away from C_8 chemistry, there will be an increase in the use of nonthermal plasma treatment. It has been shown in this chapter that plasma modification is a versatile and varied tool for imparting both hydrophobic and oleophobic functionality, as well as other functionalities. The availability of pilot-scale machinery for industry and academia aids in the development and refinement of such technology. Bulk commercial processing machinery is also available worldwide from companies such as Europlasma and Tenka Plasma Systems. Plasma treatment of textiles indeed has a bright future, aiming towards zero discharge of hazardous chemicals in the textile and footwear industry in the near future (ZDHC, 2016).

9.12 Sources of further information and advice

Plasma Technologies for Textiles (Shishoo, 2007), edited by R. Shishoo gives an overview of plasma technology with more detailed information. *The Encyclopedia of Color Science and Technology* (Goswami, 2014), edited by R. Luo gives an outline of textile finishing.

References

Alderson, A., 2015. Sports tech outdoor fabrics [Technology Wearables]. Eng. Technol. 10, 84–85.

ATSDR, 2016. Health Effects of PFAS [Online]. Agency for Toxic Substances and Disease Registry (ATSDR). Available from: http://www.atsdr.cdc.gov/pfc/health_effects_pfcs.html.

Bachmann, J., Ellies, A., Hartge, K.H., 2000. Development and application of a new sessile drop contact angle method to assess soil water repellency. J. Hydrol. 231–232, 66–75.

Bahners, T., Textor, T., Opwis, K., Schollmeyer, E., 2008. Recent approaches to highly hydrophobic textile surfaces. J. Adhes. Sci. Technol. 22, 285–309.

Balu, B., Breedveld, V., Hess, D.W., 2008. Fabrication of "roll-off" and "sticky" superhydrophobic cellulose surfaces via plasma processing. Langmuir 24, 4785–4790.

Banks, R.E., Smart, B.E., Tatlow, J.C., 1994. In: Organofluorine Chemistry: Principles and Commercial Applications. Plenum, New York, NY.

Bertaux, E., et al., 2009. Effects of siloxane plasma coating on the frictional properties of polyester and polyamide fabrics. http://www.sciencedirect.com/science/article/pii/S0257897209005763.

Blackburn, R.S., 2009. Environmentally friendly plasma technologies for textiles. Sustainable Textiles – Life Cycle and Environmental Impact, Woodhead Publishing, Cambridge (Chapter 7).

Borra, J.P., 2008. Charging of aerosol and nucleation in atmospheric pressure electrical discharges. Plasma Phys. Controlled Fusion 50, 124036.

Brinker, C., Scherer, G., 1990. Sol-Gel Science: The Physics and Chemistry of Sol-Gel Processing. Academic Press, Boston.

Buyle, G., 2009. Nanoscale finishing of textiles via plasma treatment. Mater. Technol. 24, 46–51.

Carneiro, N., Souto, A.P., Silva, E., Marimba, A., Tena, B., Ferreira, H., Magalhaes, V., 2001. Dyeability of corona-treated fabrics. Color. Technol. 117, 298–302.

Daoud, W.A., Xin, J.H., Tao, X., 2004. Superhydrophobic silica nanocomposite coating by a low-temperature process. J. Am. Ceram. Soc. 87, 1782–1784.

Davis, R., El-Shafei, A., Hauser, P., 2011. Use of atmospheric pressure plasma to confer durable water repellent functionality and antimicrobial functionality on cotton/polyester blend. Surf. Coat. Technol. 205, 4791–4797.

de Vietro, N., Belforte, L., Lambertini, V., Placenza, B., Fracassi, F., 2015. Plasma treatment for preparing durable water repellent and anti-stain synthetic fabrics for automotive applications. J. Surf. Eng. Mater. Adv. Technol. 5, 103–109.

Donnelly, V.M., Kornblit, A., 2013. Plasma etching: yesterday, today, and tomorrow. J. Vac. Sci. Technol. A 31, 050825.

Edwards, N.W.M., Best, E.L., Connell, S.D., Goswami, P., Carr, C.M., Wilcox, M.H., Russell, S.J., 2017. Role of surface energy and nano-roughness in the removal efficiency of bacterial contamination by nonwoven wipes from frequently touched surfaces. Sci. Technol. Adv. Mater. 18, 197–209.

Frisbee, S.J., Brooks, A.P., Maher, A., Flensborg, P., Arnold, S., Fletcher, T., Steenland, K., Shankar, A., Knox, S.S., Pollard, C., Halverson, J.A., Vieira, V.M., Jin, C., Leyden, K.M., Ducatman, A.M., 2009. The C8 health project: design, methods, and participants. Environ. Health Perspect. 117, 1873–1882.

Fromme, H., Tittlemier, S.A., Völkel, W., Wilhelm, M., Twardella, D., 2009. Perfluorinated compounds – exposure assessment for the general population in western countries. Int. J. Hyg. Environ. Health 212, 239–270.

Gao, L., Mccarthy, T.J., 2006. Contact angle hysteresis explained. Langmuir 22, 6234–6237.

Goswami, P., 2014. Finish, textile. In: LUO, R. (Ed.), Encyclopedia of Color Science and Technology. Springer New York, New York, NY.

Greenpeace, 2012. Chemistry for Any Weather. Greenpeace, Amsterdam.

Hearle, J.W.S., Morton, W.E. (Eds.), 2008. Physical Properties of Textile Fibres, fourth ed. Woodhead Publishing Ltd, Cambridge, UK.

Hegemann, D., 2005. Stain repellent finishing on fabrics. http://onlinelibrary.wiley.com/doi/10.1002/adem.200500063/abstract.

Heywood, D., 2003. Textile Finishing. Society of Dyers and Colourists, Bradford, UK.

Hodak, S.K., Supasai, T., Paosawatyanyong, B., Kamlangkla, K., Pavarajarn, V., 2008. Enhancement of the hydrophobicity of silk fabrics by SF6 plasma. Appl. Surf. Sci. 254, 4744–4749.

Hunter, R.J., 2001. Foundations of Colloid Science, second ed. Oxford University Press, Oxford.

Inagaki, N., et al., 1991. Hydrophobic polymer films plasma-polymerized from perfluoromethane/hydrocarbon and hexafluoroacetone/hydrocarbon mixtures. http://onlinelibrary.wiley.com/doi/10.1002/app.1991.070430320/abstract.

Iriyama, Y., Yasuda, T., Cho, D.L., Yasuda, H., 1990. Plasma surface treatment on nylon fabrics by fluorocarbon compounds. J. Appl. Polym. Sci. 39, 249–264.

Jang, J., Kim, H., 1997. Improvement of carbon fiber/PEEK hybrid fabric composites using plasma treatment. Polym. Compos. 18, 125–132.

Jelil, R.A., 2015. A review of low-temperature plasma treatment of textile materials. J. Mater. Sci. 50, 5913–5943.

Ji, Y.-Y., Hong, Y.-C., Lee, S.-H., Kim, S.-D., Kim, S.-S., 2008. Formation of superhydrophobic and water-repellency surface with hexamethyldisiloxane (HMDSO) coating on polyethyleneteraphtalate fiber by atmosperic pressure plasma polymerization. Surf. Coat. Technol. 202, 5663–5667.

Kale, K.H., Desai, A.N., 2011. Atmospheric pressure plasma treatment of textiles using nonpolymerizing gases. Indian J. Fibre Text. Res. 36, 289–299.

Kalin, M., Polajnar, M., 2014. The wetting of steel, DLC coatings, ceramics and polymers with oils and water: the importance and correlations of surface energy, surface tension, contact angle and spreading. Appl. Surf. Sci. 293, 97–108.

Kalliala, E., Talvenmaa, P., 2000. Environmental profile of textile wet processing in Finland. J. Clean. Prod. 8, 143–154.

Keller, M.A., Fortunato, G., Körner, E., Hegemann, D., 2007. Continuous coating of synthetic fibers using hexamethyldisiloxane. Plasma Process. Polym. 4, S1063–S1067.

Khelifa, F., Ershov, S., Habibi, Y., Snyders, R., Dubois, P., 2016. Free-radical-induced grafting from plasma polymer surfaces. Chem. Rev. 116, 3975–4005.

Kogelschatz, U., Eliasson, B., Egli, W., 1997. Dielectric-barrier discharges. Principle and applications. J. Phys. IV 7, C4/47–C4/66.

Kou, R.-Q., Xu, Z.-K., Deng, H.-T., Liu, Z.-M., Seta, P., Xu, Y., 2003. Surface modification of microporous polypropylene membranes by plasma-induced graft polymerization of α-allyl glucoside. Langmuir 19, 6869–6875.

Kovalchuk, N.M., Trybala, A., Starov, V., Matar, O., Ivanova, N., 2014. Fluoro- vs hydrocarbon surfactants: why do they differ in wetting performance? Adv. Colloid Interf. Sci. 210, 65–71.

Lafuma, A., Quere, D., 2003. Superhydrophobic states. Nat. Mater. 2, 457–460.

Laguardia, L., Ricci, D., Vassallo, E., Cremona, A., Mesto, E., Grezzi, F., Dellera, F., 2007. Deposition of super-hydrophobic and oleophobic fluorocarbon films in radio frequency glow discharges. Macromol. Symp. 247, 295–302.

Levalois-Grützmacher, J., Tsafack, M.-J., Kamlangkla, K., Prinz, K., 2012. Multifunctional coatings on fabrics by application of a low-pressure plasma process. In: 13th International Conference on Plasma Surface Engineering, Garmisch-Partenkirchen, Germany.

Li, S., Jinjin, D., 2007. Improvement of hydrophobic properties of silk and cotton by hexafluoropropene plasma treatment. Appl. Surf. Sci. 253, 5051–5055.

Li, S., Huang, J., Chen, Z., Chen, G., Lai, Y., 2017. A review on special wettability textiles: theoretical models, fabrication technologies and multifunctional applications. J. Mater. Chem. A 5, 31–55.

Liu, Y.C., Xiong, Y., Lu, D.N., 2006. Surface characteristics and antistatic mechanism of plasma-treated acrylic fibers. Appl. Surf. Sci. 252, 2960–2966.

Masuoka, T., Yasuda, H., 1982. Plasma polymerization of hexafluoroethane. J. Polym. Sci. Polym. Chem. Ed. 20, 2633–2642.

McCord, M., et al., 2003. Surface analysis of cotton fabrics fluorinated in radio-frequency plasma. http://onlinelibrary.wiley.com/doi/10.1002/app.11896/abstract.

Miller, J.D., Veeramasuneni, S., Drelich, J., Yalamanchili, M.R., Yamauchi, G., 1996. Effect of roughness as determined by atomic force microscopy on the wetting properties of PTFE thin films. Polym. Eng. Sci. 36, 1849–1855.

Mognetti, B.M., Yeomans, J.M., 2010. Modelling receding contact lines on superhydrophobic surfaces. Condens. Matter, 26 (23), 18162–18168. https://doi.org/10.1021/la103539m.

Morent, R., de Geyter, N., Verschuren, J., de Clerck, K., Kiekens, P., Leys, C., 2008. Non-thermal plasma treatment of textiles. Surf. Coat. Technol. 202, 3427–3449.

Plasmatreat, 2014. Plasmatreat [Online]. Available from: http://www.plasmatreat.co.uk/?gclid=Cj0KEQjwmIrJBRCRmJ_x7KDo-9oBEiQAuUPKMnqUX-Zhi5id7jmi2cWtQiBAJUljDFQngPVxYdhr1u8aAm8X8P8HAQ (Accessed 22 May 2017).

Poletti, G., Orsini, F., Raffaele-Addamo, A., Riccardi, C., Selli, E., 2003. Cold plasma treatment of PET fabrics: AFM surface morphology characterisation. Appl. Surf. Sci. 219, 311–316.

Russell, S., 2004. Handbook of Nonwovens. CRC Press, Boca Raton, FL.

Ryu, J., Wakida, T., Takagishi, T., 1991. Effect of corona discharge on the surface of wool and its application to printing. Text. Res. J. 61, 595–601.

Sahun, H.T., et al., 2002. Surface fluorination of paper in CF4-RF plasma environments. https://link.springer.com/article/10.1023/A:1020119109370.

Sajadinia, S.H., Sharif, F., 2010. Thermodynamic analysis of the wetting behavior of dual scale patterned hydrophobic surfaces. J. Colloid Interface Sci. 344, 575–583.

Sardella, E., Intranuovo, F., Rossini, P., Nardulli, M., Gristina, R., D'Agostino, R., Favia, P., 2009. Plasma enhanced chemical vapour deposition of nanostructured fluorocarbon surfaces. Plasma Process. Polym. 6, S57–S60.

Schutze, A., Jeong, J.Y., Babayan, S.E., Park, J., Selwyn, G.S., Hicks, R.F., 1998. The atmospheric-pressure plasma Jet. A review and comparison to other plasma sources. IEEE Trans. Plasma Sci. 26, 1685–1694.

Schwartz, L.W., Garoff, S., 1985. Contact angle hysteresis on heterogeneous surfaces. Langmuir 1, 219–230.

Shahidi, S., Wiener, J., Ghoranneviss, M., 2013. Surface Modification Methods for Improving the Dyeability of Textile Fabrics. InTech, Leeds, pp. 33–52.

Shishoo, R., 2007. Plasma Technologies for Textiles. Woodhead Publishing Ltd., Cambridge, UK

Sigurdsson, S., Shishoo, R., 1997. Surface properties of polymers treated with tetrafluoromethane plasma. J. Appl. Polym. Sci. 66, 1591–1601.

Sing, K.S.W., Williams, R.T., 2012. Historical aspects of capillarity and capillary condensation. Microporous Mesoporous Mater. 154, 16–18.

Skinner, F.K., Rotenberg, Y., Neumann, A.W., 1989. Contact angle measurements from the contact diameter of sessile drops by means of a modified axisymmetric drop shape analysis. J. Colloid Interface Sci. 130, 25–34.

Smiley, R.J., Delgass, W.N., 1993. AFM, SEM and XPS characterization of PAN-based carbon fibres etched in oxygen plasmas. J. Mater. Sci. 28, 3601–3611.

Sparavigna, A., 2008. Plasma treatment advantages for textiles. Los Alamos Nat. Lab. Prepr. Arch. Phys. 1, 1–16 arXiv:0801.3727v1 [physics.pop-ph].

Spori, D.M., Drobek, T., Zürcher, S., Ochsner, M., Sprecher, C., Mühlebach, A., Spencer, N.D., 2008. Beyond the lotus effect: roughness influences on wetting over a wide surface-energy range. Langmuir 24, 5411–5417.

Sun, D., Stylios, G.K., 2006. Fabric surface properties affected by low temperature plasma treatment. J. Mater. Process. Technol. 173, 172–177.

Takahashi, K., Tachibana, K., 2001. Solid particle production in fluorocarbon plasmas. I. Correlation with polymer film deposition. J. Vac. Sci. Technol. A 19, 2055–2060.

Temmerman, E., Leys, C., 2005. Surface modification of cotton yarn with a DC glow discharge in ambient air. Surf. Coat. Technol. 200, 686–689.

Thorsen, W.J., 1971. Improvement of cotton spinnability, strength, and abrasion resistance by corona treatment. Text. Res. J. 41, 455–458.

Wang, H.Z., Rembold, M.W., Wang, J.Q., 1993. Characterization of surface properties of plasma-polymerized fluorinated hydrocarbon layers: surface stability as a requirement for permanent water repellency. J. Appl. Polym. Sci. 49, 701–710.

Wang, C.X., Ren, Y., Lv, J.C., Zhou, Q.Q., Ma, Z.P., Qi, Z.M., Chen, J.Y., Liu, G.L., Gao, D.W., Lu, Z.Q., Zhang, W., Jin, L.M., 2017. In situ synthesis of silver nanoparticles on the cotton fabrics modified by plasma induced vapor phase graft polymerization of acrylic acid for durable multifunction. Appl. Surf. Sci. 396, 1840–1848.

White, L.R., 1977. On deviations from Young's equation. J. Chem. Soc. Faraday Trans. 1 73, 390–398.

Yeh, J.T., Chen, C.L., Huang, K.S., 2007. Preparation and application of fluorocarbon polymer/SiO_2 hybrid materials, part 2: water and oil repellent processing for cotton fabrics by sol–gel method. J. Appl. Polym. Sci. 103, 3019–3024.

Yuan, Y., Lee, T.R., 2013. Contact angle and wetting properties. Springer Ser. Surf. Sci. 51, 3–34.

ZDHC, 2016. Zero Discharge of Hazardous Chemicals [Online]. Available from: http://www.roadmaptozero.com/.

Zhang, J., France, P., Radomyselskiy, A., Datta, S., Zhao, J., Van Ooij, W., 2003. Hydrophobic cotton fabric coated by a thin nanoparticulate plasma film. J. Appl. Polym. Sci. 88, 1473–1481.

Zille, A., Oliveira, F.R., Souto, A.P., 2015. Plasma treatment in textile industry. Plasma Process. Polym. 12, 98–131.

Sol–gel-based treatments of textiles for water repellence

10

Ningtao Mao, Miyu Du
University of Leeds, Leeds, United Kingdom

Since the discovery of the secrets of excellent hydrophobicity and/or superhydrophobicity (Nosonovsky and Bhushan, 2009) in biological surfaces (e.g. lotus leaves, butterfly wings, mosquito eyes, red rose petals, desert beetle and fish scales) (Wagner et al., 1996), a series of new technologies (Latthe et al., 2012) (e.g. versatile electrochemical deposition, electrospraying/electrospinning method, chemical etching, hydrothermal synthesis, phase separation, self-assembly and layer-by-layer (LBL) deposition methods, solution immersion method, chemical vapor deposition (CVD) plasma treatment, laser surface treatment and sol–gel method) have been developed to form superhydrophobic surfaces (see Fig. 10.1) (Kreder et al., 2016) by incorporating both nanoscale rough surface and hydrophobic facile finishing treatment.

Those technologies are also applied on textile fabrics to improve their hydrophobicity. For example, laser surface treatment and plasma surface modification (Kan, 2016), surface nanocoating (Ramaratnam et al., 2008; Xue, 2011), electrospraying (Meirowitz, 2016; Su et al., 2016), LBL deposition (Qing et al., 2013), CVD (Meirowitz, 2016) and sol–gel coating (Gulrajani, 2013) were all used to alter surface morphology and surface energies of textile materials.

The sol–gel process is a well-recognized method of synthesizing nanoporous gels and nanoparticles. It enables the production of thin films of nanocrystalline particles, or multilayered films of porous pillars, rough coatings and nanocomposites (Montarsolo et al., 2013) on textile materials, and has wide applications for producing fabrics having superhydrophobic properties (Mahltig and Böttcher, 2003; Yu et al., 2007; Xue et al., 2008; Yin et al., 2011), antimicrobial properties (Mahltig et al., 2004; Mahltig et al., 2005; Xing et al., 2007), UV radiation protection (Abidi et al., 2007; Xing and Ding, 2007; Aķ·it and Onar, 2008), flame retardancy (Alongi et al., 2011; Aksit et al., 2016; Grancaric et al., 2017), dye fastness (Aķ·it and Onar, 2008), biomolecule immobilization (Li et al., 2007) and photocatalytic features (Colleoni et al., 2012).

10.1 Fundamentals of hydrophobicity and superhydrophobicity

The hydrophobicity of a physically smooth and chemically homogeneous solid surface (Xue, 2011) is normally expressed as the contact angle using the Young–Laplace

Waterproof and Water Repellent Textiles and Clothing. https://doi.org/10.1016/B978-0-08-101212-3.00009-5

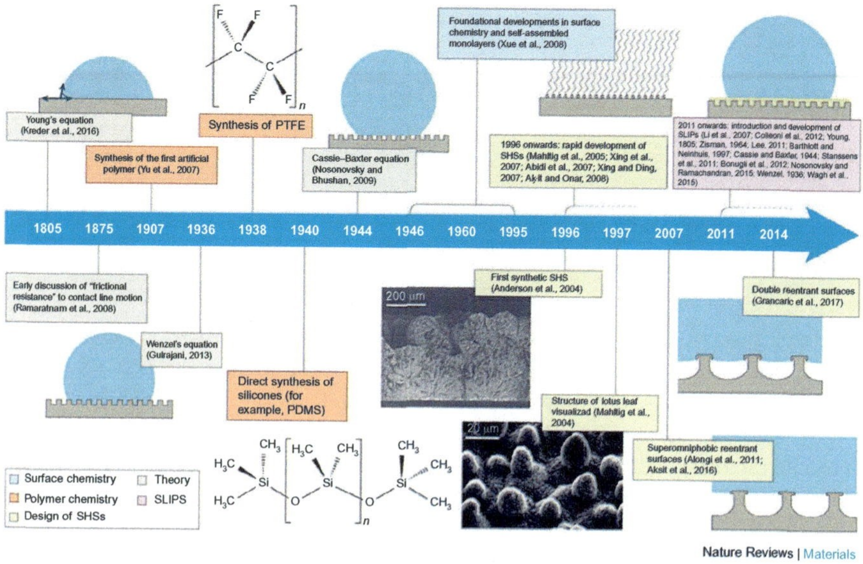

Fig. 10.1 Timeline of major advances in the area of liquid repellency (Kreder et al., 2016).

equation (Young, 1805) based in thermodynamics theory (Zisman, 1964). The contact angle, θ_c, in the Young–Laplace equation is expressed and explained in Fig. 10.2 as follows (Young, 1805).

$$\cos\theta_c = \frac{\gamma_{sg} - \gamma_{sl}}{\gamma_{\mathrm{lg}}} \tag{10.1}$$

where γ_{sg} is the surface tension of solid and gas, γ_{sl} is the surface tension of solid and liquid and γ_{lg} is the surface tension of liquid and gas.

Generally, the solid surface is considered hydrophobic if the contact angle is greater than 90 degrees and hydrophilic when less than 90 degrees.

A superhydrophobic surface exhibits a contact angle of greater than 150 degrees (Ramaratnam et al., 2008; Xue, 2011; Lee, 2011). A water droplet on a superhydrophobic surface always forms an almost perfect sphere and a slight tilting might

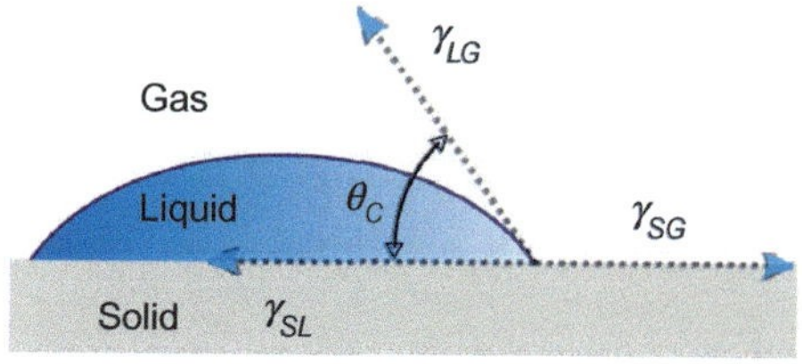

Fig. 10.2 Schematic of liquid drop on solid surface in Young's equation (Zisman, 1964).

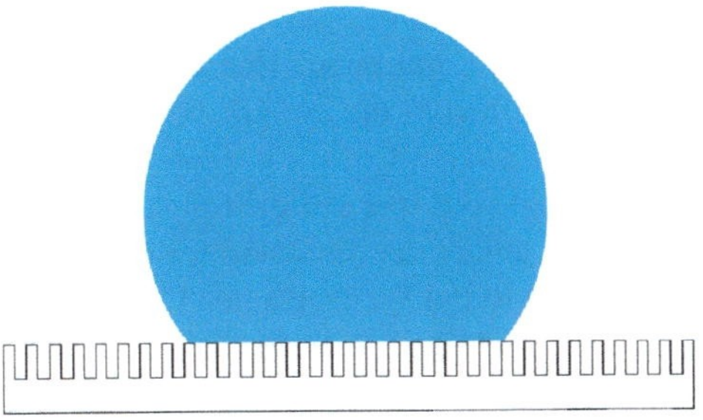

Fig. 10.3 Cassie–Baxter model (Nosonovsky and Ramachandran, 2015).

be sufficient to cause the water drop to roll off. In addition, the ability of the surface to make water droplets bounce off of it constitutes the third property of a superhydrophobic surface that is important for technical applications (Xue, 2011).

The roughness of solid surfaces is related to the materials' hydrophobicity (Ramaratnam et al., 2008). The lotus-effect (Barthlott and Neinhuis, 1997) due to microscopic roughness of plant surface exhibits excellent water repellence and self-cleaning properties. The hydrophobic principle of such special micro- to nano-scale surface structure of the lotus-effect is different from Young's equation (see Eq. 10.1) and can be explained by using Wenzel or Cassie–Baxter (Cassie and Baxter, 1944) models (Stanssens et al., 2011; Bonugli et al., 2012), as shown in Figs 10.3 and 10.4.

In the Cassie–Baxter model, the drop sits on top of the rough surface with trapped air underneath. θ_R is the apparent contact angle on a rough solid surface which is dependent on both Young's contact angle, θ_c, and the fraction of liquid/solid contact area, f. The Cassie–Baxter model indicates that the greater porosity of a solid surface would lead to smaller contact area between solid and liquid (i.e. smaller f value) and result in a higher apparent water contact angle (Bonugli et al., 2012).

$$\cos\theta_R = f\cos\theta_c + f - 1 \tag{10.2}$$

The Wenzel model (Wenzel, 1936) describes the homogeneous wetting regime as shown in Fig. 10.4, for the apparent contact angle formed by a liquid wetting on a rough surface, θ_W:

$$\cos\theta_W = \bar{r}\cos\theta_Y, \tag{10.3}$$

Fig. 10.4 Wenzel model (Nosonovsky and Ramachandran, 2015).

where $|\bar{r}\cos\theta_Y| \leq 1$, and θ_Y is the Young contact angle defined for an ideal surface which corresponds to the stable equilibrium state (i.e. minimum free energy state for the system); $\bar{r}$ is the average roughness ratio, defined as the ratio of the true area of the solid surface to the apparent area.

When a water droplet gradually penetrates into the void space of the rough surface, the air pockets underneath the water droplet are no longer thermodynamically stable, the surface wetting transitions from the Cassie state to the Wenzel state. The critical contact angles of the water droplet on the rough surface, θ_C, are expressed in the following equation (see Eq. 10.4):

$$\cos\theta_C = \frac{\phi - 1}{r - \phi} \tag{10.4}$$

where θ_C is the critical contact angle, ϕ is the fraction of solid/liquid interface where the drop is in contact with the surface and r is solid roughness (for a flat surface, $r = 1$).

When the critical contact angles are between 0 and $\pi/2$, the surface wetting transition from the Cassie state to the Wenzel state begins.

The most popular methods used to measure water hydrophobicity and the repellence property on textile materials are contact angle measurement, drop tests and spray tests. Measurements of the contact angle of the fabric surface are widely used to assess the hydrophobicity of materials, and they can be performed in a goniometer or tensiometer. In a goniometer, a sessile droplet of water is placed on the fabric surface, and the contact angle between the water droplet and the fabric on the image is measured as shown in Fig. 10.5 (Wagh et al., 2015). Contact angle hysteresis and roll-off angle could also be evaluated in this measurement.

The hydrophobicity of textile materials containing aerogel could also be examined by using the water uptake method (Aegerter et al., 2011) and weight difference method (Wagh et al., 2015). In these methods, textile materials to be tested are placed in a humid environment to allow water adsorption, then the increased mass is monitored to characterize hydrophobicity. The less the weight increases, the better the hydrophobicity. However, a larger sample is necessary to obtain the required accuracy. Two quick ways to test the water repellence property of textile fabrics are drop tests and spray tests.

Fig. 10.5 Contact angle measurement (Wagh et al., 2015).

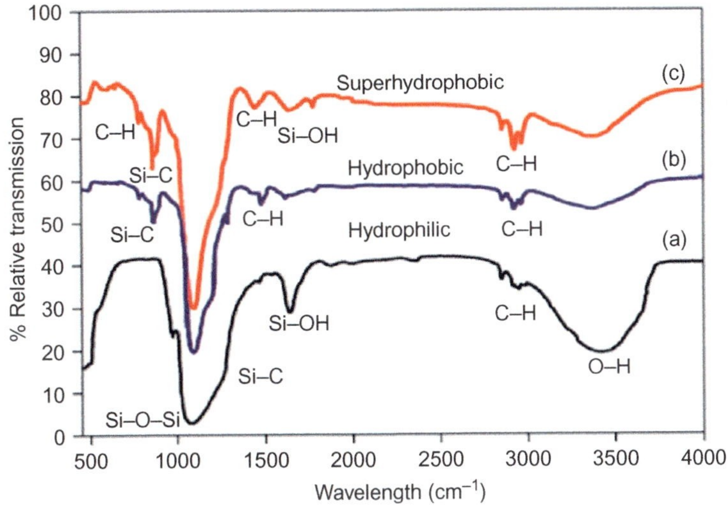

Fig. 10.6 FTIR spectra of silica aerogel. (a) Hydrophilic, (b) hydrophobic and (c) superhydrophobic (Wagh et al., 2015).

Spectroscopic methods such as Fourier transform infrared spectroscopy (FTIR) and solid-state nuclear magnetic resonance (NMR) can also be used to assess the hydrophobicity of a material's surface according to the surface chemical structure and components. They can provide information about chemical bonds for the tested material's surface as observed from the FTIR and NMR spectra, so hydrophobic and hydrophilic bonds could be identified (Wagh et al., 2015) as shown in Fig. 10.6.

10.2 Sol–gel process

The sol–gel process is a process of producing inorganic polymers or ceramic materials (or organically modified materials, ORMOSILs or CERAMERs) through a transformation from liquid precursors to a sol (or more generally, a colloidal suspension), gelation of the sol to form a gel (a nonfluid 3D network structure extending through a fluid phase), and removal of the solvent (Brinker and Scherer, 1991). A sol is more generally defined as a colloidal system which is a dispersion of one phase in another where, 'the molecules or polymolecular particles dispersed in a medium have at least in one direction a dimension roughly between 1 nm and 1 μm' (IUPAC, 2001). The sol–gel process is able to produce complex inorganic materials such as ternary and quaternary oxides from a chemically homogeneous precursor at lower processing temperatures and shorter synthesis times by ensuring atomic-level mixing of reagents; it also enables greater control over particle morphology and size (Danks et al., 2016).

Traditionally, there are three groups of sol–gel formation processes (Kakihana, 1996): (1) the sol–gel route based upon hydrolysis and condensation of metal-alkoxides (alkoxide gels), (2) the gelation route based upon concentration of aqueous solutions involving metal-chelates, often called a 'chelate gel' or 'amorphous chelate' route and (3) the organic polymeric gel route.

The sol–gel process is divided into three stages in relation to the formation of the microstructure of nanoparticles or aerogels/xerogels: sol stage before wet gel formation, gelation stage and drying stage (Brinker and Scherer, 1990). In a sol–gel process, precursors could be both inorganic and organic (e.g. metal nitrates and alkoxides). They could be colloidal metal oxides or hydroxides (e.g. alkoxide, water glass, PZT ($PbZr_{1-x}Ti_xO_3$)) and metal complex, metal oxane polymer precursors and polymer complex, as well as inorganic–organic hybrid materials. The examples of frequently used alkosilane in sol–gel process are tetramethoxysilane (TMOS), tetraethyoxysilane (TEOS), hexadecyltrunethoxysilane (HTEOS), γ-chloropropyltriethoxysilane (CPTS), trimethylethoxysilane (TMES), hexamethyldisilazane (HMDS) and methyltrimethoxysilane (MTMS), aluminium isopropoxide and titanium tetraisopropoxide.

The solvent used could be either water to form an aqueous sol–gel system or non-aqueous solvent (e.g. methanol, ethanol, etc.) to form a nonaqueous system. The precursors are first dissolved in an organic solvent that is miscible with water, and then the acid and base catalysts are added separately to speed up the hydrolysis and condensation reactions to obtain the wet gel, which is then aged and strengthened in a solvent; the strengthened wet gel finally is dried to form porous aerogel (Hüsing and Schubert, 2000).

Alkoxides, one of the most popular precursors, can be produced in many ways. For example, a large number of alkoxides such as Ta(OR)5 (R = Me, Et, nBu) are produced via metal chlorides and highly reducing metals (i.e. alkali metals and lanthanides) reacting with alcohols (Brown, 1970), and via anodic dissolution of the metal in alcohol with an electro-conductive additive such as LiBr (Turova et al., 1996).

Hydrolysis of silicon alkoxides is a typical example of forming sols in the sol–gel process. It results in the replacement of an alkoxy group with a hydroxyl in both acid and base-catalysed systems, and the inductive and steric effects of the R group in different silicon alkoxides (e.g. $SiR(OR)_3$, $SiR_2(OR)_2$ or SiR_3OR) impacts the rates of their hydrolysis reactions. The hydrolysis reactions are slower under acidic conditions and faster under basic conditions because the rate of each hydrolysis step depends on the relative electron withdrawing or donating power of –OH versus –OR groups. This leads to the different structures of resultant silica gels (Danks et al., 2016). The presence of solvents in a silica sol–gel process can either enhance mixing (many silicon alkoxides are immiscible with water) or direct interaction of solvent molecules with the silicon centre, while water itself is important, inappropriate alkoxide:water ratio can be tuned to limit hydrolysis (Fig. 10.7; Danks et al., 2016).

$$Si(OR)_4 + nH_2O \rightarrow Si(OR)_{4-n}(OH)_n + nROH$$

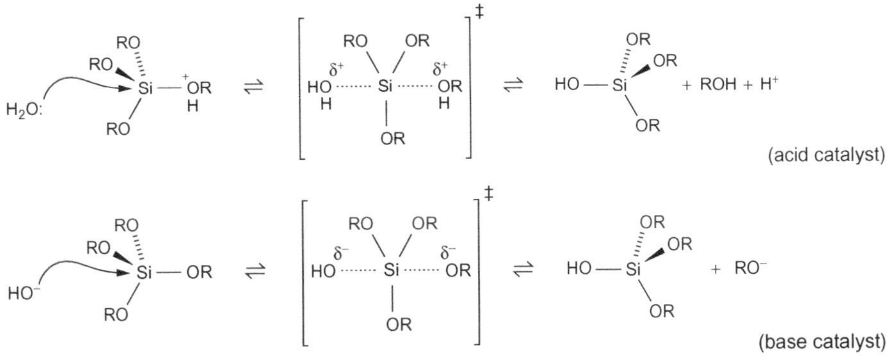

Fig. 10.7 Acid and base catalysed hydrolysis of silicon alkoxides (Danks et al., 2016)

The chelating agents containing bidentate or multidentate ligands such as acetylacetone may have stereochemical effects by directing hydrolysis and condensation to certain sites (Schubert, 2005), and the strength of binding of the chelating ligand, as well as the ligand:alkoxide ratio both impact the reactivity of the titanium precursor and affect the structure of the resultant gel. The presence of chelating agents in the sol–gel process of titanium alkoxides can be a method to reduce hydrolysis and condensation rates to slow down their sol–gel reactions (Babonneau et al., 1988), because most titanium alkoxides react vigorously with water to produce ill-defined titanium-oxo/hydroxo precipitates.

A sol is an amorphous or crystalline colloidal particle suspension including inorganic polymer particles in a liquid, and these particles might interact with each other by Van der Waals force, hydrogen bonds and covalent bonds, as well as the solvation and subsequent ionic crosslinking of biopolymers (Zha and Roggendorf, 1991). There are two stages for the formation of sol particles in a sol state: nucleation and growth. For growth of silica nanoparticles (Bogush et al., 1988), two growth mechanisms could be described by two models, monomer addition (Rao et al., 2003; Rao et al., 2007; Hegde and Rao, 2007) and controlled aggregation (Matsoukas and Gulari, 1988; Matsoukas and Gulari, 1987), respectively. In the monomer addition model, particle growth happens on the surface of primary particles which initially form nucleation; in the controlled aggregation model, particle growth occurs between primary particles, which means the nuclei occur throughout the reaction process (Eahman and Padavettan, 2012). In this aggregation process, particle growth (aggregation) primarily happens by aggregating between smaller particles and newly formed nuclei rather than interacting within larger particles (Eahman and Padavettan, 2012). In the case of existence of already formed large particles and low concentration of hydrolysed monomers, the main process is the growth of monomers on the surface of grown particles. Under high concentration of hydrolysed monomers, when the rate of monomer reaction is slower than that of hydrolysis, the dominant process is nucleation of hydrolysed monomers and the formation of more particles having smaller sizes which, results from the high concentration of hydrolysed monomers (Bogush et al., 1988).

A gel consists of a three-dimensional continuous solid network with liquid enclosed. The network is formed by aggregation of colloidal particles and agglomeration of formed aggregations (Wang and Yin, 2016). The gelation process is a transition process for a silicone precursor changing from sol to gel; it involves the condensation of formed particles and aggregates until the aggregates/clusters collide when the last link between the larger clusters is formed (Bogush et al., 1988; Matsoukas and Gulari, 1987; Bogush and Zukoski, 1991). When a precursor condenses from sol to gel in the gelation process, nanoparticles are formed and grow into particle clusters and 3D nanoparticle network structures gradually, which results in the formation of ceramic nanoparticle powders, a thin layer of nanoparticles/xerogels or nanoporous aerogels/xerogels. The microstructure of resultant aerogel/xerogels might be determined by the size of the particles/aggregations formed in the gelation process, which is also known as the particle sizes of the solid skeleton of the gel, and the drying process of wet gels. The surface roughness obtained with the sol–gel method can be tuned by changing the protocol of the method and the composition of the reaction mixture.

Aerogel is a highly nanoporous material consisting of more than 90% pores filled with air and less than 10% solid as a skeleton; it could be incorporated into textiles to obtain aerogel-textiles composite having excellent thermal insulation property (Shaid et al., 2016). Its particles are in nanoscale, and the pores are a mixture of micro-, meso- and macropores. Usually the porosities of the aerogels formed in the ambient temperature and pressure are smaller than 90%, and this type of aerogel is frequently called xerogel (Kornprobst and Plank, 2013). Organic, inorganic and organic–inorganic hybrid aerogels/xerogels could be formed based on the materials made of their solid skeleton.

The particle size distribution of nanoparticles having spherical and irregular shapes could be characterized in terms of diameter and equivalent spherical diameter, respectively. Various characterization techniques including sieving, image analysis and laser diffraction methods, were employed to examine the particle size distribution. The traditional sieving method uses different mesh sizes of a series of sieves from small to large to separate particles by size into different groups to obtain particle size distribution in terms of their masses (Seville and Wu, 2016). This method cannot provide accurate particle size distribution for particle samples and is not suitable for the particle sizes smaller than 38 μm (Ulusoy and Igathinathane, 2016).

The image analysis method (Welker, 2012) is a direct visualization method used to characterize particle size distribution, as well as pore size distribution, on a 2D surface. Two-dimensional microscopic images can be analysed using image analysis software (Hu and Stroeven, 2003), the particle size distribution, pore size distribution and porosity of the material can be quantified (Aydilek et al., 2002). The images can be obtained by using scanning electron microscopy (SEM) (Sarawade et al., 2006; Gurav et al., 2009; Kang and Choi, 2000; Rao and Bhagat, 2004), field emission scanning electron (FeSEM) (Jung et al., 2003; Lin et al., 2011), transmission electron microscopy (TEM) (Gurav et al., 2008; Rao et al., 2006), atomic force microscope (AFM) (Wei et al., 2006) and X-ray tomography. SEM can provide two-dimensional images of the microstructure of aerogel under a limited magnification. FeSEM provides much higher magnification and adequate resolution to examine the microstructure of the aerogels up to 10 nm. TEM is usually used to evaluate the substructure of particle skeletons or additional particles incorporated in aerogel skeletons and is helpful to characterize the crystalline orientation or metal nanoparticles incorporated in aerogel skeletons (Aegerter et al., 2011). X-ray tomography can provide volumetric information including both the tested surface and the thickness direction (Sondej et al., 2016). However, because only an extremely small amount of targeted samples can be examined in microscopy techniques, preparation of a representative sample of nanoparticles is key to obtaining an accurate result (Seville and Wu, 2016).

Pore size and pore size distribution of nanoporous structures obtained from the sol–gel process can also be characterized by using gas adsorption–desorption (Gurav et al., 2009; Kang and Choi, 2000; Hwang et al., 2007), mercury porosimetry (Pirard et al., 1998; Brown and Lard, 1974) and light scattering technique (Galkina et al., 2011; Yang et al., 2006; Tseng et al., 2009; Kamiyama et al., 1994; Eshel et al., 2004). Gas (helium or nitrogen gases) adsorption–desorption technique is the most widely used nondestructive approach to characterize the porous structure of aerogels. Its measurable pore sizes are in the size range of 2–300 nm. According to the obtained adsorption and desorption isotherm lines, surface area can be calculated by a universal Brunauer–Emmitt–Teller (BET) method (Aegerter et al., 2011; Lowell et al., 2004); average pore size and pore size distribution are mostly calculated by density functional theory (DFT) and Barrett–Joyner–Halenda (BJH) method based on either adsorption or desorption branch of the isotherm (Rao et al., 2004; Rassy and Pierre, 2005). DFT method is suitable to characterize materials mainly containing micropores and mesopores, and this method must be carefully used to ensure accuracy of the experiment data (Aegerter et al., 2011). BJH model is the most widely used approach to characterize material mainly containing mesopores and macropores (Sarawade et al., 2010; Aravind and Soraru, 2011). Mercury porosimetry (Lowell et al., 2004) can theoretically probe a larger range of pore size from nanometers to tens of microns, but due to the mercury used, this method is toxic, and the method is restricted for use in many labs.

Chemical components of a sol–gel-treated surface can be examined by using X-ray photoelectron spectroscopy (XPS), Rutherford backscattering spectrometry (RBS) and energy dispersive X-ray analysis (EDX) (Wei et al., 2006; Gorzalski et al., 2017). XPS analysis can identify and quantify the surface elements presented by the tested material down to several nm in depth (Guascito et al., 2015). RBS can provide an element distribution map of the tested surface at a micrometre scale. EDX can provide information regarding chemical elements and their distribution on the surface. To detect the chemical components and chemical structure of a surface, spectroscopic techniques such as FTIR and Raman spectroscopy can also be used (Erra et al., 1997). The smallest particles which can be identified for FTIR is in the range of 35–50 μm, and in the range of 1–2 μm for Raman (Welker, 2012).

10.3 The influence of sol–gel processing parameters on the structure of resultant nanoparticles and nanoporous aerogels/xerogels

Structure/morphology is determined by several sol–gel processing parameters, e.g. type of (semi)metal atom and alkyl/alkoxide groups involved (see Table 10.1), water/alkoxide ratio and presence of cosolvents.

There are many sol–gel processing parameters which affect the microstructure and properties of the resulting oxidic networks (Cireli et al., 2007), including the types of precursors and catalysts, temperature and pH, reaction time, the ratio of precursor to solvent, the ratio of precursor to water, the concentration of catalysts, ageing time and the solvent used for gel surface modification, as well as drying methods (Al-Oweini and El-Rassy, 2010; Barrett et al., 1951; Sing et al., 1985). Controlling these processing parameters significantly affects the structure and properties of crystalline and amorphous transition metal oxide nanoparticles in the form of thin films (Ohya et al., 2004), monodisperse particles (Scolan and Sanchez, 1998) or aerogels/xerogels (Du et al., 2016). The affected structure and properties of resultant gels and corresponding aerogels/xerogels include density, pore distribution, surface area, thermal stability and hydrophobicity (Fan et al., 2004; Sinko, 2010; Uhlmann et al., 1986; MacKenzie, 1986; Pajonk, 1989; Chang and Ring, 1992; Pajonk et al., 1990; Einarsrud et al., 2001). Various drying methods are adopted to avoid the cracking and shrinking of wet gels during the drying process to achieve the required nanoporous structure in the resultant aerogels/xerogels.

10.3.1 Parameters in the hydrolysis of alkoxides

Hydrolysis and condensation reactions in the sol formation process are strongly affected by processing parameters such as the nature of the R-group (e.g. inductive effects), the ratio of water to alkoxide and the presence and concentration of catalysts (Danks et al., 2016). Examples of this influence of several processing parameters in sol stage (such as molar ratio of solvent to precursor, molar ratio of precursor to water and types and concentrations of acid and base catalysts in silica sol–gel processes) are explained in the following sections.

10.3.1.1 Molar ratio of solvent to the precursor

A greater molar ratio of solvent to precursor means that the precursor was diluted with much more of the solvent, and the concentration of silica/silicone was relatively decreased. It has been reported that a smaller molar ratio of solvent to precursor leads to a higher precursor concentration and will result in a denser aerogel with smaller pores (Hegde and Rao, 2007; Gurav et al., 2008; Nadargi and Rao, 2009); usually, the greater the molar ratio, the smaller the bulk density of the aerogel would be (Matsoukas and Gulari, 1987). When the molar ratio is greater than a certain limit (e.g. 16.6:1 for silica aerogel from TEOS (Matsoukas and Gulari, 1987) and 42:1 for silicone aerogel from MTMS; Gurav et al., 2008), dispersed nanoparticles rather than nanoporous aerogel/xerogel structures would be formed.

Table 10.1 **Five types of gels relevant in sol–gel synthesis of materials (Danks et al., 2016)**

Type of gel	Bonding	Source	Gel schematic
Colloidal (Xue, 2011)	Particles connected by Van der Waals or hydrogen bonding	Metal oxide or hydroxide sols	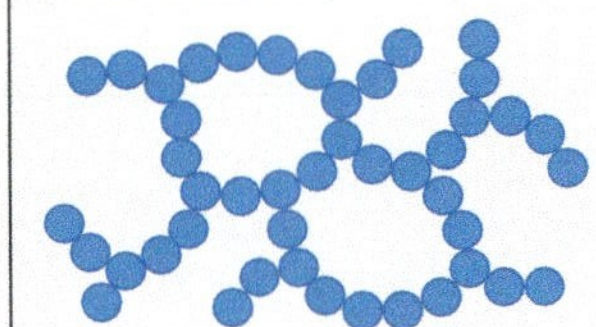
Metal-oxane polymer (Kan, 2016)	Inorganic polymers interconnected via covalent or intermolecular bonding	Hydrolysis and condensation of metal alkoxides, e.g. SiO_2 from tetramethyl orthosilicate	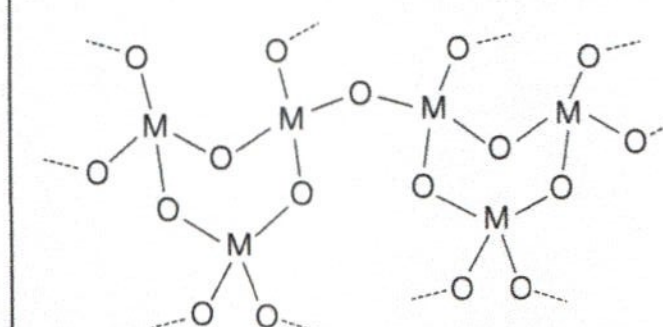
Metal complex (Meirowitz, 2016)	Weakly interconnected metal complexes	Concentrated metal complex solution, e.g. aqueous metal citrate or ethanolic metal urea Often form resins or glassy solids rather than gels	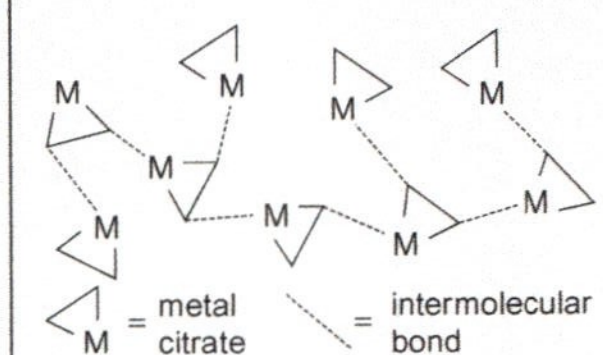

Continued

Table 10.1 Continued

Type of gel	Bonding	Source	Gel schematic
Polymer complex I In situ polymerizable complex ('Pechini' method) (Su et al., 2016; Qing et al., 2013)	Organic polymers interconnected by covalent and coordinate bonding	Polyesterification between polyhydroxy alcohol (e.g. ethylene glycol) and carboxylic acid with metal complex (e.g. metal-citrate)	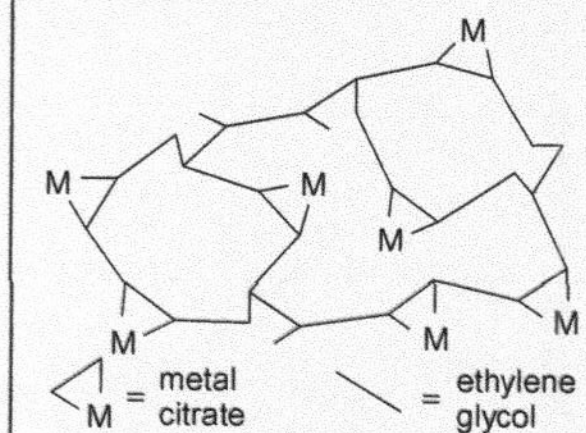
Polymer complex II Coordinating and crosslinking polymers (Gulrajani, 2013)	Organic polymers interconnected by coordinate and intermolecular bonding	Coordinating polymer (e.g. alginate) and metal salt solution (typically aqueous)	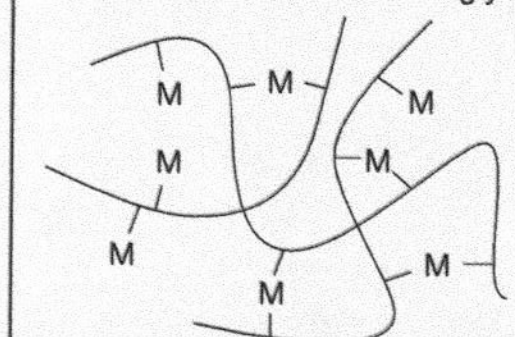

10.3.1.2 Molar ratio of precursor to water

While the amount of water determines both rate and degree of hydrolysis of precursor in the sol–gel process, research findings as to how the molar ratio of precursor to water influences the size of the nanoparticles were mixed. It was reported (Xu et al., 2012) that the size of nanoparticles in silica aerogel made from TEOS changed nonlinearly with a decrease in the molar ratio of precursor to water, ranging from 1:1.7 to 1:100 (or the concentration of water from 0.5 M to 17 M), particle size increased when the molar ratio was greater than 1:7 and decreased when the molar ratio was smaller than 1:7. However, other researchers (Bogush et al., 1988) found that particle size decreased when the molar ratio of TEOS to water decreased from 1:1.3 to 1:55.6.

10.3.1.3 Types and concentrations of acid and base catalysts

Oxalic acid, a kind of organic acid, is frequently selected as an acidic catalyst. Ammonia solution is commonly used as a basic catalyst. Oxalic acid can effectively control the hydrolysis rate to produce more uniform pore size and narrower pore size distribution (Rahman et al., 2007), and it is also a type of drying control chemical additive (DCCA), which allows faster drying. Concentrations of both acid and base catalysts can affect the speed of hydrolysis and condensation, respectively (Brinker and Scherer, 1990; Fardad, 2000). It is reported that a slow condensation speed may produce smaller particles (Rao and Bhagat, 2004; Rao et al., 2006), and greater concentration and feed rate of base catalysts result in larger particles in aerogels (Rahman et al., 2007).

10.3.2 Parameters in gelation and ageing processes

The gelation process is a transition process for hydrolysed precursors changing from sol to gel, and it involves the condensation of formed particles and aggregates until the aggregates/clusters collide when the last link between the larger clusters is formed (Bogush et al., 1988; Matsoukas and Gulari, 1987; Bogush and Zukoski Iv, 1991). When a precursor condenses from sol to gel in the gelation process, nanoparticles are formed and grow into particle clusters gradually. It is thus suspected that the size of the particles formed in sol state before they are collided into big particle frames as a gel might be influenced by sol–gel processing parameters such as the molar ratio of precursor to solvent.

The measurement of particles' size distribution in a gelation process could play an important role in guiding the products' formation in various industrial areas. The methods commonly used for characterization of particle size distribution in different stages of the sol–gel process during aerogel formation are the light scattering technique (includes dynamic light scattering) (Eshel et al., 2004; Yamamoto et al., 2002), SAXS (small-angle X-ray scattering) (Galkina et al., 2011; Yang et al., 2006; Tseng et al., 2009) and laser diffraction (Kamiyama et al., 1994). Also ultrasound techniques such as acoustic attenuation spectroscopy are used to monitor the particles size distribution in a solution with the size range from

Table 10.2 The measurable range of particles in a liquid for different techniques (Yang et al., 2006; Tseng et al., 2009; Kamiyama et al., 1994; Eshel et al., 2004; Yamamoto et al., 2002; Alba et al., 1999; Jacobsen and SUllivan, 1946)

Technique	Range of particle size measurable (μm)
Dynamic light scattering	0.001–6
Laser diffraction	0.01–3000
SAXS (small-angle X-ray scattering)	0.001–0.2
Ultrasound technique	0.01–1000
Centrifugal sedimentation	0.005–75

0.01 to 1000 μm and solution concentration from 0.5%V to 50%V (Alba et al., 1999). Traditionally, particle-size distribution measurement is done using centrifugal sedimentation, which is based on Stokes law, for measuring the required time for particles to settle down in a solution of known viscosity and density, in which the measurable particle size ranges from 5 nm to 75 μm (Jacobsen and Sullivan, 1946). Different techniques suitable for detecting certain range of particle sizes are summarized in Table 10.2.

It has been reported that gelation time relates to the temperature, molar ratio of TEOS to water and molar ratio of TEOS to ethanol, the increase of temperature and molar ratio of TEOS:ethanol; also, decrease in the molar ratio of TEOS:water results in the reduction of gelation time (Brinker and Scherer, 1990). The measurement of viscosity was used to detect the gelation process (Estella et al., 2007). The most significant noticeable change in the sol due to nanoparticle formation and growth is the abrupt increase of its viscosity over a very short interval of gelation time (Anderson et al., 2004; Sacks and Sheu, 1987); this trend of viscosity change in the gelation process for the sol of silica aerogel was studied by Sacks et al. (Sacks and Sheu, 1987) and Gauthier-Manuel et al. (Anderson et al., 2004).

Some researchers (Brinker and Scherer, 1990) monitored the sol–gel transition process using dynamic and static light-scattering techniques and found that, in the early stage of the sol-to-gel transition process, the colloidal particles first gradually grew as individual colloidal particles, then abruptly increased to aggregations of colloid particles. In the final stage, which was close to the end of gelation time, the gel network structure was formed by aggregations of colloid particles/aggregates.

It was reported that a slower, low-temperature gelation process led to smaller pores in the gel because faster gelation involved accelerated chemical reactions and severe particle aggregations, resulting in greater pores in the gel (Brinker and Scherer, 1990). Gommes et al. (2007) pointed out that, apart from the chemical principle of balancing hydrolysis and condensation reactions, the physical inter-forces might speed up the aggregation process and lead to phase separation of hybrid gel in the gelation process (Xu et al., 2012).

While the wet gel is formed, the structure and strength of the wet gel are still weak, and the process of ageing the wet gel is necessary to obtain a stronger gel (Brinker and

Scherer, 1990). The ageing process serves to mechanically reinforce the solid skeleton structure and reduce shrinkage of the wet gel during the drying process (Gommes et al., 2007).

The condensation reaction continues, and the microstructure of the wet gel continues to change throughout the ageing process. The microstructural changes in the wet gel during the ageing process include three processes: polymerization, coarsening and phase transformation (Deshpande et al., 1992; Davis et al., 1992a; Omranpour and Motahari, 2013; Davis et al., 1992b). Polymerization might increase the connectivity of the solid network by further condensation and hydrolysis (Brinker and Scherer, 1990). Coarsening is a process of dissolution/reprecipitation where smaller particles dissolve and solute precipitates onto larger particles, and the necks between particles grow through material accumulation, resulting in a strengthened network (Brinker and Scherer, 1990). Phase transformation refers to a process in which solid phase either separates from liquid or leads to crystallization of particles (Brinker and Scherer, 1990).

The effects of ageing time duration, temperature and solvents on compression strength and modulus were studied by Brinker and Scherer (1990), who concluded that the influence of ageing time and temperature on compression strength and modulus differed depending on the different solvents used, such as water, methanol and *n*-hexane). The porosity and average pore size of tetramethoxysilane (TMOS) based aerogel increased when the ageing temperature increased from 54°C to 70°C (Omranpour and Motahari, 2013). The influence of duration of ageing time (ranging from 1 day to 3 weeks) on the porous structure of a nanoporous glass film was also studied, and it was found that the porosity, median pore diameter, and surface area of the resultant porous glass film significantly increased with increases in ageing time (Brinker and Scherer, 1990).

The link between the dimensions of the nanoparticles/nanoparticle aggregations formed in the gelation process and the porous structure of resultant aerogels were investigated by Du (2014). It was found that the particle size formed in the gelation process was closely related to the porous structure of resultant aerogels when the molar ratio of methanol to MTMS changed from 15:1 to 30:1, while there was little effect on the porous structure when the ammonia feed rate changed from 0.05 to 0.2 mL min^{-1}. As shown in Figs 10.8–10.11, the peak particle size in gelation process, which is the greatest percentage of the particle size in sol, was found to affect the

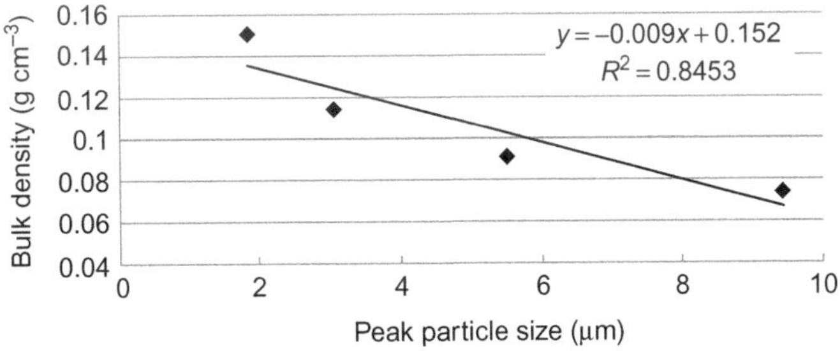

Fig. 10.8 The effect of particle size distribution in the sol on the bulk density of the silicone aerogels obtained from different molar ratios of methanol to MTMS (Du, 2014).

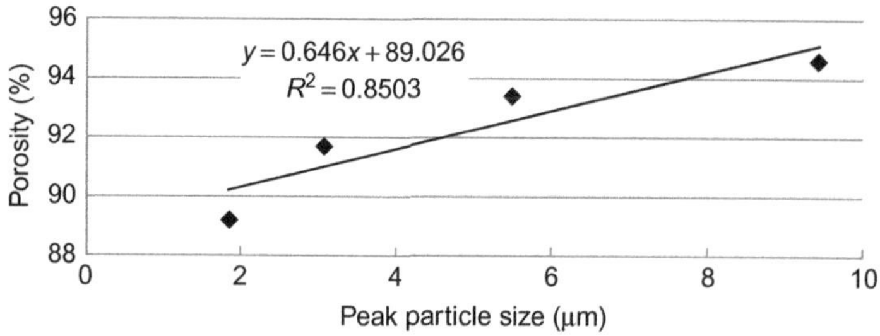

Fig. 10.9 The effect of particle size distribution in the sol on the porosity of silicone aerogels obtained from different molar ratios of methanol to MTMS (Du, 2014).

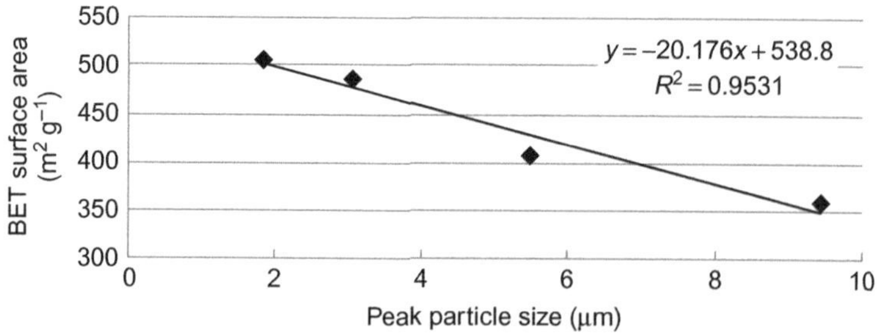

Fig. 10.10 The effect of particle size distribution in the sol on the BET surface area of silicone aerogels obtained from different molar ratios of methanol to MTMS (Du, 2014).

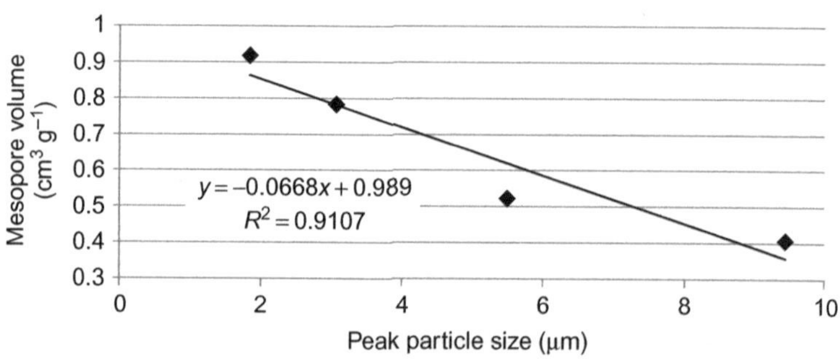

Fig. 10.11 The effect of particle size distribution in the sol on the mesopore volume of silicone aerogels obtained from different molar ratios of methanol to MTMS (Du, 2014).

characteristic pore structure of the resultant aerogels in ways such as bulk density, porosity, BET surface area and mesopore volumes. The larger particles formed in the sols lead to more macropores and less mesopores in the resultant silicone aerogels (Du, 2014).

The microstructure of aerogels resultant from different molar ratio of methanol to precursor (MTMS) is shown in Fig. 10.12.

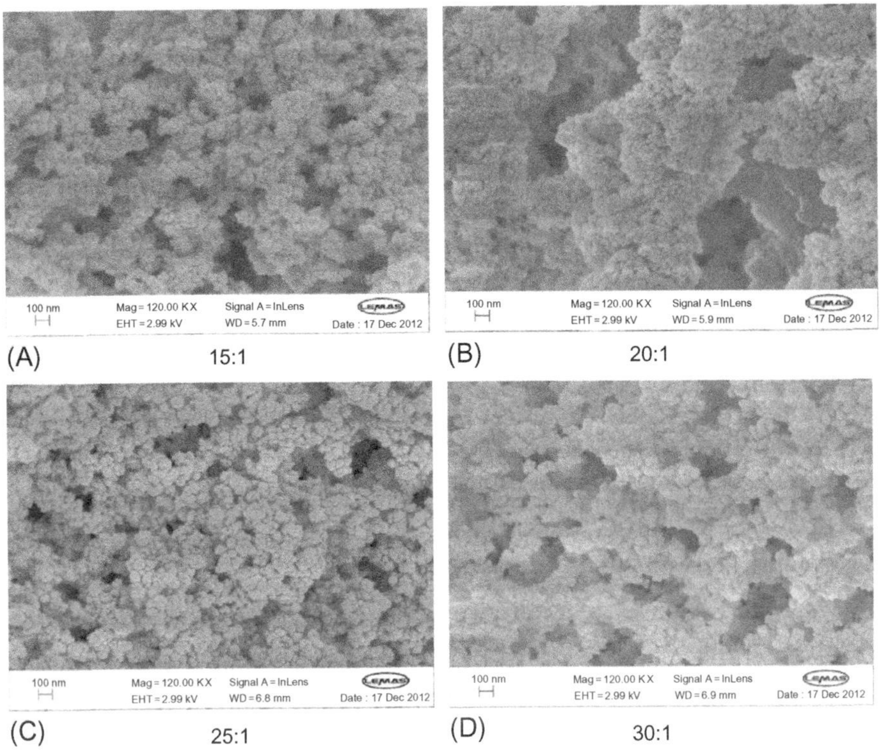

Fig. 10.12 Microstructure of silicone aerogels obtained at various molar ratios of methanol to MTMS (Du, 2014). (A) 15:1; (B) 20:1; (C) 25:1; (D) 30:1.

10.4 Applications of sol–gel treatment on textiles for water repellence

Usually, superhydrophobicity of fabrics made from inherently hydrophilic natural fibres such as cotton, silk and cellulosic fabrics is achieved via sol–gel treatments, these fabrics' inherent hydrophilic nature limits their applications; the same applies to fabrics made from synthetic fibres (e.g., polyester and polyamide fibres) due to the requirement of further improved hydrophobicity. Based on the Cassie–Baxter model, the hydrophobicity of hydrophobic textile materials can be further enhanced if nanoscopic roughness is introduced onto the textile surfaces (Xue, 2011); The hydrophobicity of hydrophilic textile materials could be achieved (Holmquist et al., 2016) by both coating nanoparticles onto their surfaces to change their surface roughness and adding functional chemical groups (e.g. fluorinated side-chains, hydrocarbon chains and organo-silicon-based groups) to alter their surface energies (Li et al., 2016; Duan et al., 2016).

Various metal oxide nanoparticles including TiO_2 (Dastjerdi et al., 2010; Dastjerdi and Montazer, 2010; Radetic, 2013), ZnO (Dastjerdi and Montazer, 2010; Staneva et al., 2015; Shaheen et al., 2015), SiO_2 (Xue, 2011) and other types of nanoparticles including organic poly(styrene-maleic anhydride) (Stanssens et al., 2011), chitosan-based particles (Ivanova and Philipchenko, 2012) and Ag/AgBr-TiO_2 nanoparticles (Rana et al., 2016) have been coated onto textile materials by the impregnation process (Xue et al., 2009), spraying coating process (Aslanidou et al., 2016) and dip coating in the sol–gel process to achieve superhydrophobicity (Kornprobst and Plsnk, 2013). Porous PZT ($PbZr_{1-x}Ti_xO_3$) and other ceramic fibres could also be produced through the sol–gel process (Kitaoka et al., 1998; Meyer et al., 1998) from the solution of lead acetate trihydrate, lanthanum isopropoxide, zirconium *n*-propoxide and titanium isopropoxide in 2-methoxyethanol as the solvent.

The sol–gel process can produce a thin film of nanoparticle clusters or nanoporous aerogels/xerogels onto substrate textile materials in a single coating step (Textor, 2009). Researchers have demonstrated that the sol–gel process offered an effective way for achieving textile material hydrophobic coating, and even a small amount (1–2 vol.%) of TEOS-based sol will significantly improve textile materials' water repellence property (Textor and Mahltig, 2010).

Composites of aerogel/xerogel and textile fabric can be obtained by using two methods, i.e. aerogel/xerogel particles are preformed first then incorporated into fabric structures as illustrated in Fig. 10.13, or aerogel/xerogels are formed with the existence of fibrous material in the sol–gel process as demonstrated in Fig. 10.14. Both of these methods are used in hydrophobic treatment of textile materials.

In the first method, impregnating and spraying were used to coat ultra-thin layers of silicon nanoparticles around fibre surfaces of needle-punched and hydroentangled nonwovens made from meta-aramide (Nomex), polyphenylene sulphide (Torcon) and polyimide (P84) fibres to improve their hydrophobic, oleophobic, antistatic and abrasion resistance properties (Schmalz, 2008).

In the second method, common technical routes used to apply sols onto textile materials to coat aerogel/xerogel particles and thin films onto the surface of fibres are dip coating, spin coating and spray coating (Textor, 2009). This method could also be used to improve the strength of aerogel substrate obtained by using textile structure as

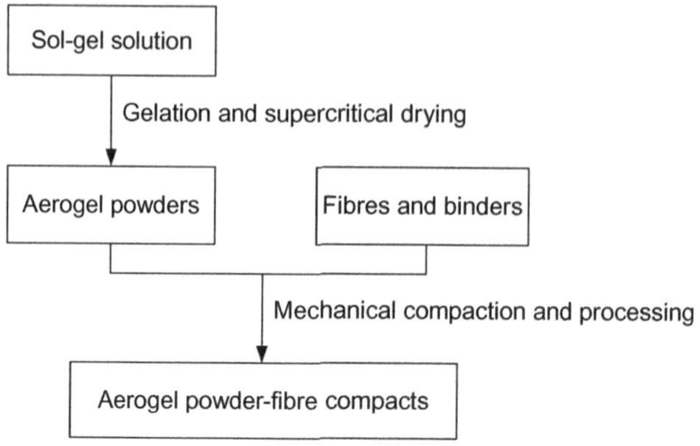

Fig. 10.13 Flowchart for incorporating aerogel particles with textile fabrics (Sangeeta and Mukund, 1994).

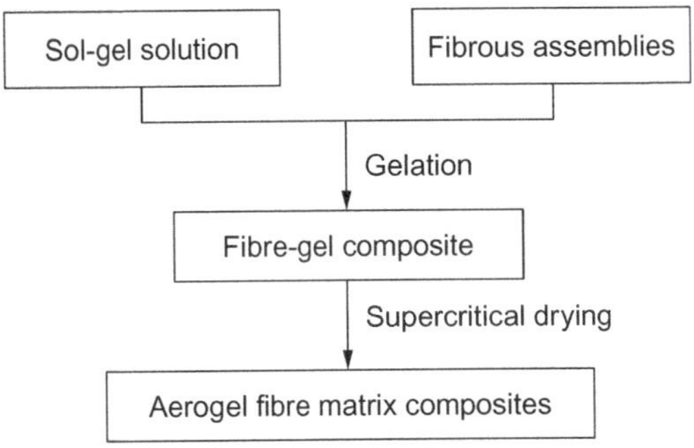

Fig. 10.14 Flowchart for coating aerogels onto fibre surfaces during sol–gel process (Sangeeta and Mukund, 1994).

a matrix, depending on the mass ratios between aerogel and fibres. For example, natural fibres of 2 mm in length up to 5% in weight (Finlay et al., 2008), ceramic fibres (e.g. SiO_2 and Al_2O_3) of 34–40 mm in length and 4–5 μm in diameter, about 7% in weight (Yang et al., 2011) and Mullite fibres (Zhang et al., 2012) were added in the sol solution to act as reinforcement fibres to improve the mechanical properties of aerogels. Polyurethane nanofibres were directly electrospun into sol of silica solution prior to its gelation to obtain nanofibre-reinforced aerogel composite (Li et al., 2009), the weight percentage of added nanofibres in terms of silica aerogel is about 1.9%.

Most existing studies on the hydrophobicity of textile fabrics treated with the sol–gel process are focused on the influence of the height difference of nanoparticles on the roughness of fabric surfaces and corresponding water repellence properties. However, aspects of the porous structure of aerogels/xerogels such as pore size distribution, porosity and the solid skeleton's structure might impact water repellence of the resultant textile materials. Currently, there is little research on the effects of porous structure of the coated aerogel/xerogel textiles on the hydrophobicity of the treated fabrics.

10.4.1 Addition of nanoparticles in sol–gel coating of textile materials

A method of adding various nanoparticles into the sol–gel process was also employed in the hydrophobic treatment of textile fabrics to achieve durable hydrophobicity. MTMS-based silicone aerogel/xerogel (Du et al., 2016) was flexible and inherently hydrophobic, and cotton fabric was treated using a MTMS-based-sol–gel process during which multiwall carbon nanotubes were added (Nasirizadeh et al., 2015). The treated cotton fabric exhibited superhydrophobicity with a water contact angle of up to 146 degrees. It was shown that the water repellence of treated cotton fabric is related to the concentration of carbon nanotubes added to the sol solution (Nasirizadeh et al., 2015). Silica and kaolin nano-clays were coated onto cotton fabric using padding and LBL coating sol–gel methods to create nano-roughness on the surface and the reported water contact angle of the treated cotton fabric was greater than 150 degrees (Joshi et al., 2012). Microstructures of cotton fibre before and after coating shows a dense coating created on the cotton fibres.

10.4.2 Assistive technologies in sol–gel treatment of textile materials

Plasma treatment could be used either as a pretreatment of textile materials prior to sol–gel treatment (Montarsolo et al., 2013; Mahltig, 2011), or as an enhanced hydrophobic treatment of sol–gel film coatings (Yang et al., 2009a; Chaivan et al., 2005; Yang et al., 2009b). Plasma pretreatment of textile materials prior to sol–gel treatment process is frequently employed to activate the fabric surface to improve the sol–gel coating's adhesion (Montarsolo et al., 2013) to facilitate its resistance to abrasion and fastness when exposed to washing stresses. The chemical composition and structure of fabric were believed to be the key factors influencing the adhesion between a silica coating and the surface of fabrics after analysis of changes in the morphology of surface, specific surface area and the microstructure of untreated and treated fibres (Hribernik et al., 2007). Cotton fabric was treated using low-pressure water vapour plasma followed by the fluoroalky-functional water-born siloxane (FAS) based sol–gel process, and it was reported that the water contact angle of treated fabric was up to 154 degrees (Vasiljević et al., 2013). The hydrophobicity of the treated cotton gradually decreased with multiple washes because of the gradual loss of nanocoatings.

Many techniques including UV (Yu et al., 2002; Liu et al., 2003), microwave (Hart et al., 2004) and plasma discharge (Mascia and Zhang, 1997; Ohsaki et al., 2009) were used to transform the gel layer coated onto substrates into ceramic materials. For example, ultraviolet light irradiation has been used as an assistive tool in the preparation of TiO_2 colloid via a photoassisted sol–gel method (Liu et al., 2003). It was found that the amorphous titania nanoparticles from this photoassisted sol–gel method can be transformed into crystalline anatase phase at a lower calcination temperature in comparison with those obtained in a conventional sol–gel process.

A novel device (Pietrzyk et al., 2013) was developed to integrate substrate preparation, aerosol–gel nanoparticle film deposition and plasma treatment of the nanoparticle film in a single reactor. The effect of deposition and plasma discharge conditions on morphology and chemical structure of the films was studied (Pietrzyk et al., 2013). All of these techniques in conjunction with sol–gel treatment of textile materials might be able to achieve ceramic coatings on textiles.

10.4.3 Superhydrophobic sol–gel coating using single and combined precursors

Nanocoatings from nanoparticles and aerogel/xerogels, made from a single alkoxide component as a precursor in sol–gel process, were widely used in the hydrophobic treatment of textile fabrics. Silica aerogel/xerogel made from tetraethoxysilane (TEOS) as a precursor in the sol–gel process is a typical inorganic aerogel and could be made via both acid catalyst (Li et al., 2008a) and base catalyst (Hribernik et al., 2007) methods. Coating silica xerogels/aerogels onto cellulosic fabrics and synthetic fabrics improved their surface hydrophobicity as well as flame retardant property (Hribernik et al., 2007; Li et al., 2008a). The surface of cotton and polyester fabrics were deposited with a layer of sol–gel silica-based film using TEOS as the precursor and treated with *N*-propyltrimethoxysilane (C3), hexadecyltrimethoxysilane (C16) and 1H,1H,2H,2H-fluorooctyltriethoxysilane (FOS) as hydrophobic additives to create a water repellent finishing (Textor and Mahltig, 2010). Cotton and polyester fabrics (Boukhriss et al., 2015), as well as polyamide fabrics (Boukhriss et al., 2015), modified by using chloropropyltriethoxysilane (CPTS) as the precursor in a sol–gel process showed excellent hydrophobicity. Their water repellence property was found to be closely related to the concentration of CPTS used.

Mixture of multiple precursors of different reactivities in the sol–gel process could result in different surface morphologies and surface free energies due to the demixing of the forming phases. Hydrophobic sol–gel coatings that offer antistatic properties for appropriate finishing of textiles contain both hydrophilic (amino-functionalized alkoxysilanes) and hydrophobic components (e.g. alkoxysilanes modified with alkyl chains) simultaneously (Textor and Mahltig, 2010).

Superhydrophobic coatings were produced using a mixture of tetraethoxysilane, tetraethyl orthotitanate and tetra-*n*-propyl zirconate as precursors in a sol–gel process (Taurino et al., 2008). Transparent and durable superhydrophobic cotton fabric was obtained via cohydrolysis and polycondensation of a mixture of hexadecyltrimethoxy-silane, tetraethoxyorthosilcate and 3-glycidoxypropyltrimethoxysilane (Daoud et al., 2004). Superhydrophobic cotton fabrics coated with a hybrid SiO_2/HTEOS/CPTS nanoparticle coating layer having a contact angle of 140 degrees were prepared by using tetraethoxysilane (TEOS), chloropropyltriethoxysilane (CPTS) and hexadecyltrimethoxysilane (HTEOS) as precursors in sol–gel process (Yin and Wang, 2013). The superhydrophobic surface was achieved by the combination of low surface energy chemical compositions (HTEOP and CPTS) and rough surface geometrical structures coated with nanoparticles (Yin and Wang, 2013). It was reported that water repellent film with a water contact angle of up to 151 degrees was created using a precursor mixture of trimethylethoxysilane (TMES)/tetraethoxysilane (TEOS). The molar ratio of TMES:TEOS significantly affected the film's hydrophobicity (Latthe et al., 2009); the greater the molar ratio, the better the hydrophobic property (Fig. 10.15).

The contact angle of up to 145 degrees was achieved in cotton fabrics dipping coated in a sol made from the mixture of TEOS-based $Zn(CH_3COO)_2 \cdot 2H_2O$

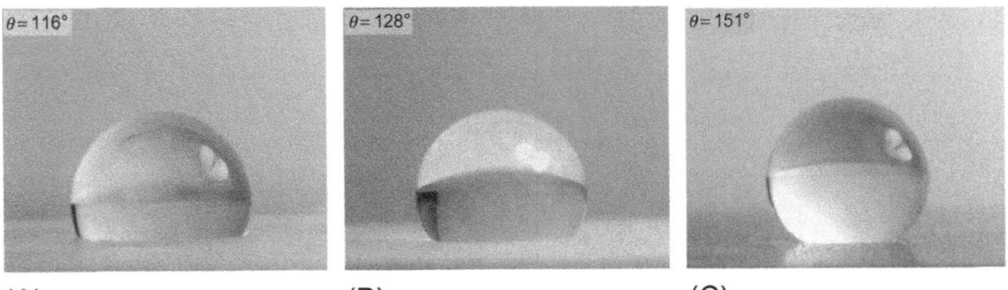

(A) (B) (C)

Fig. 10.15 Photographs of water drops on the hydrophobic silica film, molar ratio of TMES:TEOS (A) 0.5, (B) 0.8 and (C) 1.1 (Latthe et al., 2009).

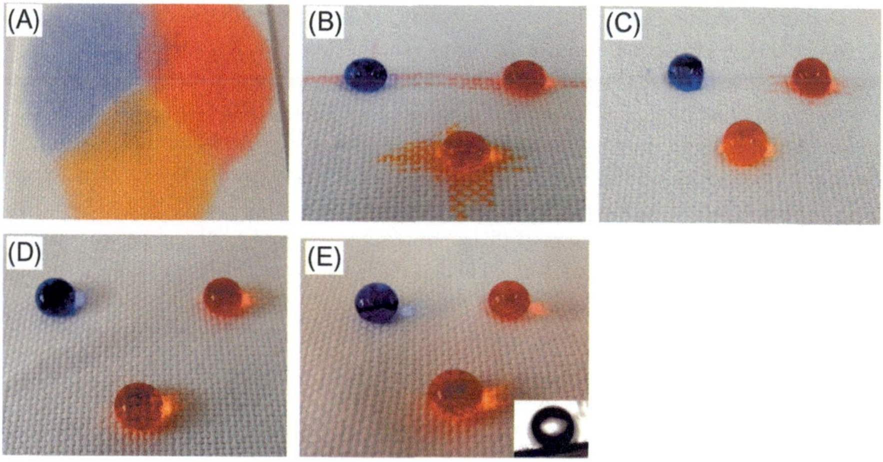

Fig. 10.16 Contact between water drops and fabric surface (A) untreated polyester fabric and treated polyester fabric with the molar ratio of TEOS:HMDS, (B) 1:0.5, (C) 1:0.75, (D) 1:1.25 and (E) 1:1.75 (Xue et al., 2013).

(Vihodceva et al., 2015). The contact angles of sol–gel SiO_2 nanoparticle-coated PET fabrics are in the range of 140 and 170 degrees (see Fig. 10.16). Its precursor is a mixture of tetraethoxysilane (TEOS) and hexamethyldisilazane (HMDS), and the PET fabric was pretreated by using NaOH solution to rough the fabric surface in order to improve the compatibility between silica sol and the fabric fibres (Xue et al., 2013). It was found that the water repellence property of treated polyester fabric is gradually improved when the proportion of HMDS increases.

The water contact angle of cellulose fibres having hydrophobic coating using SiO_2/HTEOS (hexadecyltrunethoxysilane)/CPTS (γ-chloropropyltriethoxysilane) as precursors in sol–gel treatment was up to 140 degree. The greater contact angle was believed to be due to the combined effect of low surface energy chemicals (HTEOS and CPTS) and rough surface geometrical structure (Yin and Wang, 2013). Various liquid drops including liquid soap were placed on the treated fabric to examine the liquid repellence property, and the images of contact between liquid and fabric were shown in Fig. 10.17.

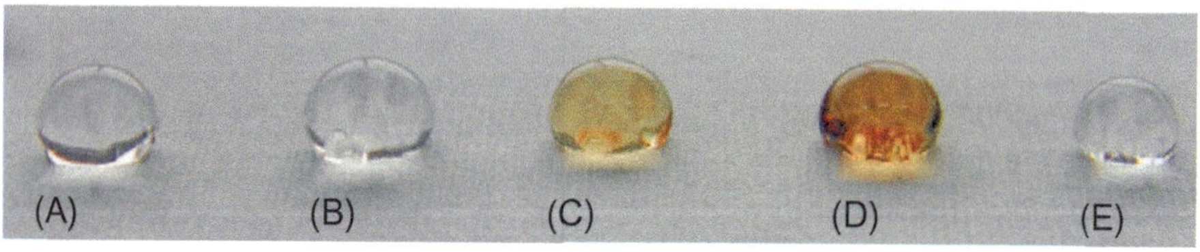

Fig. 10.17 Various liquid drops on treated fabric (A) deionized water, (B) rainwater, (C) tea, (D) coke and (E) liquid soap (Yin and Wang, 2013).

Hexadecyltrimethoxysilane (HDTMS) (Gao et al., 2009) and its mixture (e.g. water glass and hexadecyltrimethoxysilane (Li et al., 2008b) were used as a precursor in a silica sol–gel coating process to fabricate highly hydrophobic cotton and polyester fabrics. The processed textile fabrics exhibited coarse surface morphology (Xing et al., 2007) and showed excellent water repellency with water contact angles as high as 155 degrees on cotton and 143 degrees on polyester (Li et al., 2008b). The fabrics could also have their durable antimicrobial properties achieved by further treatment with silver nitrate solution (Xing et al., 2007).

10.4.4 Fabric coating with graphene

Graphene is the thinnest yet strongest material created so far (Geim and Novoselov, 2007; Lee et al., 2008). It has remarkable mechanical, thermal and electrical properties. It has high Young's modulus (~1100 GPa) and fracture strength (125 GPa) (Lee et al., 2008), thermal conductivity (~5000 $W\ m^{-1}\ K^{-1}$) (Balandin et al., 2008), mobility of charge carriers (200,000 $cm^2\ V^{-1}\ s^{-1}$) (Bolotin et al., 2008; Stoller et al., 2008) and specific surface area (calculated value, 2630 $m^2\ g^{-1}$).

Graphene is also a wonder material in handling water. It is hydrophobic and repels water, and a monolayer graphene membrane is impermeable to standard gases including helium. As a nanomembrane, it has a remarkable ability to restrict the passage of molecules (Bunch et al., 2008). However, narrow capillaries made from graphene vigorously suck in water, allowing its rapid permeation, if the water layer is only one atom thick, that is, as thin as graphene itself (Joshi et al., 2014; Cohen-Tanugi and Grossman, 2014). For example, graphene-based membranes having limited swelling exhibit high water flux and 97% rejection for NaCl (Abraham et al., 2017).

The graphene-based nanoporous membranes are in two distinct types (Yoon et al., 2016): (i) creating pores in graphene basal plane and (ii) engineering nanochannels in a graphene layers. Each has its own characteristics. Graphene nanoporous membranes have atomic thickness and present minimal resistance to fluid or ion flow, rapidly passing small molecules with little energy expenditure and retaining larger ones, while retaining high structural integrity. Therefore graphene membranes would provide a different mechanism of waterproofness than all the water repellent treatments discussed above. It was suggested (Cohen-Tanugi and Grossman, 2014) that ions smaller in size than the graphene oxide (GO) nanochannel in nanoporous graphene membranes can permeate the GO membrane at a speed orders of magnitude faster than would occur through simple diffusion, while providing excellent salt rejection to ions; so multiple-layered GO membranes with two-dimensional nanochannels have shown very intriguing separation properties and are widely used as sieving/filtration membranes for gas separation and desalination.

Graphene applied onto textile materials is usually in the form of suspension/colloid of GO nanosheets which is chemically reduced from graphite oxide powder; GO particles are abundant in hydroxyl and carboxyl groups and thus exhibit two-dimensional amphiphilic properties. GO suspension/colloid could lower the interfacial energy of the fabrics and are frequently applied onto textile fabrics in a sol–gel method

(Wang et al., 2012) and dip-drying method (Krishnamoorthy et al., 2012). Graphene-modified textiles exhibit many unique properties such as photocatalytic activity (Krishnamoorthy et al., 2012), hydrophobicity and antibacterial properties (Krishnamoorthy et al., 2012) and improved conductivity (Gu and Zhao, 2011).

GO membranes are widely used to produce tunable hydrophobic/hydrophilic smart fabrics. Thin graphene sheets applied onto a cotton fabric surface make the fabric have smart switchable surfaces, either superhydrophobic or superhydrophilic (Hasan et al., 2010). These alternate arrangements, termed 'rug' and 'brick', make the water bead up and run off or spread out and form incredibly thin sheets.

A highly sensitive amphiphilic copolymer-based nanocomposite incorporating with GO exhibits a low-intensity UV light-triggered sol–gel transition (Hu et al., 2016). Microchannels formed spontaneously within the GO-incorporated UV gel could expedite sustained drug release due to GO's unique photothermal properties.

A self-cleaning fabric was obtained from coating fabrics with composited graphene/titanium dioxide (Karimi et al., 2014); graphene/titanium dioxide nanocomposites were obtained with chemical reduction using titanium trichloride. A novel, smart, stimuli-responsive, superhydrophobic cotton fabric was produced based on the hierarchical structure of graphene and TiO_2 nanofilm with dual roughness. Upon manipulating the UV-induced hydrophilic conversion of TiO2 on a graphene/TiO_2 surface, smart surface features, such as tunable adhesiveness, wettability and directional water transport, can be obtained. The surface exhibits tunable wetting and directional water transport properties, which provides a general protocol for applications such as moisture management, microfluidic control, self cleaning and water–oil separation (Liu et al., 2015).

Hydrophobic properties of a cotton fabric with a water contact angle of up to 143 degrees were achieved by grafting graphene oxide on its surface using a dip dry method. GO was obtained through the exfoliation of graphite oxides, which was synthesized by oxidizing natural flake graphite using Hummer's method, in water using ultrasound energy. It was found that fabric hydrophobicity varies with the amount of graphene oxide loaded onto the surface of cotton fibres (Tissera et al., 2015).

10.5 Summary

Sol-gel technology that employs a wide range of precursors could be applied in the treatment of textile materials to achieve excellent water repellence properties, especially when mixture and hybrid precursors are used as well as addition of various nanoparticles. However, it is worth to note that continuous improvement in sol-gel technologies to achieve better water repellence in textile fabrics is only one aspect of the story. The success of applying sol-gel technology onto textile fabrics for clothing products relies largely on how to solve the following two problems, (1) to improve the durability of nanoparticles (xerogel or aerogel) binded on the surface of the textile fibres (e.g., especially during washing and laundry process); (2) to address the health and safety concerns of the nanoparticles, which might fall off from the sol-gel coated textile fibres, imposed on human body.

References

Abidi, N., et al., 2007. Cotton fabric surface modification for improved UV radiation protection using sol–gel process. J. Appl. Polym. Sci. 104 (1), 111–117.

Abraham, J., et al., 2017. Tunable sieving of ions using graphene oxide membranes. Nat. Nanotechnol. 12, 546–550.

Aegerter, M.A., Leventis, N., Koebel, M.M., 2011. In: Advances in sol-gel derived materials and technologies. Aerogels Handbook, Springer New York Dordrecht Heidelberg London, London p. 932.

Akş·it, A.C., Onar, N., 2008. Leaching and fastness behavior of cotton fabrics dyed with different type of dyes using sol-gel process. J. Appl. Polym. Sci. 109 (1), 97–105.

Aksit, A., et al., 2016. Synergistic effect of phosphorus, nitrogen and silicon on flame retardancy properties of cotton fabric treated by sol-gel process. Int. J. Cloth. Sci. Technol. 28 (3), 319–327.

Alba, F., et al., 1999. Acoustic spectroscopy as a technique for the particle sizing of high concentration colloids, emulsions and suspensions. Colloids Surf. A Physicochem. Eng. Asp. 153 (1–3), 495–502.

Alongi, J., Ciobanu, M., Malucelli, G., 2011. Sol–gel treatments for enhancing flame retardancy and thermal stability of cotton fabrics: optimisation of the process and evaluation of the durability. Cellulose 18 (1), 167–177.

Al-Oweini, R., El-Rassy, H., 2010. Surface characterization by nitrogen adsorption of silica aerogels synthesized from various Si(OR)4 and R″Si(OR′)3 precursors. Appl. Surf. Sci. 257 (1), 276–281.

Anderson, A.M., Bakrania, D.B., Konecny, J., Gauthier, B.M., Carroll, M.K., 2004. Detecting sol-gel transition using light transmission. J. Non-Cryst. Solids 350, 259–265.

Aravind, P., Soraru, G., 2011. High surface area methyltriethoxysilane-derived aerogels by ambient pressure drying. J. Porous Mater. 18, 159–165.

Aslanidou, D., Karapanagiotis, I., Panayiotou, C., 2016. Superhydrophobic, superoleophobic coatings for the protection of silk textiles. Prog. Org. Coat. 97, 44–52.

Aydilek, A., et al., 2002. Digital image analysis to determine pore opening size distribution of nonwoven geotextiles. J. Comput. Civil Eng., 16 (4), 280–290.

Babonneau, F., et al., 1988. XANES and EXAFS study of titanium alkoxides. Inorg. Chem. 27 (18), 3166–3172.

Balandin, A.A., et al., 2008. Superior thermal conductivity of single-layer graphene. Nano Lett. 8 (3), 902–907.

Barrett, E.P., Joyner, L.G., Halenda, P.P., 1951. The determination of pore volume and area distributions in porous substances. I. Computations from nitrogen isotherms. J. Am. Ceram. Soc. 73, 373–380.

Barthlott, W., Neinhuis, C., 1997. Purity of the sacred lotus, or escape from contamination in biological surfaces. Planta 202, 1–8.

Bogush, G.H., Zukoski Iv, C.F., 1991. Studies of the kinetics of the precipitation of uniform silica particles through the hydrolysis and condensation of silicon alkoxides. J. Colloid Interface Sci. 142 (1), 1–18.

Bogush, G.H., Tracy, M.A., Zukoski IV, C.F., 1988. Preparation of monodisperse silica particles: control of size and mass fraction. J. Non-Cryst. Solids 104, 95–106.

Bolotin, K.I., et al., 2008. Ultrahigh electron mobility in suspended graphene. Solid State Commun. 146 (9), 351–355.

Bonugli, L.O., et al., 2012. Superhydrophobic polyethylcyanoacrylate coatings. Contact area with water measured by Raman spectral images, contact angles and Cassie-Baxter model. J. Colloid Interface Sci. 388 (1), 306–312.

Boukhriss, A., et al., 2015. Sol–gel based water repellent coatings for textiles. Cellulose 22 (2), 1415–1425.

Brinker, C., Scherer, G., 1990. In: Sol-Gel Science: The Physics and Chemistry of Sol-Gel Processing. Academic: Elsevier's Science and Technology, Boston/London.

Brinker, C.J., Scherer, G.W., 1991. Sol–gel science, the physics and chemistry of sol–gel processing, Academic Press, Boston 1990, vol. xiv, 908 pp., bound—ISBN: 0-12-134970-5. Advanced Materials, vol. 3. WILEY-VCH Verlag GmbH, Weinheim, p. 522.

Brown, K.M.A.L., 1970. Synthesis and some properties of yttrium and lanthanide isopropoxides. Inorg. Chem. 9 (12), 3.

Brown, S., Lard, E., 1974. Comparison of nitrogen and mercury pore-size distributions of silicas of varying pore volume. Powder Technol. 9, 187–190.

Bunch, J.S., et al., 2008. Impermeable atomic membranes from graphene sheets. Nano Lett. 8 (8), 2458–2462.

Cassie, A.B.D., Baxter, S., 1944. Wettability of porous surfaces. Trans. Faraday Soc. 40, 546–551.

Chaivan, P., et al., 2005. Low-temperature plasma treatment for hydrophobicity improvement of silk. Surf. Coat. Technol. 193 (1), 356–360.

Chang, S.Y., Ring, T.A., 1992. Map of gel times for three phase region tetraethoxysislane, ethanol and water. J. Non-Cryst. Solids 147–148, 56–61.

Cireli, A., et al., 2007. Development of flame retardancy properties of new halogen-free phosphorous doped SiO_2 thin films on fabrics. J. Appl. Polym. Sci. 105 (6), 3748–3756.

Cohen-Tanugi, D., Grossman, J.C., 2014. Water permeability of nanoporous graphene at realistic pressures for reverse osmosis desalination. J. Chem. Phys. 141 (7), 074704.

Colleoni, C., Massafra, M.R., Rosace, G., 2012. Photocatalytic properties and optical characterization of cotton fabric coated via sol–gel with non-crystalline TiO_2 modified with poly(ethylene glycol). Surf. Coat. Technol. 207, 79–88.

Danks, A.E., Hall, S.R., Schnepp, Z., 2016. The evolution of 'sol-gel' chemistry as a technique for materials synthesis. Mater. Horiz. 3 (2), 91–112.

Daoud, W.A., Xin, J.H., Tao, X., 2004. Superhydrophobic silica nanocomposite coating by a low-temperature process. J. Am. Ceram. Soc. 87 (9), 1782–1784.

Dastjerdi, R., Montazer, M., 2010. A review on the application of inorganic nano-structured materials in the modification of textiles: focus on anti-microbial properties. Colloids Surf. B: Biointerfaces 79 (1), 5–18.

Dastjerdi, R., Montazer, M., Shahsavan, S., 2010. A novel technique for producing durable multifunctional textiles using nanocomposite coating. Colloids Surf. B: Biointerfaces 81 (1), 32–41.

Davis, P., Brinker, C., Smith, D., 1992a. Pore structure evolution in silica gel during aging/drying. I. Temporal and thermal aging. J. Non-Cryst. Solids 142, 189–196.

Davis, P., et al., 1992b. Pore structure evolution in silica gel during aging/drying. II. Effect of pore fluids. J. Non-Cryst. Solids 142, 197–207.

Deshpande, R., et al., 1992. Pore structure evolution in silica gel during aging/drying. III. Effects of surface tension. J. Non-Cryst. Solids 144, 32–44.

Du, M., 2014. The thermal insulation properties of a flexible methyltrimethoxysilane (MTMS) based silicone aerogel and its application on aerogel-spacer fabric composites. In: School of Design. University of Leeds, Leeds, p. 299.

Du, M., Mao, N., Russell, S., 2016. Control of porous structure in flexible silicone aerogels produced from methyltrimethoxysilane (MTMS): the effect of precursor concentration in sol–gel solutions. J. Mater. Sci. 51 (2), 719–731.
Duan, Y., et al., 2016. Hydrophobic silica aerogels by silylation. J. Non-Cryst. Solids 437, 26–33.
Eahman, I., Padavettan, V., 2012. Synthesis of silica nanoparticles by sol-gel: size-dependent properties, surface modification, and applications in silica-polymer nanocomposites – a review. J. Nanomater. 2012, 1–15.
Einarsrud, M.A., et al., 2001. Strengthening of silica gels and aerogels by washing and aging processes. J. Non-Cryst. Solids 285, 1–7.
Erra, P., et al., 1997. FTIR analysis to study chemical changes in wool following a sulfitolysis treatment. Text. Res. J. 67 (6), 397–401.
Eshel, G., et al., 2004. Critical evaluation of the use of laser diffraction for particle-size distribution analysis. Soil Sci. Soc. Am. J. 68, 736–743.
Estella, J., et al., 2007. Silica xerogels of tailored porosity as support matrix for optical chemical sensors. Simulataneous effect of pH, ethanol:TEOS and water:TEOS molar ratios, and synthesis temperature on gelation time, and textural and structural properties. J. Non-Cryst. Solids 353, 286–294.
Fan, J., et al., 2004. An improved model of heat and moisture transfer with phase change and mobile condensates in fibrous insulation and comparison with experimental results. Int. J. Heat Mass Transf. 47 (10–11), 2343–2352.
Fardad, M., 2000. Catalysts and the structure of SiO_2 sol-gel films. J. Mater. Sci. 35, 1835–1841.
Finlay, K., Gawryla, M., Schiraldi, D., 2008. Biologically based fiber-reinforced/clay aerogel composites. Ind. Eng. Chem. Res. 47, 615–619.
Galkina, O.L., et al., 2011. Surfactant-assisted sol-gel synthesis of TiO_2 with uniform particle size distribution. Int. J. Inorg. Chem. 2011, 1–8.
Gao, Q., et al., 2009. Formation of highly hydrophobic surfaces on cotton and polyester fabrics using silica sol nanoparticles and nonfluorinated alkylsilane. Ind. Eng. Chem. Res. 48 (22), 9797–9803.
Geim, A.K., Novoselov, K.S., 2007. The rise of graphene. Nat. Mater. 6 (3), 183–191.
Gommes, C., et al., 2007. Branching, aggregation, and phase separation during the gelation of tetraethoxysilane. J. Non-Cryst. Solids 353, 2495–2499.
Gorzalski, A.S., Donley, C., Coronell, O., 2017. Elemental composition of membrane foulant layers using EDS, XPS, and RBS. J. Membr. Sci. 522, 31–44.
Grancaric, A.M., et al., 2017. Thermal behaviour and flame retardancy of monoethanolamine-doped sol-gel coatings of cotton fabric. Prog. Org. Coat. 103, 174–181.
Gu, W.L., Zhao, Y.N., 2011. Graphene modified cotton textiles. Adv. Mater. Res. 331, 93–96.
Guascito, M.R., et al., 2015. XPS surface chemical characterisation of atmospheric particles of different sizes. Atmos. Environ. 116, 146–154.
Gulrajani, M.L., 2013. The use of nanotechnology in the finishing of technical textiles. In: Advances in the Dyeing and Finishing of Technical Textiles. Woodhead Publishing, Cambridge, UK, pp. 280–308.
Gurav, J., Nadargi, D., Rao, A., 2008. Effect of mixed catalysts system on TEOS-based silica aerogels dried at ambient pressure. Appl. Surf. Sci. 255, 3019–3027.
Gurav, J., et al., 2009. Physical properties of sodium silicate based silica aerogels prepared by single step sol-gel process dried at ambient pressure. J. Alloys Compd. 476 (1–2), 397–402.
Hart, J.N., et al., 2004. Formation of anatase TiO_2 by microwave processing. Sol. Energy Mater. Sol. Cells 84 (1), 135–143.

Hasan, S.A., et al., 2010. Transferable graphene oxide films with tunable microstructures. ACS Nano 4 (12), 7367–7372.
Hegde, N., Rao, A., 2007. Physical properties of methyltrimethoxysilane based elastic silica aerogels prepared by the two-stage sol–gel process. J. Mater. Sci. 42, 6965–6971.
Holmquist, H., et al., 2016. Properties, performance and associated hazards of state-of-the-art durable water repellent (DWR) chemistry for textile finishing. Environ. Int. 91, 251–264.
Hribernik, S., et al., 2007. Flame retardant activity of SiO_2-coated regenerated cellulose fibres. Polym. Degrad. Stab. 92 (11), 1957–1965.
Hu, J., Stroeven, P., 2003. Application of image analysis to assessing critical pore size for permeability prediction of cement paste. Image Anal. Stereol. 22, 97–103.
Hu, H., et al., 2016. Graphene oxide-enhanced sol-gel transition sensitivity and drug release performance of an amphiphilic copolymer-based nanocomposite. Sci. Rep. 6, 31815.
Hüsing, N., Schubert, U., 2000. Aerogels. In: Ullmann's Encyclopedia of Industrial Chemistry. Wiley-VCH Verlag GmbH & Co. KGaA, Weinheim, pp. 621–646.
Hwang, S., et al., 2007. Effective preparation of crack-free silica aerogels via ambient drying. J. Sol-Gel Sci. Technol. 41, 139–146.
IUPAC, 2001. Manual of Symbols and Terminology for Physicochemical Quantities and Units, Appendix II, Definitions, Terminology and Symbols in Colloid and Surface Chemistry, Part I. The International Union of Pure and Applied Chemistry (IUPAC).
Ivanova, N.A., Philipchenko, A.B., 2012. Superhydrophobic chitosan-based coatings for textile processing. Appl. Surf. Sci. 263, 783–787.
Jacobsen, A.E., SUllivan, W.F., 1946. Centrifugal sedimentation method for particle size distribution. Ind. Eng. Chem. 18 (6), 360–364.
Joshi, M., et al., 2012. Nanostructured coatings for super hydrophobic textiles. Bull. Mater. Sci. 35 (6), 933–938.
Joshi, R.K., et al., 2014. Precise and ultrafast molecular sieving through graphene oxide membranes. Science 343 (6172), 752–754.
Jung, S.-B., et al., 2003. Effect of solvent on the preparation of ambient pressure-dried SiO_2 aerogel films. Microelectron. Eng. 65 (1–2), 113–122.
Kakihana, M., 1996. Invited review "sol-gel" preparation of high temperature superconducting oxides. J. Sol-Gel Sci. Technol. 6 (1), 7–55.
Kamiyama, T., Yoshida, N., Suzuki, K., 1994. A SAXS characterization of particle aggregation in TiO_2 sol-gel system. Bull. Inst. Chem. Res. Kyoto Univ. 72 (2), 225–230.
Kan, C.W., 2016. Plasma surface treatments for smart textiles. Active Coatings for Smart Textiles. Woodhead Publishing, Cambridge, UK, pp. 221–241.
Kang, S., Choi, S., 2000. Synthesis of low-density silica gel at ambient pressure: effect of heat treatment. J. Mater. Sci. 35, 4971–4976.
Karimi, L., et al., 2014. Using graphene/TiO_2 nanocomposite as a new route for preparation of electroconductive, self-cleaning, antibacterial and antifungal cotton fabric without toxicity. Cellulose 21 (5), 3813–3827.
Kitaoka, K., Kozuka, H., Yoko, T., 1998. Preparation of lead lanthanum zirconate titanate (PLZT, (Pb, La)(Zr, Ti)O_3) fibers by sol-gel method. J. Am. Ceram. Soc. 81 (5), 1189–1196.
Kornprobst, T., Plank, J., 2013. Synthesis and properties of magnesium carbonate xerogels and aerogels. J. Non-Cryst. Solids 361, 100–105.
Kreder, M.J., et al., 2016. Design of anti-icing surfaces: smooth, textured or slippery? Nat. Rev. Mater. 1, 15003.
Krishnamoorthy, K., et al., 2012. Graphene oxide nanostructures modified multifunctional cotton fabrics. Appl. Nanosci. 2 (2), 119–126.

Latthe, S.S., et al., 2009. Superhydrophobic silica films by sol-gel co-precursor method. Appl. Surf. Sci. 256 (1), 217–222.
Latthe, S.S., Gurav, A.B., Shridhar Maruti, C., Shrikant Vhatkar, R., 2012. Recent progressin preparation of superhydrophobic surfaces: a review. J. Surf. Eng. Mater. Adv. Technol. 2 (2), 76–94.
Lee, H.J., 2011. Improving superhydrophobic textile materials. In: Functional Textiles for Improved Performance, Protection and Health. Woodhead Publishing, Cambridge, UK, pp. 339–359.
Lee, C., et al., 2008. Measurement of the elastic properties and intrinsic strength of monolayer graphene. Science 321 (5887), 385–388.
Li, F.-Y., Xing, Y.-J., Ding, X., 2007. Immobilization of papain on cotton fabric by sol–gel method. Enzym. Microb. Technol. 40 (7), 1692–1697.
Li, F., Xing, Y., Ding, X., 2008a. Silica xerogel coating on the surface of natural and synthetic fabrics. Surf. Coat. Technol. 202 (19), 4721–4727.
Li, Z., Xing, Y., Dai, J., 2008b. Superhydrophobic surfaces prepared from water glass and non-fluorinated alkylsilane on cotton substrates. Appl. Surf. Sci. 254 (7), 2131–2135.
Li, L., et al., 2009. Flexible nanofiber-reinforced aerogel (xerogel) synthesis, manufacture, and characterization. Appl. Mater. Interfaces 1 (11), 2491–2501.
Li, M., et al., 2016. Low density and hydrophobic silica aerogels dried under ambient pressure using a new co-precursor method. J. Non-Cryst. Solids 452, 187–193.
Lin, Y., et al., 2011. Superhydrophobic functionalized graphene aerogels. ACS Appl. Mater. Interfaces 3, 2200–2203.
Liu, H., et al., 2003. Synthesis and characterization of titania prepared by using a photoassisted sol–gel method. Langmuir 19 (7), 3001–3005.
Liu, Y., et al., 2015. Bioinspired, stimuli-responsive, multifunctional superhydrophobic surface with directional wetting, adhesion, and transport of water. Adv. Funct. Mater. 25 (31), 5047–5056.
Lowell, S., Shields, J.E., Thomas, M.A., Thommes, A., 2004. Characterization of porous solids and powders: surface area, pore size and density. In: Scarlett, P.B. (Ed.), ParticleTechnology SeriesKluwer Academic Publishers, The Netherlands.
MacKenzie, K.J.D., 1986. Applications of the sol-gel method: some aspects of initial processing. In: Hench, L.L.U., Donald, R. (Eds.), Science of Ceramic Processing. Wiley, New York, NY, pp. 113–122.
Mahltig, B., 2011. Hydrophobic sol–gel-based coating agent for textiles: improvement by solvothermal treatment. J. Text. Inst. 102 (5), 455–459.
Mahltig, B., Böttcher, H., 2003. Modified silica sol coatings for water-repellent textiles. J. Sol-Gel Sci. Technol. 27 (1), 43–52.
Mahltig, B., Fiedler, D., Böttcher, H., 2004. Antimicrobial sol–gel coatings. J. Sol-Gel Sci. Technol. 32 (1), 219–222.
Mahltig, B., Haufe, H., Bottcher, H., 2005. Functionalisation of textiles by inorganic sol-gel coatings. J. Mater. Chem. 15 (41), 4385–4398.
Mascia, L., Zhang, Z., 1997. Dense outer layers formed by plasma treatments of silica coatings produced by the sol–gel method. J. Mater. Sci. 32 (3), 667–674.
Matsoukas, T., Gulari, E., 1987. Dynamics of growth of silica particles from ammonia-catalyzed hydrolysis of tetra-ethyl-orthosilicate. J. Colloid Interface Sci. 124, 252–261.
Matsoukas, T., Gulari, E., 1988. Monomer-addition growth with a slow initiation step: a growth model for silica particles from alkoxides. J. Colloid Interface Sci. 132, 13–21.
Meirowitz, R.E., 2016. Coating processes and techniques for smart textiles. In: Active Coatings for Smart Textiles. Woodhead Publishing, Cambridge, UK, pp. 159–177.

Meyer, R., Shrout, T., Yoshikawa, S., 1998. Lead zirconate titanate fine fibers derived from alkoxide-based sol-gel technology. J. Am. Ceram. Soc. 81 (4), 861–868.

Montarsolo, A., et al., 2013. Hydrophobic sol-gel finishing for textiles: improvement by plasma pre-treatment. Text. Res. J. 83 (11), 1190–1200.

Nadargi, D.Y., Rao, A.V., 2009. Methyltriethoxysilane: new precursor for synthesizing silica aerogels. J. Alloys Compd. 467 (1–2), 397–404.

Nasirizadeh, N., Dehghani, M., Yazdanshenas, M.E., 2015. Preparation of hydrophobic and conductive cotton fabrics using multi-wall carbon nanotubes by the sol–gel method. J. Sol-Gel Sci. Technol. 73 (1), 14–21.

Nosonovsky, M., Bhushan, B., 2009. Superhydrophobic surfaces and emerging applications: non-adhesion, energy, green engineering. Curr. Opin. Colloid Interface Sci. 14 (4), 270–280.

Nosonovsky, M., Ramachandran, R., 2015. Geometric interpretation of surface tension equilibrium in superhydrophobic systems. Entropy 17 (7), 4686–4700.

Ohsaki, H., et al., 2009. Room-temperature crystallization of amorphous films by RF plasma treatment. Thin Solid Films 517 (10), 3092–3095.

Ohya, T., et al., 2004. Aqueous titanate sols from Ti alkoxide-α-hydroxycarboxylic acid system and preparation of titania films from the sols. J. Sol-Gel Sci. Technol. 30 (2), 71–81.

Omranpour, H., Motahari, S., 2013. Effects of processing conditions on silica aerogel during aging: role of solvent, time and temperature. J. Non-Cryst. Solids 379, 7–11.

Pajonk, G.M., 1989. Drying methods preserving the textural properties of gels. Rev. Phys. Appl. 24, 4–13.

Pajonk, G.M., et al., 1990. From sol-gel to aerogels and crypgels. J. Non-Cryst. Solids 121, 66–67.

Pietrzyk, B., et al., 2013. Plasma enhanced aerosol–gel method: a new way of preparing ceramic coatings. J. Sol-Gel Sci. Technol. 68 (3), 455–463.

Pirard, R., Heinrichs, B., Van, C., 1998. Mercury porosimetry applied to low density xerogels: relation between structure and mechanical properties. J. Sol-Gel Sci. Technol. 13, 335–339.

Qing, Y., et al., 2013. Facile approach in fabricating superhydrophobic ZnO/polystyrene nanocomposite coating. Appl. Surf. Sci. 285, 583–587.

Radetic, M., 2013. Functionalization of textile materials with TiO_2 nanoparticles. J. Photochem. Photobiol. C: Photochem. Rev. 16, 62–76.

Rahman, I.A., et al., 2007. An optimized sol-gel synthesis of stable primary equivalent silica particles. Colloids Surf. A Physicochem. Eng. Asp. 294 (294), 102–110.

Ramaratnam, K., et al., 2008. Ultradrophobic textiles using nanoparticles: lotus approach. J. Eng. Fibers Fabrics 3 (4), 1–10.

Rana, M., et al., 2016. Development of multi-functional cotton fabrics with Ag/AgBr-TiO_2 nanocomposite coating. Compos. Sci. Technol. 122, 104–112.

Rao, A., Bhagat, S., 2004. Synthesis and physical properties of TEOS-based silica aerogels prepared by two step (acid-base) sol-gel process. Solid State Sci. 6 (9), 945–952.

Rao, A.V., et al., 2003. Superhydrophobic silica aerogels based on methyltrimethoxysilane precursor. J. Non-Cryst. Solids 330 (1–3), 187–195.

Rao, A., et al., 2004. Comparative studies on the surface chemcial modification of silica aerogels based on various organosilane compounds of the type $RnSiX_{4-n}$. J. Non-Cryst. Solids 350, 216–223.

Rao, A., et al., 2006. Synthesis of flexible silica aerogels using methyltrimethoxysilane (MTMS) precursor. J. Colloid Interface Sci. 300 (1), 279–285.

Rao, V.A., Hegde, N.D., Shewale, P.M., 2007. Imperviousness of the hydrophobic silica aerogels against various solvents and acids. Appl. Surf. Sci. 253 (9), 4137–4141.
Rassy, H., Pierre, A., 2005. NMR and IR spectroscopy of silica aerogels with different hydrophobic characteristics. J. Non-Cryst. Solids 351, 1603–1610.
Sacks, M., Sheu, R., 1987. Rheological properties of silica sol-gel materials. J. Non-Cryst. Solids 92, 383–396.
Sangeeta, D.R., Mukund, R., 1994. Aerogel Matrix Composites, PCT Patent WO1993006044. Battelle Memorial Institute, Columbus, OH.
Sarawade, P., et al., 2006. Influence of solvent exchange on the physical properties of sodium silicate based aerogel prepared at ambient pressure. Aerosol Air Qual. Res. 6 (1), 93–105.
Sarawade, P., et al., 2010. Production of low-density sodium silicate-based hydrophobic silica aerogel beads by a novel fast gelation process and ambient pressure drying process. Solid State Sci. 12, 911–918.
Schmalz, I.E., 2008. Sol-Gel-Based Nanoparticulate Finishing for Functionalisation of Nonwoven. Nonwovens Research Academy, Germany.
Schubert, U., 2005. Chemical modification of titanium alkoxides for sol-gel processing. J. Mater. Chem. 15 (35–36), 3701–3715.
Scolan, E., Sanchez, C., 1998. Synthesis and characterization of surface-protected nanocrystalline titania particles. Chem. Mater. 10 (10), 3217–3223.
Seville, J., Wu, C., 2016. Particle characterization. In: Particle Technology and Engineering. Butterworth-Heinemann, Oxford, pp. 39–65.
Shaheen, T.I., et al., 2015. Durable antibacterial and UV protections of in situ synthesized zinc oxide nanoparticles onto cotton fabrics. Int. J. Biol. Macromol. 83, 426–432.
Shaid, A., et al., 2016. Preparation of aerogel-eicosane microparticles for thermregulatory coating on textile. Appl. Therm. Eng. 107, 602–611.
Sing, K.S.W., et al., 1985. Reporting physisorption data for gas/solid system with special reference to the determination of surface area and porosity. Pure Appl. Chem. 57, 603–619.
Sinko, K., 2010. Influence of chemical conditions on the nanoporous structure of silicate aerogel. Materials 2010 (3), 704–740.
Sondej, F., Buck, A., Tsotsas, E., 2016. Comparative annalysis of the coating thickness on single particles using X-ray micro-computed tomography and confocal laser-scanning microscopy. Power Technol. 287, 330–340.
Staneva, D., et al., 2015. A cotton fabric modified with a hydrogel containing ZnO nanoparticles. Preparation and properties study. Appl. Surf. Sci. 345, 72–80.
Stanssens, D., et al., 2011. Creating water-repellent and super-hydrophobic cellulose substrates by deposition of organic nanoparticles. Mater. Lett. 65 (12), 1781–1784.
Stoller, M.D., et al., 2008. Graphene-based ultracapacitors. Nano Lett. 8 (10), 3498–3502.
Su, C., et al., 2016. Fabrication of three-dimensional superhydrophobic membranes with high porosity via simultaneous electrospraying and electrospinning. Mater. Lett. 170, 67–71.
Taurino, R., et al., 2008. Facile preparation of superhydrophobic coatings by sol–gel processes. J. Colloid Interface Sci. 325 (1), 149–156.
Textor, T., 2009. Modification of textile surfaces using the sol-gel technique. In: Surface Modification of Textiles. Woodhead Publishing, Cambridge, UK, pp. 185–213.
Textor, T., Mahltig, B., 2010. A sol-gel based surface treatment for preparation of water repellent antistatic textiles. Appl. Surf. Sci. 256 (6), 1668–1674.
Tissera, N.D., et al., 2015. Hydrophobic cotton textile surfaces using an amphiphilic graphene oxide (GO) coating. Appl. Surf. Sci. 324, 455–463.

Tseng, Y., et al., 2009. Polymer-assisted synthesis of hydroxyapatite nanoparticle. Mater. Sci. Eng. C 29, 819–822.

Turova, N.Y., et al., 1996. Tantalum(V) alkoxides: electrochemical synthesis, mass-spectral investigation and oxoalkoxocomplexes. Polyhedron 15 (21), 3869–3880.

Uhlmann, D.R., et al., 1986. Kinetic processes in sol-gel processing. In: Hench, L.L.U., Donald, R. (Eds.), Science of Ceramic Processing. Wiley-Interscience, New York, USA, pp. 173–183.

Ulusoy, U., Igathinathane, C., 2016. Particle size distribution modeling of milled coals by dynamic image analysis and mechanical sieving. Fuel Process. Technol. 143, 100–109.

Vasiljević, J., et al., 2013. The surface modification of cellulose fibres to create super-hydrophobic, oleophobic and self-cleaning properties. Cellulose 20 (1), 277–289.

Vihodceva, S., Kukle, S., Bitenieks, J., 2015. Durable hydrophobic sol-gel finishing for textiles. Mater. Sci. Eng. 77, 1–4.

Wagh, P.B., et al., 2015. Hydrophobicity measurement studies of silica aerogels using FTIR spectroscopy, weight difference method, contact angle method and K-F titration method. J. Chem. Biol. Phys. Sci. 5 (3), 2350–2359.

Wagner, T., Neinhuis, C., Barthlott, W., 1996. Wettability and contaminability of insect wings as a function of their surface sculptures. Acta Zool. 77 (3), 213–225.

Wang, C., Yin, Y., 2016. Functional modification of fiber surface via sol-gel technology. In: Active Coatings for Smart Textiles. Woodhead Publishing, Cambridge, UK, pp. 301–328.

Wang, X., et al., 2012. Fabrication and characterization of graphene-reinforced waterborne polyurethane nanocomposite coatings by the sol–gel method. Surf. Coat. Technol. 206 (23), 4778–4784.

Wei, Q., et al., 2006. Surface characterisation of functional nanostructures sputtered on fiber substrates. Surf. Coat. Technol. 201, 1821–1826.

Welker, R.W., 2012. Size analysis and identification of particles. In: Developments in Surface Contamination and Cleaning. William Andrew Publishing, Oxford (Chapter 4).

Wenzel, R.N., 1936. Resistance of solid surfaces to wetting by water. J. Ind. Eng. Chem. 28, 6.

Xing, Y., Ding, X., 2007. UV photo-stabilization of tetrabutyl titanate for aramid fibers via sol–gel surface modification. J. Appl. Polym. Sci. 103 (5), 3113–3119.

Xing, Y., Yang, X., Dai, J., 2007. Antimicrobial finishing of cotton textile based on water glass by sol–gel method. J. Sol-Gel Sci. Technol. 43 (2), 187–192.

Xu, B., et al., 2012. An improved method for preparing monolithic aerogels based on methyltrimethoxysilane at ambient pressure, part I: process development and macrostructures of the aerogels. Microporous Mesoporous Mater. 148 (1), 145–151.

Xue, C.H., 2011. Improving superhydrophobic coatings for textiles through chemical modifications. In: Functional Textiles for Improved Performance, Protection and Health. Woodhead Publishing, Cambridge, UK, pp. 320–338.

Xue, C.-H., Jia, S.T., Chen, H.Z., Wang, M., 2008. Superhydrophobic cotton fabrics prepared by sol–gel coating of TiO_2 and surface hydrophobization. Sci. Technol. Adv. Mater. 9 (3), 035001.

Xue, C.H., et al., 2009. Superhydrophobic surfaces on cotton textiles by complex coating of silica nanoparticles and hydrophobization. Thin Solid Films 517 (16), 4593–4598.

Xue, C.-H., et al., 2013. Fabrication of superhydrophobic and superoleophilic textiles for oil–water separation. Appl. Surf. Sci. 284, 464–471.

Yamamoto, T., et al., 2002. Dynamic and static light scattering study on the sol-gel transition of resorcinol-formaldehyde aqueous solution. J. Colloid Interface Sci. 245, 391–396.

Yang, L., et al., 2006. Particle size control in sol-gel synthesis of nano hydroxyapatite. Adv. Mater. Res. 11–12, 231–234.

Yang, S.-H., et al., 2009a. Atmospheric-pressure plasma deposition of SiO_x films for super-hydrophobic application. Thin Solid Films 517 (17), 5284–5287.

Yang, S.H., et al., 2009b. Preparation of super-hydrophobic films using pulsed hexafluorobenzene plasma. Surf. Coat. Technol. 203 (10), 1379–1383.

Yang, X., et al., 2011. Experimental investigation on mechanical properties of fiber-reinforced silica aerogel composite. Mater. Sci. Eng. A 528, 4830–4836.

Yin, Y., Wang, C., 2013. Water-repellent functional coatings through hybrid SiO_2/HTEOS/CPTS sol on the surfaces of cellulose fibers. Colloids Surf. A Physicochem. Eng. Asp. 417, 120–125.

Yin, Y., et al., 2011. Hydrophobic properties and color effects of hybrid silica spin-coatings on cellulose matrix. J. Mater. Sci. 46 (20), 6682–6689.

Yoon, H.W., Cho, Y.H., Park, H.B., 2016. Graphene-based membranes: status and prospects. Philos. Trans. R. Soc. A: Math. Phys. Eng. Sci. 374 (2060) 20150024.

Young, T., 1805. An essay on the cohesion of fluids. Philos. Trans. R. Soc. Lond. 95, 65–87.

Yu, J.J., Zhang, J.Y., Boyd, I.W., 2002. Formation of stable zirconium oxide on silicon by photo-assisted sol–gel processing. Appl. Surf. Sci. 186 (1), 190–194.

Yu, M., et al., 2007. Superhydrophobic cotton fabric coating based on a complex layer of silica nanoparticles and perfluorooctylated quaternary ammonium silane coupling agent. Appl. Surf. Sci. 253 (7), 3669–3673.

Zha, J., Roggendorf, H., 1991. Sol–gel science, the physics and chemistry of sol–gel processing, Ed. by C.J. Brinker and G.W. Scherer, Academic Press, Boston 1990, vol. xiv, 908 pp., bound—ISBN: 0-12-134970-5. Adv. Mater. 3 (10), 522.

Zhang, Z., et al., 2012. Mechanical property of silica aerogel fibers composite insulation. Rare Metal Mater. Eng. 41, 415–418.

Zisman, W.A., 1964. Relation of the equilibrium contact angle to liquid and solid constitution. Advances in Chemistry. American Chemical Society, Washington, DC, pp. 1–51.

Superhydrophobicity

11

Jooyoun Kim, Seong-O Choi*[†]
*Seoul National University, Seoul, Republic of Korea, [†]Kansas State University, Manhattan, KS, United States

11.1 Introduction

With the discovery of self-cleaning property of lotus leaves (Barthlott and Neinhuis, 1997), known as the lotus effect, significant research efforts have been made to mimic the lotus leaf to produce superhydrophobic surfaces (Nishimoto and Bhushan, 2013; Bhushan, 2009, 2012). Surfaces that exhibit water contact angle higher than 150 degree and the contact angle hysteresis lower than 5–10 degrees are typically referred to superhydrophobic surfaces (Kota et al., 2013a; Wong et al., 2013). Barthlott and Neinhuis (Neinhuis and Barthlott, 1997; Barthlott and Neinhuis, 1997) found such superhydrophobic properties of the lotus leaf resulted from the proper interdependence between surface roughness and surface energy, which prevents the surface from being contaminated (Neinhuis and Barthlott, 1997; Barthlott and Neinhuis, 1997).

The interdependence between roughness and surface energy on wettability has been theoretically explained by the Wenzel model (Wenzel, 1936) and the Cassie–Baxter model (Cassie and Baxter, 1944). According to those models, the presence of roughness amplifies the intrinsic wettability, where intrinsic wettability is determined by the surface energy of materials. For instance, if the surface roughness is implemented to an intrinsically hydrophobic material (with low surface energy), the hydrophobicity is amplified. On the other hand, creation of roughness to an intrinsically hydrophilic material (with high surface energy) enhances the wettability. Basic wetting theories are discussed in Section 11.2.

Other wetting models to explain the wetting behavior have followed, taking into account the geometric factors and hierarchical scale of roughness (Tuteja et al., 2008; Kota et al., 2013a,b; Liu and Kim, 2014). Noticing that surface roughness is the critical factor that influences extreme wettability or anti-wettability, significant researches have been performed to fabricate roughness in various scales and patterns by employing advanced techniques (Celia et al., 2013; Du et al., 2014; Huang et al., 2015; Fujii et al., 2011; Kang et al., 2008; Kim et al., 2005; Kota et al., 2012; Kwon et al., 2014; Si et al., 2015; Liu and Kim, 2014). In Section 11.3, the focus is on techniques and methods that have been employed to control wettability by implementing surface roughness and reducing surface energy.

In Section 11.4, characterization methods to measure surface wettability and self-cleaning ability are overviewed, including static contact angle, contact angle hysteresis, sliding angle, shedding angle and the number of bouncing (Artus et al., 2006; Zimmermann et al., 2009; Park et al., 2016). Lastly, possible applications of

Waterproof and Water Repellent Textiles and Clothing. https://doi.org/10.1016/B978-0-08-101212-3.00010-1

superhydrophobic surface are introduced (Bhushan, 2012; Du et al., 2014; Si et al., 2015; Park et al., 2016; Lee et al., 2004; Liu et al., 2013; Wong et al., 2011; Shim et al., 2014; Kwon et al., 2015; Zhou et al., 2013; Nishino et al., 1999).

11.2 Wetting theories

11.2.1 Young's model for flat surfaces

A water droplet falling from a faucet (tap) is spherical, and water striders can walk on the surface of water. These phenomena occur due to the surface tension, which is the resultant of the unbalanced forces of liquid at the surface. Liquid molecules in the bulk, experience attractive forces caused by neighboring molecules in every direction, resulting in a net force of zero. Molecules at the surface, however, do not have neighboring molecules in all directions so a net force exerted on the surface molecules is not zero. Because the attraction between liquid molecules is greater than between liquid and air molecules, the net force at the liquid surface is inward, creating an internal pressure. This unbalanced force makes liquids contract to minimize the surface area (spherical in the air). Surface tension, represented by the symbol γ, is measured in force per unit length (Newton per meter, N/m) or energy per unit area (joule per square meter, J/m^2). For solids, energy per unit area is commonly used and termed surface energy.

When a liquid droplet is placed on a solid surface in the air, the shape of a liquid droplet is determined by the interactions between molecules at the interfaces. The angle formed by the solid surface and the tangent of the liquid droplet is called the contact angle, which represents the wettability of a solid by a liquid. Generally, a solid surface is considered hydrophilic if the water contact angle is smaller than 90 degree and is considered hydrophobic if the water contact angle is larger than 90 degree. If a liquid droplet on an ideal solid surface (smooth and chemically homogeneous) is considered, there are three interfaces: liquid–vapor, solid–vapor and solid–liquid interfaces (Fig. 11.1).

At static equilibrium, a liquid droplet maintains its shape due to the balance of three surface tensions at a solid–liquid–vapor contact point (γ_{lv}, γ_{sv} and γ_{sl}). The net force exerted on a liquid droplet should be zero at equilibrium, and the relationship between surface tensions can be expressed as:

$$\gamma_{sv} = \gamma_{sl} + \gamma_{lv} \cos\theta_Y \tag{11.1}$$

where γ_{sv}, γ_{sl} and γ_{lv} are the surface tension of the solid/vapor interface, of the solid/liquid interface and of the liquid/vapor interface, respectively, and θ_Y is the contact

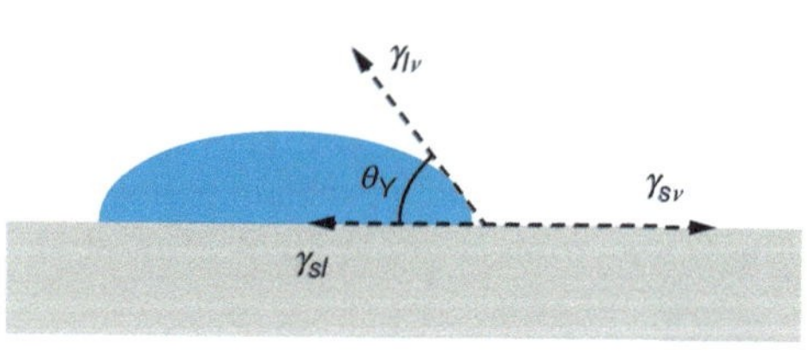

Fig. 11.1 Contact angle of a liquid droplet on a flat surface.

angle at the three-phase contact line (intrinsic contact angle). By rearranging Eq. (11.1), the contact angle of a liquid on a perfectly smooth and chemically homogeneous solid surface is obtained, which is known as Young's equation (Young, 1805):

$$\cos\theta_Y = \frac{\gamma_{sv} - \gamma_{sl}}{\gamma_{lv}} \tag{11.2}$$

Young's equation applies to flat and smooth surfaces, and the contact angle defined by the surface tensions at three interfaces reflects the intrinsic wettability of the surface.

11.2.2 Wenzel and Cassie–Baxter theories applied to rough surfaces

Because solid surfaces are not always perfectly smooth or homogeneous, the contact angles of liquid droplets measured on typical solid surfaces are significantly different from those obtained by Young's equation and are affected by the morphology and material composition of a solid surface. Wenzel (1936) considered the effect of surface roughness on wettability of solid surfaces assuming that the liquid totally wets the roughness grooves (Fig. 11.2A), and introduced a roughness factor r to describe real solid surfaces. The roughness factor is defined as the ratio of the actual surface area (A_{rough}) its projected surface area (A_{flat}) of the solid:

$$r = A_{rough}/A_{flat} \tag{11.3}$$

It should be noted that r is always greater than 1 because the surface area of a rough surface is always larger than that of a flat surface. Therefore, a rough surface has increased energy at the solid/vapor interface and at the solid/liquid interface by a factor r, compared to a flat surface. In other words, the Wenzel equation can be derived by multiplying r to γ_{sv} and γ_{sl}, respectively. Substituting surface tensions with $r{\cdot}\gamma_{sv}$ and $r{\cdot}\gamma_{sl}$ in Young's equation results in the Wenzel's equation:

$$\cos\theta_W = \frac{r \cdot (\gamma_{sv} - \gamma_{sl})}{\gamma_{lv}} = r \cdot \cos\theta_Y \tag{11.4}$$

Fig. 11.2 Schematic representation of different wetting states. (A) the Wenzel state, (B) the Cassie–Baxter state and (C) the Cassie–Baxter state with air pockets (Wenzel, 1936; Cassie and Baxter, 1944).

where θ_y is the contact angle in Young's equation (intrinsic contact angle), and θ_W is the apparent contact angle on a rough surface that follows the Wenzel model. From the Wenzel's equation, the wettability of a solid surface can be modified by introducing the roughness factor, r. For $\theta_y < 90$ degree, θ_w is always smaller than θ_y, indicating that introduction of roughness will enhance the wettability. On the other hand, θ_w becomes larger than θ_y by introducing roughness for $\theta_y > 90$ degree, indicating that the tendency of anti-wetting will also be enhanced. Therefore, initially hydrophilic surfaces will become more hydrophilic and hydrophobic surfaces will become more hydrophobic by the introduction of roughness according to the Wenzel model.

The wettability of a solid surface is influenced not only by surface roughness but also by material compositions. Cassie and Baxter (1944) considered the effect of surface heterogeneity on the wettability of solid surfaces by assuming the solid is composed of randomly distributed, different types of materials as shown in Fig. 11.2B. The surface is flat and smooth, and consists of n different materials. The surface energies at the interface with a solid and surface fraction of ith material are denoted by $\gamma_{i,sv}$, $\gamma_{i,sl}$ and f_i, respectively. Because f_i is the fraction of each surface area on the solid surface, $f_1 + f_2 + \cdots + f_i + \cdots + f_n = 1$ and the total surface energy of the solid/vapor interface (γ_{sv}) and of the solid/liquid interface (γ_{sl}) can be expressed as:

$$\gamma_{sv} = f_1 \cdot \gamma_{1,sv} + \cdots + f_i \cdot \gamma_{i,sv} + \cdots + f_n \cdot \gamma_{n,sv} = \sum_{i=1}^{n} f_i \cdot \gamma_{i,sv} \tag{11.5}$$

$$\gamma_{sl} = f_1 \cdot \gamma_{1,sl} + \cdots + f_i \cdot \gamma_{i,sl} + \cdots + f_n \cdot \gamma_{n,sl} = \sum_{i=1}^{n} f_i \cdot \gamma_{i,sl} \tag{11.6}$$

The contact angle on a heterogeneous solid surface can be obtained by putting the above surface energies in Young's equation as follows:

$$\cos\theta_{CB} = \frac{\sum_{i=1}^{n} f_i \cdot (\gamma_{i,sv} - \gamma_{i,sl})}{\gamma_{lv}} = \sum_{i=1}^{n} f_i \cdot \cos\theta_{i,Y} \tag{11.7}$$

For a surface composed of two materials, Eq. (11.7) becomes:

$$\cos\theta_{CB} = f_1 \cdot \cos\theta_{1,Y} + f_2 \cdot \cos\theta_{2,Y} \tag{11.8}$$

where $f_1 + f_2 = 1$, and $\theta_{1,Y}$ and $\theta_{2,Y}$ are the intrinsic contact angles of materials 1 and 2, respectively.

Because this equation is the general form, it can be applied to a situation where one of the materials is air. In this case, a liquid droplet sits on a surface having grooves filled with air (air-pockets) as shown in Fig. 11.2C. For the solid–liquid fraction f_{sl}, the contact angle is θ_Y because now there is only one solid component in the system. For the liquid–vapor fraction (air pockets) f_{lv}, the contact angle will be assumed as 180 degree because the fraction is completely non-wetting. Now the equation becomes:

$$\cos\theta_{CB} = f_{sl}\cdot\cos\theta_Y + f_{lv}\cdot\cos 180° = f_{sl}\cdot\cos\theta_Y - f_{lv}$$
$$= f_{sl}\cdot\cos\theta_Y - (1 - f_{sl}) \tag{11.9}$$

$$\therefore \cos\theta_{CB} = f_{sl}\cdot(\cos\theta_Y + 1) - 1 \tag{11.10}$$

If the solid–liquid fraction is partially wet by liquids due to the existence of surface roughness, the roughness factor r can be incorporated into the Cassie–Baxter equation:

$$\cos\theta_{CB} = r\cdot f_{sl}\cdot\cos\theta_Y + f_{lv}\cdot\cos 180° = r\cdot f_{sl}\cdot\cos\theta_Y - f_{lv}$$
$$= r\cdot f_{sl}\cdot\cos\theta_Y - (1 - f_{sl})$$

$$\therefore \cos\theta_{CB} = r\cdot f_{sl}\cdot\cos\theta_Y + f_{sl} - 1 \tag{11.11}$$

The above equation becomes the Wenzel equation when $f_{sl} = 1$ (i.e. surface is fully wet by liquids).

The Cassie–Baxter equation (Eq. 11.11) suggests that the contact angle θ_{CB} will increase by minimizing the solid–liquid fraction (i.e. maximizing the air-pockets). Also, θ_{CB} could exceed 90 degree even when the intrinsic contact angle θ_Y is less than 90 degree (when $0 < \cos\theta_Y < \frac{1-f_{sl}}{f_{sl}}$) which cannot be achieved in the Wenzel model. Therefore, a surface exhibiting the Cassie–Baxter state is preferred to minimize wetting and attain superhydrophobicity.

Both the Wenzel and the Cassie–Baxter models show that the introduction of roughness on a hydrophobic surface ($\theta_Y > 90$ degree) will further enhance the hydrophobicity of the surface. Theoretically, for a liquid droplet that is placed onto a rough hydrophobic surface, the favored state between the Wenzel state and the Cassie–Baxter state can be predicted by assuming an equilibrium angle (θ_c) where θ_w and θ_{CB} show the same value (Bico et al., 2002). This critical contact angle θ_c can be obtained by equating the Wenzel and the Cassie–Baxter equations, and θ_c for the transition is expressed as:

$$\cos\theta_c = \frac{f_{sl} - 1}{r - f_{sl}} \tag{11.12}$$

where f_{sl} and r are the solid–liquid fraction and the roughness factor, respectively.

From Eq. (11.12), the Cassie–Baxter state is preferred if $\theta_Y > \theta_C$ and a droplet will exhibit the Wenzel state if $\theta_Y < \theta_C$. When the roughness exists, θ_C is always greater than 90 degree because $r > 1 > f_{sl}$. Therefore, a solid surface with θ_C, where $\theta_Y > \theta_C > 90$ degree, would exist in the Cassie–Baxter state and a surface with $\theta_C > \theta_Y > 90$ degree would exist in the Wenzel state. However, droplets in the Cassie–Baxter state have also been observed on rough surfaces when $\theta_C > \theta_Y > 90$ degree, indicating that there exists a metastable state where wetting transitions can occur. Studies have demonstrated that a transition between the Cassie–Baxter and the Wenzel state can occur under certain conditions, including droplet volume, pressure and vibration (Lafuma and Quéré, 2003; Jung and Bhushan, 2009), and both states may even coexist on the same surface (Koishi et al., 2009; Jung and Bhushan, 2007; Ran et al., 2008).

11.2.3 Theories applied to surfaces with hierarchical roughness

Observations of natural surfaces such as the lotus leaf suggest that hierarchically structured surfaces, which possess multi-scale roughness, are beneficial to create superhydrophobic surfaces compared to single-textured surfaces. This is because the solid–liquid contact area in hierarchically structured surfaces is significantly smaller than that in a single-textured surface. Also, recent studies (Xue et al., 2012; Kota et al., 2012; Su et al., 2010; Nosonovsky, 2007) have demonstrated that the stability of the superhydrophobic state can be enhanced by structuring surfaces with micro- and nanoscale roughness. The micro-scale roughness reduces the solid–liquid contact area, while the nano-scale roughness helps the surface sustain the pressure that is required to maintain the Cassie–Baxter state. In addition, it was shown that it would be possible to construct a superhydrophobic surface from a material that has the contact angle below 90 degree using multi-level hierarchical structures and the macroscopic contact angle on a multi-level hierarchical structure can be given by (Herminghaus, 2000):

$$\cos\theta_{n+1} = (1 - f_{lv,n})\cos\theta_n - f_{lv,n} \tag{11.13}$$

where n, $f_{lv,n}$ and θ_n are the number of hierarchical levels, the liquid–vapor area fraction and the apparent contact angle at the nth level, respectively. From the above equation, it is evident that the apparent contact angle increases as the number of hierarchical levels increases. Because $f_{sl,n}+f_{lv,n}=1$, we can replace $f_{lv,n}$ with $(1-f_{\mathrm{sl,n}})$, which leads to:

$$\cos\theta_{n+1} = f_{sl,n}\cdot(\cos\theta_n + 1) - 1 \tag{11.14}$$

For $n=0$, the equation represents the Cassie–Baxter equation (single roughness). With the dual-scale roughness, the apparent contact angle can be calculated as follows:

$$\cos\theta_1 = f_{sl,0}\cdot(\cos\theta_0 + 1) - 1 \tag{11.15}$$

$$\cos\theta_2 = f_{sl,1}\cdot(\cos\theta_1 + 1) - 1 = f_{sl,1}\cdot f_{sl,0}\cdot(\cos\theta_0 + 1) - 1 \tag{11.16}$$

Since $f_{sl,n} < 1$,

$f_{sl,1}\cdot f_{sl,0} < f_{sl,0}$,

and $\cos\theta_2 < \cos\theta_1 (\text{i.e.}, \theta_2 > \theta_1)$.

The Wenzel state and the Cassie–Baxter state can be metastable and coexist. Rahmawan et al. (2010) described four possible wetting states on dual-scale hierarchical structures. These potential states are: Cassie (m)-Cassie (n), Wenzel (m)-Wenzel (n), Cassie (m)-Wenzel (n) and Wenzel (m)-Cassie (n), where m and n denote micro-scale and nano-scale, respectively. Schematic representation of the four states is shown in Fig. 11.3.

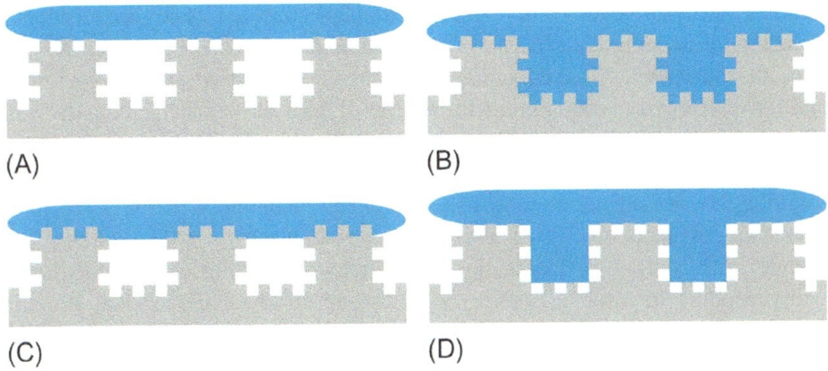

Fig. 11.3 Four possible wetting states on dual-scale hierarchical structures. (A) Cassie (m)-Cassie (n) state, (B) Wenzel (m)-Wenzel (n) state, (C) Cassie (m)-Wenzel (n) state and (D) Wenzel (m)-Cassie (n) state.

To derive theoretical contact angles for each wetting state, the effective solid–vapor and solid–liquid interfacial energies for the four states are expressed considering the roughness factors (r^m and r^n for micro- and nano-scale roughness, respectively) and wetting area fractions (f^m and f^n for micro- and nano-scale fractions, respectively) at each scale. The effective solid–vapor interfacial energy $\overline{\gamma_{sv}}$ and the effective solid–liquid interfacial energy $\overline{\gamma_{sl}}$ or each state are summarized in Table 11.1.

The apparent contact angles for the four different states ($\cos\theta_{CC}$, $\cos\theta_{WW}$, $\cos\theta_{CW}$, $\cos\theta_{WC}$) can be obtained by applying the calculated interfacial energies into Young's equation, $\cos\theta_{app} = (\overline{\gamma_{sv}} - \overline{\gamma_{sl}})/\gamma_{lv}$ is follows:

$$\text{Cassie}\,(m) - \text{Cassie}\,(n)\,\text{state}: \cos\theta_{CC} = f^m \cdot f^n \cdot (\cos\theta_Y + 1) - 1 \quad (11.17)$$

$$\text{Wenzel}\,(m) - \text{Wenzel}\,(n)\,\text{state}: \cos\theta_{WW} = r^m \cdot r^n \cdot \cos\theta_Y \quad (11.18)$$

$$\text{Cassie}\,(m) - \text{Wenzel}\,(n)\,\text{state}: \cos\theta_{CW} = r^n \cdot \{f^m \cdot (\cos\theta_Y + 1 - 1\} \quad (11.19)$$

$$\text{Wenzel}\,(m) - \text{Cassie}\,(n)\,\text{state}: \cos\theta_{WC} = r^m \cdot \{f^n \cdot (\cos\theta_Y + 1) - 1\} \quad (11.20)$$

Table 11.1 Effective interfacial energies for dual-scale hierarchical structures

Wetting state	$\overline{\gamma_{sv}}$	$\overline{\gamma_{sl}}$
Cassie (m)–Cassie (n)	$\gamma_{sv} \cdot f^m \cdot f^n$	$\gamma_{sl} \cdot f^m \cdot f^n + \gamma_{lv} \cdot (1 - f^m \cdot f^n)$
Wenzel (m)–Wenzel (n)	$r^m \cdot r^n \cdot \gamma_{sv}$	$r^m \cdot r^n \cdot \gamma_{sl}$
Cassie (m)–Wenzel (n)	$\gamma_{sv} \cdot f^m \cdot r^n$	$\gamma_{sl} \cdot f^m \cdot r^n + \gamma_{lv} \cdot (1 - f^m)$
Wenzel (m)–Cassie (n)	$\gamma_{sv} \cdot r^m \cdot f^n$	$\gamma_{sl} \cdot r^m \cdot f^n + \gamma_{lv} \cdot (r^m + r^m \cdot f^n)$

The derived equations become either a conventional Wenzel or Cassie–Baxter equation by replacing the wetting area fraction f or the roughness factor r for the surface with single-scale roughness. Also, the equation for the Cassie–Cassie state is the same as the one derived from the recursive equation for $n=1$, indicating that the recursive equation is only applicable to the situation where all the hierarchical levels exhibit the Cassie–Baxter state.

From the theories discussed in this section, it can be deduced that it is possible to construct superhydrophobic surfaces by the introduction of proper roughness on hydrophobic materials or by the creation of hierarchical structures on either hydrophobic or hydrophilic materials. Theories regarding multi-level hierarchical structures suggest that the apparent contact angle increases as the number of hierarchical level increases, if the Cassie–Baxter state is maintained. For a lotus leaf surface, the waxy surface itself is not very hydrophobic (the water contact angle is about 74.0 ± 8.5 degree) (Cheng and Rodak, 2005; Cheng et al., 2006), yet the hierarchical roughness structure at the surface produced the superhydrophobic nature with the water contact angle of ~162 degree. Other superhydrophobic hierarchical surfaces include the leaves of *Leymus arenarius* (lyme grass, static water contact angle 161 degree) and *Colocasia esculenta* (taro, static water contact angle 164 degree).

11.2.4 Re-entrant model

A study by Neinhuis and Barthlott (Neinhuis and Barthlott, 1997; Barthlott and Neinhuis, 1997) showed that most of the water repellent plant species have hierarchical surface structures, formed by densely arranged epicuticular waxes of papillose cells or in different shapes. Water droplets on those plant surfaces, however, can spread over the leaf surface under certain conditions, indicating that those surfaces exhibit the metastable Cassie–Baxter state. The metastable Cassie–Baxter state on a hydrophilic surface could be maintained or be turned into the Wenzel state. To maintain the Cassie–Baxter state, the air entrapped in the rough structures should act as an 'energy barrier' that inhibits the transition to the Wenzel state. Cao et al. (2007) hypothesized that this energy barrier might be provided by a capillary force and demonstrated that overhanging structures (re-entrant structures) with proper designs could gain superhydrophobicity on intrinsically hydrophilic materials. As shown in Fig. 11.4, the most critical parameter that determines the direction of a capillary force

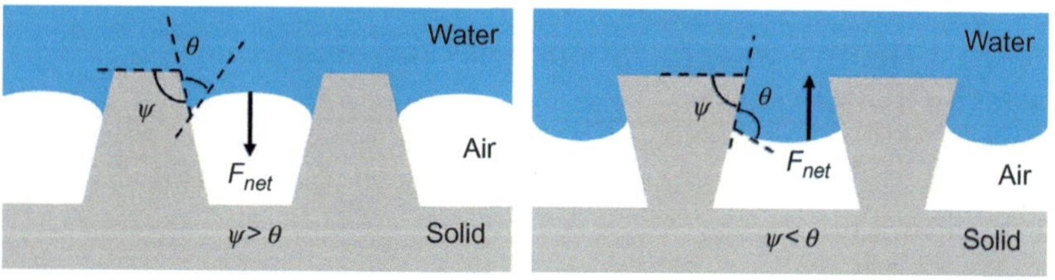

Fig. 11.4 Schematic diagrams of the expected liquid–vapor interface on two geometries with different equilibrium contact angles. The capillary force can resist the wetting of the grooves if $\psi < \theta$ (Cao et al., 2007; Tuteja et al., 2007).

is the angle formed between the sidewalls of the groove and the top surface, denoted by ψ (Tuteja et al., 2007, 2008).

When ψ is larger than the equilibrium contact angle θ, the meniscus (water–air interface) in the groove is concave (seen from the water side). In this case, the net force F_{net} generated by the meniscus is downward so that water will enter the grooves and fully wet the surface. For $\psi < \theta$, on the other hand, the meniscus is convex and F_{net} acts upward, thereby obstructing the penetration of water into the groove. Cao et al. (2007) demonstrated that re-entrant structures are desirable to construct more stable superhydrophobic surfaces by showing that an originally hydrophilic silicon surface increased the water contact angles to approximately 139–162 degrees when re-entrant structures were fabricated on the surface (Cao et al., 2007). Other studies have also demonstrated that properly designed re-entrant structures, such as dual layer re-entrant structures, can provide superoleophobic surfaces, repelling extremely low surface energy liquids (Tuteja et al., 2007, 2008). In summary, low surface energy and surface roughness (including re-entrant structure) are the key design parameters to attain superhydrophobic surfaces.

11.3 Fabrication of superhydrophobic surfaces

Except for the surface with re-entrant geometry that exhibits inherently non-wetting characteristics (regardless of the surface energy of solid), the wettability of a material is primarily dependent on the surface energy of materials. However, having a low surface energy is often not sufficient to obtain a superhydrophobic surface (water contact angle >150 degree) and the role of surface roughness becomes important to achieve superhydrophobicity, as demonstrated in theories in Section 11.2. Many superhydrophobic surfaces occurring in nature, such as the lotus leaf (Bhushan, 2009; Barthlott and Neinhuis, 1997), butterfly wing (Nishimoto and Bhushan, 2013; Bhushan, 2012) and cicada wing (Lee et al., 2004; Nishimoto and Bhushan, 2013), have certain roughness structures on waxy materials. With such observations, many studies have attempted to fabricate bioinspired superhydrophobic surfaces (Liu et al., 2013; Nishimoto and Bhushan, 2013; Wong et al., 2011; Bhushan, 2012), employing advanced techniques to lower the surface energy and to form surface roughness. Section 11.3 gives an overview of design strategy and techniques to fabricate superhydrophobic surfaces.

11.3.1 Modification of surface energy

Decreasing the surface energy of a solid surface is the primary step to reduce wettability, while this process itself may not be sufficient to attain superhydrophobicity. Textile materials often have inherent roughness resulting from the curvature of fibers/yarns, weave and knit patterns. Thus, the modification of surface energy by itself can significantly change the wetting characteristics. Indeed, superhydrophobic woven fabrics can be obtained primarily by coating the textile surface with fluorinated chemicals (Shim et al., 2014; Kwon et al., 2015). In the following sections, wet and vapor coating processes are overviewed as common application methods to modify the surface energy of materials.

11.3.1.1 Wet process

Wet processes, including dip-pad-cure coating, knife coating and foam coating, have been employed in the textile industry as conventional finishing processes to modify the material properties. For water-resistant and water repellent finishing, chemicals in low surface tension, such as siloxanes and fluorinated compounds, have been frequently applied to textiles.

When fluorinated compounds are coated on a surface, it is desirable for the fluorine-containing groups to be arranged outward from the surface. Depending on the orientation and distribution of fluoroalkyl groups, the surface energy can be lowered to 5–20 dyne/cm (Zhou et al., 2013), and a flat surface packed with $-CF_3$ groups gives water contact angles about 120 degree (Nishino et al., 1999). Among the fluorinated compounds, perfluorocarbon compounds with C8 or longer chains have been used as effective repellent agents. However, the use of C8 fluorocarbon agents is being restricted because they produce perfluorooctanoic acid (PFOA) in the decomposition process of C8 fluorocarbons. PFOA is known to be potentially carcinogenic and hazardous to the body (Armitage et al., 2009).

The dip-pad-cure method is used for a wide range of textile finishing; in this method, a textile is immersed into a chemical solution and then fixed through drying and cure (Shim et al., 2014; Kwon et al., 2015). This wet coating process can be easily applied in mass production of textiles. The thickness of coating produced by the dip-pad-cure method is relatively thick compared to that produced by the vapor deposition process (Kwon et al., 2014, 2017). If a thick coating is applied, the coating agent can block pores of the treated textile, possibly leading to undesirable physical properties such as lower air and vapor permeability. Wet processing usually leaves toxic waste water; therefore, it is necessary to explore environmentally more responsible processes. As an alternative approach, chemical vapor deposition processes have been employed (Park et al., 2016; Kwon et al., 2014, 2015; Yuan et al., 2016).

11.3.1.2 Chemical vapor deposition (CVD)

Chemical vapor deposition (CVD) is the process where chemical precursors are transported in the vapor phase to react on a substrate to form a thin layer of coating. CVD can be applied to deposit hydrophobic compounds on a surface or to form rough surfaces (Celia et al., 2013). As the CVD produces a submicron- scale thin layer, it generally allows the bulk properties of substrates to be maintained. Also, compared to typical wet processes, CVD process leaves less toxic wastes; thus, this process has been discussed as an environmentally more sustainable approach to granting functionality (Kwon et al., 2014; Park et al., 2016).

As a vapor deposition method, plasma enhanced chemical vapor deposition (PECVD) has also been used to fabricate superhydrophobic surfaces (Kim et al., 2005, 2016; Zhang et al., 2003; Kwon et al., 2014; Park et al., 2016). In this process, a vaporized chemical precursor is ionized by a strong electromagnetic field, polymerized and then deposited on a substrate. The deposited polymer adheres to the surface chemically and/or physically. Fluorine compounds have been frequently used to produce superhydrophobic textiles by the plasma process (Kim et al., 2005; Zhang et al., 2003). When SF_6 radio frequency plasma was applied to cotton, silk and polyethylene terephthalate (PET) fabrics, the resulting surfaces exhibited the water contact angle between 130 and 150 degrees (Riccardi et al., 2001). Hexamethyldisiloxane (HMDSO) has also been commonly used in the PECVD process to form a thin layer of hydrophobic coating on a substrate (Park et al., 2016; Kim et al., 2016; Kwon et al., 2014).

PECVD forms a thin film in several to tens of nanometers, thus the intrinsic bulk properties of the treated substrates are minimally influenced by the treatment. In Kwon et al.'s (2014) study, a single-faced superhydrophobic lyocell fabric was produced by the PECVD process with HMDSO chemical. The coating layer was thin enough not to influence the intrinsic hydrophilicity of the untreated side (lyocell), but the coated side (with HMDSO) exhibited superhydrophobicity (Kwon et al., 2014, 2015). Air and vapor permeability of the treated fabric did not change significantly, which demonstrates negligible influence of thin coating on comfort properties (Kwon et al., 2014, 2017). With such advantages of the vapor deposition techniques, the scalable vapor process needs to be further studied.

11.3.2 Formation of surface roughness

A lotus leaf with a hierarchical dual scale roughness has been regarded as a typical example of superhydrophobic surface, and many researches have put substantial effort into fabricating such a roughness structure by employing various techniques (Liu et al., 2013; Nishimoto and Bhushan, 2013). In this section, fabrication methods that have been employed to implement surface roughness to create superhydrophobic surfaces are overviewed. The roughness types are classified as follows: inherent roughness resulting from the fibrous structure; roughness created by adding materials onto a surface; roughness created by etching the top surface (Table 11.2).

11.3.2.1 Inherent roughness of textile fabrics

Textiles are composed of multiple strands of micro-sized fibers, thereby holding microscale roughness. Though it is challenging to achieve superhydrophobicity solely by the microscale roughness attributed to fibers and yarns, woven fabrics coated by fluorinated finishing agents can show water contact angle close to or greater than

Table 11.2 Methods to form surface roughness for superhydrophobic textile/polymer materials

Inherent fabric roughness	Bottom up roughness creation	Top down roughness creation
Weave/knit pattern Texture from yarns/fibers	Electrospinning Colloidal assembly Layer-by-layer deposition Sol–gel method	Lithography Plasma etching Chemical etching

150 degree (Shim et al., 2014; Kwon et al., 2015). The fluorinated fabric surface produced the higher level of hydrophobicity when the yarn consisted of more fibers and when the fibers were textured to give additional roughness (Shim et al., 2014). The following section presents the theoretical explanation on wettability of woven fabrics.

Fabric roughness explained by the Wenzel model

In the wetting state that perfectly follows the Wenzel model, fibers that comprise a yarn are completely wet and are in contact with liquid. The surface area that is in contact with liquid on a woven fabric can be calculated from the yarn radius, fiber radius and the number of fibers that comprise a yarn. From the calculated surface area, the roughness factor r in the Wenzel model can be estimated by the equations given in Table 11.3 (Shim et al., 2014; Michielsen and Lee, 2007). When a yarn is composed

Table 11.3 Calculation of roughness factor from the woven textile surface (Shim et al., 2014; Michielsen and Lee, 2007)

Woven with monofilament yarns		
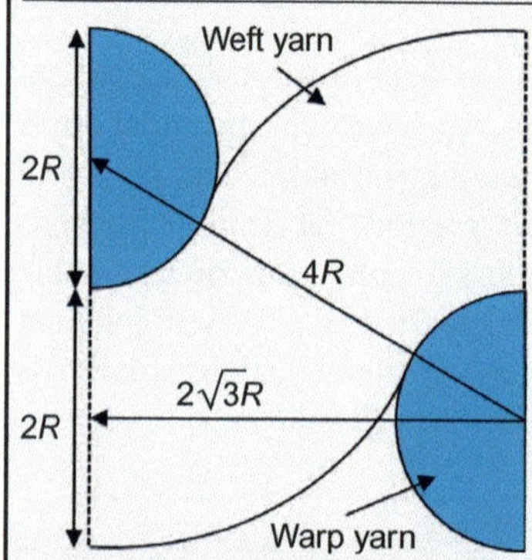	Projected area $2\sqrt{3}R \times 2\sqrt{3}R = 12R^2$ Area in contact with water $(2\pi R \times 4R) \times 2 = 16\pi R^2$	$r = 4.19$
Woven with multifilament yarns		
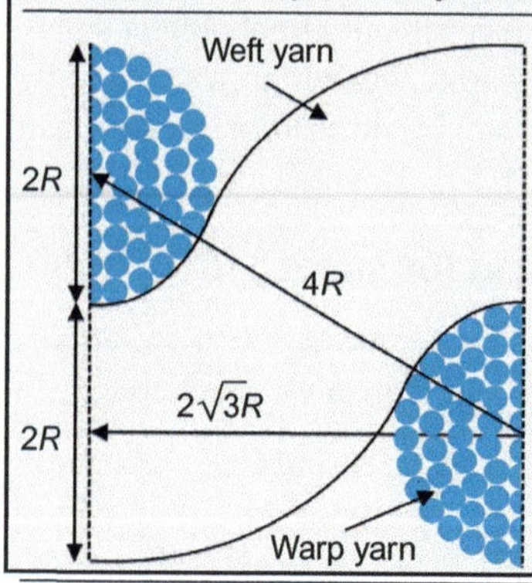	Projected area $2\sqrt{3}R \times 2\sqrt{3}R = 12R^2$ Area in contact with water $N(2\pi R_f) \times 4R \times 2 = 16\pi N R_f R$	$r = 4.19\frac{NR_f}{R}$

R, yarn radius; R_f, fiber radius; N, number of fibers comprising the yarn.

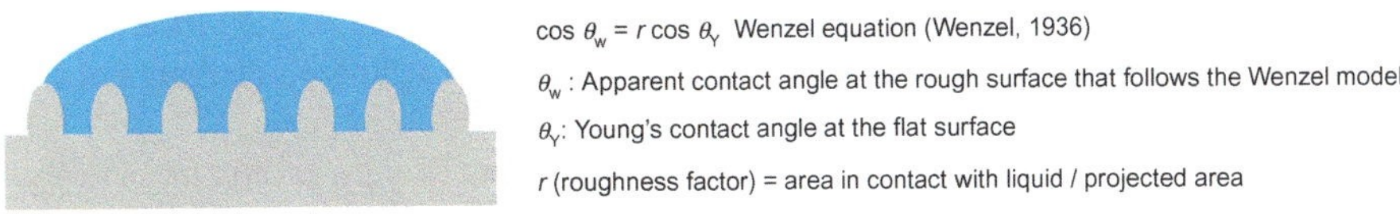

Fig. 11.5 Wenzel's wetting state.

of multi-filament fibers, the roughness factor r increases as the yarn radius decreases and the number fibers in a yarn increases. Knowing the roughness factor r, the apparent contact angle in the Wenzel's wetting state can be calculated by the following Wenzel equation (Fig. 11.5):

$$\cos\theta_w = r\cos\theta_y$$

$$r(\text{roughness factor}) = \frac{\text{area that is in contact with liquid}}{\text{projected area}}$$

Shim et al. (2014) estimated the roughness factor r by the model suggested by Michielsen and Lee (2007); where, r from the polyester woven fabric ranged from 10 to 43. Due to the porosity of the fabric, water is quickly wicked through the pores and spread on the fabric surface. The wicking occurred more quickly when the capillary action was facilitated by a larger number of channels (or a larger number of filaments) in a yarn (Shim et al., 2014).

Fabric roughness explained by the Cassie–Baxter model

The wettability of superhydrophobic fabrics can be explained by the Cassie–Baxter model. Michielsen and Lee (2007) elaborated the Cassie–Baxter equation to be applied to woven fabrics, by incorporating the structural parameters of yarns comprising a fabric. In this model, yarns are modeled as separated cylinders in parallel by a distance of $2d$, as illustrated in Fig. 11.6. Liquid sits on the surface of cylindrical yarns,

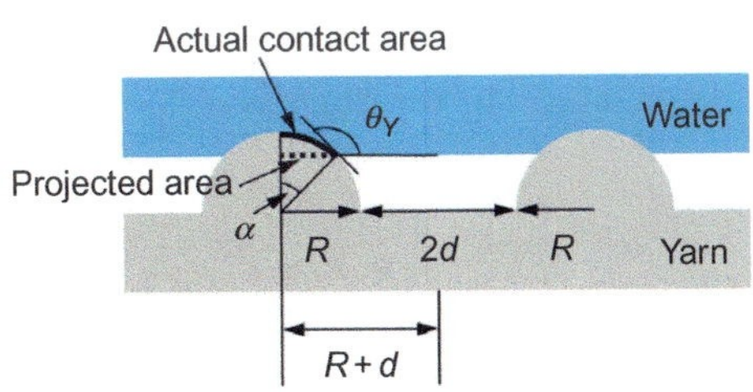

Fig. 11.6 Illustration of the Cassie–Baxter wetting state on a fibrous surface (Michielsen and Lee, 2007).

and the bold arc around the yarn cross-section represents the actual contact area of a yarn that is in contact with liquid. The dotted line represents the projected area of the yarn surface that is in contact with liquid.

The Cassie–Baxter equation can be rewritten in terms of yarn radius (R) and the half of the yarn to yarn distance (d) as follows:

$$\cos\theta_{CB} = f_1\cos\theta_Y + f_2\cos\theta_{Y'}, \tag{11.21}$$

θ_{CB}: apparent contact angle at the rough surface that follows the Cassie–Baxter wetting state
θ_Y: Young's contact angle at the flat surface
$\theta_{Y'}$: contact angle of air, assumed as 180 degree.

f_1 (area fraction that liquid is in contact with solid) $= \dfrac{\text{area that liquid is in contact with solid}}{\text{projected area}}$

f_2 (area fraction that liquid is in contact with air) $= \dfrac{\text{area that liquid is in contact with air}}{\text{projected area}}$

where the projected area is $R+d$.

The equation leads to:

$$\cos\theta_{CB} = f_1\cos\theta_Y - f_2 \tag{11.22}$$

Then f_1 and f_2 can be expressed in terms of f and r_f, which are defined as follows: $f_1 = r_f f$, and $f_2 = 1 - f$

$$r_f = \frac{\text{area that liquid is in contact with solid}}{\text{projected area of solid surface that is in contact with liquid}}$$

$$f = \frac{\text{projected area of solid surface that is in contact with liquid}}{\text{whole projected area}}$$

$$\cos\theta_{CB} = r_f f\cos\theta_Y + f - 1 \tag{11.23}$$

$$r_f f = \frac{R\alpha}{R\sin\alpha} \times \frac{R\sin\alpha}{R+d} = \frac{R\alpha}{R+d}$$

where $f = \dfrac{R\sin\alpha}{R+d}$, $\alpha = \pi - \theta_Y$, R is yarn radius, d is half the distance between two yarns.

The Cassie–Baxter equation can be rewritten as:

$$\cos\theta_{CB} = \left(\frac{R}{R+d}\right)(\pi - \theta_Y)\cos\theta_Y + \left(\frac{R}{R+d}\right)\sin(\pi - \theta_Y) - 1 \tag{11.24}$$

where $f_1 = \left(\dfrac{R}{R+d}\right)(\pi - \theta_Y)$, and $f_2 = 1 - \left(\dfrac{R}{R+d}\right)\sin(\pi - \theta_Y)$.

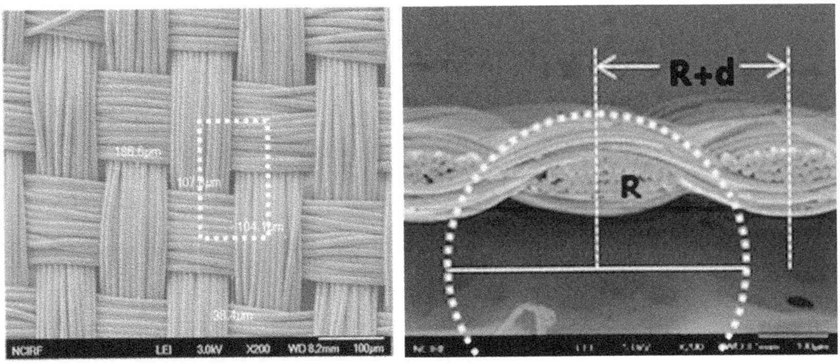

(A) (B)

Fig. 11.7 Estimation of yarn radius, R and yarn to yarn distance, d from a woven fabric. From Shim, M.H., Kim, J., Park, C.H., 2014. The effects of surface energy and roughness on the hydrophobicity of woven fabrics. Text. Res. J. 84, 1268–1278. https://doi.org/10.1177/0040517513495945, with permission from SAGE Publishing.

Shim et al. (2014) measured R and d from a hydrophobic woven fabric as demonstrated in Fig. 11.7, and calculated f_1, f_2 and θ_{CB} predicted by the extended Cassie–Baxter equation by Michielsen and Lee (Cassie and Baxter, 1944; Michielsen and Lee, 2007). On a surface that follows the perfect Cassie–Baxter wetting state (Cassie and Baxter, 1944), f_1+f_2 is 1. However, calculation of f_1+f_2 from Shim et al.'s study produced slightly greater than 1; this implies that a water drop was slightly pinned through the roughness structure, and the wetting state was in the transition state between the Cassie–Baxter and the Wenzel states. The measured contact angle approximately corresponded to the theoretically predicted θ_{CB} on a fabric that was composed of non-textured filament fibers. When textured filaments with artificial crimps were used for fabrics, the measured contact angle was larger than the predicted one, possibly due to the additional roughness that was not considered in the model (Michielsen and Lee, 2007; Shim et al., 2014). As noted, roughness attributed to the curvature of yarns and fibers can contribute to anti-wettability of fabrics, and the additional roughness on fibers and yarns such as crimp can further amplify the anti-wetting property of fabrics. To conclude, the fabric can produce superhydrophobic property by its intrinsic roughness coming from the yarns and fibers if the surface energy can be properly adjusted.

11.3.2.2 Creation of surface roughness by bottom up approach

Electrospinning

Electrospinning technique can be employed to create nanoscale roughness and attain superhydrophobic materials by a one-step process (Lin et al., 2010; Lee et al., 2013; Sas et al., 2012; Kang et al., 2008). Electrospinning is a process that produces submicron to micron size fibers from the electrostatically driven jets of a polymer solution or polymer melt. The basic set-up consists of a feeding system, a high voltage power

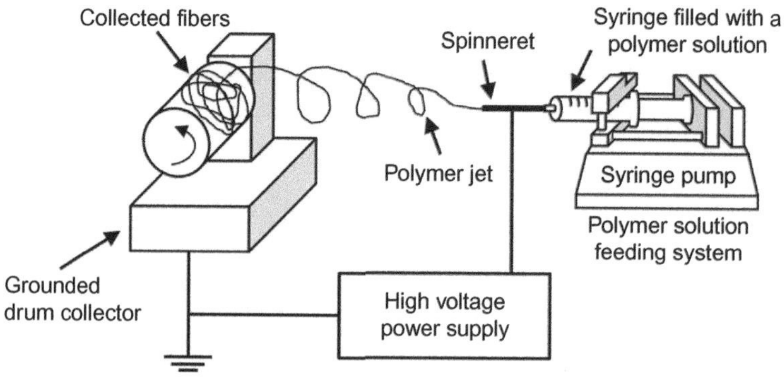

Fig. 11.8 Schematic of electrospinning set-up.

supply and a grounded collector (Fig. 11.8). The feeding system is composed of (1) a precursor polymer solution or melt, (2) a pump that allows the control of feeding rate of a precursor and (3) a spinneret that is connected to a power supply. The electric field is created between the spinneret and the grounded collector, forcing the precursor liquid to overcome the surface tension and to eject from the spinneret. As the polymer jet travels towards the collector, it is elongated and solidified to form thin fibers, being deposited on the collector as a nonwoven form (Sas et al., 2012).

As the morphology and roughness of an electrospun web can be controlled by the electrospinning conditions (Sas et al., 2012), if a polymer in low surface energy is chosen, the electrospinning process by itself can produce a superhydrophobic nonwoven web unlike other methods that require two-step processes to modify both the surface energy and the roughness. For example, polystyrene, with its low surface energy (about 33 mN/m), has been used to fabricate superhydrophobic electrospun web (Yuan et al., 2016; Sas et al., 2012; Kang et al., 2008; Lin et al., 2010). To attain superhydrophobic nonwoven webs, rough surface structures such as beads, wrinkles and pores can be useful, though in general a uniform fiber is the preferred morphology for many textile applications.

Morphology of electrospun webs can be controlled by the combination of the following parameters: (1) properties of polymer solution or melt, such as viscosity, conductivity, solvent volatility, surface tension and polymer molecular weight; (2) process parameters including voltage, tip to collector distance and feeding rate; and (3) ambient conditions including relative humidity and temperature (Sas et al., 2012). Fig. 11.9 shows the morphological changes of fibers with the concentration of a precursor polymer solution. When the viscosity of the precursor solution is too low (at a low concentration), a jet collapses into droplets and beads are formed. Above the critical viscosity, fibers are formed, and then their thickness becomes larger as the viscosity increases (and concentration increases) (Yuan et al., 2016).

In addition, the phase separation driven by the difference in volatilities for a mixture of solvents can produce irregular secondary roughness on fibers and beads. In Fig. 11.10, polystyrene precursor solution was prepared with a mixture of tetrahydrofuran (high volatility) and dimethylformamide (low volatility). Due to the

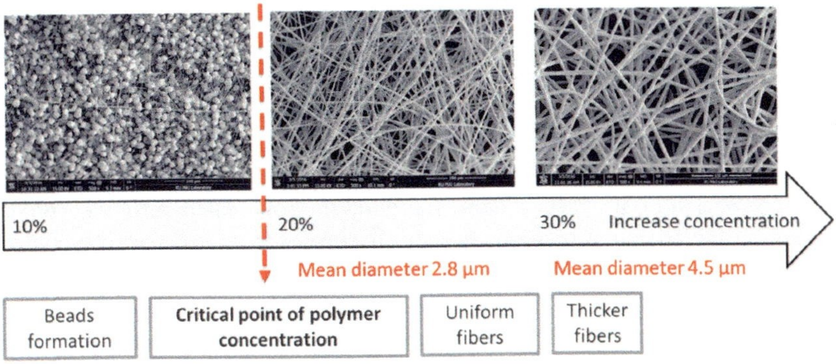

Fig. 11.9 Morphology of electrospun web with concentration changes of polymer solution.

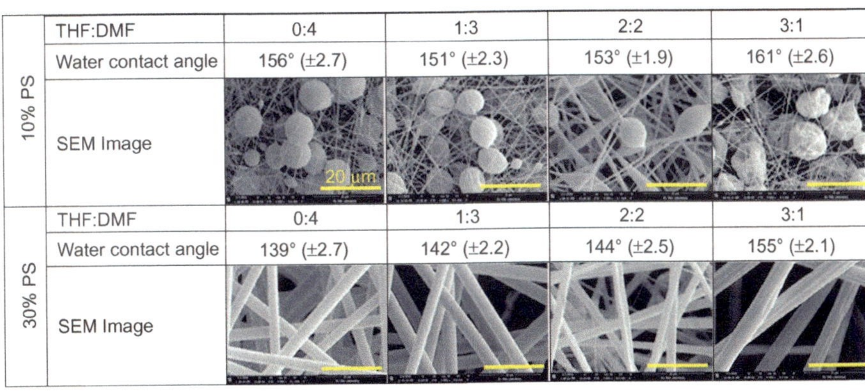

Fig. 11.10 Morphology of the PS electrospun webs produced under different solvent mixing ratios: THF, tetrahydrofuran; DMF, dimethylformamide. *All inserted *yellow bars* represent a length of 20 µm.
From Yuan, Y., Choi, S.-O., Kim, J., 2016. Analysis of contact area between water and irregular fibrous surface for prediction of wettability. RSC Adv. 6, 73313–73322. https://doi.org/10.1039/c6ra15389e.

differences in the rate of solvent evaporation, phase separation into polymer-rich and polymer-poor regions occurred during fiber formation. The resulting webs generated pores and wrinkles on the surface of beads and fibers. This secondary roughness appeared to contribute to the enhanced anti-wettability (shown as higher contact angles) (Yuan et al., 2016). The morphologies generated by the solvent mixture and the resulting contact angles are presented in Fig. 11.10 (Yuan et al., 2016).

Electrospinning is the facile, economic and scalable process. The fine-tuned morphology from this process may not be durable enough for a long-time use, but application for disposable nonwoven fabrics that require repellent property would be relevant.

Colloidal assembly

To attain superhydrophobicity by creating surface roughness, inorganic or polymeric particles can be arrayed in an ordered or non-ordered form on a substrate. Among various methods, colloidal assembly is easy to apply and a cost-effective method. Colloidal assembly is the process that forms the assemblies of dispersed particles through chemical bonding or van der Waals forces, commonly by spin coating or dip coating (Roach et al., 2008). Thus, the durability of treated surfaces would be dependent on the chemical bonding strength between the treated particles and surfaces. The process can be readily applied to textiles by the conventional wet processes.

Self-assembled monolayers (SAMs) can be driven by specific interactions favored by colloidal particles and a solid substrate, and the resulting assembled layer or colloidal aggregates are strongly bonded to the substrate. Particularly, multilayered roughness can be attained by using different sizes of colloidal particles in the self-assembly process, and this technique can be utilized to fabricate superhydrophobic surfaces (Yan et al., 2011; Zhang et al., 2005). Zhang et al. (2005) formed binary colloidal assemblies with $CaCO_3$-loaded hydrogel spheres and polystyrene particles by dip-coating to produce hierarchical roughness on a silicon substrate. The subsequent surface energy modification was performed to produce a superhydrophobic surface. Athauda and Ozer (2012) attached silica nanoparticles with different sizes (7–40 nm) onto the cotton substrate to produce a dual scale roughness, thereby obtaining superhydrophobic fabrics.

Layer-by-layer deposition (LBL)

Layer-by-layer (LBL) deposition utilizes the electrostatic interactions between the polyelectrolyte layers of polyanion and polycation to construct multilayered films (Chen and McCarthy, 1997). This method, like colloidal assembly method, does not require specific preparation for template replication, thus it is a relatively facile and economical process. Employing LBL technique, multilayers of SiO_2 nanoparticles were formed by immersing the substrates repetitively into cationic and anionic aqueous electrolytes with particles (Zhao et al., 2010; Bravo et al., 2007). As polyelectrolytes are hydrophilic, post-treatment of hydrophobization is necessary to yield superhydrophobic properties. When fluoroalkylsilane was coated after the LBL process, the resulting substrate exhibited superhydrophobicity with a low sliding angle (Zhao et al., 2010).

Sol–gel process

The sol–gel process uses a colloidal chemical solution, sol, as a precursor to form a gel-like integrated network of particles on a surface. In this process, a chemical solution (sol) is deposited on a surface by spin coating, spraying or immersion, to form a gel that contains both liquid and solid phases. The gel experiences the phase separation of the solvent-rich and solid-rich region, forming a structured solid surface (Yan et al., 2011). Using the sol–gel method, TiO_2 nanoparticles were aggregated on a fabric surface, yielding a dual scale roughness. The surface was then coated with 1H,1H,2H,2H-perfluorodecyltrichlorosilane (PFDTS) to lower the surface energy. The resulting fabric exhibited superhydrophobic characteristics (Xue et al., 2008). In another study, ZnO nanorods were grown using a seed layer of ZnO nanoparticles on a fabric surface by the sol–gel method. The subsequent hydrophobization with the silane compound on a roughened surface produced the superhydrophobic fabric (Xu and Cai, 2008). Durability of the resulting fabrics needs to be further tested for clothing applications.

11.3.2.3 Surface roughness by top down approach

Lithography and template

Lithography using UV, X-ray or electron beam allows the accurate control of patterns in various shapes, and it has been employed to construct micro- and nanostructures to test the relationship between the surface structure and wetting state (Spori et al., 2010; Cerrina, 2000; Xia and Whitesides, 1998; Lee and Kwon, 2007; Cho and Choi, 2008). Using lithographic methods, an ordered array of micro- and nanostructures can be made, and the subsequent hydrophobization of the patterned surface can produce the superhydrophobic surface (Shieh et al., 2010; He et al., 2011). In Liu and Kim's study (Liu and Kim, 2014), nail-like micro-posts in five different geometric patterns were fabricated employing the photolithography. In this study (Liu and Kim, 2014), geometric parameters of the posts were analyzed with regard to the wettability, leading to the conclusion that the re-entrant geometric structure gave the inherently non-wetting surface to most liquids.

Template-based method involves lithography and imprinting to create a patterned template master, from which a patterned structure is replicated by molding (Spori et al., 2010; Cerrina, 2000; Xia and Whitesides, 1998; Lee and Kwon, 2007; Cho and Choi, 2008). Employing this method, a gecko-mimetic superhydrophobic surface with nanopillar structures was fabricated on a polydimethylsiloxane (PDMS) substrate (Cho and Choi, 2008). Lithography is complex and costly, thus commercial applications for clothing may not be relevant.

Plasma etching

Reactive ion etching (RIE) by the plasma process has been used to form nanoscale roughness in preparing superhydrophobic surfaces (Kwon et al., 2014; Park et al., 2016). Anisotropic etching is particularly helpful to create high-aspect-ratio roughness. Anisotropic etching occurs when an electromagnetic field generates the chemically reactive ions that carry a high energy in a straight direction, causing physical collisions with the surface from the vertical direction (Tsougeni et al., 2009; Tawfick et al., 2012). The chemical etching can occur in all directions but at the slower pace. This etching mechanism produces an anisotropic etching. As the etching time increases, the etching in vertical direction occurs at a faster pace, while the width of nanopillar reduces at a relatively slower pace; as a result, the aspect ratio of nanopillar increases with time (Ramaratnam et al., 2008; Xue et al., 2009).

Fig. 11.11 (Park et al., 2017) shows the fabric treated by oxygen plasma with different etching duration. The uniform and dense nanopillars were formed, and the aspect ratio was increased with longer etching duration. The creation of high-aspect-ratio roughness was attributed to: (1) the anisotropic etching by the linear collision of

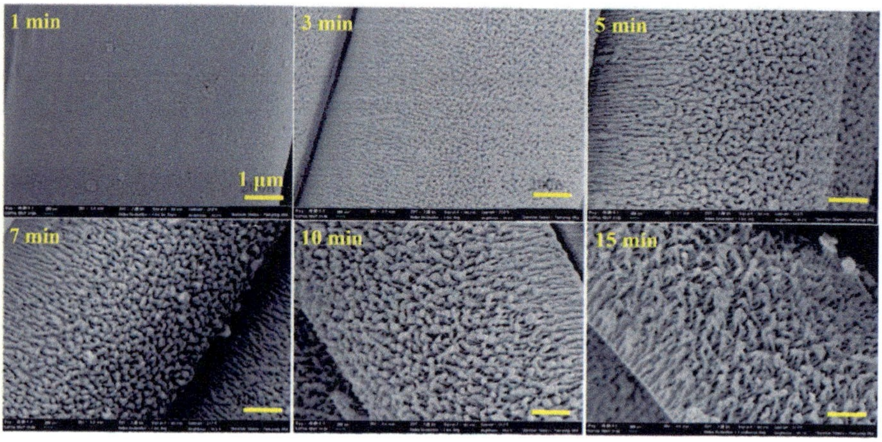

Fig. 11.11 Plasma etching of polyester fiber with different etching duration: oxygen plasma for 1, 3, 5, 7, 10 and 15 min. Scale bars represent 1 μm.
From Park, S., Kim, J., Park, C.H., 2016. Influence of micro and nano-scale roughness on hydrophobicity of a plasma-treated woven fabric. Text. Res. J. Online first published. https://doi.org/10.1177/0040517515627169, with permission from SAGE Publishing.

oxygen plasma, and (2) the inhibition of etching by the metal deposits coming from the stainless steel chamber (Tsougeni et al., 2009; Tawfick et al., 2012). It is explained that the metal surface of the plasma chamber is partially ionized by the reactive oxygen, resulting in the deposition of metal particles onto the substrate. Due to the masked area by the deposited metal, selective etching occurs in the unmasked area of substrate, assisting the anisotropic etching (Park et al., 2016). As the reactive oxygen and deposited metals would make the surface hydrophilic, the surface energy of the etched surface should be lowered by hydrophobic coating. When the etched surface was treated with HMDSO by PECVD, a superhydrophobic fabric was obtained with water contact angle close to 180 degree (Kim et al., 2016; Park et al., 2016; Kwon et al., 2014).

When the anisotropic etching was performed on a fibrous substrate, etching occurred mostly at the top layer of the fiber from submicron to micron depth, leaving the inner layer intact. Thus, the intrinsic bulk properties of the substrate were not significantly affected by the plasma etching process (Kwon et al., 2014, 2017; Park et al., 2016). Kwon et al. (2014, 2017) developed a single-faced superhydrophobic fabric from hydrophilic lyocell fabric by combining the oxygen plasma etching and the subsequent plasma enhanced chemical vapor deposition (PECVD) with hexamethyldisiloxane (HMDSO). The resulting fabric exhibited asymmetric wettability (Fig. 11.12); the treated surface turned into superhydrophobic, and the untreated surface remained hydrophilic. On this treated surface, water vapor was easily condensed into a liquid droplet, which then rolled off the surface (Kwon et al., 2014, 2017) (Fig. 11.13).

While this process has been reported as being effective to fabricate superhydrophobic fabrics, the nanopillars created by this process would not be very durable

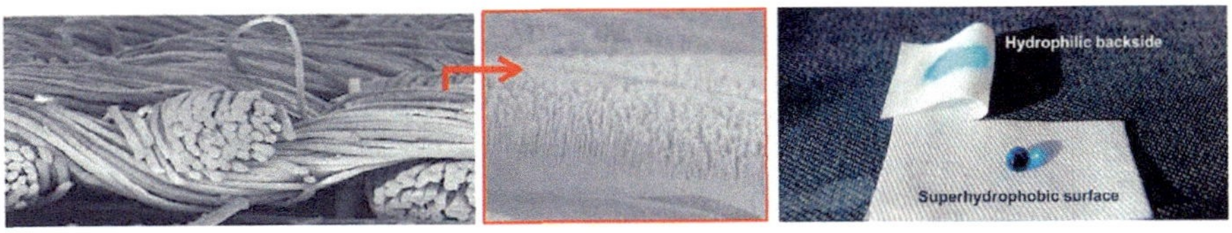

Fig. 11.12 A single-faced superhydrophobic lyocell fabric: left, lyocell fabric treated by oxygen etching for 20 min; right, asymmetric wettability of surfaces on the treated fabric. From Kwon, S.O., Ko, T.J., Yu, E., Kim, J., Moon, M.W., Park, C.H., 2014. Nanostructured self-cleaning lyocell fabrics with asymmetric wettability and moisture absorbency (part I). RSC Adv. 4, 45442–45448. https://doi.org/10.1039/C4RA08039D.

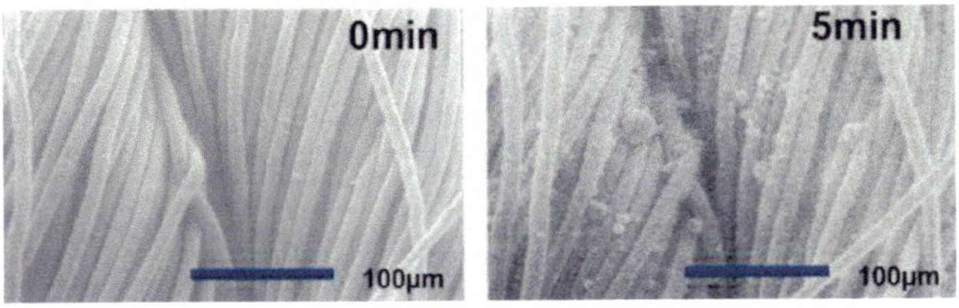

Fig. 11.13 Water condensation for superhydrophobic lyocell surface: water drops were observed after 5 min in the saturated condition by an environmental scanning electron microscope.
From Kwon, S.O., Kim, J., Moon, M.W., Park, C.H., 2017. Nanostructured superhydrophobic lyocell fabrics with asymmetric moisture absorbency: moisture managing properties. Text. Res. J. Online first published. https://doi.org/10.1177/0040517516639832, with permission from SAGE Publishing.

to abrasion or harsh laundering conditions. Furthermore, the process is relatively slow and costly, and the application of this process has been limited in laboratory-scale.

Chemical etching

Chemical etching is a facile process to create roughness. In Qian and Shen's study (Qian and Shen, 2005), metal surfaces of Zn, Al and Cu were etched by acids followed by fluoroalkylsilane coating to make superhydrophobic surfaces. In another studies (Xue et al., 2014; Mazrouei-Sebdani and Khoddami, 2011), poly(ethylene terephthalate) (PET) textile surface was etched by the alkali hydrolysis process to form nanoroughness; the etched surface was coated by polydimethylsiloxane (PDMS) (Xue et al., 2014) and fluorocarbon (Mazrouei-Sebdani and Khoddami, 2011) for hydrophobization to attain superhydrophobicity with washing durability and wear-resistance (Xue et al., 2014).

While the accurate control of surface morphology by this method would be challenging, this process would be readily applied to textiles. As this process involves caustic chemicals, efforts should be made to minimize the environmental impacts associated with the process.

11.4 Characterization of superhydrophobicity

11.4.1 Static water contact angle

Wettability can be evaluated by measurement of static contact angle (CA), contact angle hysteresis (CAH), sliding angle and shedding angle. CA refers to the angle between the surface and the liquid meniscus near the line of contact (Yan et al., 2011). CA measurement is a convenient way of evaluating wettability, but the measurement can be influenced by several factors including: the size and weight of the liquid drop, height that the liquid is dropped on the substrate; and the determination of baseline at the contacting surface (Zimmermann et al., 2009). They demonstrated that the measurement of CA can give a high deviation of more than 10 degree when there is inconsistency in lighting, focus and contact baseline. From the experiment, a shift of the substrate baseline by a single pixel in the image resulted in an apparent difference of almost 5 degree (Zimmermann et al., 2009).

The deviation of measurement can be large when a textile substrate with an irregular and non-reflective surface is used as shown in Fig. 11.14. From such a substrate, a baseline can be determined by capturing the image at the moment when the liquid contacts the substrate, though this method can still generate inaccuracy. In the case of superhydrophobic surfaces that have very high water contact angles, it is often difficult to discriminate the level of hydrophobicity by measuring the static contact angle.

11.4.2 Contact angle hysteresis (CAH)

As a measure of stickiness or roll-off ability of liquid from the surface, contact angle hysteresis (CAH) can be measured. CAH is determined by measuring the difference between the advancing and receding angles of a liquid drop. The superhydrophobic surface shows a CAH lower than 10 degree (Kota et al., 2013a; Wong et al., 2013), and this low CAH makes easier roll-off of liquid drops from the surface. The tendency of roll-off ability is also measured by sliding angle or shedding angle, which are explained in the following sections (Artus et al., 2006).

Fig. 11.14 Example of a water drop on a nonwoven textile substrate.

11.4.3 Sliding angle

Non-wetting surfaces can give a similar range of CAs (as close as 180 degree), making it difficult to differentiate the level of wettability by CA measurements alone. To discriminate the wettability of such extremely hydrophobic surfaces, alternative measurement such as sliding angle and shedding angle can be useful. Both sliding angle and shedding angle can be referred to as a roll-off angle because they measure the minimum tilting angle of the stage where a water drop begins to roll off the substrate placed on the stage (Artus et al., 2006; Zimmermann et al., 2009). When the evidence lacks for identifying whether a surface follows the Cassie–Baxter wetting state (with entrapped air pockets) or not, often a low roll-off angle is suggested as an indirect indication that the surface follows the Cassie–Baxter wetting state, where a liquid drop is held by the entrapped air instead of being pinned through the rough structures. If pinning had occurred, the drop would be attached more firmly on the surface, thereby hindering the roll-off (Park et al., 2016).

For the measurement of sliding angle, a water drop is placed on a staged substrate, and the stage is tilted gradually from the horizontal position until the water drop slides down spontaneously. The tilted angle at which the water drop slides is determined as the sliding angle. The higher the hydrophobicity is, the lower the sliding angle tends to be (Artus et al., 2006; Furmidge, 1962). In this method, a water droplet can be pinned through the roughened surface as time passes. Especially for fabric surfaces, fiber ends projecting from the fabric surface can be attached to the liquid drop, preventing it from rolling off. On a rough surface, a liquid drop deposited on a depressed area is likely to give higher sliding angle than on a flat area (Zimmermann et al., 2009).

11.4.4 Shedding angle

For shedding angle measurement, a liquid drop is released onto the tilted substrate from a defined height, and the minimum tilting angle from which the drop slides down or roll off is determined (Fig. 11.15). The tilting begins at 85 degree and is reduced until the drop does not roll off the surface. The lowest tilting angle at which the drop completely rolls off from the surface is determined as the shedding angle (Zimmermann et al., 2009). The surface with low wettability and good self-cleaning ability gives a low shedding angle.

The shedding angle can provide a sensitive measurement for roll-off ability of superhydrophobic surfaces, but it is influenced by several factors including the drop size, releasing height and the roughness and porosity of substrates (Zimmermann et al., 2009). The shedding angle is generally lower with a larger drop size (Carre and Shanahan, 1995; Zimmermann et al., 2009) and with a longer needle-to-substrate distance; a larger impact leads to pinning of liquid, thereby hindering roll-off (Zimmermann et al., 2009). Also, the substrate with pliable open structures can cushion the drop upon impact, giving a higher shedding angle (Zimmermann et al., 2009).

11.4.5 Bouncing of water drops

Water drops can bounce off from the superhydrophobic surface. Crick and Parkin (2011) measured the number of droplets bouncing as an alternative characterization for superhydrophobicity. In general, at the fixed droplet-releasing height, the number of droplets

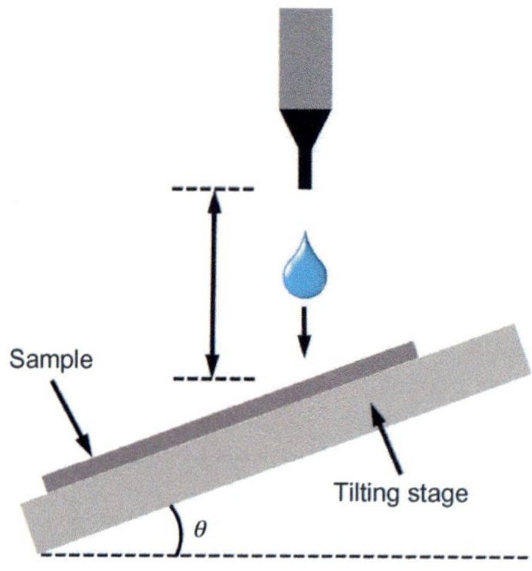

Fig. 11.15 Illustration of shedding angle measurement.

bouncing increased as the CA of surface increased. This study demonstrated the relationship between the bouncing behavior and the level of repellency for a superhydrophobic surface that follows the Cassie–Baxter state.

11.4.6 Self-cleaning property

A low roll-off angle is commonly regarded as an indication of the self-cleaning property (Kwon et al., 2014), yet there is no standard method or specification available to test the self-cleaning ability. Park et al. (2016) observed the self-cleaning effect of superhydrophobic polyester woven fabrics that had water shedding angle less than 1 degree. For the test of self-cleaning ability, 0.5–0.8 g of hydrophilic (silicon carbide) and hydrophobic (sudan black B) particles were sprinkled over a superhydrophobic fabric, and then water drops were subjected to flow on the fabric surface that was staged on a tilted cradle at 1 degree. In this experiment, both hydrophilic and hydrophobic particles were effectively adsorbed to the rolling water and were washed away from the superhydrophobic fabric, clearly demonstrating the self-cleaning ability (Fig. 11.16). This self-cleaning phenomenon can be attributed to the reduced work of adhesion resulting from the reduced contact area between the solid particles and rough superhydrophobic surface.

11.5 Applications

A wide range of potential applications of superhydrophobic materials has been studied; these include self-cleaning textiles (Park et al., 2016, 2017), liquid filtration/separation (Lee et al., 2013; Si et al., 2015), anti-icing (Howarter and Youngblood, 2007), anti-biofouling (Zhang et al., 2013; Liu et al., 2015), corrosion control (Liu et al.,

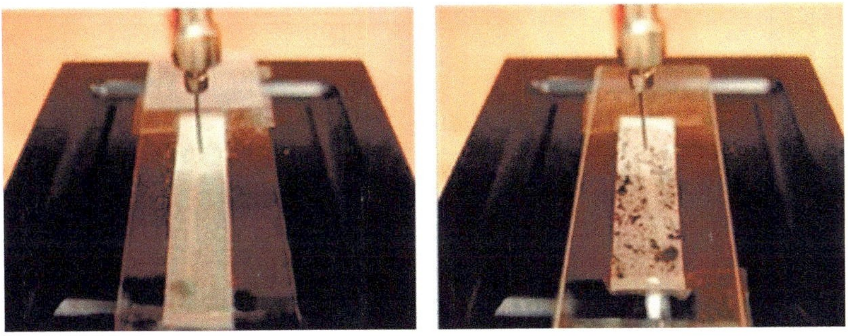

Fig. 11.16 Self-cleaning effect of superhydrophobic polyester woven fabric by water precipitation: left, silicon carbide particles; right, sudan black B particles.
From Park, S., Kim, J., Park, C.H., 2016. Analysis of wetting state for super-repellent fabrics to liquids in varied surface tensions. RSC Adv. 6, 45884–45893. https://doi.org/10.1039/c5ra27281e.

2007) and atmospheric water collection (Lo et al., 2014). On a superhydrophobic surface that follows the Cassie–Baxter state, the contact area between the particles and the surface can be reduced due to the entrapped air. Thus, the adhesion of particles to the surface is weakened. When water droplets roll on the surface, the weakly adhered particles are attached to the rolling water and are detached together with the droplets from the superhydrophobic surface (Park et al., 2016), thereby keeping the surface clean. This characteristic can be applied to clothing and interior textiles that require easy removal of dirt and soil.

The probability of icing can be significantly reduced on a superhydrophobic surface because water droplets deposited on the surface can be easily removed. The presence of entrapped air on a superhydrophobic surface that favors the Cassie–Baxter state can reduce the thermal conductivity, lowering the freezing point (Darmanin and Guittard, 2014). Consequently, superhydrophobic surfaces delay or reduce the adhesion of wet frost and ice when water droplets freeze in the Cassie–Baxter state (Kim et al., 2012). The entrapped air on a superhydrophobic surface can also be beneficial for improving anti-corrosion properties of metal surfaces. A superhydrophobic surface that follows the Cassie–Baxter model showed high impedance ($10^{10}\ \Omega\ cm^2$) with entrapped air inhibiting ion transport from the electrolyte to the metal surface (Ejenstam et al., 2013). Superhydrophobic materials, therefore, have great potential to replace toxic chromium (VI)-containing anti-corrosives.

One of the actively sought-out environmental applications of superhydrophobic materials is oil-water separation. Wettability-based separation utilizes the differential affinity of superhydrophobic materials to oil and water. A porous superhydrophobic/lipophilic membrane selectively passes oily liquids through the membrane while it prevents water from passing; consequently, the membrane will effectively separate water and oil (Lee et al., 2013; Si et al., 2015; Wang et al., 2016). Superhydrophobic nonwoven fabrics could be used for this application.

Anti-bioadhesion or biofouling is another application. Accumulation of biological matters on a surface can be a big challenge to various industrial sectors, including water treatment, marine vessel and other engineered systems (Wang et al., 2016). For instance, bio-adhesion of marine vessel surfaces can increase the drag resistance and accelerate corrosion (Gittens et al., 2013). Research has shown that superhydrophobic surfaces with entrapped air can reduce the attachment of bacteria and biological substances, thus mitigating bio-adhesion or biofouling (Zhang et al., 2013; Liu et al., 2015). Because bio-adhesion is a complex phenomenon, mechanism for bio-adhesion needs more studies in consideration of the chemistry and morphology of adhering bio-substances. The controlled adhesion of biological substances can be applied to antibacterial textiles that resist bacterial adhesion and biofouling. With potential risks of using antibiotics in developing antibiotic-resistance, anti-biofouling textiles could be a promising alternative to use of antibiotics and provide protection from bacterial infection. The anti-bioadhesion property could be well applied to medical textiles and military clothing.

A superhydrophobic surface can enhance the vapor condensation. Due to the water repellency of the surface, the vapor phase of moisture will not be absorbed much into the surface; instead, vapor will effectively condense at the surface. When this property is applied to the outer surface of the clothing, the fabric would not be easily wet in high humidity. This property can be also utilized for water collection from dew/fog and for efficient heat management in thermal power generation (Wang et al., 2016; Gittens et al., 2013; Kwon et al., 2017). The superhydrophobic surfaces with entrapped air can induce the liquid slippage due to the presence of a shear-free interface between air and liquid. By implementing directionally aligned topography, directional drag in microfluidic surfaces can be designed (Darmanin and Guittard, 2014).

11.6 Summary

Surface wettability is dependent on surface energy and surface roughness. The Young's wetting model explains the wettability as the relationship between surface tension of liquid and surface energy of solid on a flat and smooth surface. The Wenzel and Cassie–Baxter models consider the role of surface roughness in surface wettability by incorporating the roughness factor, which counts the solid–liquid contact area on the surface, into the models. For the fabrication of superhydrophobic surfaces, various technologies have been used to modify surface energy and implement surface roughness, including dipping method, chemical vapor deposition, colloidal assembly, layer-by-layer deposition, sol–gel process, lithography, plasma and chemical etching. The characteristics of superhydrophobic surfaces have been analyzed by static and dynamic contact angles, sliding/shedding angles and rebound of droplets. The applications of superhydrophobic surfaces include self-cleaning, anti-corrosion, anti-icing, anti-biofouling, liquid separation and facilitated vapor condensation, which are attributed to the Cassie–Baxter wetting state.

Despite the advanced technologies for designing superhydrophobic surfaces, fabrication processes are often complex, limiting scale-up productions of superhydrophobic materials. Also, it is challenging to achieve robust superhydrophobic properties against refurbishing procedures and environmental threat. For textile materials, a single process to lower surface energy may produce superhydrophobic property, as the surfaces usually have microscale roughness attributed to fibers and yarns. As perfluorinated compounds with C8 are being restricted for use, alternative chemicals as effective as C8 chemical should be identified or developed. Also, simple processes that allow precise control of nanoscale topography need to be further explored to enable mass productions of robust superhydrophobic materials.

References

Armitage, J., Macleod, M., Cousins, I., 2009. Comparative assessment of the global fate and transport pathways of long-chain perfluorocarboxylic acids (PFCAs) and perfluorocarboxylates (PFCs) emitted from direct sources. Environ. Sci. Technol. 43 (15), 5830–5836.

Artus, G., Jung, S., Zimmermann, J., Gautschi, H., Marquardt, K., Seeger, S., 2006. Silicone nanofilaments and their application as superhydrophobic coatings. Adv. Mater. 18 (20), 2758–2762.

Athauda, T., Ozer, R., 2012. Investigation of the effect of dual-size coatings on the hydrophobicity of cotton surface. Cellulose 19 (3), 1031–1040.

Barthlott, W., Neinhuis, C., 1997. Purity of the sacred lotus, or escape from contamination in biological surfaces. Planta 202 (1), 1–8.

Bhushan, B., 2009. Biomimetics: lessons from nature–an overview. Phil. Trans. R. Soc. A 367 (1893), 1445–1486.

Bhushan, B., 2012. Bioinspired structured surfaces. Langmuir 28 (3), 1698.

Bico, J., Thiele, U., Quéré, D., 2002. Wetting of textured surfaces. Colloids Surf. A 206 (1), 41–46.

Bravo, J., Zhai, L., Wu, Z., Cohen, R., Rubner, M., 2007. Transparent superhydrophobic films based on silica nanoparticles. Langmuir 23 (13), 7293–7298.

Cao, L., Hu, H., Gao, D., 2007. Design and fabrication of micro-textures for inducing a superhydrophobic behavior on hydrophilic materials. Langmuir 23 (8), 4310–4314.

Carre, A., Shanahan, M., 1995. Drop motion on an inclined plane and evaluation of hydrophobia treatments to glass. J. Adhes. 49 (3–4), 177–185.

Cassie, A., Baxter, S., 1944. Wettability of porous surfaces. Trans. Faraday Soc. 40, 1905–1971.

Celia, E., Darmanin, T., Taffin de Givenchy, E., Amigoni, S., Guittard, F., 2013. Recent advances in designing superhydrophobic surfaces. J. Colloid Interface Sci. 402, 1–18.

Cerrina, F., 2000. X-ray imaging: applications to patterning and lithography. J. Phys. D. Appl. Phys. 33, R103–R116.

Chen, W., McCarthy, T., 1997. Layer-by-layer deposition: a tool for polymer surface modification. Macromolecules 30 (1), 78–86.

Cheng, Y., Rodak, D., 2005. Is the lotus leaf superhydrophobic? Appl. Phys. Lett. 86, 144101.

Cheng, Y., Rodak, D., Wong, C., Hayden, C., 2006. Effects of micro- and nano-structures on the self-cleaning behaviour of lotus leaves. Nanotechnology 17 (5), 1359–1362.

Cho, W., Choi, I., 2008. Fabrication of hairy polymeric films inspired by geckos: wetting and high adhesion properties. Adv. Funct. Mater. 18 (7), 1089–1096.

Crick, C., Parkin, I., 2011. Water droplet bouncing-a definition for superhydrophobic surfaces. Chem. Commun. 47 (44), 12059–12061.

Darmanin, T., Guittard, F., 2014. Recent advances in the potential applications of bioinspired superhydrophobic materials. J. Mater. Chem. A 2, 16319–16359.

Du, C., Wang, J., Chen, Z., Chen, D., 2014. Durable superhydrophobic and superoleophilic filter paper for oil–water separation prepared by a colloidal deposition method. Appl. Surf. Sci. 313, 304–310.

Ejenstam, L., Ovaskainen, L., Rodriguez-Meizoso, I., Wågberg, L., Pan, J., Swerin, A., Claesson, P.M., 2013. The effect of superhydrophobic wetting state on corrosion protection—the AKD example. J. Colloid Interface Sci. 412, 56–64.

Fujii, T., Aoki, Y., Habazaki, H., 2011. Fabrication of super-oil-repellent dual pillar surfaces with optimized pillar intervals. Langmuir 27 (19), 11752–11756.

Furmidge, C., 1962. Studies at phase interfaces. I. The sliding of liquid drops on solid surfaces and a theory for spray retention. J. Colloid Sci. 17 (4), 309–324.

Gittens, J., Smith, T., Suleiman, R., Akid, R., 2013. Current and emerging environmentally-friendly systems for fouling control in the marine environment. Biotechnol. Adv. 31 (8), 1753–1783.

He, Y., Jiang, C., Yin, H., Chen, J., Yuan, W., 2011. Superhydrophobic silicon surfaces with micro–nano hierarchical structures via deep reactive Ion etching and galvanic etching. J. Colloid Interface Sci. 364 (1), 219–229.

Herminghaus, S., 2000. Roughness-induced non-wetting. Europhys. Lett. 52 (2), 165–170.

Howarter, J., Youngblood, J., 2007. Self-cleaning and anti-fog surfaces via stimuli-responsive polymer brushes. Adv. Mater. 19 (22), 3838–3843.

Huang, J., Li, S., Ge, M., Wang, L., Xing, T., Chen, G., Liu, X., Al-Deyab, S., Zhang, K., Chen, T., Lai, Y., 2015. Robust superhydrophobic TiO2@fabrics for UV shielding, self-cleaning and oil–water separation. J. Mater. Chem. A 3 (6), 2825–2832.

Jung, Y., Bhushan, B., 2007. Wetting transition of water droplets on superhydrophobic patterned surfaces. Scr. Mater. 57 (12), 1057–1060.

Jung, Y., Bhushan, B., 2009. Dynamic effects induced transition of droplets on biomimetic superhydrophobic surfaces. Langmuir 25 (16), 9208–9218.

Kang, M., Jung, R., Kim, H., Jin, H., 2008. Preparation of superhydrophobic polystyrene membranes by electrospinning. Colloids Surf. A 313, 411–414.

Kim, S., Kim, J., Kang, B., Uhm, H., 2005. Superhydrophobic CFx coating via in-line atmospheric RF plasma of He-CF4-H2. Langmuir 21 (26), 12213–12217.

Kim, P., Wong, T., Alvarenga, J., Kreder, M., Adorno-Martinez, W., Aizenberg, J., 2012. Liquid-infused nanostructured surfaces with extreme anti-ice and anti-frost performance. ACS Nano 6 (8), 6569–6577.

Kim, J., Kim, H., Park, C., 2016. Contribution of surface energy and roughness to the wettability of polyamide 6 and polypropylene film in the plasma-induced process. Text. Res. J. 86 (5), 461–471.

Koishi, T., Yasuoka, K., Fujikawa, S., Ebisuzaki, T., Zeng, X., Widom, B., 2009. Coexistence and transition between Cassie and Wenzel state on pillared hydrophobic surface. Proc. Natl. Acad. Sci. U. S. A. 106 (21), 8435–8440.

Kota, A., Li, Y., Mabry, J., Tuteja, A., 2012. Hierarchically structured superoleophobic surfaces with ultralow contact angle hysteresis. Adv. Mater. 24 (43), 5838–5843.

Kota, A., Choi, W., Tuteja, A., 2013a. Superomniphobic surfaces: design and durability. MRS Bull. 38 (5), 383–390.

Kota, A., Mabry, J., Tuteja, A., 2013b. Superoleophobic surfaces: design criteria and recent studies. Surg. Innov. 1 (2), 71–83.

Kwon, S., Ko, T., Yu, E., Kim, J., Moon, M., Park, C., 2014. Nanostructured self-cleaning lyocell fabrics with asymmetric wettability and moisture absorbency (part I). RSC Adv. 4 (14), 45442–45448.

Kwon, S., Park, C.H., Kim, J., 2015. Breathable, antistatic and superhydrophobic PET/lyocell fabric. J. Eng. Fibers Fabr. 10, 46–56.

Kwon, S., Kim, J., Moon, M., Park, C., 2017. Nanostructured superhydrophobic lyocell fabrics with symmetric moisture absorbency: moisture managing properties. Text. Res. J. https://doi.org/10.1177/0040517516639832 (Online First Published).

Lafuma, A., Quéré, D., 2003. Superhydrophobic states. Nat. Mater. 2 (7), 457–460.

Lee, S., Kwon, T., 2007. Effects of intrinsic hydrophobicity on wettability of polymer replicas of a superhydrophobic lotus leaf. J. Micromech. Microeng. 17 (4), 687–692.

Lee, W., Jin, M., Yoo, W., Lee, J., 2004. Nanostructuring of a polymeric substrate with well-defined nanometer-scale topography and tailored surface wettability. Langmuir 20 (18), 7665–7669.

Lee, M., An, S., Latthe, S., Lee, C., Hong, S., Yoon, S., 2013. Electrospun polystyrene nanofiber membrane with superhydrophobicity and superoleophilicity for selective separation of water and low viscous oil. ACS Appl. Mater. Interfaces 5 (21), 10597–10604.

Lin, J., Ding, B., Yu, J., 2010. Direct fabrication of highly nanoporous polystyrene fibers via electrospinning. ACS Appl. Mater. Interfaces 2 (2), 521–528.

Liu, T., Kim, C., 2014. Turning a surface superrepellent even to completely wetting liquids. Science 346 (6213), 1096–1100.

Liu, T., Chen, S., Cheng, S., Tian, J., Chang, X., Yin, Y., 2007. Corrosion behavior of superhydrophobic surface on copper in seawater. Electrochim. Acta 52 (28), 8003–8007.

Liu, M., Wang, S., Jiang, J., 2013. Bioinspired multiscale surfaces with special wettability. MRS Bull. 38 (5), 375–382.

Liu, L., Ercan, B., Sun, L., Ziemer, K., Webster, T., 2015. Understanding the role of polymer surface nanoscale topography on inhibiting bacteria adhesion and growth. ACS Biomater. Sci. Eng. 2 (1), 122–130.

Lo, C., Wang, C., Lu, M., 2014. Scale effect on dropwise condensation on superhydrophobic surfaces. ACS Appl. Mater. Interfaces 6 (16), 14353–14359.

Mazrouei-Sebdani, Z., Khoddami, A., 2011. Alkaline hydrolysis: a facile method to manufacture superhydrophobic polyester fabric by fluorocarbon coating. Prog. Org. Coat. 72 (4), 638–646.

Michielsen, S., Lee, H., 2007. Design of a superhydrophobic surface using woven structures. Langmuir 23 (11), 6004–6010.

Neinhuis, C., Barthlott, W., 1997. Characterization and distribution of water-repellent, self-cleaning plant surfaces. Ann. Bot. 79 (6), 667–677.

Nishimoto, S., Bhushan, B., 2013. Bioinspired self-cleaning surfaces with superhydrophobicity, superoleophobicity, and superhydrophilicity. RSC Adv. 3 (3), 671–690.

Nishino, T., Meguro, M., Nakamae, K., Matsushita, M., Ueda, Y., 1999. The lowest surface free energy based on – CF3 alignment. Langmuir 15 (13), 4321–4323.

Nosonovsky, M., 2007. Multiscale roughness and stability of superhydrophobic biomimetic interfaces. Langmuir 23 (6), 3157–3161.

Park, S., Kim, J., Park, C., 2016. Analysis of the wetting state of super-repellent fabrics with liquids of varying surface tension. RSC Adv. 6 (51), 45844–45893.

Park, S., Kim, J., Park, C., 2017. Influence of micro and nano-scale roughness on hydrophobicity of a plasma-treated woven fabric. Text. Res. J. 87 (2), 193–207.

Qian, B., Shen, Z., 2005. Fabrication of superhydrophobic surfaces by dislocation-selective chemical etching on aluminum, copper, and zinc substrates. Langmuir 21 (20), 9007–9009.

Rahmawan, Y., Moon, M., Kim, K., Lee, K., Suh, K., 2010. Wrinkled, dual-scale structures of diamond-like carbon (DLC) for superhydrophobicity. Langmuir 26 (1), 484–491.

Ramaratnam, K., Iyer, S., Kinnan, M., Chumanov, G., Brown, P., Luminov, I., 2008. Ultrahydrophobic textiles using nanoparticles: lotus approach. J. Eng. Fibers Fabr. 3 (4), 1–14.

Ran, C., Ding, G., Liu, W., Deng, Y., Hou, W., 2008. Wetting on nanoporous alumina surface: transition between Wenzel and Cassie states controlled by surface structure. Langmuir 24 (18), 9952–9955.

Riccardi, C., Barni, R., Fontanesi, M., Marcandalli, B., Massafra, M., Selli, E., Mazzone, G., 2001. A SF6 RF plasma reactor for research on textile treatment. Plasma Sources Sci. Technol. 10, 92–98.

Roach, P., Shirtchliffe, N., Newton, M., 2008. Progress in superhydrophobic surface development. Soft Matter 4, 224–240.

Sas, I., Gorga, R., Joines, J., Thoney, K., 2012. Literature review on superhydrophobic self-cleaning surfaces produced by electrospinning. J. Polym. Sci. B Polym. Phys. 50 (12), 824–845.

Shieh, J., Hou, F., Chen, Y., Chen, H., Yang, S., Cheng, C., Chen, H., 2010. Robust airlike superhydrophobic surfaces. Adv. Mater. 22 (5), 597–601.

Shim, M., Kim, J., Park, C., 2014. The effects of surface energy and roughness on the hydrophobicity of woven fabrics. Text. Res. J. 84 (12), 1268–1278.

Si, Y., Fu, Q., Wang, X., Zhu, J., Yu, J., Sun, G., Ding, B., 2015. Superelastic and superhydrophobic nanofiber-assembled cellular aerogels for effective separation of oil/water emulsions. ACS Nano 9 (4), 3791–3799.

Spori, D., Drobek, T., Zürcher, S., Spencer, N., 2010. Cassie-state wetting investigated by means of a hole-to-pillar density gradient. Langmuir 26 (12), 9465–9473.

Su, Y., Ji, B., Zhang, K., Gao, H., Huang, Y., Hwang, K., 2010. Nano to micro structural hierarchy is crucial for stable superhydrophobic and water-repellent surfaces. Langmuir 26 (7), 4984–4989.

Tawfick, S., De Volder, M., Copic, D., Park, S., Oliver, C., Polsen, E., Roberts, M., Hart, A., 2012. Engineering of micro- and nanostructured surfaces with anisotropic geometries and properties. Adv. Mater. 24 (13), 1628–1674.

Tsougeni, K., Vourdas, N., Tserepi, A., Gogolides, E., Cardinaud, C., 2009. Mechanisms of oxygen plasma nanotexturing of organic polymer surfaces: from stable super hydrophilic to super hydrophobic surfaces. Langmuir 25 (19), 11748–11759.

Tuteja, A., Choi, W., Ma, M., Mabry, J., Mazzella, S., Rutledge, G., McKinley, G., Cohen, R., 2007. Designing superoleophobic surfaces. Science 318 (5856), 1618–1622.

Tuteja, A., Choi, W., Mabry, J., McKinley, G., Cohen, R., 2008. Robust omniphobic surfaces. Proc. Natl. Acad. Sci. U. S. A. 105 (47), 18200–18205.

Wang, Z., Elimelech, M., Lin, S., 2016. Environmental applications of interfacial materials with special wettability. Environ. Sci. Technol. 50 (5), 2132–2150.

Wenzel, R., 1936. Resistance of solid surfaces to wetting by water. Ind. Eng. Chem. 28, 988–994.

Wong, T., Kang, S., Tang, S., Smythe, E., Hatton, B., Grinthal, A., Aizenberg, J., 2011. Bioinspired self-repairing slippery surfaces with pressure-stable omniphobicity. Nature 477 (7365), 443–447.

Wong, T., Sun, T., Feng, L., Aizenberg, J., 2013. Interfacial materials with special wettability. MRS Bull. 38 (5), 366–371.

Xia, Y., Whitesides, G., 1998. Soft lithography. Angew. Chem. 37 (5), 550–575.

Xu, B., Cai, Z., 2008. Fabrication of a superhydrophobic ZnO nanorod array film on cotton fabrics via a wet chemical route and hydrophobic modification. Appl. Surf. Sci. 254 (18), 5899–5904.

Xue, C., Jia, S., Chen, H., Wang, M., 2008. Superhydrophobic cotton fabrics prepared by sol–gel coating of TiO2 and surface hydrophobization. Sci. Technol. Adv. Mater. 9 (3), 035001.

Xue, C., Jia, S., Zhang, J., Tian, L., 2009. Superhydrophobic surfaces on cotton textiles by complex coating of silica nanoparticles and hydrophobization. Thin Solid Films 517 (16), 4593–4598.

Xue, Y., Chu, S., Lv, P., Duan, H., 2012. Importance of hierarchical structures in wetting stability on submersed superhydrophobic surfaces. Langmuir 28 (25), 9440–9450.

Xue, C., Li, Y., Zhang, P., Ma, J., Jia, S., 2014. Washable and wear-resistant superhydrophobic surfaces with self-cleaning property by chemical etching of fibers and hydrophobization. ACS Appl. Mater. Interfaces 6 (13), 10153–10161.

Yan, Y., Gao, N., Barthlott, W., 2011. Mimicking natural superhydrophobic surfaces and grasping the wetting process: a review on recent progress in preparing superhydrophobic surfaces. Adv. Colloid Interf. Sci. 169 (2), 80–105.

Young, T., 1805. An essay on the cohesion of fluids. Philos. Trans. R. Soc. Lond. 95 (1), 71.

Yuan, Y., Choi, S., Kim, J., 2016. Analysis of contact area between water and irregular fibrous surface for prediction of wettability. RSC Adv. 6 (77), 73313–73322.

Zhang, J., France, P., Radomyselskiy, A., Datta, S., Zhao, J., van Ooij, W., 2003. Hydrophobic cotton fabric coated by a thin nanoparticulate plasma film. J. Appl. Polym. Sci. 88, 1473–1481.

Zhang, G., Wang, D., Gu, Z., Möhwald, H., 2005. Fabrication of superhydrophobic surfaces from binary colloidal assembly. Langmuir 21 (20), 9143–9148.

Zhang, X., Wang, L., Levänen, E., 2013. Superhydrophobic surfaces for the reduction of bacterial adhesion. RSC Adv. 3 (30), 12003–12020.

Zhao, Y., Tang, Y., Wang, X., Lin, T., 2010. Superhydrophobic cotton fabric fabricated by electrostatic assembly of silica nanoparticles and its remarkable buoyancy. Appl. Surf. Sci. 256 (22), 6736–6742.

Zhou, H., Wang, H., Niu, H., Gestos, A., Lin, T., 2013. Robust, self-healing superamphiphobic fabrics prepared by two-step coating of fluoro-containing polymer, fluoroalkyl silane, and modified silica nanoparticles. Adv. Funct. Mater. 23 (13), 1664–1670.

Zimmermann, J., Seeger, S., Reifler, F., 2009. Water shedding angle: a new technique to evaluate the water-repellent properties of superhydrophobic surfaces. Text. Res. J. 79 (17), 1565–1570.

Further reading

Shi, F., Niu, J., Liu, J., Liu, F., Wang, Z., Feng, X., Zhang, X., 2007. Towards understanding why a superhydrophobic coating is needed by water striders. Adv. Mater. 19 (17), 2257–2261.

Part Three

Water repellent textiles in practice: Performance, testing and applications

Designing waterproof and water repellent clothing for wearer comfort—A paradigm shift

12

Jeni Bougourd, Jane McCann†*
*Consultant, London, United Kingdom, †Design Consultant, Northern Ireland, United Kingdom

12.1 Introduction

The design of water resistant, water repellent and waterproof clothing has always challenged innovators. This has never been more important than at present as designers seek to balance the need for effective waterproofing with the urgent need to establish sustainable practice.

The traditional clothing design practice of take, make, use and dispose has led to this industry being one of the most wasteful in the world. While this philosophy of consumption was inherent in the industrial revolution, if the planet and the economy are to survive there is a need to move towards a circular economy.

This need for the design of water repellent and water resistant clothing has been accelerated by concern expressed in the first Greenpeace (1997) on the hazardous nature of polyfluorinated chemicals (PFCs) used in waterproofing membranes and durable waterproof repellent (DWR) finishes for this type of protective clothing.

The concept of the circular economy is not new for companies who were producing waterproof clothing without these hazardous materials (e.g. Paramó) but since this time several have adopted a circular economy strategy (e.g. Vaude) and others have moved or are seeking to move towards it by eliminating these hazardous chemicals from their products (Greenpeace, 2015).

It is within this context that the design of water repellent and water resistant clothing is explored in this chapter.

Some of the changes in policy and process that are being implemented by designers of protective clothing to adopt this concept are considered. A co-design strategy is used where the consumers' active participation and requirements are integrated into the design and development process. Four aspects of comfort are identified and their relevant criteria used as a framework within which to discuss key stages of the sustainability design process. New product range concepts for design briefs are outlined and factors affecting the selection and testing of materials presented. Key processes affecting the design of water repellent clothing, their components and features are

Waterproof and Water Repellent Textiles and Clothing. https://doi.org/10.1016/B978-0-08-101212-3.00011-3

outlined prior to discussing garment laboratory testing and criteria for field studies. It concludes with a review of current sustainable practices, outlines some of the benefits of emerging technologies, and reaffirms the role of designers in the circular economy.

12.2 The circular economy: Avoiding waste and damage

Half a century ago seminal texts warned that man's exploitation of the planet cause waste and damage—for example, *Silent Spring* (Carson, 1962) and *Limits to Growth* (Meadows, et al., 1972). The Massachusetts Institute of Technology (MIT) team that published the second of these declared, 'If the present growth trends in world populations, industrialization, pollution, food production and resource depletion continues unchanged the limits to growth on the planet will be reached within the next 100 years'. These conclusions were supported by further studies in 1992 and 2004 (MIT) and 2014 (Turner, 2014). There is now further evidence that mankind is living at a rate requiring resources (materials, water, energy) beyond those of the entire planet (materials, water, energy—Fig. 12.1).

Despite scepticism, the United Nations established its Environment Program in 1972, when the concept of sustainable development was mooted: 'Sustainable

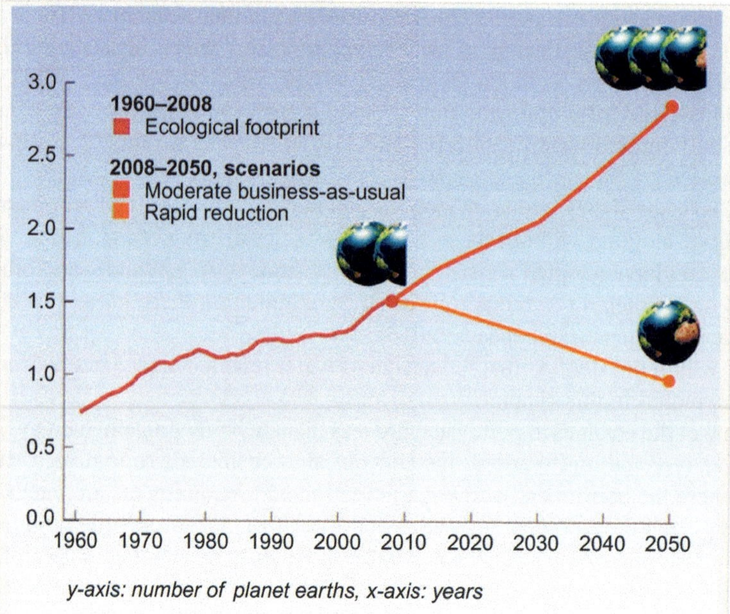

Fig. 12.1 A representation of the ways in which resource use and the generation of waste outpace the capacity of the planet—the 'global footprint of the human race' since 2008 and a scenario for the period up to 2050 (Global Footprint Network, 2017).

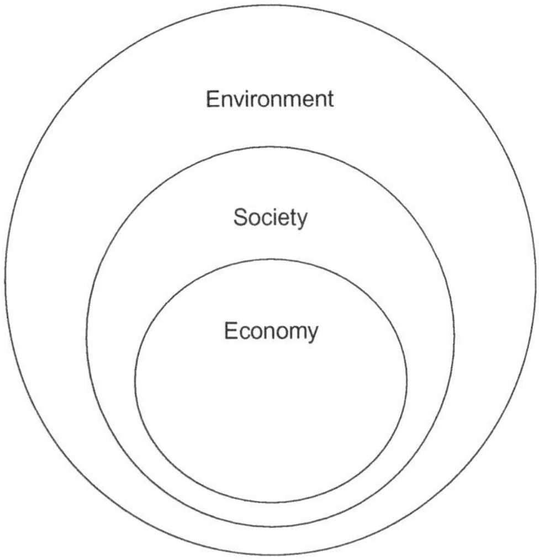

Fig. 12.2 The 'three pillars of sustainability', where both economy and society are constrained by environmental limits (Scott Cato, 2009).

development is development that meets the needs of the present without compromising the ability of future generations to meet their own needs' (Bruntland, 1987).

Sustainability has been characterized as a homeostatic ecosystem, with three interdependent components, or pillars: environmental, social and economic (see Fig. 12.2).

These three pillars of sustainability have been widely adopted by companies to describe core values and sustainable strategies (PUMA, 2017); to discuss ethical movements in fashion (Csanák, 2014); and to form the basis of standards for achieving and measuring sustainability, such as ISO 26000:2010 (Guidance on social responsibility). Metrics include benchmarks, audits, indexes and accounting, as well as assessment, appraisal and other reporting systems.

The pillars also form a framework for Sustainability Development Goals (SDGs), which came into effect on Jan. 1, 2016 and states that the 'fundamental changes in the way societies consume and produce are indispensable for achieving global sustainable development' (UN, 2012). SDGs offer guidance for the decisions to be taken over the next 15 years. 'All of us will work to implement the Agenda within our own countries and at the regional and global levels, taking into account different national realities, capacities and levels of development and respecting national policies and priorities' (UN, 2015).

Many issues identified by the SDGs are relevant to the clothing industry, highlighting the need for sustainable consumption and prompting innovation. Consumption has been described as an underlying driver of direct human impacts on the environment (Michaelis and Lorek, 2004). Industrialization has brought consumerism with it as an integral part of the economy; growth demands the marketing of new products and corresponding disposal of old ones. The clothing industry, worth trillions of dollars, is one of the most resource-intensive and one of the most polluting (Muthu, 2014). In the United Kingdom the processing of raw materials into finished products generates a third of the waste and over three quarters of the carbon and water footprint produced by the sector (WRAP, 2012).

Around 30% of clothes in the household wardrobe typically have been unused for at least a year—worth over $1200 per household or $38 billion across the United Kingdom. This means that extending the life of clothes by 9 months of active use would reduce the carbon, water and waste impacts by 20%–30% (WRAP, 2012). In addition, 30% or 700,000 tonnes of unwanted clothes go into landfill, which, if reused or recycled, would have a value of $175 million every year (WRAP, 2012).

Observers suggest the economy cannot survive the 'take, make, use, and throw away' approach of the clothing industry, which depends on complex linear supply chains. The adoption of a circular economy is essential if industry's impact on the environment is to be reduced. The circular economy described by WRAP as 'an alternative to a traditional linear economy (make, use, dispose) in which resources are kept in use for as long as possible, extract the maximum value from them whilst in use, then recover and regenerate products and materials at the end of each service life' (WRAP, 2017). Hence 'transition to the circular economy may be the biggest revolution and opportunity for how we organize production and consumption in our global economy' (Lacy and Rutqvist, 2015).

The clothing and textile industry has been encouraged to embrace the circular economy. That encouragement has come from, for example, NGOs such as WRAP (2017) and ECAP (2015); companies (e.g. Páramo, Patagonia, M&S, Nike); independent and campaigning organizations such as the Ellen MacArthur Foundation, Greenpeace, the Sustainable Apparel Coalition (SAC) and its associated Higg Index. This pressure has prompted the identification and measurement of best practice and its certification (e.g. Cradle to Cradle, Life Cycle Assessment, Corporate Social Responsibility (CSR)), with some companies already training their designers to use these sustainability measurement tools.

Priorities for attention within the circular economy have varied. Some clothing outerwear companies have chosen to concentrate on CSR, the supply chain and the production components. However, though it has been argued that 'sustainability is the ultimate design brief' and that 'sustainability needs to be design led', the design process has, by comparison, received less attention (Sherwin, 2012; Angel, 2013). The EU Eco-innovation Observatory nevertheless sees design as a cornerstone of the circular economy, as it can determine whether the product can be rented, repaired, reused, remanufactured or recycled when its initial life span is complete (Eco-innovation Observatory, 2017, see a generic overview of the circular economy adapted from Buttin and Brieuc (2017) in Fig. 12.3).

This section has placed the design of clothing for protection from water in the context of waste and damage caused by a clothing industry that relies on marketing huge volumes of ready-to-wear (RTW) clothing. Emphasis has been placed on broadening understanding of the design process to embrace the circular economy.

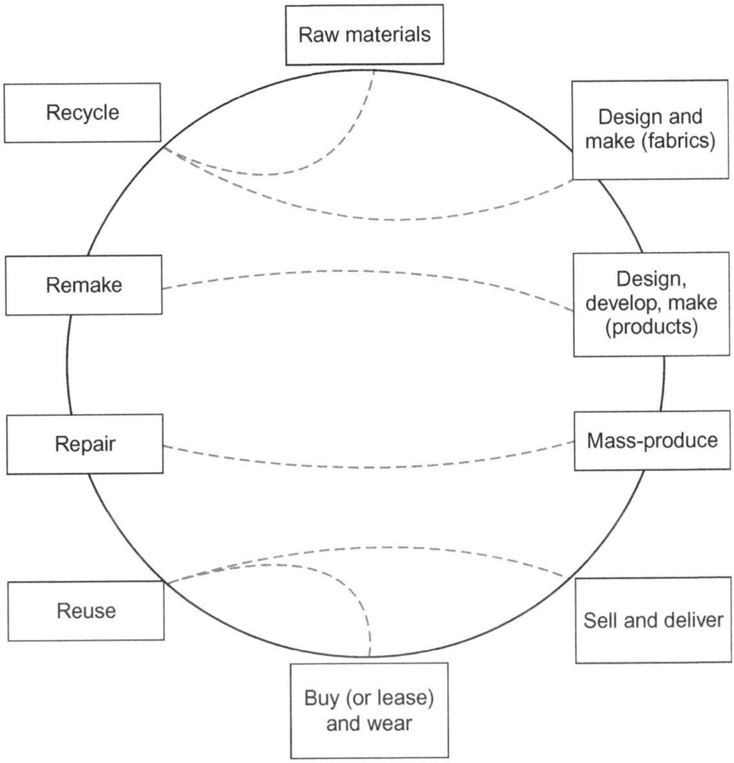

Fig. 12.3 A generic overview of the circular economy for the design and development of water-resistant and water repellent clothing.
Adapted from Buttin, N., Brieuc, S., 2017. Innovation + Économie Circulair, from WITHAA website. https://www.slideshare.net/wiithaa/business-model-and-circular-economy (Accessed July 2017).

12.3 Design of waterproof and water repellent clothing within a circular economy

Many companies that design waterproof and water repellent clothing have embraced the circular economy at their inception; one such is Páramo. Some have undertaken the transition from a linear to a circular economy (e.g. Houdini, 2017a; Vaude, 2017a,b); others already have eco certified materials (e.g. Halti, 2017) or are planning to develop increasingly sustainable practices (e.g. Rapha, 2016). A consequence of a very full involvement of Greenpeace with significant outdoor companies that produce this protective clothing will be to stimulate others to review their priorities (see Fig. 12.4).

The circular economy created by Vaude (2017b), echoes that of the generic process described in Fig. 12.3, and although it references CSR aspects for manufacturing, the emphasis on a design-led sustainable practice is paramount (Vaude, 2017a). The key processes affecting design of waterproof and water repellent clothing (indicated in this

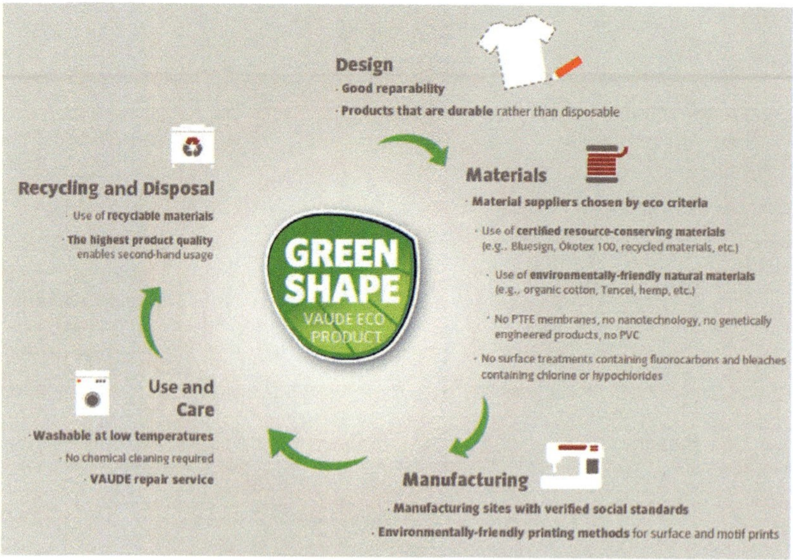

Fig. 12.4 A sustainable cycle in use in the wet weather clothing industry by a company who had a very early commitment to sustainable practice (Vaude, 2017a).

structure), suggests that products need to be designed and manufactured to support end-of-life usage, and made from eco-certified natural materials (not from materials, membranes and finishes, and after-care processes that are known to be harmful to human health, to the environment, or to both).

12.4 Policies and goals for designing sustainable waterproof clothing

Company goals comprise a need to make consumers comfortable and safe in the most extreme wet conditions, ultimately allowing people to stay outside for longer, while ensuring key sustainable needs to achieve zero negative impact. Patagonia's goal (since the recession in 1991) has been to build the best product, to cause no unnecessary harm, and to use business to inspire and implement solutions to the environmental crisis; while for Houdini it is to seek answers to the following questions:

- Does this product deserve to exist?
- Will it last long enough?
- Is it versatile enough?
- Will it age with beauty and be a lifelong companion?
- Nothing unnecessary added? (Houdini, 2017b).

Following on from these key sustainable goals are those companies that are design-driven, encompassing a need to offer the consumer comfort along with the highest level of performance, innovation and quality, and also to produce protective rainwear that is attractive, simple and desirable, using modern materials and construction techniques. Overall, the aim is to produce innovative clothes that leave the smallest possible ecological footprint.

12.5 The design concept

A fundamental change from traditional practice of clothing design for protection from water is to ensure that the design concept, the selection of materials, colours and components, and the assembly processes are suitable for sustainable aftercare, sufficiently durable for renting and resale, and appropriate for future repair or remake before final recycling. In addition to Vaude, these design concepts have been adopted by other rainwear designers, e.g. Paramó in 2017 (for their polyester based products) and Patagonia—the latter launching a Common Threads Partnership in 2005, which included a category of 'reduce' where consumers agree to buy only what they need and what will last (Patagonia, 2011).

12.5.1 Sell, rent, resell and repair

Companies are committed to making products durable, i.e. making products last so that they can be reused (either rented or resold) or repaired. Some companies already provide garments for renting alongside those that are new and second-hand, as well as offering comprehensive online information on how consumers can repair products themselves (e.g. waterproof waders) or ways in which they can access a company's own repair service. Patagonia (2017) goes a step further, offering a program of mobile services to sell 'worn wear' and to repair products, while companies who use branded material assemblies, such as Gore-Tex, have appointed specialists who provide repairs (Scottish Mountain Gear, 2017).

An interesting issue arising from a recent review was that, while longevity is a key goal, the proposed lifespan of products ranged from 1 year to a lifetime guarantee. While longevity is welcome and there is some evidence to support varying understanding between suppliers and consumers (Burman, 2014), the implication of ensuring sustainable aftercare practice for maintaining durability and waterproofing remains uncertain.

12.5.2 Remake and alter

Remake and alterations to waterproof and water repellent clothing is a sustainable practice but one that can be difficult for companies to provide and, although many offer repairs and replacement parts for their shells (draw cords, cord endings, etc.) and dry suits (pockets, etc.) (Vaude, 2017a,b; Houdini, 2017b; Typhoon, 2017b). These are not at present full remake or alteration offers. Full remake services are being actively encouraged by WRAP, but an increase in the provision of mass-customized or made-to-measure waterproof clothing could help alleviate some of the acknowledged problems of poor fitting garments and choice of colour.

12.5.3 Recycle

An important decision at the design concept stage for a growing number of companies is the need to design waterproof garments that can be recycled either as a new fabric or as a new product.

Design for recycling requires either (a) that all items in a garment be constructed from the same materials or, (b) have components that can easily be disassembled.

There are several companies at the forefront of recycling:

- one that set out at the onset to establish its own closed loop recycling system was Páramo;
- some who have introduced practices where every garment that is made can now be recycled—(which as stated by Chouinard in 2013, a practice which was unthinkable10 years ago).
- those who recycle their own garments to their own materials supplier's closed loop system (such as Houdini and Teijin);
- others who have recognized the need and are actively engaged in assessing their use of resources and those researching opportunities—especially companies whose current products involve using identified toxic materials or need to overcome challenges arising from a mixture of fibres and dyes.

Key to the success of material selection for recycling is the collection of unwanted clothes. There are several routes for reclamation of clothing (the charity bag, clothing banks and return of garments to the store in exchange for cash or credit towards a future purchase, etc.). Problems may nevertheless arise when seeking to achieve a circular economy through reclamation. While there are materials that are suitable for current recycling systems, others may be needed for new systems, such as those suitable for some waterproof hard shells.

12.6 Co-design

As sustainable design practice has gained recognition it has become apparent that the traditional design practice (of presenting RTW products to the consumer, without consultation) continues to swell the volume of waste, and that a co-design process (where the consumer's needs are addressed at a much earlier stage of design) can help to reduce adverse environmental impact (Eco-innovation Observatory, 2017).

Designers of sustainable wet weather clothing support this view and consider it essential to involve 'expert consumers' throughout the iterative cycle of design, development and testing. For many companies these individuals or groups of experts (available now through an initiative supported by ISPO) are not only evaluators of process and products but also brand ambassadors.

A clear advantage of the co-design process for sustainable industrial practice (as opposed to applications for research projects) is that the needs encapsulated in company goals for the product and the consumer can be articulated to help compile a design brief, aid the design, development and evaluation of products designed to meet the end of life usage. These stages broadly reflect those of traditional stages of co-design: comprising the specification of requirements, design and evaluation (ISO 13407).

12.6.1 Specifying requirements

Frameworks for identifying user needs for performance and protective clothing have been well documented (McCann, 1999; Timmins and McCann, 2015).

The performance design tree proposed by McCann was created to ensure that garments are fit for purpose. The components relating to function and form are a useful guide and echo some of the general requirements identified for protective clothing in EN 340; for example, user profile (size, shape and movement), the recreational or occupational activities, the climatic conditions within which a user may be engaged, and sustainable practice (Fig. 12.5).

The outdoor industry has collaborated with SAC in promoting the HIGG Design and Development Module (DDM) as a sustainable design tool.

However, while such directives are helpful, there is little published guidance on criteria for the user to judge the effectiveness of clothing design in maintaining comfort while protecting the user from the ingression of water whether on sea or land.

12.6.1.1 Comfort

Comfort has been described as multidimensional, something that is both difficult to define and, among all the aspects associated with human feelings and desires, represents a central concern (Kilinc-Balci, 2011). This description and one proposed by

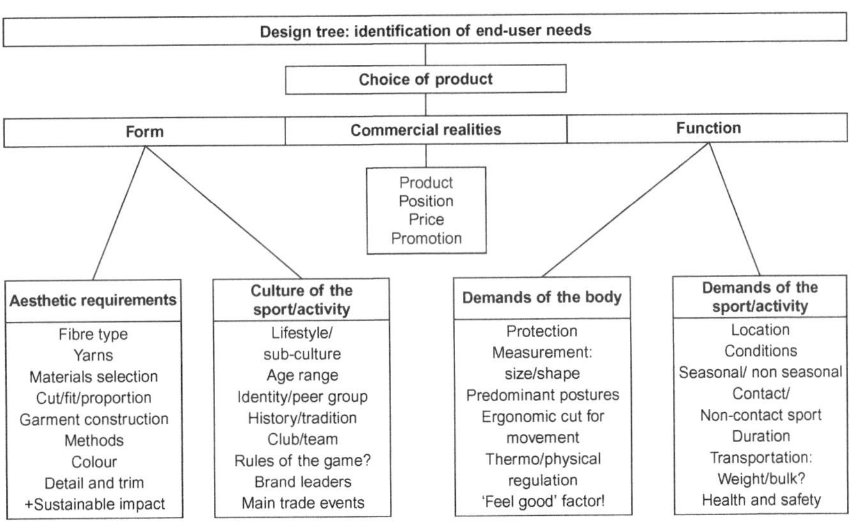

Fig.12.5 A design tree; a hierarchy of design requirements to consider in drawing up a brief.

Slater (1985), 'a pleasant state of psychological, physiological, and physical harmony between a human and the environment' has been extensively explored in a publication on improving comfort in clothing (Song, 2011). These types of criteria for comfort in clothing have been defined and extended by Saville (1999), Rossi (2005) and Bartels (2011): thermal physiological comfort; sensorial comfort; ergonomic comfort; and psychological comfort.

(a) Thermal physiological comfort

This comfort category is seen as the heat and moisture transportation properties of the clothing and the way in which clothing can help to maintain the heat balance of the body during various levels of activity. The factors affecting this type of comfort are seen by Rossi as heat balance—the exchange of heat by radiation, conduction and convection; the loss of heat by evaporation of sweat; the physical activity being undertaken by the person; and the environment (ambient temperature, air humidity and movement).

(b) Skin sensorial comfort

This type of comfort concerns the sensation of how the fabric feels when worn on or near the skin. It can relate to fabric softness and pliability in movement, and to thermo physiological comfort should a fabric become wetted through with sweat and cling to the dampened skin. And, more recently as a category within some sustainable certification indexes, where levels of toxicity are determined according to the amount of a garment that touches the skin.

(c) Ergonomic comfort

This level of comfort can be directly related to the design, cut, fit and style of a garment; the ease of movement within it; and the range and type of activity. The weight and lightness of material can also affect the microclimate within a garment or its system layers. Poor fit (e.g. too tight, too loose or the incorrect length for the activity) is also a key contributor to nonuse by the consumer and consequent waste.

(d) Psychological comfort

The comfort considered in this category deals with the aesthetics of clothing and the influence a wearer's attitude can have towards a brand image which could include a company's sustainability policy, as well as the suitability of a protective garment (where the level of protection may be inadequate).

These criteria reflect some of those set out in relevant standards for general and specific protective clothing (e.g. BS EN ISO 13688:2013, BS EN 343 2003+A1 2007) and can be seen as providing users with a level of comfort consistent with:

- climatic conditions;
- user morphology;
- activity—dynamic protection against water afforded by the selection and evaluation of materials;
- protection against water through overall design, layering systems, design features and evaluation of the product;
- extended and repeated use;

- preparation for production and product lifespan.

These levels of comfort will be used to guide the discussion within this chapter.

It is further suggested that clothing is, 'to be made from materials with low water vapor resistance and/or high air permeability and/or shall be sufficiently ventilated to minimize discomfort and thermal stress' (BS EN ISO 13688:2013).

12.6.1.2 Climatic conditions

Climatic conditions on land or sea present different kinds of wetness. These can take the form of water collected in clouds, fog, freezing ice, rain, hail, sleet, snow and storms. Each has a wetness descriptor. For example, rain, apart from having different droplet sizes and velocities, can (like snow) be described as having weight (light, medium, heavy). Also, snow is referred to as being dry or wet. In addition, these different experiences of wetness are experienced across a range of temperatures (an example being heat in association with monsoon rain). Another climatic feature is wind velocity, which can influence the experience of air temperature (wind chill) and on the direction of some of the types of water in the air and in the sea.

Water may be considered a hazard and may adversely affect the health and survival of users, particularly if it occurs during extreme conditions.

An overview of clothing requirements for different environmental conditions was reported by Rengasamy (2011). Other authors have addressed clothing needs for extreme conditions; they deal, for example, with low temperatures (Holmér, 2009); heat and fire (Makinen, 2005); and radiation protection (Zhou et al., 2005). Holmér (2011) has looked at protection from wet in cold but insufficient attention has been given to other conditions for protective clothing for ingression of water.

Understanding of varying climatic conditions may need to change as global warming increases the frequency and intensity of some types of extreme weather. Thus '… warming is causing more rain to fall in heavy downpours. There are also longer dry periods between rainfalls. This, coupled with more evaporation due to higher temperatures, intensifies drought. Wet places have generally become wetter, while dry places have become drier. Heat waves have become more frequent and intense, while very cold days have decreased'. (Climate Communication, 2017). Wind speeds are also predicted to increase in some northern global areas.

The implications of these changes for wet weather clothing suggest that (if there is no reversal of global climatic changes), protection from ingression of water and protection from wind may need to increase, which may require a sustainable fabric of a higher water and wind resistance than is currently available.

12.6.1.3 The 'expert' consumer

The provision of sustainable clothing for protection from water can range between consumers requiring traditional classic rainwear, leisurewear and activity or occupational wear. Many companies design clothing for several activities in all types of wet weather (hiking, skiing, multisport and active lifestyle). Others specialize in many aspects of a single activity in all weathers (cycling, sailing) and some for a single occupation in specified extreme conditions (oil workers, sea and mountain rescue). There are consequently a wide range of consumers of varying ages and body sizes engaged in different types of outdoor activities and weather conditions that can guide the co-design process. For example:

- general population for active lifestyle;
- expensive products for high socioeconomic consumers 30+ years;
- active professional and leisure pursuits 30–40 years;
- 'Best-agers' 55–70+ years.

Those consumers not defined by age may be:

- those who took part in a specific outdoor activity;
- 'global sophisticated experts', or 'mountain types';
- groups with recognized cultural needs, for example, preferred colours.

Anthropometric shape and size databases are available for some countries and regions and are mined for age-related 3D, 2D and 1D data according to selected market segments for consumers in the general population. 3D body shapes are used to help create size tables, automatically generate block and graded patterns and aid the development of mannequins, for example, SizeUK/Rohan (Sizemic, 2017).

While these data can be used for companies that produce classic rainwear, functional urban and some occupational protective clothing in a wide range of sizes, many leisure and other activities require clothing that reflects the anatomical development and range of dynamic movements related to a particular activity.

Working with expert consumers who represent participants of identified outdoor activities or job performance to verify morphologies related to a particular activity is common practice. For example, a practitioner's body characteristics with broad shoulders and/or large arm or leg muscles. The capture of these various sizes and shapes can be achieved through the application of 3D shape analysis, using 3D scanning technology, traditional manual methods (e.g. video recordings) or independent company methods, as the shapes of professional cyclists are derived from an 'on bike' position (Rapha, 2016), a process which echoes earlier work by Kryzwinski et al. (2005) and Ashdown (2011). However, while expert representatives can be used throughout a co-design process (and in some cases combined with reference to industry-standard mannequins to help maintain consistency), the generalization of these individual or small group body shapes to their wider populations is not ideal. Either a 3D survey of individual populations representative of a particular activity or the introduction of a mass customized process would be preferable to help ensure accurate static shape capture.

Capturing changing sizes and shapes for a range of dynamic activities is a complex exercise. Daanen and Reffeltrath (2007) reported on several postures in which the extremes of a range of motions were reached during activities required for protective clothing, while some of the difficulties of describing body dimensional changes during human body movement were highlighted by Ashdown (2011). The application of new 3D manual methods (as described by Houdini (2017c) in their press release in

Fig. 12.6) and digital dynamic ergonomic analysis for assessing the range of movement related to individual activities is recognized and, as with shape analysis, these new approaches are now being adopted and regarded by some designers as essential tools.

Notwithstanding the usefulness of these 3D static applications, 360-degree motion-capture devices and 4D temporal-based systems for documenting movement, function and pose are now available (e.g. 3dMD 2017). It is to be hoped that these applications for the design of protective clothing will lead to a more accurate capture of body movements and subsequently a greater understanding of clothing comfort for a range of activities.

Fig. 12.6 'Made to move', a protective garment designed to offer a high level of waterproofing that gives the wearer full freedom of movement, using new 3D manual applications. Courtesy of Houdini, 2017c. Press Release. The International Fachmesse für Sportartikel und Sportmode. The International Sporting Goods Trade Fair (ISPO), Munich.

12.6.2 Selection of materials

Waterproof, water repellent and water-resistant clothing has evolved from simple garments using locally available materials which would now be recognized as part of a circular economy; for example, dried seal gut, woven bark and oilskins made from recycled sails (see Figs 12.7 and 12.8). A more recent innovation 'incorporating activated carbon derived from recycled coconut shells' is being used in nylon and polyester polymers (Rengasamy, 2011).

The design and development of early military and mass-produced clothing by Macintosh, Aquascutum (derived from the Latin for water and shield) and Burberry whose iconic trench coat (developed for the military in the First World War) is a classic rainwear design for many other companies (see Fig. 12.9).

Material development is now a highly technical process involving the provision of clothing for a multiplicity of activities taking place in a variety of climatic environments.

Over time, the terms waterproof, water repellent and water resistant have become to some extent interchangeable, while their meanings have been extended to include complex materials and processes. Two main categories have been identified by Wilson (2010): waterproof and water repellent (the second includes water resistant and shower-proof). An overview, together with some of their advantages and disadvantages for clothing, are listed below.

Waterproof—a fabric that is impervious to water.

There are three groups:

- naturally waterproof, for example, rubber;
- protective coatings, for example, rubber, polyurethane (PU) and polyvinyl chloride (PVC), wax and oil compounds. Fabrics coated with these materials have several end uses although

Fig. 12.7 Summer dried seal gut, designed with drawstring hoods and cuffs (sometimes called Kamleikas) Aleutian Islands, Yupik, 20th century (Yupic, 2017).

Fig. 12.8 Coir raincoat, Chejing province, southern China (Mallett, 2017).

Fig. 12.9 Belted trench with Stormwear, made from sustainable nonperfluorinated material (Marks and Spencer, 2017).

when some are used for clothing they are not 'breathable', i.e. they do not allow water vapour or perspiration to evaporate away from the inside of the garment. Nonporomeric and microporous coatings with pores fine enough to allow water vapour and air to pass out and prevent larger water droplets from penetrating from the outside.
- laminated fabrics: many of these fabrics are usually completely waterproof. Their advantage is that they can have various built in systems that allow them to be both waterproof and breathable. (Although it is noted that microporous films are not completely waterproof, i.e. Gore-Tex Windstopper.)

Water repellent: a fabric not easily penetrated by water. These can also be produced in several ways:

- naturally water repellent, for example, wool, nylon and polyester;
- densely woven materials, for example, canvas, Ventile, Gabardinè and more recent microfiber constructions;
- protective finishes can use various natural and chemical processes. Unlike coatings water repellent finishes usually coat just the surface fibres on the fabric and do not completely fill the gaps in the fabric construction enabling the fabrics to be breathable, i.e. allow water vapour and air to pass through the structure. Chemical finishes include silicone compounds and fluorocarbon finishes.

In comparison with the history of fashion, technical performance clothing emerged within a relatively short time span. Manmade fibres and polymers had a dramatic effect on the design and comfort of clothing (e.g. PVC, neoprene, polyurethane, nylon, expanded polytetrafluoroethylene—PTFE and polyester materials); some of which are now regarded as unsustainable.

The first effective nondeteriorating all-weather clothing was made from PVC and although considered a nonbreathable material, the garments were loosely cut to allow air circulation. PVC also had the advantage of being a material that can have high-frequency welded seams and, although toxic properties have been identified in production and it does not biodegrade (Greenpeace, 1997), several work wear and fashion companies nevertheless continue to offer PVC rainwear. However, some agreement appears to have been reached regarding the use of recycled PVC (VinyLoop, 2017).

Wax and oil compounds allow water to bead up and disperse and are used in some of waterproof garments today (e.g. Paramó and Barbour; see Fig. 12.10), but it is suggested that these material finishes can be easily contaminated by oil and dirt.

The process of producing polyurethane coatings has also been found to contain toxic compounds (organotins) that are harmful to the body and has been banned in Europe as part of the Zero Discharge of Hazardous Chemicals (ZDHC) program (Ethical Fashion Forum, 2015). Such materials have a range of design applications and it is proposed that calcium-zinc stabilizers can be used, as they offer similar qualities for clothing (they are lightweight, flexible, provide adequate waterproofing on lightweight fabrics, and resist abrasion).

Wetsuit design has developed so that different types or weights of materials can be mapped onto areas of the body to aid movement, compression or moisture transmission, but many are still made from sandwiched neoprene, a material where the manufacturing process has been identified as toxic (Neoprene, 2017). Yulex and Patagonia have developed a sustainable alternative to neoprene from 'Seed to Suit', a plant-based bio rubber

Fig. 12.10 Paramó Velez Jacket. Paramó 'Velez' jacket with Nikwax waterproof badge denotes that a garment has shown no water ingress during an independent laboratory test. This, biomimicry, nonperfluorinated, fabric system remains comfortable during high-aerobic activity without noticeable condensation build-up. It is recoverable as it dries out quickly by body heat alone, a distinct benefit in extreme environments where prolonged exposure can endanger health (Paramó, 2017a,b).

made from guayule; a renewable, nonfood crop that requires little water and uses no pesticides—an echo from the 1960s when Rachel Carson campaigned to prevent synthetic pesticides being used in agriculture (Carson, 1962). Patagonia has since offered open access to their new material (Sustainable Brands, 2014).

There is wide concern regarding the 2012 Greenpeace report and the use of Durable Water Repellents (DWR) coatings in the outer layer and in waterproof membranes. There are available alternatives but there is some doubt as to whether the durability will equal those of the original C8 materials, in particular, for water, snow and protection from extreme conditions (Davies, 2014). C8 materials are also oleophobic offering good soil resistance. Meanwhile, there is continued use of C6 but the need to seek alternative materials with the same durable protection given by the original C8 materials and less harmful DWR finishes is being actively pursued. Extensive public discussions have also been ongoing regarding the adverse effects of plastic (fibres and micro-beads) in the ocean, in fish and for human beings and, although some designers are already in conversation with washing machine manufacturers, the European outdoor group (EOG) has formed a consortium to consider:

- how much microfibre shedding is being produced;
- how are the fibres being shed;
- fabric types that may be the key offenders.

The EOG plans to place the research findings on an open source database (European Outdoor Group, EOG, 2017).

12.6.2.1 Tools to guide the selection of waterproof and sustainable materials

Guidance for the selection of sustainable materials in available standards for protective clothing includes a flow chart for considering acceptability of materials and mentions that they should be selected to minimize environmental impact (e.g. EN 340: 2003). However, while this is useful, it appears to rely on received manufacturer guidance.

Tools that give direct guidance to designers on the selection of sustainable materials are emerging. There is now a comprehensive list following the introduction of an EU regulation concerning the 'registration, evaluation, authorisation and restriction of chemicals' (REACH); the Higg Material Sustainability Index (MSI) provides opportunity for designers to compare and select sustainable materials; and Nike updated its 2013 open-source MAKING application in Dec. 2016. This App was created for designers to better understand the environmental impact of materials (Nike, 2016).

WRAP provides specific information on the impact of fibre processing and general guidance on selection of materials for sportswear but does not offer specific information for waterproof materials or those specific to layering systems.

Training and education for sustainable design practice is under development, and some companies, such as Vaude (2017b) are already giving their own design teams training in sustainable practices and introducing them to in-house rules for material selection—helping to ensure that all components in material assemblies, layering systems and garment accessories for protective clothing meet sustainable criteria.

Apart from being sustainable, the initial criteria for selection of material for protective garments are that they be waterproof, windproof, breathable, thermo-regulating, durable and hardwearing. Also, for today's active consumers, garments need to be lightweight and flexible, with UV protection, long-lasting and with sustainable aftercare. There is useful guidance for ultraviolet radiation (UVR) factors for fibre, fabric and colour (Zhou et al., 2005).

12.6.2.2 Testing materials for resistance to water penetration

Materials may be tested in independent and fabric supplier laboratories, universities and company-owned facilities, or any combination.

There are numerous standards for assessing individual material components and material assemblies for their resistance to water penetration. Methods have evolved into company-specific, national, regional and international standards, for example, Gore-Tex, Patagonia's 'Killer wash' (a process used to simulate years of use and abuse in a short period of time and based on JIS L 1092 B).

Standards for measuring performance of materials to be used for waterproof and water resistant clothing include ISO/DIS 811:2016 Determination of resistance to water penetration (currently under review); BS EN ISO 13688:2013 Protective clothing: General requirements: BS EN 343: 2003+A1: 2007 Protective Clothing: Protection against rain. Examples of key performance requirements from the latter standard are set out in Tables 12.1–12.3.

Table 12.1 **Application of performance tests on the components (British Standards, 2007).**

Property	Reference clause	Outer shell material	Liner or thermal liner	Lining
Resistance to water penetration (before and/or after pretreatment)	4.1	X	X	
		(In combination if applicable)		
Water vapour resistance	4.2	(In combination if applicable)		
Tensile strength	4.3	X		
Tensile resistance	4.4	X		
Dimensional change	4.5	X	X	X
		(In combination if applicable)		
Seam strength	4.6	X		

Note: Includes test for seam strength in accordance with the principles of EN ISO 13935-2.

Table 12.2 **Classification of resistance to water penetration (British Standards, 2007)**

	Class		
Water penetration (*Wp*)	1	2	3
Specimen to be tested – Material before pretreatment	$Wp \geq 8\ 000$ Pa	No test required[a]	No test required[a]
– Material after each pretreatment (see 5.1.3.2–5.1.3.5)	No test required	$Wp \geq 8\ 000$ Pa	$Wp \geq 13\ 000$ Pa
– Seams before pretreatment	$Wp \geq 8\ 000$ Pa	$Wp \geq 8\ 000$ Pa	$Wp \geq 13\ 000$ Pa

Note: For each class several requirements shall be met.
Notes:

- Class 3 with the highest value would provide the best result.
- Tests for seams are included which should take account of any stitching holes, seam adhesives and tapes (e.g. lifting of seam tapes).
- References to 5.1.3.2–5.1.3.5 include tests for dry-cleaning and/or washing; abrasion; longitudinal and cross direction flexing; influence of fuel and oil.

[a]No test required because the worst case situation for class 2 and class 3 is after pretreatment.

The results of these evaluations can help guide designers in the selection of materials and, in conjunction with general requirements listed in BS EN ISO 13688:2013, aid in the preparation of point-of-sale information, for example, after-care, size designations and labelling.

Table 12.3 Classification of water vapour resistance (British Standards, 2007)

Water vapour resistance R_{et}	Class		
	1[a]	2	3
m^2 Pa/W	R_{et} above 40	$20 < R_{et} \leq 40$	$< R_{et} \leq 20$

Notes:

- Class 3 with the smallest value would provide the best result.
- Test in accordance with BS EN 31092:1994 (replaced by BS EN ISO 11092: 2014).

[a]*Warning*: Class 1 has a restricted wearing time.

This standard also includes guidance for practical performance tests (see Annex C) and, although this information is helpful (and supports guidance from WRAP, 2012), it does not take account of the need to ensure consumer satisfaction and longevity of garment use by integrating the expert consumer throughout the co-design process.

The standard referred to above is only one example of testing and there is a wide range of tests, for example, tests for water vapour resistance, which (depending on test conditions) can give different numbers and units for the same fabric. Some of these alternative standards are additive, and could aid designers in the selection of multiple layers.

12.6.2.3 Testing materials for thermal properties

In addition to the standards identified above procedures for evaluating thermal properties together with water vapour resistance are available. A relevant example is BS EN ISO 11092:2014, Textiles—determination of physiological properties—measurement of thermal and water-vapour resistance under steady state conditions (sweating guarded-hot plate test).

12.7 Product ranges and the design brief

Traditional protective clothing ranges tend not to be seasonal. They include professional, leisure, core and basic ranges for specified activities (such as skiing, cycling, bird watching and mountaineering), as well as multifunctional activities for lifestyle.

The development of clothing product systems that formed part of ranges, and the need for compatibility of design and materials used for garments in those systems continue to be important. In particular, when products are differentiated by climatic conditions and activity and are not designed as all-season products. This is highlighted where companies not only attach a label with a graphic image of the shell garment (with its important features identified) but also those of the other layers that had been specifically designed to be worn with the outer shell to maintain protection and comfort.

One of the most dramatic changes in the way in which some company products are now being designed and categorized is the way they are being rated according to a judgement on sustainability. Halti suggests the following two categories.
The highest product rating: 'Think ahead products'

- 90% of the textile fabric used for inner and outer layers of a garment, including any printing, to be Bluesign-approved (Bluesign is an independent system listing safe chemicals, including over 60 alternatives to PFCs) Bluesign (2017);
- the remaining 10% (i.e. not Bluesign approved fabric) must not be in positions on the garment placed directly against the skin;
- additionally, products must meet the criteria and limits of the Bluesign restricted substance list (RSL).

ECO products that are rated in 'second position' require that the garment meets one of the following requirements:

- made from Bluesign-approved fabric;
- is certified to meet Global Organic Textile Standard (GOTS) for organic certified cotton;
- meets the Responsible Down Standard (RDS);
- meets Responsible Wool Standard (RWS);
- there is no perfluorinated compound (PFC) water resistant treatment or used recycled fibre content in the material.

These new ratings suggest that clothing design for protection from water will be driven as much by environmental concerns as by the need for comfort, aesthetics, functional developments and business considerations.

The content of design briefs for clothing for the protection from water can vary considerably as they are linked to the frequency with which product concept and range reviews occur and each of the ranges is redesigned. Most companies are looking for long product availability as a key criterion, so the introduction of new styles; redesign of a range, system or a component is determined by the expected life of a material and/or product. For example, a development cycle for a new product could take three to four seasons, or up to 4 years; refining a style could take place between one season and another (e.g. redesign of a hood, see Fig. 12.11), while the updating of a core garment style might take a few months.

The range of clothing product types for the protection of the ingression of water which could be included in a design brief would be consistent with a selected activity or occupation. These could include basic outerwear garments such as suits, coats, jackets, soft shell, hard shell, pants, trousers, salopettes, wet and dry suits, and more specialist products such as waders, capes and cover-alls. These might be designed in conjunction with base and mid layers for layering or an outer and mid layers may be combined as a single garment.

A key design concern would be how products interface with existing head-, hand-, foot-wear and specialist equipment, for example, life jackets and back-packs. Although many upper body garments are designed with integral or detachable hoods it may be necessary to consider head protectors as integral body components, or, as for wet suits, design suits that include integral socks to go inside existing footwear.

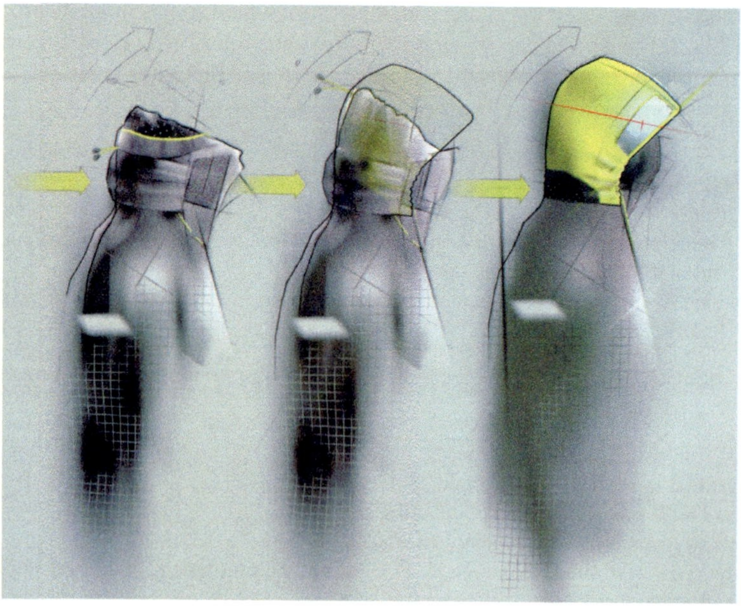

Fig. 12.11 New offshore hood system offering a storage pod, quick drying, safety features and increased field of vision (Henri Lloyd, 2017).

Colour palettes do not follow trends. Selected colours are specific to a designer or company and chosen to meet their needs. Houdini would offer an example where colours can be selected to be timeless (to fit with product longevity policy), to enhance safety such as high visibility, to be UV compatible or selected to merge with a particular seasonal surroundings but above all to be created from sustainable dyes so that consumers can wear them with a clear conscience (Houdini, 2017a).

12.7.1 Designing layering systems

As with the earlier rain and seawear examples, many fibre and fabric technologies, adopted for protective clothing, were originally developed for military preparedness and the space race. The typical layering system (made up of a wicking base layer, insulating mid layer(s) and a protective outer shell garment) was devised for US combat wear to address the harsh conditions experienced in the Korean War; although much more attention is now given to the importance of design and the integration of design elements and features throughout the layers (Bougourd and McCann, 2009).

Several authors have discussed the traditional layering system in relation to protective clothing (McCann, 2005; Black et al., 2005; Timmins and McCann, 2015). Holmér (2011) has included the layering system in his review of cold weather clothing and comfort but, as aspects of the materials used in some layers have direct relationship to overall comfort in varying temperatures and have become part of a sustainability debate, they are briefly reviewed.

12.7.1.1 Wicking base layer

A base or next to skin layers is designed to trap a warming layer of air between skin and fabric and wick moisture away from the skin, enabling the body to remain dry and comfortable. Helly Hansen developed an early base layer (polypropylene LIFA), but there are now designs in various materials and weights of material, including polyester and merino wool, for different temperatures and activities. Musto (2017) has four base layers:

- *thermal*—for use during medium energy activities in freezing conditions. It is usually lightweight and ergonomically fitted, fast wicking and low absorption fabric, which keep the body dry and comfortable during energy exertion, which can help regulate temperature in the cold
- *active*—a layer for high-energy activities in cold and freezing temperatures. An antibacterial wicking fabric with body mapped variable knits, which regulates the body temperature and reduces risk of injury by supporting muscles and joints
- *merino*—for high-energy activities in freezing to mild climates. The natural antibacterial wool fibres wick moisture away and help regulate temperature
- *quick dry*—a four-way stretch knit layer for keeping dry and comfortable in hot climates during low- to high-energy activity. It is super lightweight with welded and flat-locked seams

A very thin close-fitting base layer is preferred, as this kind of fit will speed up the base layer's drying speed, helped by its low water holding ability. It is suggested that because these layers are worn for long periods it is crucial they are comfortable, i.e. not made from thick itchy fibres and, if synthetic fibres are used, some applications may require treatment with an antimicrobial or FR finish (Stevens and Fuller, 2015). However, it is noted that there are concerns regarding the safety of some of these chemical treatments (Thomas, 2011).

12.7.1.2 Insulating mid layer

The mid layer provides an essential layer of warmer air whilst wicking may remove moisture so little water remains on the skin, evaporation from the skin removes more heat than from an outer layer. Helly Hansen, along with Calami (Italy) and Malden Mills (United States), was again at the forefront in developing mid-layer insulation using fibre pile and revolutionary lightweight polyester fleece fabrics; although there are still debates on the benefits of synthetic versus natural wool and down materials. It has been suggested that synthetics are good for warm and wet conditions, down for dry, cold temperatures and the combination of down jacket and a waterproof breathable shell for protection from cold and wet conditions. There are some disadvantages to using down:

- even when it is treated with a durable water repellent (DWR), if it gets wet the plumules clump together and it loses the ability to trap body heat—particularly important if the garment is being worn in cold wet conditions (EVO, 2017);
- some sources of down are linked to cruel animal practices and companies are implementing standards to reassure consumers of safe practices. See, for example, The global traceable down standard (NSF, 2015).

An insulation layer will usually be lightweight, relatively thin and windproof, designed to maintain a comfortable skin temperature and yet be able to transfer heat before overheating, for example, designed with a full zip and sleeves wide enough to be pulled up to the elbows.

For some activities a secondary high-loft insulation garment can be held in reserve for passive insulation as it is beneficial when stationary or when warming. This type of clothing puts lower demands on a shell.

12.7.1.3 Protective outer shell

The primary role of the outer shell is to maintain comfort by providing protection from water, wind and/or snow while allowing vapour (perspiration) to escape, i.e. being breathable. An early development in protecting outer shells using polymers was the Gore Tex PTFE membrane both for more traditional outdoor products, often in red, royal, forest green and navy but with lighter weight more design aware ski wear styles.

Notwithstanding these early improvements in design, to reiterate, concern has since been raised regarding the toxicity of some PFCs used in DWR coatings and in waterproof membranes. DWR is the finish applied to the outer surface of shells to make them water resistant (or hydrophobic). They are used to prevent the outer surface of shells from wetting out and reducing breathability (moisture transportation through the breathable membrane; Fig. 12.12). Some DWR chemicals are harmful to the body, resistant to degradation, and are being banned in Europe (Greenpeace, 2015) (see Table 12.4).

12.7.1.4 Layering for specialist activities

It is suggested that the design of layering systems for some activities demanded a different emphasis for base moisture transfer layers for winter and summer (e.g. cycling); while the choice for mid insulating layers can be between natural materials (which offer the best thermal efficiency) and synthetic insulation (which absorbed next to no moisture but rapidly transport it away from the body).

Traditional protective shell features usually have full water and wind protection, which may not be the case with all soft shells, some of which can be made from lighter weight, more flexible materials that are only water resistant. More advanced hybrid assemblies of combined mid and shell layers, offer closer fits and greater versatility.

Over all the design of a system will be guided by the intended activity and the climatic conditions within which the activities are taking place (Fig. 12.13). For example, garment systems for sailing:

- *Ocean*—designed for foul weather and the most extreme conditions;
- *Offshore*—designed and engineered for prolonged durations out at sea;
- *Race Lite*—performance clothing for racing inshore and around the coast, in wet, warm environments and in trade winds (Musto, 2017)

There is criticism of the effectiveness of a concentration on three layer systems, particular in relation to changeable wet conditions, when the number and type of layers

Table 12.4 Overview of waterproof and water resistant material suppliers, clothing brand partners and applications

System	Construction/finish	WP producers	Examples of WP brand names	Examples of brand partners	End-user applications
4–7 multilayer					Extreme end-use: firefighting, military, etc.
3 layer	Sandwich construction: outer fabric, bonded membrane with inner knit/ mesh	W.L. Gore, Toray, Teijin, eVent, Sympatex Technologies	Gore-Tex Pro Shell Gore-Tex Pro Shell Dermizax NX Apex 3 layer Hardshell, Sympatex, Pertex Shield	Musto Patagonia/Adidas/Berghaus Northface Bergens Houdini Vaude	Sailing Mountaineering Hill walking, Skiing, snowboarding Ski touring
Soft shell	Outer fabric with or without a membrane bonded to either single knit, mesh or fleece construction	W.L. Gore, Polartec, Sympatex Technologies, Teijin, eVent	Gore WINDSTOPPER, NeoShell, Windmaster, Eco Circle Motion Comfortshell	Gore Bikewear Rapha/Rab Vaude Houdini	Cycling Cycling/Alpine Outdoor, hiking Outdoor
2.5 layer	Outer fabric with membrane with inner abrasion-resistant 3D foam 'print' to protect membrane	W.L. Gore, Pertex,	Gore-Tex PacLite Pertex Shield +2.5L Ceplex Advanced membrane 3D Phaseable laminate	Gore Bikewear/Arcteryx Rab Vaude Local	General wet conditions: Cycling, climbing Hiking Cycling
2 layer	Outer fabric with coating or membrane	W.L. Gore	DRYVENT 2L, Gore-Tex 2 Layer, Altura Shield, Adidas Climaproof	The North Face Berghaus Altura Adidas	Hill walking General outdoor Cycling Urban
2 layers	2 layers of closely woven water repellent fabric without membrane	Nikwax/Páramo	Nikwax Analogy	Páramo	Hill walking Skiing

Continued

Table 12.4 Continued

System	Construction/finish	WP producers	Examples of WP brand names	Examples of brand partners	End-user applications
Double layer	Double layer fabric system with an 'Air Gap'	Keela	System Dual Protection (SDP)	Keela Maier Sports	Mountain rescue, skiing outdoor
Drop liner	A membrane bonded to lightweight fabric placed between a single detailed outer layer and an inner loose lining	W.L. Gore	Gore-Tex drop liner, Gore-Tex Performance Shell Z-Liner (Other laminating companies with other products)	Musto Whisper jacket Revit	Country clothing: shooting, bird watching Motorcycling Ski fashion
Single layer	Calendared Ripstop Pertex Shield: Lightweight durable shell +* microporous coating	Mitsui & Co., Ltd. Also products by Toray & F.O.V	Also: Pertex, Pertex Classic Pertex Quantum Pertex Equilibrium (NB: also used with pile fabrics)	Montane (Pertex Shield) Rab (Pertex Shield) Buffalo (Pertex with fibre pile inner)	Mountaineering
Single layer	Densely woven cotton using long staple fibre	Ventile sole producer	Ventile	Snugpak Howies	EG: polar expeditions (dry cold)
Single layer	Fibres of woven fabrics encapsulated with silicone 'barrier'. Suitable on cottons, Synthetics or blends	Nextec Applications, Inc.	EPIC by Nextec Suitable on cottons, Synthetics or blends	Greg Norman, Howies, LL Bean	Golf wear Jeans
Single unified layer	Membrane fabric with direct integration of proofing membrane into core of fabric during knitting process		Voormi		Skiing, Snowboarding

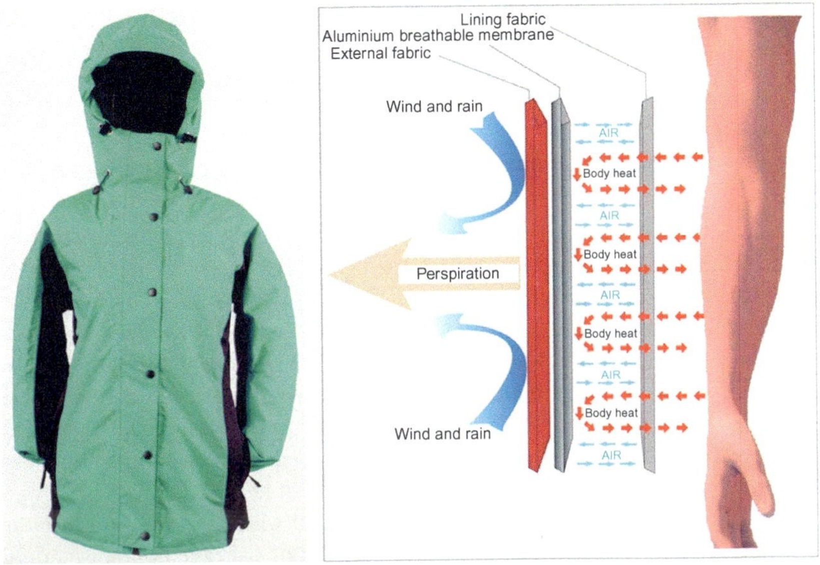

Fig. 12.12 An example of a ladies' Osprey jacket (left) incorporating a thermo-regulated, waterproof, breathable and windproof fabric system by APT (right) (Aclimatise, 2017). The system is composed of a lining fabric, an aluminium coated breathable membrane and a wind- and rainproof external fabric.

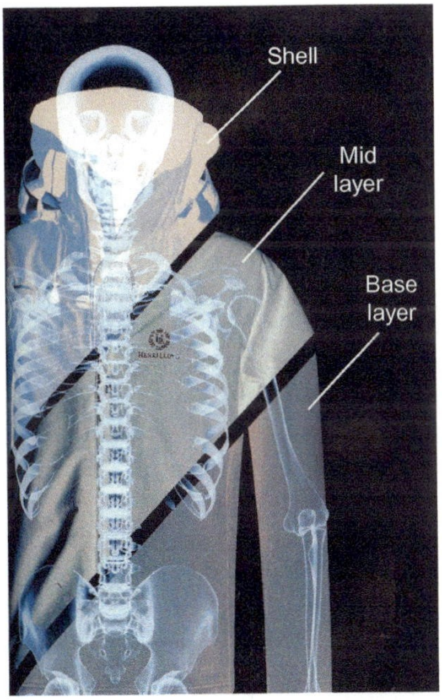

Fig. 12.13 The three-layer system; traditional layering for extreme conditions such as offshore sailing (Henri Lloyd, 2017).

need to be removed or adjusted and the intended microclimate is disrupted. Although it is argued that a softshell, a vented Pertex system, or an additional insulated layer can offset some of these uncomfortable experiences (Stevens and Fuller, 2015).

12.8 Design of waterproof and water repellent clothing

In addition to key sustainability concerns, today's design of wet weather clothing has moved beyond functional protection towards a growing demand for a balance between technical and aesthetic design requirements.

Both the International London Boat Show and ISPO (the international sports/outerwear trade show) launched more than 40 years ago have helped to accelerate the design and marketing of waterproof clothing. Francis Chichester's circumnavigation of the world in 1967 and Chris Bonnington's climb of the South Face of Annapurna in 1970 drew popular attention to the need for protection from water and the outdoor clothing industry that strives to meet that need.

The importance of design was further accelerated when, in 1984, ICI re-launched nylon (e.g. Tactel) which, together with the growth of package holidays, stimulated fashion trends that began to influence the skiwear market (M&S and C&A). In the 1990s, snowboard and skateboard styling crossed over with street wear. In the 2000s, waterproof materials became lighter (and more breathable). Design silhouettes for some sectors of the protective clothing market became more streamlined. Since that time, the emergence of new systems to capture body size and shape, novel garment cutting and joining techniques, together with refined waterproof zips and fastenings, have enabled designers of waterproof clothing to engineer relatively tailored garments, that can be co-branded with fabric suppliers.

12.8.1 Design realization

This process can comprise the creation of a sketch, mock-ups, block pattern and style pattern generation, iterations of toiles, and prototypes (number would depend on the complexity of the design) after which prototypes can then be test cut and fitted for articulation and garment functionality.

However, while these traditional manual design and development practices are undertaken by designer/makers in companies producing water protective clothing (e.g. Paramó, in discussion with authors) the overall process of realizing designs is changing as technologies become more advanced and reliable. Other practices, using a mix of manual and digital processes, can include using independent technologies for each stage of the process such as CAD design and pattern cutting systems. The outcome can either be physical toile or prototype garment as above (e.g. Rapha Racing Ltd. in discussion with authors), or a virtual prototype by wrapping a pattern around an avatar (representative of an average shape or an expert consumer) to assess the design and fit of a style (e.g. author observation at an M&S supplier).

Further advances in digital technologies allow designers to create a virtual garment directly on a representative 3D avatar and (depending on the materials) select or reject

Fig. 12.14 An e3D garment designed using ergonomic, three-dimensional pattern technology for enhanced comfort and mobility (Arc'teryx, 2017).

a design before making a physical prototype. It is clear, that (in the same way as 3D technologies are being used to capture and identify shape), designers are benefitting from these technologies. They are using them to create a required degree of fit and ease for a garment (e.g. snug fit), introduce ergonomic three-dimensional contour fitting for components and some articulated features, and also to support a sustainable design practice (e.g. Figs 12.10 and 12.14). However, while these 3D systems and avatars have beneficial applications for individual virtual garment development it is at present difficult to create and assess the fit and balance of multilayered garment systems.

12.8.2 Design and assembly of features

There is a wide range of style and design features that can be integrated into garments to aid comfort and enhance protection from the ingression of water.

As joining techniques for rainwear garments continue to improve and seam technology becomes less obtrusive, style design features have, in turn, become less restrictive and more creative; although some of their placements (e.g. areas over the shoulder, sleeve, back) still need to be appropriate to ensure comfort when wearing additional equipment. The water proofing of seams, however, still depends on the base fabric and seam type (Fig. 12.15). Bonded seams are usually waterproof; others need to be sewn and sealed with waterproofed thread and tape and (as with some materials)

Fig. 12.15 A lightweight 2.5 shell layer with a print on the inner face of the lining to protect the laminate. The mesh pocket is mounted on a border of the printed lining fabric and taped. Tapes are self-fabric.

be environmentally safe, i.e. without PFC's, and designed to be compliant with REACH and to meet Oeko-Tex Standard as well as offer optimum water resistance after launderings (WMB UK, 2017).

12.8.2.1 General features

Some design features can be common to several types of rainwear garments and integral to the design while others tend to be product and activity specific and include proprietary features or systems. Common features can include:

- attached or detachable hoods and high collars to protect the head, neck and face;
- dropped hems, detachable snow skirt to prevent rain and snow entering the jacket;
- welded or stitched outer pockets with protective storm flaps for venting, storage and/or linings to keep the hands warm;
- inner pockets for storage;
- longer lengths of jackets or extended jacket tails, sleeves and trousers to ensure full coverage of body when in activity position, raising arms and bending elbows and knees;
- varying lengths and locations of coated waterproof zips for closures, pit vents with zip pulls to aid ease of closures in high winds or during darkness (see Fig. 12.16);
- wrist, ankle and neck closures—depending on the activity these closures can have Velcro fastenings, be elasticated, overlapped, etc. for ventilation, or for secure protection from water ingression have flexible contoured seals (see Fig. 12.19);
- cinching of hoods around the face, across and over the head, waist and hem to aid fit and protection of the garment microclimate but to be adjustable to ensure ventilation and breathability (see Fig. 12.14);
- venting features and systems for maintaining air flow directly related to the activity and/or integral to the design (see Fig. 12.17);
- decorative features can be incorporated for safety and breathability requirements, for example, the use of colour for venting (see Fig. 12.17), and 'High-Vis' glass bead wrist prints (Typhoon, 2017c);

Fig. 12.16 A lightweight 2.5 shell layer with bonded zip on the mouth of a pocket and a centre front stitched zip with inner guard.

Fig. 12.17 A classic rain jacket that introduces innovative laser-cut venting slats to the back panel for an on-bike position to improve airflow and regulate body temperature during extreme action. The jacket is made from 2.5 Polartec Neoshell bodyfabric, soft, 3-ply forearms, taped seams and neoprene cuffs (Rapha, 2017).

12.8.2.2 *Activity specific features*

Some designs, such as the Eco-jacket (Fig. 12.18), have their own proprietary breathable system as well as a set of unique features for the specified activity.

The upper torso of this waterproof garment has an engineered Breathable Construction System (BCS) with a laser-cut stretch insert to maximize freedom of movement. The inside of the jacket has a knitted 3D tape combined with airflow panels for extra space and ventilation.

Fig. 12.18 Halti's Autumn 2017 Legenda ski jacket, designed by Studio TINT, is filled with innovative, functional technologies. It uses DrymaxX, with a customized membrane, which is an approved Bluesign eco material that is fluorine free and water repellent. It has a 4-way stretch outer fabric and is insulated with Primaloft Silver Eco 80 with a body mapped lining.

The outer waterproof and breathable material has 4-way stretch and a Bionic-Finish Eco, i.e. a fluorocarbon-free water repellent treatment. Traditional features include water repellent zips on front, sleeve and hand warmer pockets and air ventilations, waterproof seams and a detachable, adjustable hood with integrated pocket to stow a beanie in lining. In addition to the BSC system, the unique features designed for ski-wear to aid comfort, protection and support activity requirements comprise:

- mechanical venting on inside back panel for maximum comfort;
- zip-out stretch snow gaiter;
- Lycra cuffs;
- integrated glove hook;
- audio cable holder;
- detachable goggle wiper;
- ski pass pocket;
- multiple inner pockets for goggles, valuables and electronics.

12.8.2.3 Prototypes and fitting

The physical fitting of waterproof prototypes at the end of a co-design process has a distinct advantage over a traditional process used for RTW garments. Initial block patterns which may have some articulation to allow movement already included or toiles can be fitted during their development and do not necessarily rely on a final fitting at the end of the design process. Nevertheless, if the shell garment has been designed as part of a layering system it is important to assess each garment, as well as all garments comprising a layered system before formal laboratory and field-testing is undertaken.

There are many definitions of fit (Yu, 2004) and both subjective and objective assessment methods have been widely debated. Criteria for assessing fit on a person in a static posture usually comprises grain, set, line, balance and ease and although, plus or minus ease, is design and material dependent, it is an element of pattern technology that needs to be carefully addressed to ensure comfort and enhance breathability in both single and multilayered systems. A rainwear garment or system that is too tight, too loose or poorly fitted between layers can inhibit range of movement and performance during an activity, which could ultimately impair health (Rossi, 2005).

Criteria for assessing aspects of movement in EN 340:2003 includes, donning or doffing and/or movement and postures that the wearer could experience during the course of their activity or work. While donning, doffing and a limited assessment of movements can be reviewed during a physical prototype fitting a full evaluation undertaken during the expected environmental conditions needs to be completed.

12.8.2.4 Donning and doffing

Donning and doffing is an important functional feature for some occupational activities, for example, sea rescue, where ease and speed of dressing in waterproof jackets and trousers is an essential requirement.

Dry suits designed for sailing are made from breathable fabrics with selected reinforcements for high-wear garment areas. They usually have several unique design features for keeping dry on land and at sea, aiding mobility and maintaining safety. For example:

- fully articulated arms, legs and body for improved fit and movement;
- 3D dry hood with spray protector;
- detachable fleece collar for adding/subtracting warmth;
- High-Vis glass bead wrist prints and reflective hood tape for increased visibility;
- glide skin seals at ankle, wrist and neck designed specifically for salt-water use;
- extra long lower leg covers to work with full height boots.

A donning and doffing feature is key for all-in-one designs, but the way in which zips are applied can vary. Some designs rely on a diagonal insertion while others, with a built-in X Hinge zip and a double waist with internal braces, can give easier access (see Fig. 12.19).

12.8.3 Testing waterproof garment and garment systems

The evaluation of a final garment and of a layered garment system and its features is of paramount importance for all rainwear garments, whether classic, multi functional or activity specific. There are two basic means of evaluating performance: laboratory facilities and field studies.

12.8.3.1 Laboratory testing

Several laboratory tests are used to assess the protective qualities of a garment and the effectiveness of clothing ensembles designed to be worn in the rain. Some tests are set out in international, regional or national standards and others exist as unpublished company-specific tests. They can be applied to static or dynamic mannequins, be water specific. or focused on thermal comfort.

Fig. 12.19 The PS330 Extreme Dry suit has been designed by Typhoon to cope with 'the harshest conditions' and facilitate donning and doffing. The arms, legs and body are articulated with an all-round waist zip to improve fit and comfort and the wrist, ankle and neck are sealed (Typhoon, 2017a,b). They are usually worn over an all-in-one fleece.

Rain room tests

As with the testing of waterproof materials, testing for water ingression on garments may be undertaken in independent laboratories, university and company-specific facilities. There is, however, variation within the tests, some of which are set out in standards (e.g. BS EN 14360:2004) or published by companies (e.g. Paramó, 2005), and others that are not available in the public domain that may in turn, limit comparison between outcomes.

The example of a standard given in the previous paragraph—Protective clothing against the rain-test method for ready made garments—Impact from above with high-energy droplets. This standard describes testing procedure for the:

- rain tower, its droplet size and relative climatic conditions; setting the air and water temperature (to ensure compatibility);
- shape, size and posture of an adult male mannequin (leaning backwards to assess the waterproof zip and its insertion);
- clothing to be placed on the mannequin (water absorbent materials);
- critical locations for the placement of humidity sensors to detect the timing of the water ingression (e.g. shoulders and zip position).

In addition to establishing critical locations, garments may be tested for the integrity of the placement and assembly of seams and closures, design features, leakage through the creases in the garment due to any limb articulation (e.g. bent arms), and for any interfaces between the neck and hood, sleeves and gloves, trousers and footwear as well as the length of garments.

The duration of tests can vary and, while a minimum of 1 h is recommended in this EU standard, some companies test products for more than 4 h (e.g. Paramó, 2005).

While this standard is useful, it is limited to static mannequins and specified rain conditions. Other, more sophisticated systems, can test the clothing on expert consumers in static or dynamic postures and have side nozzles as well as overhead simulated rain that not only simulates the direction of rain but also the velocity of associated wind.

Thermal tests

In addition to garment tests directly related to the ingression of water there are tests designed to link to the evaluation of thermal performance for ergonomic and physiological comfort. Two such standards are outlined below:

- BS EN ISO 9920:2009, Ergonomics of the thermal environment—Estimation of thermal insulation and water vapour resistance of a clothing ensemble. Note that this standard does not take account of the influence of rain or snow on the thermal characteristics. It is usually undertaken on a wetted or sweating mannequin;
- BS EN ISO 15831:2004, Clothing—Physiological effects—measurement of thermal insulation by means of a thermal mannequin. Measurement is usually performed with a stationary or moving mannequin although the movements are limited to those of the arms and legs and a defined stride length.

While the results from the use of standards are seen by Rossi (2005) as providing a realistic simulation of heat and moisture transfer from the body through the clothing to the environment, mannequins are made to a standard size and shape and not necessarily suitable for users with activity related morphology. An alternative, suggested by Bartels (2011), is to conduct a trial with a wearer who is an expert consumer, using a climatic chamber. Such a trial can also be undertaken using the ASTM Standard F-1154, 'a set of eight standard range of movement body positions and actions, and a set of seven work related activities', discussed by Ashdown (2011). Ashdown argues that studies that investigate the activities in which the user of clothing engages will give results that are more valid.

More recently, thermal imaging systems have been used by some companies to assess breathability and comfort (e.g. Musto, 2017; see Fig. 12.20).

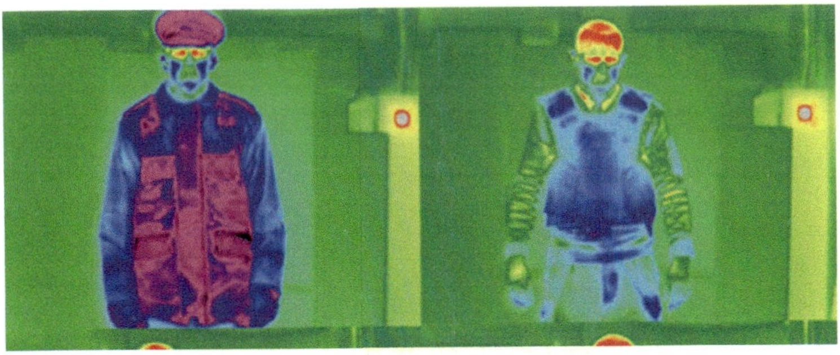

Fig. 12.20 Thermal imaging used in testing breathability and comfort in layered protective clothing (SATRA, 2017).

Other, more advanced systems propose a simulation of thermoregulation under transient and environmental conditions and include different activity levels (TAIThermIR, 2017). Meanwhile, some companies have set up biophysical laboratories that can recreate many of the environments found on the earth's surface. Key components comprise both environmental chambers and rain towers designed to measure the performance, protection, breathability and comfort benefits for the end user (Gore, 2016).

12.8.3.2 Field testing

There is wide agreement that field tests can offer valuable feedback on the performance of clothing designed to protect the user from the ingression of water. However, the procedures and criteria for undertaking such tests can vary according to the type of garment, activity and expected climatic conditions.

Some company garments are subject to extensive and iterative field-testing in a variety of climatic conditions by users of different occupations and professions, for example, British Mountain Rescue teams, Greenpeace and the British Antarctic Survey, whose tests ranged over 4 months (Paramó, 2005). Other companies, such as Typhoon, test garments and fabrics in the harshest of environments to produce the best possible combination of durability, performance and safety. Tests have taken place during the Volvo Ocean Race, the Sydney Hobart race, circumnavigations of the British Isles and extreme kayaking in the arctic.

Not all companies publish their procedures for testing but it is suggested by Havenith (2009) that tests should, where possible, be designed within set periods to minimize expert consumer fatigue, and that an expert panel of assessors be recruited to conduct a follow-up questionnaire (with rated/scales). This structure, together with laboratory tests, could provide an objective, subjective and correspondence evaluation.

Guidance for the selection of criteria, included in BS EN 340:2007 and provided by Patagonia (2015), is primarily based on ergonomics of comfort, with a few references to sensorial criteria while other sources propose criteria which could be classified within all four categories: thermal physiological comfort; sensorial comfort; ergonomic comfort; psychological comfort. Examples are shown below.

(a) Thermal physiological comfort

Do any of the following items affect thermal physiological comfort while undertaking the activity in wet (windy/cold) climatic conditions?

- Weight of layers.
- Insulation and breathability.
- Venting.
- Interaction of layers with each other.
- Interaction of layers with body.
- Waterproofing and insulation.
- Amount of layers.

The moisture level and temperature between the mid layer and shell can also be measured with micro sensors.

(b) Skin sensorial comfort

Do any of the following items affect the skins sensorial comfort while undertaking the activity in wet (windy/cold) climatic conditions?

- Rough, sharp, hard surfaces.
- Fabric worn on or near the skin.
- Fabric prickle/itch.
- Pressure from garment/layers/components.
- Softness/smoothness.
- Pliability in movement.
- Stiffness.
- Fabric cling to dampened skin.
- Fabric wetted through with sweat.
- Leaking and wetting of inside of the garment.
- Leaking and wetting of other layers.
- Interaction between fabric, human skin and atmospheric conditions.

(c) Ergonomic comfort

Do any of the following items affect ergonomic comfort while undertaking the activity in wet (windy/cold) climatic conditions?

- Freedom and ease of movement (within range and type of activity, for example, standing/sitting/walking/stairs/hands above head/bending).
- Body coverage during activity
- Overall fit of ensemble
- Cut of garment and components
- Too tight (for deep breathing)
- Too loose
- Overall, fit of individual garment within ensemble.
- Fit of individual garment component.
- Ease of donning and doffing (without assistance).
- Clarity and field of vision.
- Ease of accessing zipped/cinched components.
- Ease of opening/closing zipped components.
- Ease of opening/closing cinched components.
- Length of clothing and component parts.
- Size and position of storm flaps.
- Potential weakness in construction.
- Compatible with equipment.

(d) Psychological comfort

Do any of the following items affect psychological comfort while undertaking the activity in wet (windy/cold) climatic conditions?

- Overall external and internal attitude to sustainability and clothing brand.
- Overall well-being.
- Perception of overall system.
- Perception of protective shell.
- External personal preference of garment aesthetics (design, materials, colour, components, shape).
- Internal personal preferences (attitude/sensitivity to situation).

Note: Responses to the foregoing could be positive or negative and vary between and within populations and consumer ages.

A useful framework for drawing up conclusions to field studies for protective garments is proposed in BS EN ISO 13688:2013, but there is scope for a more comprehensive and reliable conclusion through the combination of quantitative and qualitative assessment.

12.9 Preparation for production, labelling and point of sale

12.9.1 Preparation for production

Many companies prepare tech packs, supported by quality manuals. Typical packs comprise information including prototype garments, patterns, style notes, initial specifications, corrections and product data (all information on materials and lab dip approval, trims, colours, sizing, construction details, labelling and packaging). Specifications for protective clothing are often extensive, covering all internal and external processes and, in some cases, expanded drawings. Some designers and QA personnel visit both local and overseas production facilities during the introduction of new styles. In some cases, counter prototype samples are created. At the production stage, factory-ready samples for all sizes and technical documents are combined with an accurate sample reference. There is, however, little reference to sustainable packaging.

12.9.2 Labelling

There are wide variations in the type and number of labels associated with rainwear garments. Labels range from sewn-in size and care guidance, through branding for outer shells (such as Gore-Tex) to extremely detailed multilabelling on swing tickets (as discussed earlier) covering not only technical drawings for the product but also technical drawings for all other garments designed for the layered system (Musto, 2017). There is a growing range of sustainability labels relating to materials (Eco-tex, Bluesign, Oeko tex) and also for sustainable products (Halti's ECO, and VAUDE's Green shape). WRAP has made proposals for increasing information on labels to cover durability and longevity of garment use (WRAP, 2015). Laundering procedures are given but consumers of wet weather clothing are unclear how often they should clean and reproof a garment (Burman 2014). Length of wash cycles varies between machine manufacturers settings as does the temperature of a tumble dryer.

12.9.2.1 Garment size and protection labelling

Guidance is given in ISO 340: 2003 for size designation, for example, EN13402-3 (under revision). This standard outlines the method of presenting various combinations of measurements to the consumer on a pictogram with either numerical or letter indicators. Equivalent key body sizes are usually provided in online charts.

ISO 343:2003+A12007 provides information for 'marking' protective occupational rainwear (represented by the image of an umbrella), which can include results of testing for water resistance to water penetration and water vapour resistance.

12.9.3 Point of sale

Many companies provide online information, in either textual or photographic form. These can include product ranges, materials, components and features as well the results of testing for resistance to water and water vapour penetration. In addition to general information on washing and dry cleaning many companies provide extensive directions on how to maintain and prolong the protective quality of the DWR. That is, to keep the outer fabric from becoming saturated so that the breathable barrier can continue to maintain its effectiveness; more research is required to assess the impact on the environment of these recommendations. Guidance is widely published for end users on care and maintenance, and it is common practice to recommend washing products to maintain waterproof finish. However non-C8 products are more easily soiled and need reproofing more often. Many companies augment their textual and photographic information with video clips, especially for the donning and doffing of dry suits, for example, Typhoon (2017c).

12.10 Looking ahead

The transition from a linear to a circular economy presents a challenge. This chapter has addressed some of the issues affecting the design and development of waterproof and water repellent clothing to provide wearer comfort within the context of that transition. In their publication Waste to Wealth, Lacy and Rutqvist (2015) suggest that, 'from a historical perspective, resource stewardship has only been in vogue for a short time'. However, time is short if designers are to help arrest and reverse the overuse of such resources.

The early history of wet-wear clothing illustrated the possibility of a compatible relationship with nature. Moreover, while the introduction of polymers created opportunities for complete protection from all aspects of wet climatic conditions, verification from Greenpeace suggests that there are still issues to address, with regard to textile toxicity and microfibre release, in particular, if consumers are going to be comfortable buying and wearing water repellent clothing. In a survey by Burman (2014), consumers commonly keep water proof garments for 10 years and indicated they don't want to pay more for an inferior garment. In addition, there are issues to be addressed related to the sustainability of product aftercare and the maintenance of waterproof finishes. More generally, the Greenpeace initiative has had a significant impact on the textile and clothing companies in the Outdoor market and on consumers whose commitment to a sustainable future has been helping to drive the need to find alternative materials. There has been a surge of new materials and, though there are several companies whose products do not contain PFC's, others are being obliged to make changes, many making commitments for a PFC-free future by 2020. For example, W. L. Gore is committed to eliminating 85% of its hazardous PFCs by 2020 and from its specialized outdoor waterproofing laminates by 2023 (Greenpeace, 2015).

In addition, if global warming is not arrested and the frequency of storms and wind velocity continues to increase, then perhaps the quest for a more water repellent and windproof material (and subsequently more stringent standards for testing) may be required to achieve a higher level of protection and comfort for the consumer.

Comfort in clothing, whilst defined and debated, remains difficult to achieve. In comparison with RTW processes, a co-design process, testing and field studies, can help achieve relatively successful protective rainwear products for a range of varying morphologies and activities. Nevertheless, as reported above, some technologies are helping to aid sustainable practice and reduce the number of product iterations. The creation of virtual designs, improved contour 3D fit and emerging technologies for the capture of shape with motion may help designers to visualize and assess comfort for activities without making initial physical garments.

Some generic criteria for assessing comfort in the field test were explored but no attempt was made to link specific garment components to a specific comfort category. For example, a hand-warmer pocket or a fleece-lined collar could be identified and evaluated as sensorial comfort criteria. These links could, however, be considered for evaluating clothing that had been designed for a specific activity and climatic condition.

Many designers and many companies now start out in the knowledge that everything is produced at a cost to the environment. They work to lower the use of environmental resources at every phase of a product's life cycle, anticipating the need to reuse, repair and recycle products within a circular economy. There is emerging evidence of good practice in design and development, the introduction of training in design for sustainability and the assembly of rules to aid the selection of sustainable materials. An outcome of these practices is a change in the way products are rated. It is now suggested that clothing, designed to resist water penetration, will be driven as much by environmental concerns, as by the need for comfort, aesthetics, functional developments and business considerations. Perhaps confirming the belief that 'sustainability is the ultimate design brief' and that 'sustainability for water repellent clothing needs to be design led'.

Acknowledgments

The authors would like to acknowledge and thank the following for their advice and assistance:
Apt and Aclimatise: Stanley Russell
Arc'teryx: Merle Schubert
Halti and Studio TINT: Maija Kiiski and Mari Mattila-Kamarainen
Henri Lloyd: Craig Prest
Houdini: Mia Grankvist
Marks and Spencer: Matt Linford
Musto: Alex Russell and Nick Houchin
Páramo: Alasdair Leighton-Crawford
Patagonia: Glen Morden
Rapha: Simon Huntsman and Olga Galikowska
SATRA: Jake Lock
Typhoon: Samantha Leonard
Vaude: Hilke Patzwall, Benedikt Troester and Bettina Roth

References

Aclimatise, 2017. Product and Incorporated Fabric (by APT). From Website, http://aclimatise.co.uk/waterproof-ladies-osprey-jacket-pink/ (February 2017).

Angel, S., 2013. Sustainable fashion needs to be design led. The Guardian. (20 February).

Arc'teryx, 2017. wet Weather Products Advertised Online. http://www.arcteryx.com/product.aspx?country=gb&language=en&model=Beta-AR-Jacket (Accessed February 2017).

Ashdown, S.P., 2011. Improving body movement comfort in apparel. In: Song, G. (Ed.), Improving Comfort in Clothing. Woodhouse, Cambridge.

Bartels, V.T., 2011. Factors affecting comfort: human physiology and the role of comfort. In: Song, G. (Ed.), Improving Comfort in Clothing. Woodhead, Cambridge.

Black, S., Kapsali, V., Bougourd, J., Geesin, F., 2005. Fashion and function: factors affecting the design and use of protective clothing. In: Scott, R. (Ed.), Textiles for Protection. Woodhead, Cambridge.

Bluesign, 2017. What Is the 'Bluesign' Standard for Textiles? http://business-ethics.com/2012/11/11/what-is-the-bluesign-standard-for-textiles/ (Accessed May 2017).

Bougourd, J., McCann, J., 2009. Factors affecting the design of cold weather performance clothing. In: Williams, J. (Ed.), Textiles for Cold Weather Apparel. Woodhead, Cambridge.

British Standards, 2004. Protective Clothing Against Rain. Test Method for Ready-Made Garments. Impact from Above With High Energy Droplets. BS EN 14360:2004.

British Standards, 2007. Protective Clothing—Protection Against Rain. BS EN 343:2003 +A1:2007.

British Standards, 2009. Ergonomics of the Thermal Environment. Estimation of Thermal Insulation and Water Vapour Resistance of a Clothing Ensemble. BS EN ISO 9920:2009.

British Standards, 2013. Protective Clothing—General Requirements. BS EN ISO 13688:2013.

Bruntland, 1987. Report of the World Commission on Environment and Development: Our Common Future. Oxford University Press, London.

Burman, G., 2014. Expectations of durable water repellent fabric finishes. In: Durable Water Repellency—Study Phase 1. De Montfort University, Leicester.

Buttin, N., Brieuc, S., 2017. Innovation+Économie Circulair, from WITHAA. https://www.slideshare.net/wiithaa/business-model-and-circular-economy (Accessed February 2017).

Carson, R., 1962. Silent Spring. Houghton Mifflin, Boston.

Chouinard, Y., 2013. Prosperity with less: what would a responsible economy look like? Guardian. (Sustainable Business, 4 October).

Climate Communication, 2017. Online Information Published by a Science Outreach Organisation. https://www.climatecommunication.org/climate/global-warming/ (accessed May 2017).

Csanák, E., 2014. Eco-friendly concepts and ethical movements in the fashion industry. In: 7th International Textile, Clothing and Design Conference.

Daanen, H.A.M., Reffeltrath, P.A., 2007. Function, fit and sizing. In: Ashdown, S.P. (Ed.), Sizing in Clothing: Developing Effective Sizing Systems for Ready-to-Wear Clothing. Elsevier, London.

Davies, A., 2014. Durable Water Repellency—Study, Phase 1. De Montfort University, Leicester.

ECAP, 2015. European Clothing Action Plan. http://www.ecap.eu.com/about-ecap/ (accessed April 2017).

Eco-Innovation Observatory, 2017. Eco-Innovate: A Guide to Eco-Innovation for SMEs and Business Coaches. From Website, http://www.eco-innovation.eu/index.php?option=com_

content&view=article&id=638:sme-guide2&catid=79:thematic-reports&Itemid=212 (accessed February 2017).

Ellen MacArthur Foundation, 2017. The Circular Economy. 2017.

Ethical Fashion Forum, 2015. 10 Toxic Chemicals to Avoid in Your Products. From, http://source.ethicalfashionforum.com/article/10-toxic-chemicals-to-avoid-in-your-products.

European Outdoor Group (EOG), 2017. Outdoor Industry Microfiber Consortium Formed. http://www.europeanoutdoorgroup.com/news/outdoor-industry-microfibre-consortium-formed (Accessed May 2017).

EVO, 2017. How to Choose Insulated Garments. www.evo.com/guides/how-to-choose-insulated-garments (Accessed February 2017).

Global Footprint Network, 2017. From Website http://nourishtheplanet.com/2015/08/global-overshoot-day-2015-weve-spent-annual-natures-budget/ (Accessed January 2017).

Gore, W.L., 2016. Online Announcement of a Fabric Testing Facility. https://www.gore.com/news-events/press-release/fabrics-press-release-opening-of-testing-facility-and-40th-anniversary (Accessed May 2017).

Greenpeace, 1997. What's Wrong With PVC? From Website, http://www.greenpeace.org.uk/files/pdfs/migrated/MultimediaFiles/Live/FullReport/5575.pdf (Accessed January 2017).

Greenpeace, 2015. Detox Outdoors. http://detox-outdoor.org/ (Accessed February 2017).

Halti, 2017. Author Interview with Company Representative.

Havenith, G., 2009. Laboratory assessment of cold weather clothing. In: Williams, J. (Ed.), Textiles for Cold Weather Apparel. Woodhouse, Cambridge.

Holmér, I., 2009. Protective clothing for cold workplace environments. In: Williams, T. (Ed.), Textiles for Cold Weather Apparel. Woodhead, Cambridge.

Holmér, I., 2011. Cold weather clothing and comfort. In: Song, G. (Ed.), Improving Comfort in Clothing. Woodhead, Cambridge.

Houdini, 2017a. Advice on the Circular Economy. https://www.houdinisportswear.com/en/houdini-services (Accessed May 2017).

Houdini, 2017b. Author Interview with Company Representative.

Houdini, 2017c. Press Release. The International Fachmesse für Sportartikel und Sportmode. The International Sporting Goods Trade Fair (ISPO), Munich.

Kilinc-Balci, F.S., 2011. How consumers perceive comfort in apparel. In: Song, G. (Ed.), Improving Comfort in Clothing. Woodhead, Cambridge, pp. 97–113.

Kryzwinski, S., Rödel, H., Siegmund, J., 2005. Virtual product development of close-fitting garments of knitwear with Elastane yarns. Text. Process. 2, 136–140.

Lacy, P., Rutqvist, J., 2015. Waste to Wealth: The Circular Economy Advantage. Palgrave Macmillan, London.

Lloyd, 2017. Pure Sailing. Henri Lloyd, London.

Makinen, H., 2005. Fire fighters protective clothing. In: Scott, R.A. (Ed.), In: Textiles for Protection. Woodhead, Cambridge.

Mallett, M., 2017. From Website http://www.marlamallett.com/w-3955.htm (Accessed January 2017).

Marks and Spencer, 2017. Promotional Website http://www.marksandspencer.com/buttonsafe-crease-resistant-belted-mac-with-stormwear/p/p22427670 (Accessed January 2017).

McCann, 1999. Establishing the requirements for the design development of performance sportswear. Unpublished MPhil dissertation, University of Derby.

McCann, J., 2005. Material requirements for the design of performance sportswear. In: Shishoo, R. (Ed.), Textiles in Sport. Woodhead, Cambridge.

Meadows, D.H., Meadows, D.L., Randers, J., Behrens, W.W., 1972. Limits to Growth. Universe Books, New York.

Michaelis, L., Lorek, S., 2004. Consumption and the Environment in Europe: Trends and Future. Danish Environmental Protection Agency, Copenhagen.

MUSTO, 2017. The Inside Edge, Product Catalogue. MUSTO Ltd, Laindon, Essex.

Muthu, S., 2014. Assessing the Environmental Impact of Textiles and the Clothing Supply Chain. Woodhead, Cambridge.

Neoprene, 2017. See Discussion at: https://en.wikipedia.org/wiki/Neoprene (Accessed February 2017).

NIKE Making App, 2016. https://itunes.apple.com/us/app/making-of-making-powered-by-nike-msi/id662227880?mt=8 (Accessed May 2017).

NSF, 2015. Global Traceable Down Standard. See Website, http://www.nsf.org/services/by-industry/sustainability-environment/responsible-sourcing/global-traceable-down-standard (Accessed February 2017).

Paramó, 2005. How Do We Test Our Waterproof Jackets? http://www.paramo-clothing.com/blog/en-gb/rainroomtest (Accessed May 2017).

Paramó, 2017a. The Velez Jacket. From Website, http://www.paramo-clothing.com/en-gb/explore-range/product/?pk=0E33E82B-CDF8-4CCE-A46F-E66CF4545B71 (Accessed February 2017).

Paramó, 2017b. Report on Analogy Fabric Used to Waterproof the Velez Jacket. From Website, https://www.youtube.com/watch?v=yTKV_4smGS8.

Patagonia, 2011. Company Blog. https://www.patagonia.com/blog/2011/09/introducing-the-common-threads-initiative (Accessed May 2017).

Patagonia, 2015. Our DWR Problem (Updated). http://www.patagonia.com/blog/2015/09/our-dwr-problem-updated/ (Accessed February 2017).

Patagonia, 2017. Patagonia Repair & Care Guides. http://www.patagonia.com/worn-wear-repairs/ (Accessed February 2017).

Puma, 2017. Company declaration on sustainability, From Website http://about.puma.com/en/sustainability/strategy (Accessed January 2017).

Rapha, 2016. Author Interview.

Rapha, 2017. Product Information Published Online http://www.rapha.cc/gb/en/shop/collector%27s-edition-classic-rain-jacket/product/RJK07XX (Accessed February 2017).

REACH, 2017. European regulation on safe use of chemicals, From http://www.reachready.co.uk/reachthebasics.php (Accessed February 2017).

Rengasamy, R.S., 2011. Improving moisture management in apparel. In: Song, G. (Ed.), Improving Comfort in Clothing. Woodhead, Cambridge.

Rossi, R.M., 2005. Interactions between protection and thermal comfort. In: Scott, R.A. (Ed.), Textiles for Protection. Woodhead, Cambridge, pp. 233–260.

Satra, 2017. Breathability and Clothing Comfort. https://www.satra.co.uk/spotlight/article_view.php?id=295 (Accessed January 2017).

Saville, B.P., 1999. Comfort. In: Saville, B.P. (Ed.), Physical Testing of Textiles. Woodhead, Cambridge, pp. 209–243.

Scott Cato, M., 2009. Green Economics. Earthscan, London p. 36f.

Scottish Mountain Gear, 2017. Gore-Tex Repair Service. See Website, https://scottishmountaingear.com/repair-service_gore-tex.asp accessed February 2017.

Sherwin, C., 2012. Sustainability is the ultimate design brief. The Guardian. 17 September.

Sizemic, 2017. Fashion technology, size and fit solutions, From www.sizemic.eu/.

Slater, K., 1985. Human Comfort. Charles C. Thomas, Springfield, IL.

Song, G. (Ed.), 2011. Improving Comfort in Clothing. Woodhead, Cambridge.

Stevens, K., Fuller, M., 2015. Thermo regulation and clothing comfort. In: McCann, J., Bryson, D. (Eds.), Textile-Led Design for the Active Ageing Populations. Woodhead, Cambridge.

Sustainable brands, 2014. News Item in Webpage. http://www.sustainablebrands.com/news_and_views/collaboration/sustainable_brands/patagonia_sharing_proprietary_biorubber_advance_sust (Accessed February 2017).

TAIThermIR, 2017. Online Description of the Simulation of Human Thermal Sensation and Thermoregulation Under Transient and Asymmetric Environmental Conditions. http://www.thermoanalytics.com/products/taithermir/human-therma (Accessed May 2017).

Thomas, P., 2011. Antibacterial clothing—a Fashionable threat to human health. Natural Health News. 29 November.

Timmins, M., McCann, J., 2015. Overview of the design requirements of the active ageing. In: McCann, J., Bryson, D. (Eds.), Textile-Led Design for the Active Ageing Population. Woodhead Publishing, Cambridge.

Turner, G., 2014. Is Global Collapse Imminent? An Updated Comparison of The Limits to Growth with Historical Data. (Research Paper), Melbourne Sustainable Society Institute.

Typhoon, 2017a. 'Equator Hinge', Spring/Summer Catalogue 2016-17. Typhoon, London.

Typhoon, 2017b. PS330 Extreme Drysuit. From Website, https://www.typhoon-int.co.uk/products/surface/drysuits/other/100151/ps330-extreme-drysuit/ (Accessed January 2017).

Typhoon, 2017c. Promotional Video. Typhoon PS330 Drysuit. https://www.youtube.com/watch?v=cxrWI9AhZtQ (Accessed May 2017).

UN, 2012. A/RES/66/288—The Future We Want. United Nations, New York.

UN, 2015. A/RES/70/1—Transforming Our World: The 2030 Agenda for Sustainable Development. United Nations, New York.

Vaude, 2017a. Sustainable Clothing Philosophy and Practice. https://www.vaude.com/en-GB (Accessed February 2017).

Vaude, 2017b. Author Interview with Company Representative.

Vinyloop, 2017. Vinyl Loop, Promotional Leaflet. From Website, http://www.eswa.be/pdf/camb01_4002_vinylplus_brochure_en.pdf (Accessed January 2017).

Wilson, J., 2010. Classic and Modern Fabrics. Thames and Hudson, London.

WMB UK, 2017. West Bridgeford Machine. Seam-Sealing Tapes. http://www.wbm-uk.com/ (Accessed August 2017).

WRAP, 2012. Working Together for a World Without Waste. From Website, http://www.wrap.org.uk/sites/files/wrap/VoC%20FINAL%20online%202012%2007%2011.pdf (Accessed February 2017). http://www.wrap.org.uk/sustainable-textiles/scap.

WRAP, 2015. Sustainable Clothing Guide. From Website, http://www.wrap.org.uk/sites/files/wrap/WRAP%20Sustainable%20Clothing%20Guide.pdf (Accessed February 2017).

WRAP, 2017. WRAP and the Circular Economy. From Website, http://www.wrap.org.uk/about-us/about/wrap-and-circular-economy (Accessed February 2017).

Yu, W., 2004. Subjective assessment of clothing fit. In: Fan, J., Yu, W., Hunter, L. (Eds.), Clothing Appearance and Fit: Science and Technology. Woodhead, Cambridge.

Yupic, 2017. Summer dried seal gut parka, From Website https://commons.wikimedia.org/wiki/File:Summer_dried_seal_gut_parka,_Aleutian_Islands,_Yupik,_20th_century,_Honolulu_Museum_of_Art,_2014-25-01.JPG (Accessed January 2017).

Zhou, W., Reddy, N., Yang, Y., 2005. Overview of protective clothing. In: Scott, R.A. (Ed.), In: Textiles for ProtectionWoodhead, Cambridge.

Sources of further information

BS 8001: 2017 Framework for implementing the principles of the circular economy in organizations. Guide BSI London: May 2017.

Horrocks, A.R., Anand, S.C., 2016. Handbook of Technical Textiles: Technical Textile Applications. Woodhead, Cambridge.

ISO, 2016a. ISO18825: 2016. Clothing—Digital Fittings—Part 1 Vocabulary and Terminology for the Virtual Human Body.

ISO, 2016b. ISO18825: 2016. Clothing—Digital Fittings—Part 2 Vocabulary and Terminology for the Attributes of the Virtual Human Body.

ISO, 2016c. ISO18163: 2016. Clothing—Digital Fittings—Vocabulary and Terminology for the Virtual Garment.

ISO, 2016d. ISO18831: 2016. Clothing—Digital Fittings—Attributes of Virtual Garments.

ISO, 2017a. ISO 8559-1: 2017. Part 1: Anthropometric Definitions for Body Measurement.

ISO, 2017b. ISO 8559-2: 2017. Part 2: Size Designation of Clothes Part 2 Primary and Secondary Dimension Indicators.

ZHDC, 2012. Durable Water and Soil Repellent Chemistry in the Textile Industry—A Research Report November 2012 Version 1.0 P05 Water Repellency Project. http://www.roadmaptozero.com/programme/manufacturing-restricted-substances-list-mrsl-conformity-guidance/.

Further reading

Black, S., 2012. The Sustainable Fashion Handbook. Thames & Hudson, London.

Fletcher, K., Grose, L., 2012. Fashion & Sustainability: Design for Change. Lawrence King, London.

Teijin, 2017. Solutions. See Website, http://www.teijin.com/solutions/ecocircle/ (Accessed February 2017).

Performance evaluation and testing of water repellent textiles

13

Alice J. Davies
University of Leeds, Leeds, United Kingdom

13.1 Introduction

Liquid repellent finishes are commonplace in a number of sectors within the textile industry. They are used to provide a fabric with improved resistance to aqueous and sometimes oily mediums, enabling liquid droplets to bead up and roll away, leaving the textile unwetted. They are also able to improve the efficacy of soil release by domestic laundering.

A variety of standardized test methods are available to assess repellent fabrics, covering water and oil repellency and soiling, offered by a number of organizations including: British Standards Institute (BSI); International Organisation for Standardisation (ISO); American Association of Textile Chemists and Colorists (AATCC). In the majority of cases the pass/fail criteria is not specified within the standard, rather this is established by individual retailers dependent on the product end use and consumer requirements.

This chapter discusses and compares available methods for testing the initial performance of a water repellent fabric as well as those for testing the durability of repellency. The suitability of each method will also be considered. In doing so, it should be noted that the majority of tests have been devised for perfectly smooth and homogenous solid surfaces. Textiles have more complex topographies and surface roughness can significantly affect repellency. For example, super-hydrophobicity is often achieved by adding microscopic roughness to a fabric surface, which creates an effect similar to a lotus leaf, thus increasing repellency.

13.2 Static test methods

A number of static test methods exist for assessing initial repellency of a fabric. The majority of these methods assess the interaction between the fabric surface and a droplet of water or other liquid of lower surface tension. While these methods do not replicate the conditions of wear for clothing applications, they can be useful for categorizing the repellent type and providing an initial differentiation of repellency performance.

Waterproof and Water Repellent Textiles and Clothing. https://doi.org/10.1016/B978-0-08-101212-3.00012-5

13.2.1 Resistance of fabrics to water absorption: static immersion test

- BS/EN 3449:1990 Method for resistance of fabrics to water absorption (static immersion test) (BSI, 1990).

A sinker is used to submerge a weighed fabric specimen under water for a period of 20 minutes ± 10 seconds, after which time, the fabric is run in a centrifuge for 15 ± 1 seconds to remove any excess water before it is reweighed to the nearest milligram to determine how much water has been absorbed during the test.

This static test is not representative of conditions of use for water repellent clothing but could be useful to suggest fabric behaviour under prolonged exposure to water.

13.2.2 Water contact angle

- Goniometer system for measuring contact angles of static drops (Holme, 2003).
- Sessile drop method and contact angle hysteresis method (Yuan and Lee, 2013).

A water droplet of controlled size is introduced on to the fabric surface and images of the droplet are recorded using the goniometer system video microscope. These images are used to calculate the contact angle between the water droplet and the fabric surface. A mean value is calculated from multiple images to provide an indication of the water repellency of the fabric. While a higher contact angle theoretically indicates superior performance, the equations are based on ideal, smooth surfaces and the wetting of a textile surface is often more complex than indicated by this method.

Contact angle is dependent on the relative interactive tensions between solid, liquid and vapour interfaces which are described by Young's equation,

$$\gamma_{lv} \cos\theta_{\gamma} = \gamma_{sv} - \gamma_{sl}, \tag{13.1}$$

where γ_{lv}, γ_{sv} and γ_{sl} signify the liquid–vapour, solid–vapour and solid–liquid interface tensions respectively, and θ_{γ} signifies the resultant contact angle (Rouette, 2001; Yuan and Lee, 2013). In sessile drop contact angle measurement, this is the angle formed at the edges of the liquid droplet where the liquid–vapour interface contacts the solid surface (measured though the liquid phase) (Ruckman, 2005; Samanta et al., 2015). Contact angle values <90° indicate that the fabric surface is wettable/hydrophilic. Contact angle values >90° denote hydrophobicity. Materials described as super-hydrophobic are those exhibiting contact angles >150° (Fig. 13.1) (Yuan and Lee, 2013).

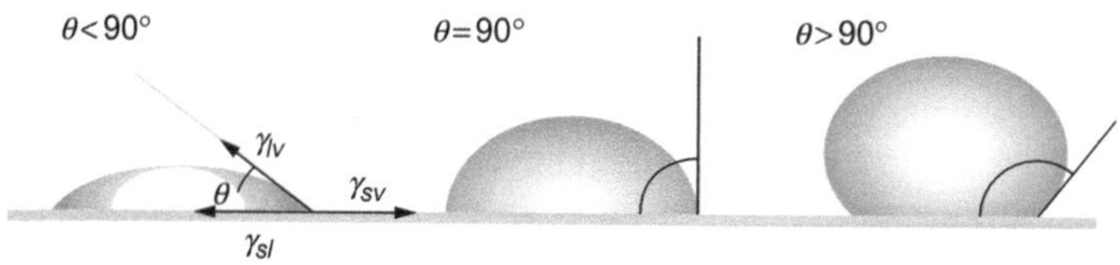

Fig. 13.1 Illustration of contact angles formed by liquid drops on homogenous solid surface. Modified from Yuan, Y., Lee, T., 2013. Contact angle and wetting properties. In: Bracco, G., Holst, B. (Eds.), Surface Science Techniques. Heidelberg: Springer, pp. 3–29.

13.2.2.1 Contact angle and hysteresis

In practice, wetting is not a static state and so to characterize wetting behaviour based on a single static contact angle may be inadequate. Contact angles formed as the liquid expands and contracts are referred to as advancing (θ_a) and receding (θ_r) respectively. The difference between the advancing and receding angle is known as contact angle hysteresis (H),

$$H = \theta_a - \theta_r. \tag{13.2}$$

It is generally accepted that this phenomenon arises due to surface roughness and heterogeneity. Young's equation does not account for surface topography and so a measurement of contact angle hysteresis, as a derivation of the standard method, more accurately reflects the relative strengths of molecular interactions between interfaces and therefore provides a more accurate measurement of contact angle (Yuan and Lee, 2013).

13.2.3 Aqueous liquid repellency: water/alcohol solution test

- BS/EN/ISO 23232:2009 Aqueous liquid repellency. Water/alcohol solution resistance test (BSI, 2009).
- AATCC 193-2007 Aqueous Liquid Repellency: Water/Alcohol Solution Resistance Test (AATCC, 2007b).

The aqueous liquid repellency test (also known as water rating method) describes a procedure where by 20 μL drops of solution with increasing concentration of isopropyl alcohol are introduced to the fabric and a rating number is assigned based on which strength of solution the fabric is able to successfully repel. Table 13.1 outlines which concentration of isopropyl alcohol solution equates to which rating number. The fabric must be able to repel the solution for 10 ± 2 seconds to be deemed successful. A visual comparison is used to aid assessment; grading examples shown in Fig. 13.2.

As a static test, this method does not replicate end use, however it could be considered to be more stringent than other water drop methods as it makes use of solutions with surface tensions lower than that of water.

13.2.4 Oil repellency: hydrocarbon resistance test

- BS/EN/ISO 14419:2010 Textiles. Oil repellency. Hydrocarbon resistance test (BSI, 2010).
- AATCC 118-1997 Oil Repellency: Hydrocarbon Resistance Test (AATCC, 1997).

The oil repellency test (also known as the oil rating method) describes a similar procedure to the aqueous liquid repellency test, where oils of decreasing density and surface tension are used in place of isopropyl alcohol solution. 20 μL of each oil is introduced to the fabric surface and a rating number assigned based on which oil the fabric is able to successfully repel. A list of standard oils and corresponding rating numbers can be found in Table 13.2. For this method the fabric should repel the oil for 30 ± 2 seconds to be assessed as successful (the same visual assessment criteria is used here as for the aqueous liquid repellency test: Fig. 13.2).

Table 13.1 Aqueous liquid repellency test standard test liquids (British Standards Institute, 2009)

Aqueous solution repellency grade number	Composition (by volume)	Surface tension at 25°C (dyn/cm)
0	Water	72.75
1	98:2/Water:isopropyl alcohol	59
2	95:5/Water:isopropyl alcohol	50
3	90:10/Water:isopropyl alcohol	42
4	80:20/Water:isopropyl alcohol	33
5	70:30/Water:isopropyl alcohol	27.5
6	60:40/Water:isopropyl alcohol	25.4
7	50:50/Water:isopropyl alcohol	24.5
8	40:60/Water:isopropyl alcohol	24

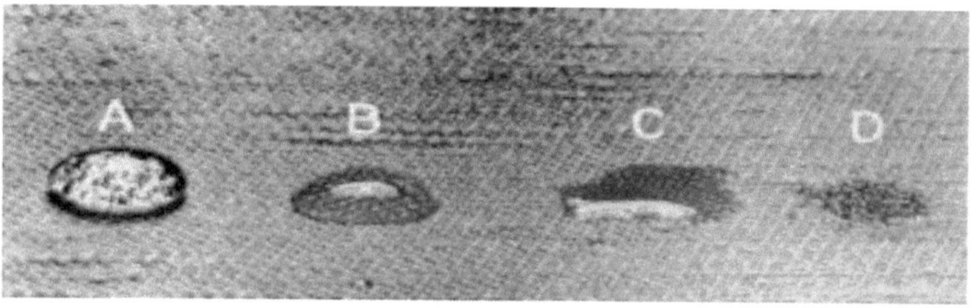

A passes; clear well-rounded drop
B borderline pass; rounding drop with partial darkening
C fails; wicking apparent and/or complete wetting
D fails; complete wetting

Fig. 13.2 Aqueous liquid repellency test grading examples (British Standards Institution, 2009).

The oil repellency method is generally only applicable for fluorochemical based finishes as alternative repellent technologies are not commonly capable of providing oil repellency. This method is most applicable for protective clothing applications, as it unlikely that true oil repellency will be required for apparel applications such as sportswear. That said, assessing repellency to oil can provide further performance differentiation than testing for water repellency alone and may also give an indication of the fabric's resistance to staining and soiling from oily substances.

13.2.5 Suitability of static test methods

While static methods can be useful for assessing the basic repellency performance of a fabric, they are not representative of wearer use of water repellent apparel. A true indication of performance cannot be obtained from the use of static methods in isolation. Water and oil rating methods are useful for identifying the category of repellent finish

Table 13.2 **Oil repellency test standard test liquids (BSI, 2010)**

Composition	Oil test liquid number	Density (kg/L)	Surface tension (N/m at 25°C)
None (fails with white mineral oil)	0	–	–
White mineral oil	1	0.84–0.387	0.0315
65:35 white mineral oil: *n*-hexadecane by volume	2	0.82	0.0296
n-Hexadecane	3	0.77	0.0273
n-Tetradecane	4	0.76	0.0264
n-Dodecane	5	0.75	0.0247
n-Decane	6	0.73	0.0235
n-Octane	7	0.70	0.0214
n-Heptane	8	0.69	0.0198

being used, providing an effective distinction between fluorochemical-based and fluorochemical-free finishes but are not suitable to provide any further meaningful differentiation (Davies, 2014).

13.3 Dynamic test methods

13.3.1 Determination of resistance to surface wetting: spray test

- BS/EN/ISO 4920:2012 Textile fabrics. Determination of resistance to surface wetting (spray test) (BSI, 2012a).
- AATCC 22-2001 Water Repellency—Spray Test (AATCC, 2001).

The spray test is the basic method for assessing water repellency of apparel fabrics. This method is intended to test a fabric's ability to resist surface water only, not resistance to water penetration. A fabric specimen is mounted at a 45° angle below a spray nozzle, with a distance of 150 ± 2 mm between the centre of the nozzle and the centre of the fabric specimen as shown in Fig. 13.3. 250 ± 2 mL of deionized water is poured into a funnel at such a rate as to create 25–30 seconds continuous flow of water from the spray nozzle onto the fabric specimen. The fabric and holder are removed from the test apparatus and tapped twice against a solid object to remove excess water drops, after which the fabric is assessed and rated.

A comparative photographic scale (Fig. 13.4) is used to rate the fabric performance. AATCC and ISO standards specify different numeric scales (AATCC 0-100; ISO 0-5) but both specify six possible ratings. For outdoor apparel applications, a rating of 80 (ISO 3) is commonly regarding as a pass by the industry (P05 Project Team, 2012).

This is the simplest method to perform and can be used with reasonable success to quantify the basic repellency performance of a fabric or finish. However, the use of subjective assessment criteria leaves the ratings open to experimenter bias that could result in inconsistencies in data.

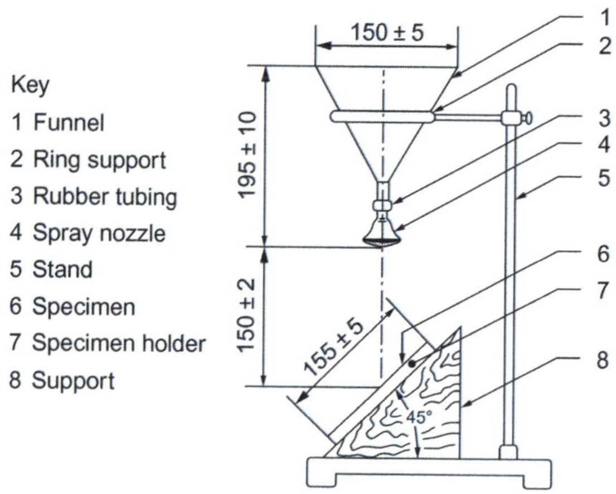

Fig. 13.3 Apparatus for spray test (BSI, 2012a).

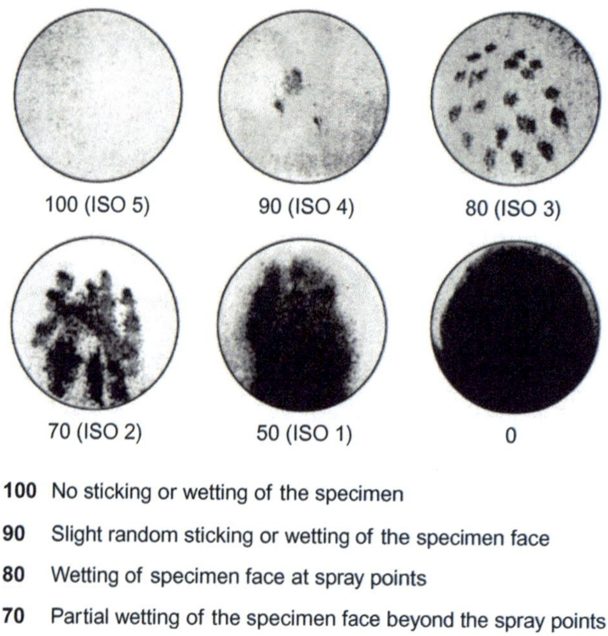

Fig. 13.4 Spray test rating chart: photographic scale (BSI, 2012a).

13.3.2 Impact penetration test

- BS/EN/ISO 18695:2007 Textiles. Determination of resistance to water penetration. Impact penetration test (BSI, 2007).
- AATCC 42-2007 Water Resistance: Impact Penetration Test (AATCC, 2007a).

The impact penetration test is used to predict the probable rain penetration resistance of a fabric. A fabric specimen is mounted at 45° angle below a spray nozzle, with a

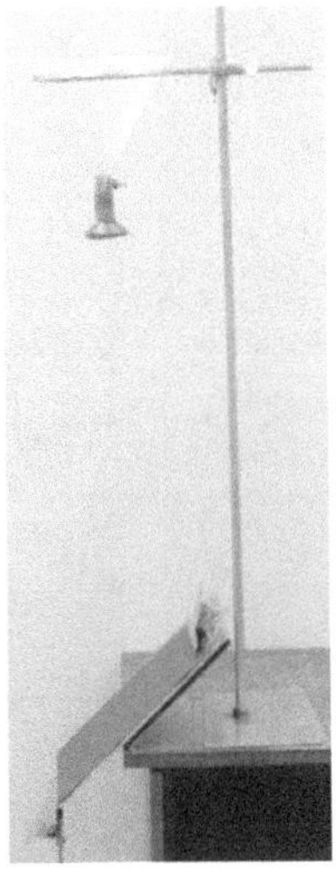

Fig. 13.5 Apparatus for impact penetration test (BSI, 2007).

distance of 610 ± 10 mm between the centre of the nozzle and the centre of the fabric specimen. The fabric is backed with a weighed blotting paper and clamped flat with spring clamps as shown in Fig. 13.5. 500 ± 10 mL of deionized water is poured into a funnel allowing a continuous flow of water from the spray nozzle onto the fabric surface.

The blotting paper is removed and reweighed after testing to determine how much water has been able to penetrate the fabric.

While this method provides a more objective assessment than the spray test, it does not consider surface wetting which is still an important factor; the fabric may still exhibit unsatisfactory surface wetting even if it is able to resist water penetration in which case the results of the impact penetration test could be misleading for assessing overall repellency performance.

13.3.3 Bundesmann rain-shower test

- BS/EN/ISO 29865:1993 Determination of water repellency of fabrics by the Bundesmann rain-shower test (BSI, 1993).

Fabric specimens are mounted on holder cups and positioned 1500 mm below a multi-nozzle drop producer. Water is allowed to flow from the nozzles onto the cups for 10 minutes, effectively subjecting the test specimens to a simulated rain shower. Cross-shaped arms or blades inside the cups also rub the underside of the fabric specimen to simulate the action of a garment rubbing against the body.

Each cup is weighed before and after testing and the difference is calculated to determine the volume of water that has penetrated the fabric. A comparative photographic scale is also used to visually assess the surface wetting (Fig. 13.6) similar to

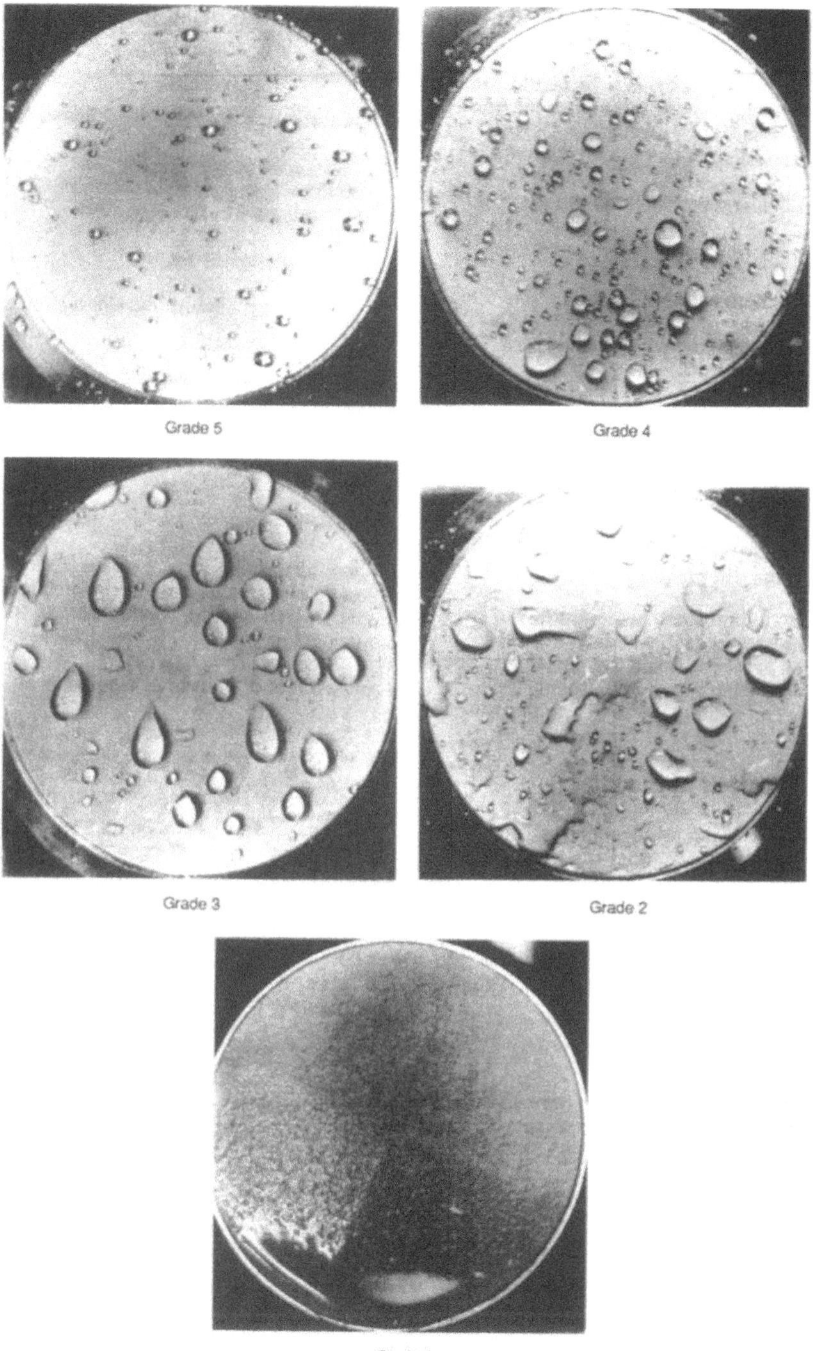

Fig. 13.6 Bundesmann rain-shower test rating chart: photographic scale (BSI, 1993).

the spray test. Visual assessments can also be made at predetermined intervals throughout the test, for example after 1, 5, 10 minutes. For outdoor apparel applications, a rating of 4 after 10 minutes is generally considered to be a pass by the industry (P05 Project Team, 2012).

The Bundesmann method is one of the most representative of conditions of use because it utilizes a greater distance between the nozzles and the test specimens as well as a prolonged flow of water. It also has the advantage of assessing both the surface wetting and water penetration of the fabric.

Although the Bundesmann rain-shower test is seemingly quite important to the industry it appears to have fallen out of favour due to the requirement of specialized equipment and larger test samples, making it both complex and costly to perform. Most independent testing laboratories contacted in the UK were unable to perform the test.

13.3.4 Hydrostatic head test

- BS/EN 20811:1992 ISO 811:1981 Textiles. Determination of resistance to water penetration—Hydrostatic pressure test (BSI, 1992)

A hydrostatic head test can be used to determine the resistance of fabrics to penetration by water. Booth (1968) noted that as the ability for water to penetrate a fabric relies heavily on the density of the fabric structure, the hydrostatic head method is unlikely to be sensitive to any differences in fabric finishes and coatings.

A fabric specimen of 100 cm^2 is clamped to the apparatus in horizontal orientation and exposed to increasing water pressure from above or below until penetration occurs in three places. The test procedure specifies that the specimen should be watched continuously for evidence of penetration and that very fine droplets or subsequent drops in the same location on the fabric should be disregarded, leaving some room for subjectivity of results.

The water used for testing should be distilled or deionized and at a temperature of either $20 \pm 2°C$ or $27 \pm 2°C$. The water pressure should be increased at a rate of either 10 ± 0.5 cm or 60 ± 3 cmH_2O/min.

According to the standard (BSI, 1992) a hydrostatic head of 1000 mm can be considered 'penetration resistant'. However, this is generally considered to be a minimum figure for waterproofness and although various figures are quoted across the industry, a hydrostatic head of 10,000 mm can be considered sufficient for preventing water penetration under normal conditions of use for waterproof apparel (Fuller and Taylor, 2012).

13.3.5 Suitability of dynamic test methods

Dynamic methods more closely replicate wearer use of water repellent clothing than do static methods, making them more suitable for assessing repellency performance. The basic spray test method can be used to provide consistent and repeatable differentiation between different repellent technologies. Its effectiveness is increased by measuring the percentage weight increase of the fabric specimen before and after test to provide an additional objective measure of performance (Davies, 2014). Rain room methods might also be considered as a dynamic test of the full garment under controlled conditions prior to wearer trials (see Section 13.4.4). The impact penetration and hydrostatic head test methods are most applicable to fabrics intended to be waterproof (resistant to water penetration) rather than just water repellent (resistant to surface wetting), such as those with microporous or hydrophilic membranes or coatings (Fuller and Taylor, 2012).

13.4 Methods for assessing durability of performance

A number of parameters can influence the durability and longevity of an item of water repellent clothing. The effect of laundering and of abrasion from continued wear are perhaps the most important and noticeable factors for the consumer, but prolonged exposure to water can also cause a reduction in repellent performance.

13.4.1 Laundering

- BS/EN/ISO 6330:2012 Textiles. Domestic washing and drying procedures for textile testing (BSI, 2012b).
- AATCC Monograph M6 Standardization of Home Laundry Test Conditions (AATCC, 2011a).

Selected repellency tests may be repeated on a fabric after it has been subjected to repeated washing and drying cycles, the number of which may vary significantly in practice throughout the industry depending on fabric type and anticipated end use (P05 Project Team, 2012). Comparing these results to those gained before laundering will provide an indication of the durability of the repellent under the specific conditions used.

The ISO standard (BSI, 2012b) describes testing procedures for three washing machine types and six possible drying methods: horizontal axis, front-loading type (Type A); vertical axis, toploading agitator type (Type B); vertical axis, top-loading pulsator type (Type C) and line dry; drip line dry; flat dry; drip flat dry; flat press dry; tumble dry (vented—Type A1; condenser—Type A2; large vented—Type A3) respectively. The standard also describes six variations of detergent, either non-phosphate powder or liquid, with or without optical brighteners and enzymes (BSI, 2012b).

The AATCC standard (2011a) describes testing procedures for top-loading and front-loading washing machines, and for five possible drying methods: tumble dry; line dry; drip dry; screen dry and flatbed press dry. The AATCC's reference detergents are recommended for performing laundering methods to this standard: 1993 AATCC Standard Reference Detergent Powder; 2003 AATCC Standard Reference Liquid Laundry Detergent (AATCC, 2012).

These methods do not describe a procedure specifically for assessing the durability of repellent finishes. Rather, this is generally determined by individual brands within the industry. As an example, a spray test may be repeated on a fabric after 10 laundering cycles and if it is still able to achieve a minimum rating of 80 (ISO 3). This would usually be considered a 'pass' by the outdoor apparel industry (P05 Project Team, 2012).

While this is certainly an important test, there is a large variety of parameters that need to be taken into consideration and it would not be realistic to test all possible combinations. It would be most practical to select a combination of laundering conditions that is reflective of consumer behaviour however, this might also be difficult to determine particularly as laundering behaviour is likely to differ internationally (Burman, 2014).

13.4.2 Abrasion

- BS/EN/ISO 12947(1-4):1998 Textiles. Determination of the abrasion resistance of fabrics by the Martindale method (BSI, 1998).
- BS/EN/ISO 12945-2:2000 Textiles. Determination of fabric propensity to surface fuzzing and to pilling. Modified Martindale method (BSI, 2000).
- AATCC 93-2011 Abrasion Resistance of Fabrics: Accelerotor Method (AATCC, 2011b).

Abrasion can occur at various points in a garment simply through everyday wear, such as the action of sleeves of the garment rubbing against the body, or by rubbing from an external source such as a backpack, all of which may contribute to reduced repellency performance. The abrasion methods described here are not in particular reference to assessing durability of repellent finishes, but can be applied in a similar way to the laundering methods; the repellency testing is repeated after abrasion and results compared to the virgin fabric performance.

ISO standard 12947 (1-4) (BSI, 1998) outlines a method for testing abrasion resistance of a fabric against worsted wool. 38 ± 0.5 mm diameter fabric specimens are mounted in a holder with a foam backing and subjected to a specified load dependent on the intended end use of the fabric: 797 ± 5 g for work wear, upholstery and fabrics for technical use (resulting pressure of 12 kPa) or 595 ± 7 g for apparel and household textiles (pressure of 9 kPa). The fabric specimen is rubbed against the standard worsted wool abradant in translational motion, either for a specified number of rotations or until breakdown of the specimen is observed.

The test specimen is assessed at specified intervals after which it is either removed or returned to test until the next interval depending on the results of the assessment. The standard suggests assessing unfamiliar fabrics at intervals of 2000 rubs to determine a specified end point for subsequent specimens. There are three possible methods for assessment of abrasion resistance:

- determination of specimen breakdown
- determination of mass loss
- assessment of appearance change

The method used may be dependent on the fabric construction and requirements for end use. A combination of methods may also be used.

Table 13.3 Visual assessment grading for surface fuzzing and pilling (BSI, 2000)

Grade	Description
5	No change
4	Slight surface fuzzing and/or partially formed pills
3	Moderate surface fuzzing and/or moderate pilling. Pills of varying size and density partially covering the specimen surface
2	Distinct surface fuzzing and/or distinct pilling. Pills of varying size and density covering a large proportion of the specimen surface
1	Dense surface fuzzing and/or severe pilling. Pills of varying size and density covering the whole of the specimen surface

As this method of abrasion is intended to degrade the fabric itself it may not be considered suitable for testing degradation of repellent finishes, as it would become difficult to determine whether any loss in repellency was caused by removal of or damage to the finish alone or by damage to the fabric itself. Additionally, the specimen dimensions for this method are smaller than those for most repellency tests, significantly limiting the range of tests that could be performed.

The modified Martindale method (BSI, 2000) for determining fabric propensity to surface fuzzing and to pilling describes a method by which test specimens of 140 ± 5mm diameter are rubbed against specimens of the same fabric, rather than against worsted wool. This method also exerts a lower pressure on the fabrics than the regular Martindale method with the total weight of specimen holder and load at 415 ± 2 g (compared to either 797 ± 5 g or 595 ± 7 g depending on end use for the regular Martindale method).

The surface change of the fabric specimen is visually assessed at set intervals according to the grading scale shown in Table 13.3, usually up to a minimum of 2000 rubs.

The use of a fabric-to-fabric set up and lighter load make this method less abrasive than the regular Martindale method and the specimen dimensions are sufficient such that they could be subjected to the full range of repellency tests. Providing the test fabric achieved a minimum rating of grade 4 for surface change assessment, testing repellency before and after this method could be suitable; to achieve a grade 4 or above the fabric must exhibit little or no change so it would not be unreasonable to assume that any loss in repellency at these levels is the result of removal of or damage to the finish alone rather than by damage to the fabric itself.

13.4.3 Prolonged exposure to water

While abrasion can certainly be considered as a contributing factor to loss of repellency performance between washes, simply repeated wetting may also be a significant problem and so testing a fabric's resistance to prolonged exposure to water may be valuable when assessing overall repellency performance. The Bundesmann rain-shower test is the only standard method available for testing this but as discussed previously, this is not a readily available or easily performed test. However, it is also possible to test resistance to prolonged exposure to water by modifying some of the simpler water repellency test methods.

Extended spray tests. Based on:

- BS/EN/ISO 4920:2012 Textile fabrics. Determination of resistance to surface wetting (spray test) (BSI, 2012a)
- BS/EN/ISO 29865:1993 Determination of water repellency of fabrics by the Bundesmann rain-shower test (BSI, 1993)

Alterations:

- Extended spray test: increased height method—standard spray test with the distance between the spray nozzle and test specimen increased from 150 ± 2 mm to 1500 mm, as is specified for the Bundesmann rain-shower test.
- Extended spray test: continuous water flow method—standard spray test with the distance between the spray nozzle and test specimen increased to 1500 mm and using a continuous flow of water for 10 minutes, as is specified for the Bundesmann rain-shower test.

The alterations suggested here are intended to replicate some of the parameters tested for by the Bundesmann rain-shower test. By increasing the height of the spray nozzle and subjecting the fabric specimen to a continuous flow of water for a prolonged time period, these methods are more closely representative of the conditions of use for water repellent apparel.

Extended immersion test. Based on:

- BS/EN 3449:1990 Method for resistance of fabrics to water absorption (static immersion test) (BSI, 1990).

Alterations:

- Static immersion test altered to move the test specimens up and down within 250 mL of distilled water at a controlled rate of 30 rpm for 20 minutes.

This method provides a more easily repeatable procedure for exposing fabrics to water for an extended time period but does not mimic the likely conditions of use for the end product (Davies, 2014). It would be necessary to repeat a standard spray test or other repellency tests after extended immersion to assess whether there has been any noticeable effect on repellency performance.

13.4.4 Wearer trials

- BS 7754:1994 Code of practice for: Garment evaluation by wearer trials (BSI, 1994)

Wearer trials are a useful method for assessing the true performance of fabrics in garment form under conditions of real use. They can help to determine the suitability of laboratory test methods as well as allowing for subjective assessment where it is not possible to easily replicate the wearer use by any other means. The BSI code of practice (1994) suggests that there should be approximately 10–12 replications of each variation within a trial for an apparel article. While larger scale trials may be preferable in some cases, it is important to note that this will likely reduce control over the wearers and the accuracy of feedback given. Laing et al. (2008) noted that it is often difficult to predict the performance of a garment under conditions of use based on fabric performance testing, particularly due to the subjectivity of user perception.

By using wearer trials to evaluate the performance of water repellent garments, it would be possible to assess the effect of launderability, abrasion and prolonged exposure to water in combination. However, it may become difficult to determine which of these factors has had the most significant effect on performance degradation and there may be additional factors to consider that are more difficult to define. For example, contact with oils or other impurities such as from the skin, sun cream or food stuffs may affect the performance of a repellent but may not necessarily be recorded by the trial participant.

As new water repellent technologies are introduced to replace fluorochemical-based finishes, it is likely to become increasingly important to correlate the results of laboratory test methods with wearer trials to successfully differentiate between the performances of the various alternatives. This is important in determining whether new technologies still meet consumer expectations; many consumers expect their water repellent clothing to last up to ten years or more as well as expecting such items to require washing only very infrequently as this is what they have become accustomed to with fluorochemical-based finishes (Burman, 2014). It is unlikely that consumers will be accepting of inferior performance, particularly if it comes at a higher cost.

The full garment may also be assessed in the laboratory prior to wearer trials by a rain room test that involves fitting the garment to a mannequin or test subject and exposing it to a simulated rain shower. This method allows for any areas of water penetration in either the fabric or garment construction to be observed under simulated conditions of use. Water penetration may be detected by sensors fitted to the mannequin or by fitting the mannequin with an absorbent inner layer underneath the test garment so that any colour change or weight increase can be easily observed.

13.4.5 Suitability of methods for assessing durability of performance

Durability of water repellency is important for apparel applications particularly as many consumers and brands expect such products to have a useful life in excess of ten years (European Outdoor Group, 2015). Laundering is certainly an important factor on durability but it is difficult to achieve a true indication of performance in use, as there are multiple variables and differences in consumer behaviour that need to be accounted for. Most outdoor apparel brands do not specify laundering recommendations and consumers launder their water repellent garments very infrequently (European Outdoor Group, 2015) so it can be assumed that other factors are affecting repellency performance in between washes and it may be more beneficial for brands to test for durability to these other parameters such as abrasion and prolonged exposure to water. There are no current industry standards specifically for testing the durability of water repellent finishes but a number of modifications to existing methods can be considered as discussed previously. Wearer trials are perhaps the most valuable method for assessing the true performance of garment provided they are conducted in conjunction with laboratory tests and with as much control of variables as is possible.

13.5 Assessing restoration of performance

13.5.1 Laundering

Laundering can also be an important factor in maintaining good levels of repellency performance. Exposure to impurities from dirt or oils can interfere with the repellent properties of the fabric, shortening its useful lifespan (Gore et al., 2016). Removal of these impurities by laundering can assist in restoring some of the performance lost but it is essential to consider that some laundry detergents might also interfere with the repellent finish; nonbiological detergent is recommended (Berghaus, 2014).

While useful as a method for removing impurities, repeated laundering has been shown to negatively impact the performance of fluorocarbon-based repellent finishes (Wakida et al., 1993; Wakida et al., 1994; Arunyadej et al., 1998). This loss in repellency is suggested to be due to a disturbance in the orientation of the fluoroalkyl groups (Wakida et al., 1994) or adsorption of laundering detergents into the fibre (Arunyadej et al., 1998), although it is likely that a combination of these among other factors are involved. Arunyadej et al. (1998) noted that repellency was fully lost (tested as fully wettable to the aqueous liquid repellency method) after 20 wash cycles. It is unlikely that an average consumer would ever see this level of performance loss in their water repellent clothing as most surveyed by Burman (2014), expected their products to last for 10 or more years but only washed them approximately once a year. That said, the majority of retailers surveyed did not provide any recommendation for frequency of laundering (Burman, 2014) so it is possible that some consumers may experience premature loss in repellency performance simply due to unclear care instructions.

13.5.2 Application of heat

Heat application has been shown to restore some of the repellency performance lost after laundering, be this by tumble drying or ironing (Wakida et al., 1993; Wakida et al., 1994; Arunyadej et al., 1998). For fluorocarbon-based finishes, the best results were recorded after heat treatment at temperatures above 80°C (Wakida et al., 1993). Exposure to heat is suggested to reorient the fluoropolymer chains of the water repellent finish, thus restoring performance (Arunyadej et al., 1998; Fuller and Taylor, 2012). Heat application has also been shown to improve performance for a number of fluorocarbon-free repellent types (dendrimer, wax, silicone, polyurethane), with all showing acceptable levels of restoration (spray rating within the range initially recorded before prolonged exposure to water) after 30 minutes tumble drying (Davies, 2014).

Although many outdoor retailers recommend to tumble dry or iron their water repellent products using a low heat setting, exact times and temperatures are not suggested so variation in machines and consumer behaviour is likely to influence the success of performance restoration by these methods. That said, the majority of consumers avoid any application of heat altogether (Burman, 2014) and so miss out on a valuable opportunity to extend the useful lifespan of their garments.

13.5.3 Reproofing products

There are a number of products on the market that can be used to restore or replace the water repellent finish on a garment either by spray or in-wash application. There is not an industry agreed recommendation for how frequently these reproofing products need to be, or should be, applied with many outdoor retailers providing no recommendation at all (Burman, 2014). Interestingly, consumer behaviour appears to follow one of two contrasting patterns; either the consumer does not use re-proofing products at all or they re-proof after every wash (Burman, 2014). The most popular of the available products are Nikwax and Grangers (Burman, 2014) both of which utilize different technologies within their products. Nikwax products are promoted as being all water-based, fluorocarbon-free and free of volatile organic compounds. All Nikwax finishes develop their repellent properties on air drying and so do not require heat activation (Nikwax, 2016), potentially making them more accessible to a wider consumer base. Grangers offer a number of C6 fluorocarbon-based products and have recently added fluorocarbon-free technology to their range. All Grangers finishes require heat activation, which is recommended to be achieved either by tumble drying or ironing at a low heat setting (Grangers, 2016).

While it stands to reason that the addition of a re-proofing finish will restore lost water repellent performance (by essentially topping up the factory-applied finish), it is unclear to what extent performance can be restored and at what level of deterioration they may be required. Additionally, consumers may be adding reproofing agents which utilize a different repellent type to the factory applied finish (for example adding wax based reproofing products to fluorocarbon-based factory finishes) which may actually reduce rather than restore performance.

13.6 Performance comparison of available types of water repellent textile finishes

13.6.1 Initial observations

As can be expected, fluorocarbon-based finishes (long and short chain) still provide the best level of repellency performance when tested using the industry standards methods discussed in this chapter. In spite of this, many commercially available alternative repellent technologies are able to achieve the minimum pass rating as specified by the industry: spray rating 80 (ISO 3) (P05 Project Team, 2012). Modifications to the standard methods may need to be considered to provide meaningful distinction for determining which fluorocarbon-free repellents offer the best performance. For example, measuring the percentage weight increase of test specimens (dendrimer, wax, silicone and polyurethane-based repellent finishes) after spray-rating testing has been demonstrated to provide greater differentiation of results than the spray rating alone (Davies, 2014).

13.6.2 Durability

A number of studies have shown that the performance of fluorochemical-based finishes is reduced by repeated laundering (Wakida et al., 1993; Wakida et al., 1994) although repellency is not fully lost until after 20 wash cycles (Arunyadej et al., 1998) which is unlikely to be reached during the average useful life of a water repellent garment. It is not clear how fluorocarbon-free finishes compare in terms of durability to laundering although prolonged immersion in water has been demonstrated to negatively impact the repellent performance of these finishes more so than fluorochemical-based finishes (Davies, 2014) so it can be expected that this would be reflected in laundering trials also.

13.6.3 Restoration of performance

Application of heat has been shown to be a successful method of restoring water repellent performance of fluorochemical-based (Wakida et al., 1993; Wakida et al., 1994; Arunyadej et al., 1998) and fluorochemical free finishes (Davies, 2014) as shown in Fig. 13.7. The efficacy of re-proofing products is unclear as are the optimal conditions

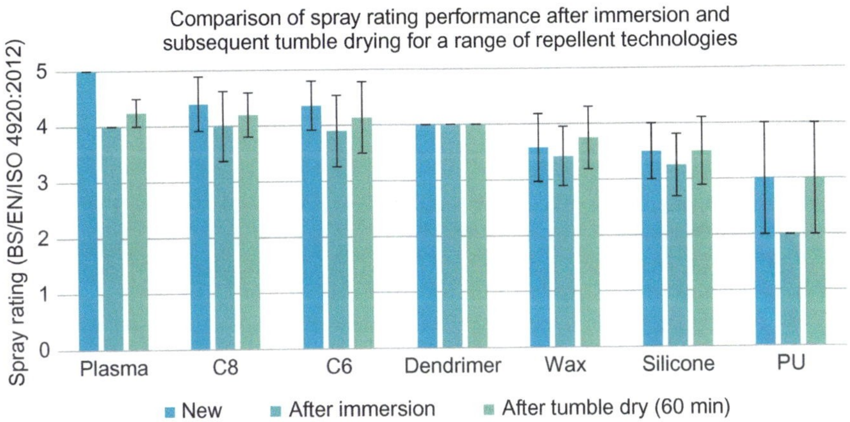

Fig. 13.7 Comparison of spray-rating performance after immersion and subsequent tumble drying for a range of repellent technologies.
Modified from Davies, A., 2014. Durable Water Repellency—Study Phase I: An Evaluation of Test Methods for Assessing Durable Water Repellent Fabrics Within the Outdoor Industry [WWW]. Available from: http://www.europeanoutdoorgroup.com/files/DWR-Study_Alice_Davies__digital_.pdf [Accessed 01 June 2017].

for their use. It should also be considered that the factory-applied water repellent type may be of a different technology to that of the re-proofing product used by the consumer, which would presumably be less effective at restoring performance than using the same technologies at both stages.

13.7 Recent developments

As new technologies are developed to replace fluorochemical-based water repellents, there is also a need for the development and use of more relevant testing methods. This is particularly true in the case of super-hydrophobicity which is often created by increasing surface roughness as the typical contact angle measurement methods are designed for perfectly smooth and homogenous surfaces. A number of more relevant methods are being explored in this area. Contact angle hysteresis accounts for the fact that wetting of a textile surface is not a static state by measuring the difference between the advancing and receding contact angles, thus providing a more accurate measurement for textile surfaces (Yuan and Lee, 2013). Sliding angle and shedding angle methods can also provide further differentiation as many super-hydrophobic finishes will display similarly high contact angles. Both methods work by measuring the minimum tilting angle of the test stage at which a liquid droplet begins to roll off the textile substrate which may be significantly influenced by the surface roughness (Zimmerman and Seeger, 2009). Water droplet bouncing has also been proposed as an alternative method for characterization of super-hydrophobic surfaces. This method measures the number of times a droplet bounces on a textile substrate, which is influenced by its surface topography. The number of bounces has been found to be linearly correlated to the contact angle (Crick and Parkin, 2011).

References

American Association of Textile Chemists and Colorists, 1997. AATCC 118-1997 Oil Repellency: Hydrocarbon Resistance Test. American Association of Textile Chemists and Colorists, Durham.

American Association of Textile Chemists and Colorists, 2001. AATCC 22-2001 Water Repellency—Spray Test. American Association of Textile Chemists and Colorists, Durham.

American Association of Textile Chemists and Colorists, 2007a. AATCC 42-2007 Water Resistance: Impact Penetration Test. American Association of Textile Chemists and Colorists, Durham.

American Association of Textile Chemists and Colorists, 2007b. AATCC 193-2007 Aqueous Liquid Repellency: Water/Alcohol Solution Resistance Test. American Association of Textile Chemists and Colorists, Durham.

American Association of Textile Chemists and Colorists, 2011a. AATCC Monograph M6 Standardization of Home Laundry Test Conditions. American Association of Textile Chemists and Colorists, Durham.

American Association of Textile Chemists and Colorists, 2011b. AATCC 93-2011 Abrasion Resistance of Fabrics: Accelerotor Method. American Association of Textile Chemists and Colorists, Durham.

American Association of Textile Chemists and Colorists, 2012. AATCC Monograph M2: 2003 AATCC Standard Reference Liquid Laundry Detergent. American Association of Textile Chemists and Colorists, Durham.

Arunyadej, S., Mitchell, R., Walton, J., Carr, C.M., 1998. An investigation into the effect of laundering on the repellency behaviour of a fluorochemical treated cotton fabric. J. Text. Inst. 89 (4), 696–702.

Berghaus (2014). Step by Step: Washing Waterproof Jackets [WWW] Available from: http://community.berghaus.com/knowledge-advice/washing-waterproof-jackets/ (Accessed 25 October 2016).

Booth, J.E., 1968. Principle of Textile Testing: An Introduction to the Physical Methods of Testing Textile Fibres, Yarns and Fabrics, third ed. Butterman Heinemann Limited, UK.

British Standards Institution, 1990. BS/EN 3449:1990 Method for Resistance of Fabrics to Water Absorption (Static Immersion Test). British Standards Institution, London.

British Standards Institution, 1992. BS/EN 20811:1992 ISO 811:1981 Textiles. Determination of Resistance to Water Penetration—Hydrostatic Pressure Test. British Standards Institution, London.

British Standards Institution, 1993. BS/EN/ISO 29865:1993 Determination of Water Repellency of Fabrics by the Bundesmann Rain-Shower Test. British Standards Institution, London.

British Standards Institution, 1994. BS 7754:1994 Code of Practice for: Garment Evaluation by Wearer Trials. British Standards Institute, London.

British Standards Institution, 1998. BS/EN/ISO 12947(1-4):1998 Textiles. Determination of the Abrasion Resistance of Fabrics by the Martindale Method. British Standards Institution, London.

British Standards Institution, 2000. BS/EN/ISO 12945-2:2000 Textiles. Determination of Fabric Propensity to Surface Fuzzing and to Pilling. Modified Martindale Method. British Standards Institution, London.

British Standards Institution, 2007. BS/EN/ISO 18695:2007 Textiles. Determination of Resistance to Water Penetration. Impact Penetration Test. British Standards Institution, London.

British Standards Institution, 2009. BS/EN/ISO 23232:2009 Aqueous Liquid Repellency. Water/Alcohol Solution Resistance Test. British Standards Institution, London.

British Standards Institution, 2010. BS/EN/ISO 14419:2010 Textiles. Oil repellency. Hydrocarbon Resistance Test. British Standards Institution, London.

British Standards Institution, 2012a. BS/EN/ISO 4920:2012 Textile Fabrics. Determination of Resistance to Surface Wetting (Spray Test). British Standards Institution, London.

British Standards Institution, 2012b. BS/EN/ISO 6330:2012 Textiles. Domestic Washing and Drying Procedures for Textile Testing. British Standards Institution, London.

Burman, G., 2014. Durable Water Repellency—Study Phase I: Expectations of Durable Water Repellent Finishes. Available from:http://www.europeanoutdoorgroup.com/files/DWR-Study_Georgina_Burman__digital_.pdf (Accessed 01 June 2017).

Crick, C.R., Parkin, I.P., 2011. Water droplet bouncing—a definition for superhydrophobic surfaces. Chem. Commun. 47, 12059–12061.

Davies, A., 2014. Durable Water Repellency—Study Phase I: An Evaluation of Test Methods for Assessing Durable Water Repellent Fabrics Within the Outdoor Industry. Available from: http://www.europeanoutdoorgroup.com/files/DWR-Study_Alice_Davies__digital_.pdf (Accessed 01 June 2017).

Fuller, M., Taylor, M., 2012. Waterproof Breathable Fabric—Explained [WWW]. Available from http://www.ukclimbing.com/articles/page.php?id=4556 (Accessed 25 October 2016).

Gore, W.L., et al., 2016. Restoring Water Repellency [WWW]. Available from https://www.gore-tex.com/en-us/support/restoring-water-repellency (Accessed 25 October 2016).

Grangers, 2016. Clothing Care [WWW]. Available from https://grangers.co.uk/clothing-care (Accessed 25 October 2016).

Holme, I., 2003. Water-repellency and waterproofing. In: Heywood, D. (Ed.), Textile Finishing. Society of Dyers and Colorists, Bradford, pp. 135–213.

Laing, R.M., Sims, S.T., Wilson, C.A., Niven, B.E., Cruthers, M., 2008. Differences in wearer response to garments for outdoor activity. Ergonomics 51 (4), 492–510.

Nikwax, 2016. Waterproofing [WWW]. Available from https://www.nikwax.com/en-gb/productselector/waterproofing.php (Accessed 25 October 2016).

P05 Project Team, 2012. Durable Water and Soil Repellent Chemistry in the Textile Industry—A Research Project. P05 Water Repellency Project, Version 1.0 (November 2012).

Rouette, H.-K., 2001. Encyclopedia of Textile Finishing. Woodhead Publishing Limited, Cambridge.

Ruckman, J., 2005. Water resistance and water vapour transfer. In: Shishoo, R. (Ed.), Textiles in Sport. Woodhead Publishing Limited, Cambridge, pp. 287–305.

Samanta, K., Basak, S.C.S., Gayatri, T., 2015. Water-free plasma processing and finishing of apparel textiles. In: Muthu, S. (Ed.), Handbook of Sustainable Apparel Production. CRC Press, Boca Raton, pp. 3–37.

Wakida, T., Li, H., Satot, Y., Kawamura, H., Ueda, M., Mizushima, H., Takekoshi, S., 1993. The effect of washing and heat treatment on the surface characteristics of fluorocarbon resin-treated polyester. J. Soc. Dyers Color. 109, 292–296.

Wakida, T., Goto, T., Li, H., Sato, T., Lee, M., Chen, J., 1994. Effect of washing and subsequent heat treatment on the water repellency of poly(ethylene terephthalate) fabric and film treated with carbon tetrafluoride and trifluoromethane low-temperature plasmas. Transaction 50 (1), 533–537.

Yuan, Y., Lee, T., 2013. Contact angle and wetting properties. In: Bracco, G., Holst, B. (Eds.), Surface Science Techniques. Springer, Heidelberg, pp. 3–29.

Zimmerman, J., Seeger, S., 2009. Water shedding angle: a new technique to evaluate the water-repellent properties of superhydrophobic surfaces. Text. Res. J. 79 (17), 1565–1570.

Sportswear

14

Zehra Evrim Kanat
Namık Kemal University, Tekirdağ, Turkey

14.1 Introduction

Obesity has become widespread in society and this creates serious health problems. With the goal of making the population leaner and fitter, governments are encouraging people to adopt more active lifestyles and play sports. Future generations will understand the importance of staying fit. Sports/fitness companies, governments, health organizations, insurance companies and community-minded groups work together to support physical activity (Buirski, 2005). 'For sports, television is the great popularizer, benefactor and dictator. It encourages the growth of organized sports at all levels; it ensures huge profits for professional owners, promoters and athletes: it has subtly changed the character of both amateur and professional sports' (Baker, 1982).

Outdoor activities allow people to feel better in a physical, mental and social sense. Also, low stress levels and high physical activity reduce the prevalence of common health problems such as high blood pressure, obesity, heart attack, cancer and mental health problems. Spending time in natural environments due to positive health, mental, social and environmental outcomes is now considered an important preventive medicine practice (Morrissey and Rossi, 2013).

Consumers expect high performance from sportswear. Developments in active sportswear fabrics are concerned with achieving high function and comfort. Sportswear demands protection from environmental conditions (rain, snow, cold, heat and UV rays), strength, drape, comfort, fit and ease of movement. Sportswear manufacturers use innovative textile technologies to meet athletic and leisure-time activities' requirements for better performance (Özdil and Anand, 2014). It is an innovative field that is investing heavily for research and development, pioneering new technologies and concepts, and improving performance and comfort. In sportswear design, it is necessary to consider both performance and aesthetic development, from second skin clothing to outerwear (Bramel, 2005).

However, for all sport clothes high functionality is not necessary. In walking or leisure wear, aesthetic demands are more important. A good water vapour permeability and water repellency, if it is used in wet conditions, could be adequate, but expectations for professional sportswear are higher. Higher water vapour and liquid moisture transmission together with properties that will enhance users' performance are desired.

In the design of waterproof and water repellent sportswear, the level of activity must be taken into consideration, along with the climatic conditions under which it

Waterproof and Water Repellent Textiles and Clothing. https://doi.org/10.1016/B978-0-08-101212-3.00013-7

will be used by the wearer. In addition to protecting the user from the weather conditions, sportswear also provides comfort by allowing sweat that occurs on the body to be removed. This chapter gives an overview of waterproof and water repellent sportswear.

14.2 Sportswear and its functional requirements

Today, sports clothes and leisure wear are overlapping each other. It is possible to classify sportswear as performance wear, outdoor wear and sports-inspired wear (Shishoo, 2015).

Textile materials are used in all sports as clothing, and in many games as equipment and footwear. Examples of sportswear include swimwear, water sports clothing, fitness apparel, skiwear, gym wear, soccer gear, cricket clothing, riding gear, rugby gear, winter active clothing, climbing and hiking clothes, cycling clothes, motor racing clothing, tennis clothes, athletic clothing (sweatshirts, shorts, jackets and trousers) and outdoor leisure wear. Fig. 14.1 shows some types of sportswear.

Athletic shoes, football boots, gym shoes, tennis shoes and walking boots are examples of sports footwear (Shishoo, 2005). While designing sports footwear, comfort, performance, protection, support and shock absorption should be considered. Footwear needs to be light and supportive, and should not become heavy in wet conditions. Different sports have different requirements for footwear. For example, running shoes need to be flexible, but footwear for outdoor activities should be strong, thick and rigid (Nebo, 2005).

Similarly, for different sports activities, expectations are different for clothing. When designing sportswear, activity level, duration of use and place of use (e.g. stadium, triathlon, a day trekking or polar expedition, etc.) should be considered. Also there are differences between training, competition and post-exercise clothes (McCann, 2005). Functional requirements of some types of sports are given in Table 14.1.

In addition, requirements of sportswear vary depending on weather conditions. For instance, cold weather sports clothing requires good thermal insulation as well as

Fig. 14.1 (1) Tennis wear (Women's Tennis Blog, 2017) (2) Gym wear (Nike, 2017) (3) Running wear (Footwear News, 2017) (4) Swimwear (Speedo, 2017) (5) Snowboard gear (Snowboarding.transworld, 2017) (6) Motor sport clothing (Kruse Motor Sport, 2017) (7) Rugby body armour (Gilbert Rugby, 2017).

Table 14.1 **Main requirements of some types of sports (Özdil and Anand, 2014; Gupta, 2011)**

Type of sport	Main requirement
Aerobic/gym	Stretchability, opacity
Ball sports	Sweat absorption and transfer, fast drying, cooling
Athletics	Sweat absorption and transfer, fast drying, cooling, stretchability
Water sports	Low water and air resistance, stretchability, opacity
Snow sports	Water proofing, water vapour permeability, quick drying and maintaining body core temperature
Motor sports	High tenacity, thermal retention, low air resistance
Contact sports	Lightweight, flexible, shock absorption and protection support

protection against adverse weather conditions. In dry conditions, windproof could be adequate, but in rainy conditions a waterproof outer shell is desired (Rossi, 2015).

On the other hand, low thermal insulation values are expected from garments that are worn in hot weather conditions, in addition to high air permeability to minimize heat stress. In hot and rainy weather conditions, water repellency is better than the blocking of air permeability.

14.3 The growing market

In recent years, there has been a significant consumption of sportswear as interest in indoor and outdoor sports and outdoor leisure pursuits has increased. More leisure time, greater awareness of health, increased indoor and outdoor sports facilities, and the introduction of well-designed functional sportswear have increased interest in playing sports (Shishoo, 2015). Since the 1980s, the great increase in participation in active sports all over the world with performance sportswear has also had an important influence on leisure wear (McCann, 2015). In the twentieth century clothing has become an important issue for the general consumer, the active athlete and those who exercise in leisure time for fitness (Uttam, 2013).

According to a report published by Euromonitor International, the global sportswear market was valued at $78 billion in 2016. Sports-inspired apparel sales were expected to grow 6% in 2017 (Kissane, 2017).

The survey, conducted by the Physical Activity Council shows participation in sports, fitness and related physical activities increased in 2015 in the United States. There were 212.6 million 'actives' taking part in a wide range of sports and fitness activities in 2015, a slight increase from 209.3 million actives in 2014. Overall inactivity dropped in 2015, going from 82.7 million people to 81.6 million. This shows that 1.2 million people who were inactive in 2014 participated in some sort of fitness activity in 2015. The same survey investigated spending trends and showed over 40% of people purchased sports/recreational footwear or clothing in 2015 (Physical Activity Council, 2016).

Besides this, the number of participants in recreational outdoor sports is growing. According to an Outdoor Foundation report nearly half of all Americans (48.4%) participated in at least one outdoor activity in 2014. This equates to 141.4 million participants, who went on a collective 11.8 billion outdoor outings. Numbers for outdoor participation in the United States are shown in Fig. 14.2. The same research shows people prefer outdoor activities such as jogging, hiking, cycling, camping and fishing (Outdoor Foundation, 2016).

Surveys conducted by Sport England also show participation in sporting activity in the United Kingdom has increased in recent years. Fig. 14.3 shows the number of adults participating in sporting activities for a minimum of 30 min at moderate intensity at least once a month has increased by 1.7 million over the last 10 years (Sport England, 2016).

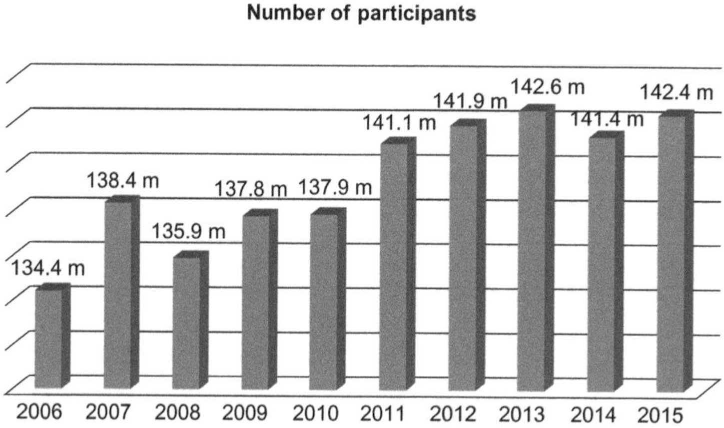

Fig. 14.2 Numbers of outdoor participation in the United States.

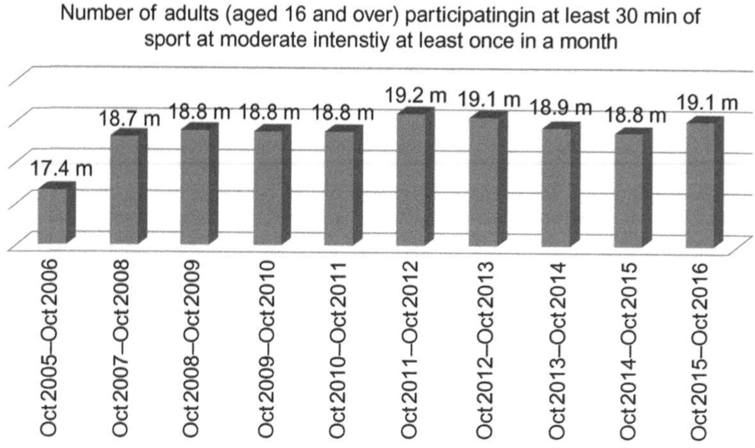

Fig. 14.3 Number of participants in sports activity in the United Kingdom.

With increased participation in outdoor sports, the use of waterproof and water repellent sportswear is growing.

14.4 Waterproof breathable and water repellent sportswear

In addition to protection from weather conditions like rain and wind, for outdoor sports, breathability is expected from sportswear. That's why waterproof breathable or water repellent fabrics are preferred.

14.4.1 Waterproof breathable sportswear

Waterproof breathable fabrics are designed to protect the user from environmental factors such as wind and rain, as well as heat loss from the body. They prevent the penetration and absorption of liquid water from outside while at the same time allowing water vapour to be transmitted to the outside of the fabric (Mukhopadhyay and Midha, 2008a; Chaudhari et al., 2016). Breathability is insignificant if there is not high waterproofness. The expected waterproof values are the initial hydrostatic head of 500 cm of water (50 kPa) for high-quality products to 130 cm of water (13 kPa) for lower-grade products (Mukhopadhyay and Midha, 2008a).

There are different breathable waterproof or water repellent textile products on the market. They can be classified as closely-woven fabrics, microporous membranes and coatings, hydrophilic membranes and coatings, combination microporous and hydrophilic membranes and coatings, smart breathable fabrics and fabrics based on biomimetics (Mukhopadhyay and Midha, 2008a; Özen, 2012; Chaudhari et al., 2016). These waterproof breathable fabrics are also used for sportswear production. Table 14.2 shows some commercial waterproof breathable applications used in sportswear and their categories.

One closely woven breathable fabric, Ventile used since WW2, is made from long staple cotton fibres. When the fabric gets wet, the cotton fibres swell and the size of the pores between the yarns decreases (10–3 μm). Ventile is now considered a 'smart' textile because its waterproof property improves its characteristics; air permeability and breathability of the fabric are also better (Morrissey and Rossi, 2013; Chaudhari et al., 2016). Water resistance properties can be enhanced with water repellency application. Ventile fabrics are now used for outdoor sportswear such as mountaineering, trekking and nature watching (Ventile, 2017).

It is also possible to produce closely woven breathable fabrics with microfilament synthetic yarns, which are individual filaments less than 10 μm in diameter. These fabrics are produced in a non-porous structure weight between 170 and 295 g/m^2 (Morrissey and Rossi, 2013; Chaudhari et al., 2016). The yarn is woven into various dense fabric constructions like taffeta, twill and oxford weave, and given a water repellent finish. This type of weaving results in a wind-proof fabric with an excellent water vapour permeability compared to laminates and coatings but with inferior water proofness (Mukhopadhyay and Midha, 2008a).

Table 14.2 **Waterproof breathable applications**

Company	Application name	Type of waterproofness	Usage area
Stotz & Co. AG	Eta Proof	Densely woven cotton fabric	Skiwear
Teijin	Microft	Densely woven microfibre fabric	Sportswear
Gore	Gore-tex	Microporous membrane	Outer wear, foot wear
Toray	Torex Entrant	Microporous coating	Skiing, snowboarding, athletics
Finetex, ınc.	Polartec_NeoShell	Electrospun nanoweb membrane	Sportswear
Teijin	Eco Storm	Hydrophilic coating	Sportswear
Sympatex Technologies	SympaTex	Hydrophilic membrane	Snowwear, Outerwear, running, biking
Toray	Torex Dermizax	Hydrophilic membrane	Mountaineering, outdoor and winter sports
Gore	Gore-tex (2nd generation)	Bi component membrane	Outerwear, footwear
Teijin	Super Microft	Biomimetic	Sportswear
X-Technologies	X- Bionic	Biomimetic	Bike, running
Akzo Nobel	Stomatex	Biomimetic	Rugby, hockey

The principles of microporous membranes and coatings are similar. Water droplets cannot penetrate micropores because the pores are smaller than a rain drop, but water vapour can pass through because the pores are larger than vapour molecules (Mukhopadhyay and Midha, 2008a; Özdil and Anand, 2014; Chaudhari et al., 2016). Diameter of a water vapour molecule is 40×10^{-6}μm and the diameter of various drop types can be seen in Table 14.3 (Holmes, 2000b).

A waterproof, breathable, microporous membrane introduced by W.L. Gore in 1976 has an important place in the performance sportswear market (McCann, 2005; Buirski, 2005; Shishoo, 2015). It is first-generation Goretex (Bartels, 2005). This membrane has approximately 1.4 billion pores per square cm, with each pore 20,000 times smaller than a drop of water (Holmes, 2000b).

In microporous membranes and coatings, the average pore size is in the range of 0.1–50 μm. The water vapour transmission rate in these constructions is also related to the porosity, thickness and pore size. In general, the decrease in pore size and thickness increase the water vapour transmission. It is stated the waterproof properties of the fabrics are sufficient when the maximum pore size of the barrier layer on the outer surface is 2–3 μm. Microporous constructions are also air permeable and

Table 14.3 **Diameters of some types of drop (Holmes, 2000b)**

Type of drop	Diameter (mm)
Drizzle	0.02
Light rain	0.045
Moderate rain	0.10
Heavy rain	0.15
Excessive rain	0.21
Cloud burst	0.30

the water vapour permeability properties are physiologically acceptable (Mukhopadhyay and Midha, 2008a). The most widely used are: polyurethanes (PUs), poly-tetrafluoroethylene (PTFE), acrylics, polyamino acids and poly-vinylidene fluoride (PVDF) (Mukhopadhyay and Midha, 2008a; Morrissey and Rossi, 2013; Chaudhari et al., 2016).

In addition, a new microporous membrane or coating category has emerged. It is produced from electrospun nanofibres (Morrissey and Rossi, 2013). Electrospinning makes it possible to produce waterproof breathable sportswear with a pore structure that provides the desired barrier and comfort properties, with membrane-like webs providing very fine fibres, high specific surface area, flexibility and lightness (Yoon and Lee, 2011; Harifi and Montazer, 2017). Lee and Obendorf (2009) and Schreuder-Gibson et al. (2002) reported waterproof breathable nanoweb textiles could be used in sportswear and leisure clothing. Multi-layered fabric systems have been made from elecrospun nano webs produced with different composite structures, substrate fabrics and nanofibre web densities in different levels for outdoor sportswear. High resistance to water penetration with high water vapour permeability and air permeability was obtained (Lee and Obendorf, 2007; Harifi and Montazer, 2017). It was determined nanoweb laminate had a higher water vapour transmission rate but lower water resistance than polytetrafluoroethylene laminate. Results from water penetration tests, however, suggest water resistance may be sufficient to prevent wetting by rain (Ahn et al., 2010). Yoon and Lee (2011)'s findings implied nanofibre webs should be incorporated into layered composite structures to achieve a combination of high barrier and comfort performance. Also, in cold weather clothing, Al-deposited electrospun nanowebs can be used for improving thermal comfort properties. They are still vapour permeable and satisfy waterproofing standards (Kim and Park, 2013).

It is commercially available as Polartec_NeoShell in the electrospun nanowebs market. Finetex manufactures this using a patented procedure (Morrissey and Rossi, 2013).

A nano-membrance consisting of nano-composite TiO_2 and ZnO material, fluoronates surfactant and perfluoroalkyl is used in water vapour single-directed nano-functional shoes (LSMZ Series), having waterproof, permeability and antibacterial properties. This nano-membrane has lotus effect (superhydrophobicity) and gas can easily pass through it (Chunyan et al., 2011).

Hydrophilic membranes are obtained through chemically modifying polyester or polyurethane by the forming of an amorphous region of polyethylene oxide in the main polymer system. Water vapour molecules can pass through these amorphous regions, because they behave like intermolecular pores. However, penetration of water is prevented by the solid structure of the membrane. Hydrophilic coatings have the same mechanism as hydrophilic membranes. The water vapour is transported through these structures by adsorption–diffusion and desorption mechanisms (Chaudhari et al., 2016). Sympatex is the best known example of this type of membrane (Bartels, 2005). In Fig. 14.4 microporous and hydrophilic transfer mechanisms are shown.

Hydrophilic WBFs are non-porous (in the conventional sense) and impermeable to air (windproof) (Morrissey and Rossi, 2013). The water vapour transmission in the hydrophilic system is based on the chemical chain reaction between the moisture molecule and the non-porous film. In hydrophilic films and coatings, gas or vapour from the highly concentrated surface is absorbed. It is diffused through the film with a difference in concentration between the two surfaces. Then it is desorbed at the low-concentration surface. In hydrophilic polymers, water vapour permeability is inversely proportional to thickness, so for hydrophilic water vapour permeable raincoats, thin polymer coatings with good physical properties at low application weights are preferred. Development of breathable fabric with good durability has been reported consisting of lightweight (15–50 g/m^2) coatings of silicone rubbers and some of their blends with polyurethane (PU) and acrylic formulation. The use of polymer blend incorporating hydrophilic polymers or pigments, e.g. mixtures of polyethyl acrylate and polyvinyl alcohol has also been advocated (Mukhopadhyay and Midha, 2008a).

Jassal et al. (2004) studied the performance of environmentally-friendly water-based polyurethane dispersions (PUD) for waterproof breathable coating. The study showed when the hydrophilic component increases, water vapour penetration increases and water resistance to water penetration decreases in the coated samples.

Fabrics may also be coated with copolymers having both hydrophilic and hydrophobic segments. In these coatings, the hydrophobic part provides waterproofness, and the hydrophilic part provides water vapour permeability (Mukhopadhyay and Midha, 2008a; Morrissey and Rossi, 2013). In the bi-component WBFs, the previously mentioned microporous and hydrophilic fabrics are laminated to atightly woven

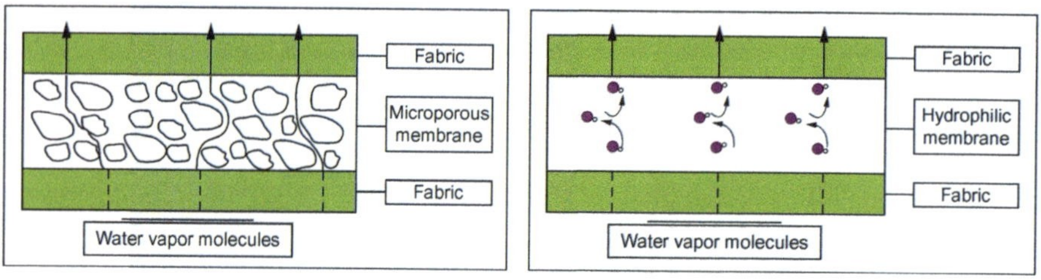

Fig. 14.4 Schematic diagram of microporous and hydrophilic transfer mechanism (Arkema, 2017).

face fabric. The best-known bi-component WBF example is second generation GORE-TEX. In bi-component WBFs, the hydrophilic coating partially fills the microporous structure (Morrissey and Rossi, 2013).

The main problem with use of WBFs in cold and rainy weather is the formation of condensation on the inner surfaces. Rossi et al. (2004) found condensation accumulation in rain, and at low temperatures, is largest in microporous WBFs, followed by hydrophilic and bi-component WBFs. Whereas microporous membrane becomes impermeable due to physical blocking of pores by rain, solid hydrophilic and bi-component WBFs continue breathing, provided there is a favourable water vapour pressure gradient (Morrissey and Rossi, 2013). Ren and Ruckman's (2003) study showed under isothermal conditions the water vapour transfer rate of polyurethane-laminated fabrics was greater, while under non-isothermal conditions the water vapour transfer rate of hydrophilic laminated fabrics was greater.

In bicomponent WBFs, water vapour permeability and air permeability decrease, and water resistance increases with the amount of coating substance (Jeong and An, 2001). The study of Bakshi (2015) showed the optimum combinations yielding the best performance in terms of waterproof properties and breathability are samples with 80%–20% PU-Si and 85%–15% PU-Si. He concluded these combinations can be used to create active wear worn in harsh conditions as well as apparel for regular wear.

It has been found that smart breathable fabrics and biomimetic-based fabrics can be highly effective in moisture transmission (Mukhopadhyay and Midha, 2008a). One of the most important developments in this area is shape-memory, polymer-based, waterproof, breathable fabric (Ding et al., 2004). This fabric reduces steam and heat transfer at low temperatures and prevents heat loss from the body while at higher temperatures it exhibits better water vapour permeability than ordinary waterproof breathable fabrics (Mukhopadhyay and Midha, 2008a).

The suitability of using Phase Change Material (PCM) in breathable fabrics was noted in the literature (Chung and Cho, 2004; Li and Zhu, 2004; Choi et al., 2004; Ghali et al., 2004; Shim et al., 2001). The use of this material does not directly affect the water vapour transmission property. However, the comfort characteristics of vapour-permeable, water repellent fabrics treated with microcapsules containing PCM are better due to improving heat and moisture management properties when worn next to the skin (Chung and Cho, 2004).

Biomimetic-based fabrics are also used for WBFs. Akzo Nobel marketed a product under the trade name of Stomatex. Stomatex is generally made from thermo-insulating, closed-cell foam materials such as Neoprene, which is a high-performance fabric made from non-porous polyester membrane that is weatherproof and breathable. Stomatex uses the concept of transpiration by leaves to provide a controlled release of water vapour to provide more comfortable wear characteristics than standard closed cell neoprene (Holmes, 2000b; Mukhopadhyay and Midha, 2016). X-BIONIC in Switzerland has marketed a unique Symbionic Membrane based on bionic research and inspired by amphibian skin. They transport both the water vapour and droplets of sweat. The ultra-fine hairs absorb it and transport it to the outside air, where the moisture evaporates (Shishoo, 2015).

14.4.2 Water repellency

The other option for protecting from weather conditions is water repellency. Water repellent fabrics, unlike waterproof fabrics, have open pores and are therefore permeable to air and water vapour. However, at high hydrostatic pressure, liquid water can pass through these fabrics. A hydrophobic material is applied to the fibre surface to render a fabric water repellent (Özen, 2012). Because gaps in the fabric weave are not covered during the process, the fabric remains porous allowing air and water vapour to pass through. Also, the fabric handle does not change as it does with waterproof coatings. However, in heavy rain conditions, water can penetrate the fabric. In addition, the opening and closing of gaps in the fabric during wear can allow water to permeate (Malik and Sinha, 2012).

According to a study by Burman (2014), end-users do not wash outdoor water repellent jackets often. When they do wash a jacket, most do not follow manufacturer's recommendations for water temperature or detergent. They do not remove dirt and oil because the washing temperature is too low. Inappropriate detergent can fill fabric pores reducing water repellency.

Also water repellency can be use with WBFs. Microporous WBFs can become physically blocked by rain, leading them to be impermeable. When a fabric wets out, the temperature difference between its two faces decreases causing condensation. Water repellency is applied to a waterproof breathable shell layer to reduce condensation accumulation and prevent the microporous structure from being blocked by rain (Morrissey and Rossi, 2013).

Repellency is imparted by a finishing treatment that reduces free energy in the fibre surface. If the adhesive interaction between the fibre and the droplets of liquid falling onto the fibre is greater than the cohesive interactions within the fluid itself, it spreads to the dripping surface. Otherwise the drop remains on the surface without spreading. Low interacting surfaces with liquids are called low-energy surfaces (Sayed and Dabhi, 2014). As the surface energy reduces, repellency increases. For example, surface tension of water is 72 mN/m and *n*-Octane oil is 22 mN/m. If a textile material is impregnated with fluoromethane CF_2 with a surface energy of 18 mN/m, it will not get wet with water and oil. This phenomena is called 'water and oil repellency'. The surface energies and tensions of some liquids and moieties are given in Table 14.4 (Posner, 2012).

The oldest water and oil repellents are simple paraffin and wax coatings, and have low resistance to washing (Åkerblom and Göranzon, 2013). Perfluorochemicals (PFCs) are the only chemicals that can repel water, oil and other liquids that cause stains (Sayed and Dabhi, 2014). The most important features of PFCs are low surface energy, and a decrease in surface energy as the carbon chain grows (up to 12) (Åkerblom and Göranzon, 2013).

Typical fluorocarbons are based on C8 chemistry, but the release of toxic perfluorooctanic acid (PFOA) and perfluorooctane sulfonate (PFOS) is also associated with C8-based fluorocarbons (Mohsin et al., 2016). Fluorocarbons may have direct or indirect environmental effects (Dimitrov et al., 2004). Studies have reported PFOA and PFOS exposure cause low fertility, cancer and damage to the immune system in children (Mohsin et al., 2016). C6 fluorocarbon applications are popular today

Table 14.4 Surface tensions and energies of different moieties and liquids (Posner, 2012)

Moiety/liquids	Surface energy (γ_c)/Surface tension (Υ_L) (mN/m)
–CF_3–	6
–CF_2H–	15
–CF_2–	18
–CH_3–	22
–CH_2–	31
–CH_2CHCL–	39
–Polyester	42
–Polyamide	46
–Cotton	44
Water	72
n-Octane	22
Olive oil	32

because they are considered less dangerous than C8 (Åkerblom and Göranzon, 2013). However, because C6 fluorocarbons have chain lengths lower than C8, less effective oil and water repellency properties can be gained (Mohsin et al., 2016). Because of the hazardous effects of fluorocarbons, many brands eliminated PFC-based durable water repellents (DWRs). They replaced them with alternative technologies such as wax based, or less hazardous C-6. Some brands maintain studies for PFOA-free and PFC-free applications (Davies, 2014).

Mohsin et al. (2016) showed the combination of acrylate-based fluorocarbon (oleophobol) and polyurethane-based fluorocarbon (phobol) exhibited a much better oil and water repellency rating than any of the fluorocarbons used alone.

It is possible to make fabric water repellent and flexible with a silicon coating that can be applied to the fibre surface or the fabric surface. With the use of concentrated silicone coatings, abrasion resistance and water repellent applications are very useful for outdoor activities. Washable or dry-cleanable fabrics can be obtained with very thin coatings and the coating is generally permanent. Silicone-finished fabrics do not deform, shrink or crease after repeated washings. A blend of polyamide with elastane can result in a fabric that is comfortable and, in a blend with Teflon, one that is water repellent and stain resistant (Özdil and Anand, 2014).

Producing water- and soil-repellent fabrics with the lotus-leaf effect is a significant and commercially successful biomimetic application. Water droplets roll like mercury on the lotus leaf, which is microscopically rough and covered with a wax-like low-surface-tension substance. In this process, the surface structure of a lotus leaf is imitated. The air trapped in the dents when the water drops to the surface forms a boundary with water and the drop is easily rolled off the textile surface (Shishoo, 2015). Durability in textile applications need further evaluation.

The surface of a characteristic wax microstructure can have a contact angle of up to 170 degrees, which is much higher than the smooth fluorochemical-based coatings.

This super hydrophobic effect allows the removal of any or all soil from the surface quickly and efficiently, which is known as a self-cleaning effect (Holme, 2007). Teijin Co., Japan, developed a water repellent fabric named Super-Microft that emulates the structure of lotus leaves by holding a layer of air on its surface with its small protrusions (Mukhopadhyay and Midha, 2008a). Scholler's water repellent and easy-care finish NanoSphere, Nano- Tex's Aquapel technology and LLC's Nano-Care and Nano- Pel treatments are examples of finishes inspired by the lotus-leaf effect (Holme, 2007). Nanotex and Nanosphere are examples of eco-friendly water repellent treatments. They are Fluorocarbon-free/PFOA-free.

Kwon et al. (2017) analysed the moisture management properties of the developed superhydrophobic lyocell fabric for sweat absorbency capacity, wetting time, spreading speed, wicking rate and drying rate. Their results showed superhydrophobic-treated lyocell maintained 80% of the moisture absorbency with faster drying.

With the spread of environmental and sustainability studies, people are increasingly concerned about the environment. Because the harm various chemicals cause is being publicized, consumers tend to pay more for products that cause less harm to humans and the environment but do not want to pay more for an inferior product.

14.5 Comfort in sportswear

In active and endurance sports, comfort is very important. Active garments used in the outdoors should keep the wearer's thermal balance and protect from environmental factors such as wind, sun, rain and snow. The increased metabolic rate caused by activity increases body heat and perspiration. Sweat evaporation through fabric is possible with good moisture management (Anbumani and Babu, 2008; Shishoo, 2005, 2015). Clothing comfort is a factor that affects the performance and efficiency of athletes (Bartels, 2005).

Slater (1985) defined comfort as, 'a pleasant state of physiological, psychological, and physical harmony between human being and environment'. The most commonly used definition for comfort is 'freedom from pain and discomfort' (Hatch, 1993). Along with improvements in textile technology, clothing comfort has become an important feature demanded by consumers.

One of the basic functions of clothing is to help the human body maintain its thermal balance and thermal comfort. Clothing is necessary to protect the body against climatic influence and to aid the body's thermal comfort functions under different environmental conditions and physical activities. A combination of clothing, climate and physical activity affects the thermal comfort of a human being (Li, 2001).

Clothing comfort is a complex phenomenon and it can be divided into four main aspects: thermophysiological wear comfort; skin sensorial wear comfort; ergonomic wear comfort; psychological wear comfort (Li, 2001; Bartels, 2005; Anbumani and Babu, 2008; Özdil and Anand, 2014).

The essential components of comfort are as follows (Li, 2001):

- Comfort is related to subjective perception of various sensations.
- Comfort involves many aspects of human senses, such as visual (aesthetic comfort), thermal (cold and warm), pain (prickle and itch) and touch (smooth, rough, soft and stiff).

- The subjective perceptions involve psychological processes in which all relevant sensory perceptions are formulated, weighed, combined and evaluated against past experiences and present desires to form an overall assessment of comfort status.
- Body clothing interactions (both thermal and mechanical) play important roles in determining the comfort of the wearer.
- External environments (physical, social and cultural) have great impact on the comfort status of the wearer.

The human body tries to maintain an operating temperature of 37°C under changing conditions (Holmes, 2000a; Özdil and Anand, 2014). The sum of metabolic heat produced and heat received from external sources must be equal to the heat lost from the body. Otherwise, if the sum of heat produced plus received heat is higher than the heat lost, the body heat will rise and vice versa. Both situations create various health risks. In hot climates it is a problem to remove heat from the body, while in cold climates it is a challenge to retain heat (Malik and Sinha, 2012).

Clothing systems contribute to comfort by regulating heat transfer between the body and the environment (Morrissey and Rossi, 2013). The selection of fabric in the design of outdoor clothes is very important. Thermal insulation of the selected fabric influences heat flow from the body to the environment or from the environment to the body (McCann, 2015). The heat and moisture exchange between the human body and the clothing system and between the clothing system and the environment is very important in the dynamic thermal comfort of clothing (Wang et al., 2003). Heat generated through exercise is removed from the body by radiation, convection, conduction and evaporation. Conduction is the transfer of heat between objects that are in contact, while heat is transferred by radiation through space. In convection, movement of the air next to skin transfers heat. When the ambient temperature is higher than the body temperature, the body sweats to cool itself and provides cooling by evaporation (McCann, 2015; Rossi, 2015; Morrissey and Rossi, 2013: Mukhopadhyay and Midha, 2008a).

As a consequence, heat balance between the body and environment can be expressed as:

$$M - W = E + R + C + K + S$$

where M is the metabolic rate of the body in W or W/m^2; W is the mechanical work; E is heat transfer by evaporation; R is heat transfer by radiation; C is heat transfer by convection; K is heat transfer by conduction, and S is heat storage. In this equation heat storage (S) should ideally equal zero, which means heat production is equal to heat loss (Rossi, 2015).

In textiles, liquid and moisture transfer mechanisms are provided by vapour diffusion (in void space), and moisture sorption by the fibres' evaporation and capillary effects. Vapour passes through the textile depending on the water vapour concentration gradient, generally from the inside to the outside. The structure and chemical properties of the fibres affect the water absorption properties. Liquid moisture transmission through the surface of the fibres is determined by surface tension and effective capillary pore distributions. The heat transfer process in textiles depends on moisture transfer processes by phase changes such as moisture absorption/desorption and evaporation/condensation (Wang et al., 2003).

To maintain the thermal comfort capacities of fabric, it is expected to have the following properties: (Malik and Sinha, 2012):

- Insulation
- Wind proofing
- Moisture vapour permeability
- Waterproofing

The garment may be wetted from the inside by perspiration or from outside by, for example, rain. Because the thermal conductivity coefficient of water is much higher than that of air, thermal insulation of the fabric falls when it gets wet. Increased body heat in hot environments and during exercise is reduced by sweat evaporation. In some cases the rate of sweat evaporation becomes lower than the rate of sweat that occurs. In this case water vapour can accumulate inside the inner garment of the wearer, especially in a cold and wet climate. The insulation value of the garment falls and it can lead to a chill after exercise or, in extreme cases, hypothermia (Anbumani and Babu, 2008; Holmes, 2000a: Jeong and An, 2001; Özdil and Anand, 2014; Rossi, 2015; Mukhopadhyay and Midha, 2008a). The garment also gets heavier due to accumulation of water.

If perspiration on the body cannot evaporate and the insulation value of the clothing does not reduce, the body continues to be protected from cold and hyperthermia can result (Holmes, 2000a). To prevent these situations the garment underlayer should wick away humidity to the outer layers. The outer layer of the garment should breathe to prevent condensation of sweat.

Climatic conditions have a significant effect on water vapour permeability and thermal properties of fabrics. For example, wind increases vapour permeability as a result of increased vapour pressure differences across the fabric from one side to the other. However, under windy and rainy conditions, water vapour transfer rates drop (Ruckman, 1997). Wind also increases convective heat loss around the body (Bougourd and McCann, 2009).

To evaluate the performance of waterproof breathable fabrics, resistance to penetration and absorption of liquid water, wind resistance and water vapour permeability are measured. Although literature suggests some vapour permeability values for breathable fabric, it is known that water vapour permeability changes too much according to the standards used, relative humidity and temperature during measurement. Especially when the performance of sportswear is being evaluated, actual weather conditions are simulated (Holmes, 2000a,b). Also, when assessing waterproof breathable sportswear, multilayer garment systems should be considered (Morrissey and Rossi, 2013).

Water vapour resistance (R_{et}) values can be used to evaluate breathable fabric performance. Lower R_{et} value expresses less resistance to moisture transfer, therefore higher breathability. Bartels's studies show waterproof breathable fabrics could achieve R_{et} values of 7–13. These values are evaluated as a good rating (Bartels, 2005).

Waterproof, windproof and breathable fabrics provide good protection in a cold environment (Rossi, 2015). Garments that are both waterproof and breathable provide the user with better thermal comfort. Consumers attach importance to comfort in performance apparel and are willing to maintain their comfort by paying more for waterproof and breathable garments (Mukhopadhyay and Midha, 2008a).

14.6 Layering system and soft shell

The use of multi-layer fabrics produced using different warp and weft knitting technologies in active sportswear is widespread, and these multi-layer systems provide better thermophysiological comfort than single layer fabrics (Özdil and Anand, 2014). Sportswear is designed as a multi-layer garment system based on an onion-skin principle, i.e., different layers provide variable thermal insulation. Thus the garment can be adapted to changing situations by wearing the layer designed to protect against wind or wet weather (Anbumani and Babu, 2008).

In this multilayer system each layer has a separate task. The first layer, which is worn next to the skin, wicks body moisture to the outer layers by transplanar wicking and spreads the moisture over a large area, thus evaporative cooling and faster drying properties are improved (Bramel, 2005; Morrissey and Rossi, 2013; Özdil and Anand, 2014; Rossi, 2015). By changing the fibre cross-section, blending hydrophilic and hydrophobic fibres, combining hydrophilic and hydrophobic yarns in double layer fabrics, or changing the surface hydrophilicity of hydrophobic fibres with chemical or plasma treatment, the wicking properties of textiles can be enhanced (Morrissey and Rossi, 2013).

Coolmax by Dupont, Coolpass by Hengligroup and Triactor by Toyoba Co Ltd. are examples of modified cross-section polyester fibres. Dri-Release is a blend yarn consisting of 85%–90% hydrophilic polyester fibre and 10%–15% hydrophilic cotton fibre (Özdil and Anand, 2014; Chaudhari et al., 2016).

Fig. 14.5 shows a two-layer fabric structure combining hydrophilic and hydrophobic yarns. In this fabric a synthetic material such as polyester, nylon, acrylic or polypropylene is used on the absorption/wicking layer for good moisture transfer properties and good capillary action. On the absorption/evaporation layer, material with good absorption properties such as cotton, wool or viscose rayon is preferred.

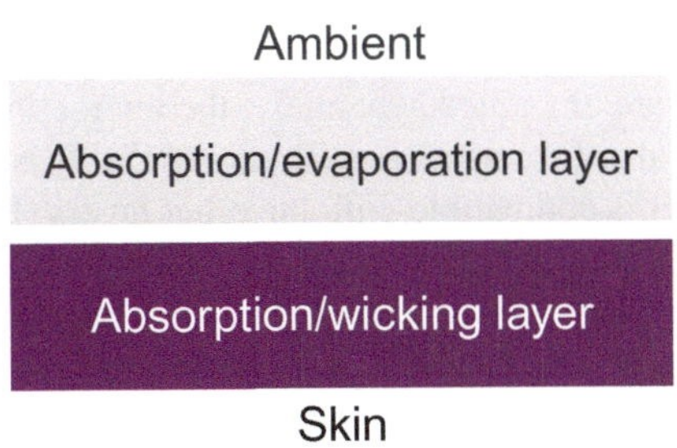

Fig. 14.5 Two-layer fabric structure.

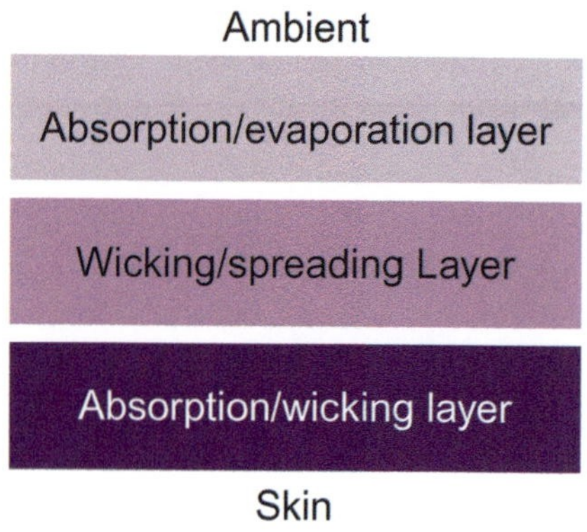

Fig. 14.6 Three-layer fabric structure.

For three-layer structures in which microfibre yarns are used, capillary action and thermophysiological regulation improves (Fig. 14.6) (Özdil and Anand, 2014).

Some researchers claim using open structure fabrics (e.g. fishnet) as the base layer has advantages in drying time and transverse wicking (Morrissey and Rossi, 2013)

The second layer in particular provides thermal insulation and also transports body moisture to the outer layer. In the second layer, synthetic- or natural-fibre knitted-fleece structures are generally preferred. Also, sandwiched fabrics created with tightly woven fabrics and synthetic or natural nonwoven battings or down fibres can be used. New generation sportswear is focused on reducing bulk without reducing insulation to provide the best warmth–to-weight ratio (Bramel, 2005; Morrissey and Rossi, 2013; Rossi, 2015).

The task of the outer shell, or protective layer, is to provide a balance between windproofness and waterproofness versus breathability at the specified range of activities (McCann, 2005). The shell layer should prevent wind and rain from entering and at the same time protect the clothing system mechanically. It should also provide comfort by allowing the water vapour to be transmitted from the clothing microclimate to the environment. The shell layer could be any type of waterproof breathable fabric as mentioned in Section 14.4 (Morrissey and Rossi, 2013). The different types of laminated systems are shown in Fig. 14.7.

Fig. 14.8 shows construction of a three-layered sports garment. While the characteristics of each layer are examined separately, the interaction of the different layers with each other must also be considered (Rossi, 2015). It is very important that the fibre used in each layer is compatible with the other layers (Buirski, 2005). The temperature distribution and the saturation pressures in the garment affect its overall insulation and the permeability of all layers. For this reason, liquid evaporation or condensation in any layer, and heat and mass transfer within each layer affect neighbouring layers. Thermal insulation and moisture management properties of garments must be evaluated not only for individual layers, but also for the temperature and relative humidity conditions the garment system can be used in Rossi (2015).

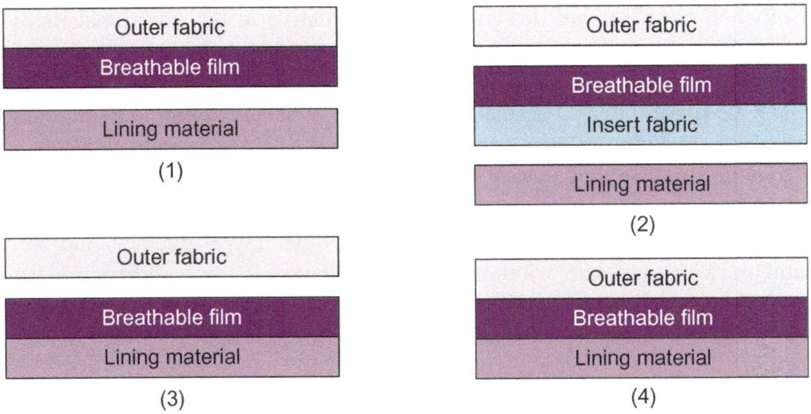

Fig. 14.7 (1) Outer fabric laminate, (2) Insert laminate, (3) Lining laminate and (4) Three-layer laminate fabric (Mukhopadhyay and Midha, 2008a).

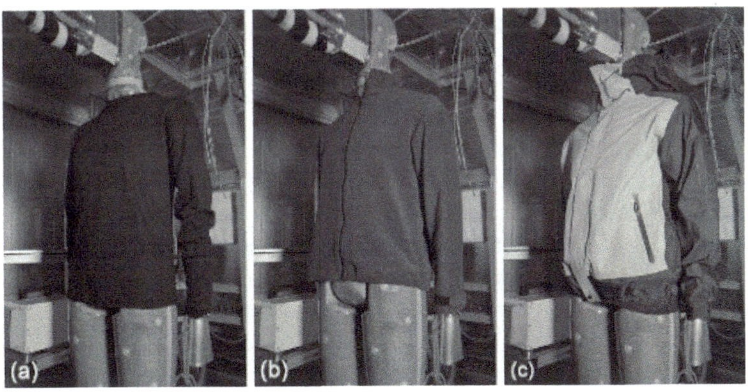

Fig. 14.8 Construction of a three-layered sports garment: (A) base layer, (B) middle (insulation) layer and (C) outer shell (Rossi, 2015).

The waterproof and breathable fabric (WBF) layer usually has the highest water vapour resistance due to its weather protection properties. In different studies it has been shown condensation can occur at low temperatures in multi-layer systems. In several studies it has been shown the accumulation of condensate at low temperatures in microporous WBF is more prominent than the hydrophilic WBF (Rossi, 2015). The durable water repellent (DWR) finishing operations on the shell layer are important to minimize condensation (Morrissey and Rossi, 2013).

Some physical properties of fabrics can be changed to solve the condensation problem of three-layer WBF. To increase water vapour permeability and reduce condensation, the thickness of the waterproof membrane and outer layer fabric can be decreased and the average diffusion coefficient across the outer layer and membrane can be increased (Mukhopadhyay and Midha, 2008b).

Soft-shell garments are designed as an alternative to the traditional multi-layer system. The soft-shell system provides protection from cold, wind, rain and overheating without the need to add or remove any layers. In doing so, they do not use a waterproof breathable membrane. This is achieved through lamination or stitching of the knitted fleece or pile fabrics, and a tightly woven shell layer from the conventional layered system normally without the membrane (Morrissey and Rossi, 2013).

Soft-shell garments serve as mid layer and outer layer together, and are more breathable with moderate wind and water resistance. They could be preferred for high-exertion activities where breathability is more important. They could also protect from light showers and most snowfall (Outdoor Conservation, 2017; Rei, 2017). Some soft-shell garment examples are given in Fig. 14.9.

The classic protective clothing system is well adapted to outdoor activities from walking to skiing. In extreme cold or damp conditions the three-layer system still gives better results. While the waterproofing feature is not always necessary, it is important. Breathability is more important in 90% of sports activities, while waterproofing is a priority in 10% of outdoor activities. Breathability becomes a key performance feature when waterproofness ebbs in importance. In this case, the second layer of outerwear is becoming the focal point. This new second layer is designed to provide flexibility, wind protection and, to some degree, thermal insulation and water resistance, depending on its positioning (Bramel, 2005).

A hard shell provides the most weather protection. Hard shells are fully windproof, waterproof and breathable. Soft shells are flexible and air permeable, meant to protect but not to provide shelter. Best suited to better weather without much precipitation, soft shells excel at maintaining a comfortable temperature during high exertion and are exceptionally mobile, moving freely and easily. They are water repellent and air permeable. Consistently wet weather is best managed with a hard shell; better conditions offer the opportunity to wear a soft shell (Arcteryx, 2016).

Fig. 14.9 Soft-shell garments (Backcountry, 2017).

Soft shells emerged as a result of new technologies that allow various textiles to be bonded, including knitting and fleece. With the developments in lamination technology, glues and fleece, manufacturers have developed permanent composite flexible textiles with various functions such as breathability, thermal insulation, ease of movement and/or—depending on face fabric abrasion—water resistance. Soft-shell clothing usually provides 30–40 min of protection against water penetration, which is usually enough time to find a shelter or put on a waterproof hard-shell jacket (McCann, 2005; Bramel, 2005).

The rise of soft-shell garments is challenging the traditional three-layer protective garment system (Bramel, 2005). In the sportswear market, various types of fabrics were introduced for special uses. Companies have developed 2-, 2½-, 3- and 4-layer laminates, various types of DWR finishing agents, and membranes with different properties to suit the specific technical requirements of users or customers.

14.7 Designing requirements

As mentioned above, according to the activities performed, requirements for sportswear are changing (Uttam, 2013; Özdil and Anand, 2014). When designing a sportswear garment, the following parameters should be considered (McCann, 2005):

1. Form/Style
   a. Aesthetic parameters (colour, materials and cut/fit/proportion)
   b. Culture of the sport
2. Commercial realities
3. Function
   a. Demands of the body (protection, anthropometry, ergonomics of movement, thermophysiological regulation, psychological considerations)
   b. Demands of the sport (duration of activity, safety/survival, range of likely sporting conditions)

In addition to providing mobile functionality and comfort, sportswear is expected to protect the wearer from environmental conditions. Sportswear may require different performance characteristics depending on the activity being performed and highly functional textiles are used for this. For example, good thermal properties are required for sports in cold weather conditions (e.g. PCMs), aerodynamic features for downhill skiing and swimming and breathable waterproofing for many outdoor pursuits (Shishoo, 2005).

Air is regarded as the most important parameter for active sportswear in all weather conditions and in all layers of the garment. It provides lightness and helps with temperature regulation (Bramel, 2005).

Waterproof breathable garments are preferred in outdoor clothing for leisure and sports. This includes rainwear, skiwear, golf suits, sport footwear linings, cycling clothing and mountaineering clothing. They provide protection from low temperatures, rain and wind (Mukhopadhyay and Midha, 2008a; Chaudhari et al., 2016), and are preferred in wet conditions (Ren and Ruckman, 2003).

Consumers mostly prefer insulated water repellent jackets for snow-related sports, trekking, mountaineering and casual use, but will not choose a non-insulated jacket for skiing and snowboarding. Also the end-user would rather wear a non-insulated water repellent jacket in mild/wet conditions, but insulated, water repellent jackets are primarily preferred in cold/dry weather conditions (Burman, 2014).

Performance, protection and comfort-related features are essential expectations for consumers' active sportswear and leisure wear (Shishoo, 2005). Professional sports clothing is generally worn for a comparatively short time and climatic conditions (temperature, relative humidity and air speed) are constant during this time. However, physical performance is at a maximum level. On the other side, leisure sports participants wear the garment from several hours to a whole day and climatic conditions may change during that time. Active and rest phases take place. Therefore there is a greater control range for leisure sportswear compared to professional sportswear (Uttam, 2013).

In the last two decades, significant developments have been made in the textile industry. These developments are in the field of yarn-spinning technologies, functional knitting and woven fabric production techniques, and functional coating and laminating technologies (Anbumani and Babu, 2008; Shishoo, 2005, 2015). It is possible to meet the aesthetic, design and functional requirements consumers expect from sportswear using new polymeric membranes and surface finishes (Shishoo, 2005; Chaudhari et al., 2016). Highly functional garments are expected to be waterproof/breathable, sweat absorbent and have high thermal insulation with low thickness (Shishoo, 2005).

14.8 Conclusion

While the sportswear industry develops high-performance products using innovations in textile technology and science, it also pioneers progress in technology. Requirements for active sportswear are high functional properties with comfort. The sportswear industry uses new technologies to provide all these requirements for users. Developments in active sportswear include new fibre, yarn and fabric technologies, as well as new finishing techniques.

Active sportswear has to be designed for special uses taking users' activity levels and weather conditions into consideration. Necessary protection with minimum weight is the key element. Beside this, users have to feel dry and comfortable while wearing sportswear. Body movement comfort has to be kept in mind during the design process. To achieve these requirements, microfibres, bi-component yarns, elastomerics and a range of knitted, woven, multi-layer and non-woven constructions, breathable membranes, coatings, insulations and water-resistant finishes are useful.

At the same time, producers aim to establish an ecologically sound and sustainable approach. One hundred percent solvent-free foam print and coatings, use of recycled polyester face fabric, PFOA-free and PTFE-free laminates and Fluorocarbon-free DWR are some of these developments.

References

Ahn, H.W., Park, C.H., Chung, S.E., 2010. Waterproof and breathable properties of nanoweb applied clothing. Text. Res. J. 81 (14), 1438–1447.

Åkerblom, D., Göranzon, E., 2013. Greener Water Repellency? Feasible Alternatives to Fluoro Chemicals for DWOR Treatments on Textiles. Bachelor in Textile Engineering Swedish School of Textiles, University of Borås, Sweden. p. 71.

Anbumani, N., Babu, B.S., 2008. Comfort properties of bi-layer knitted fabrics. Indian Text. J. 118 (11), 17–28

Arcteryx.com, 2016. ARC'TERYX official website. Available from: http://arcteryx.com/HardshellvsSoftshell.aspx?language=EN (accessed 29.10.16).

Arkema, 2017. Arkema official website. Available from: http://www.extremematerials-arkema.com (accessed 02.06.17).

Backcountry, 2017. Backcountry official website. Available from: https://www.backcountry.com/explore/what-the-hell-is-softshell# (accessed 11.06.17).

Baker, W.J., 1982. Sports in the Western World. Rowman & Littlefield, Totowa, NJ.

Bakshi, A.S., 2015. Development and Study of Waterproof Breathable Fabric Using Silicone Oil and Polyurethane Binder (Master's Thesis and Doctoral Dissertations). Eastern Michigan University. p. 611.

Bartels, V.T., 2005. Physiological comfort of sportswear. In: Shishoo, R. (Ed.), Textiles in Sports, first ed. Woodhead Publishing, Cambridge, pp. 177–202.

Bougourd, J., McCann, J., 2009. Factors affecting the design of cold weather performance clothing. In: Williams, J.T. (Ed.), Textiles for Cold Weather Apparel, first ed. Woodhead Publishing, Cambridge, pp. 152–195.

Bramel, S., 2005. Key trends in sportswear design. In: Shishoo, R. (Ed.), Textiles in Sports, first ed. Woodhead Publishing, Cambridge, pp. 25–43.

Buirski, D., 2005. Current sportswear markets. In: Shishoo, R. (Ed.), Textiles in Sports, first ed. Woodhead Publishing, Cambridge, pp. 15–23.

Burman, G., 2014. Durable Water Repellency—Study Phase I, Expectations of Durable Water Repellent Fabric Finishes, European Outdoor Group Official Website, Available at: http://www.europeanoutdoorgroup.com/files/DWR-Study_Georgina_Burman__digital_pdf (accessed 05.06.17).

Chaudhari, S.S., Chitnis, R.S., Ramkrishnan, R., 2016. Waterproof breathable active sports wear fabrics. Sasmira official website. Available at: http://www.sasmira.org/sportwear.pdf (accessed 30.05.16).

Choi, K., Cho, G., Kim, P., Cho, C., 2004. Thermal storage/release and mechanical properties of phase change materials on polyester fabrics. Text. Res. J. 74 (4), 292–296.

Chung, H., Cho, G., 2004. Thermal properties and physiological responses of vapor-permeable water repellent fabrics treated with microcapsule containing PCMs. Text. Res. J. 74 (7), 571–575.

Chunyan, L., Xingliang, L., Meng, S., Yanfen, X., 2011. Study on application and biosafety of nano-materials in sports engineering. In: International Conference on Future Computer Science and Education, Xi'an, China, pp. 131–134.

Davies, A., 2014. Durable water repellency—study phase I, an evaluation of the test methods used for assessing durable water repellent fabrics within the outdoor industry, European Outdoor Group official website. Available from: http://www.europeanoutdoorgroup.com/files/DWR-Study_Alice_Davies__digital_pdf (accessed 05.06.17).

Dimitrov, S., Kamenska, V., Walker, J.D., Windle, W., Purdy, R., Lewis, M., Mekenyan, O., 2004. Predicting the biodegradation products of perfluorinated chemicals using CATABOL. SAR QSAR Environ. Res. 15 (1), 69–82.

Ding, X.M., Hu, J.L., Tao, X.M., 2004. Effect of crystal melting on water vapor permeability of shape-memory polyurethane film. Text. Res. J. 74 (1), 39–43.

European Outdoor Conservation Association, 2017. European Outdoor Conservation Association official website. Available from: http://www.outdoorconservation.eu/conservation-section.cfm?pageid=12&sectionid=17 (accessed 11.06.17).

Footwear News, 2017. Footwear News official website. Available from: http://footwearnews.com/2016/focus/athletic-outdoor/puma-olympics-2016-running-gear-caribbean-teams-jamaica-244068/ (accessed 11.06.17).

Ghali, K., Ghaddar, N., Harathani, J., Jones, B., 2004. Experimental and numerical investigation of the effect of phase change materials on clothing during periodic ventilation. Text. Res. J. 74 (3), 205–214.

Gilbert Rugby, 2017. Gilbert rugby official website. Available from: http://www.gilbertrugby.com/eng/collection/body-armour/body-armour/triflex-womens-xp3 (accessed 11.06.17).

Gupta, D., 2011. Functional clothing—definition and classification. Indian J. Fibre Text. Res. 36, 321–326.

Harifi, T., Montazer, M., 2017. Application of nanotechnology in sports clothing and flooring for enhanced sport activities, performance, efficiency and comfort: a review. J. Ind. Text. 46 (5), 1147–1169.

Hatch, K.L., 1993. Textile Science. West Publishing Company, New York. pp. 268–269.

Holme, I., 2007. Innovative technologies for high performance textiles. Color. Technol. 123, 59–73.

Holmes, D.A., 2000a. Performance characteristics of waterproof breathable fabrics. J. Ind. Text. 29 (4), 306–316.

Holmes, D.A., 2000b. Waterproof breathable fabrics. In: Horrocks, A.R., Anand, S.C. (Eds.), Handbook of Technical Textiles, first ed. Woodhead Publishing, Cambridge, pp. 282–314.

Jassal, M., Khungar, A., Bajaj, P., 2004. Waterproof breathable polymeric coatings based on polyurethanes. J. Ind. Text. 33 (4), 269–280.

Jeong, W.Y., An, S.K., 2001. The transport properties of polymer membrane-fabric composites. J. Mater. Sci. 36, 4797–4803.

Kim, K.S., Park, C.H., 2013. Thermal comfort and waterproof breathable performance of aluminum- coated polyurethane nanowebs. Text. Res. J. 83 (17), 1808–1820.

Kissane, B., 2017. New apparel and footwear research category overview: sportswear maintains momentum, Euromonitor International official website. Available from: http://blog.euromonitor.com/2017/01/new-apparel-and-footwear-research-category-overview-sportswear-maintains-momentum.html (accessed 03.02.17).

Kruse Motor Sport, 2017. Kruse Motor Sport blog. Available from: http://krusemotorsport.com/ (accessed 11.06.17).

Kwon, S., Kim, J., Moon, M.-W., Park, C.H., 2017. Nanostructured superhydrophobic lyocell fabrics with asymmetric moisture absorbency: moisture managing properties. Text. Res. J. 87 (7), 807–815.

Lee, S., Obendorf, S.K., 2007. Transport properties of layered fabric systems based on electrospun nanofibers. Fibers Polym. 8, 501–506.

Lee, S., Obendorf, S.K., 2009. The effects of laundering on the mechanical properties of mass-produced nanofiber web for use in wear. Text. Res. J. 79 (12), 1085–1090.

Li, Y., 2001. The science of clothing comfort. Text. Prog. 31 (1–2), 1–135.

Li, Y., Zhu, Q., 2004. A model of heat and moisture transfer in porous textiles with phase change materials. Text. Res. J. 74 (5), 447–457.

Malik, T., Sinha, T.K., 2012. Clothing comfort: a key parameter in clothing. Pak. Text. J. 61 (1), 55–57.

McCann, J., 2005. Material requirements for the design of performance sportswear. In: - Shishoo, R. (Ed.), Textiles in Sports, first ed. Woodhead Publishing, Cambridge, pp. 44–69.

McCann, J., 2015. Environmentally conscious fabric selection in sportswear design. In: - Shishoo, R. (Ed.), Textiles for Sportswear, first ed. Woodhead Publishing, Cambridge, pp. 17–52.

Mohsin, M., Farooq, A., Abbas, N., Noreen, U., Sarwar, N., Khan, A., 2016. Environment friendly finishing for the development of oil and water repellent cotton fabric. J. Nat. Fibers 13 (3), 261–267.

Morrissey, M.P., Rossi, R.M., 2013. Clothing systems for outdoor activities. Text. Prog. 45 (2–3), 145–181.

Mukhopadhyay, A., Midha, K.V., 2008a. A review on designing the waterproof breathable fabrics part I: fundamental principles and designing aspects of breathable fabrics. J. Ind. Text. 37 (3), 225–262.

Mukhopadhyay, A., Midha, K.V., 2008b. A review on designing the waterproof breathable fabrics part II: construction and suitability of breathable fabrics for different uses. J. Ind. Text. 38 (1), 17–41.

Mukhopadhyay, A., Midha, K.V., 2016. Waterproof breathable fabrics. In: Horrocks, A.R., Anand, S.C. (Eds.), Handbook of Technical Textiles: Technical Textile Applications, second ed. Woodhead Publishing, Duxford, pp. 27–55.

Nebo, M., 2005. Functional sport footwear. In: Shishoo, R. (Ed.), Textiles in Sports, first ed. Woodhead Publishing, Cambridge, pp. 70–88.

Nike, 2017. Nike official website. Available from: http://store.nike.com/us/en_us/pd/dry-womens-training-t-shirt/pid-11464638/pgid-12005075?intpromo=pdp:crossSellSelected:1:2257 (accessed 11.06.17).

Outdoor Foundation, 2016. Outdoor foundation official website. Available from: http://www.outdoorfoundation.org/pdf/ResearchParticipation2015Topline.pdf (accessed 29.10.16).

Özdil, N., Anand, S., 2014. Recent developments in textile materials and products used for activewear and sportswear. Electron. J. Text. Technol. 8 (3), 68–83.

Özen, İ., 2012. Multi-layered breathable fabric structures with enhanced water resistance. J. Eng. Fibers Fabr. 7 (4), 63–69.

Physical Activity Council, 2016. Physical Activity Council official website. Available from: http://www.physicalactivitycouncil.com/pdfs/current.pdf (accessed 29.10.16).

Posner, S., 2012. Perfluorinated compounds: occurrence and uses in products. In: Knepper, T.P., Lange, F.T. (Eds.), Polyfluorinated Chemicals and Transformation Products, the Handbook of Environmental Chemistry. In: vol. 17. Springer, Berlin, Heidelberg, pp. 25–39.

Rei Coop, 2017. Rei coop official website. Available from: https://www.rei.com/learn/expert-advice/rainwear.html (accessed 11.06.17).

Ren, Y.J., Ruckman, J.E., 2003. Water vapour transfer in wet waterproof breathable fabrics. J. Ind. Text. 32 (3), 165–175.

Rossi, R.M., 2015. Cold weather sports clothing. In: Shishoo, R. (Ed.), Textiles for Sportswear, first ed. Woodhead Publishing, Cambridge, pp. 197–212.

Rossi, R., Gross, R., May, H., 2004. Water vapor transfer and condensation effects in multilayer textile combinations. Text. Res. J. 74 (1), 1–6.

Ruckman, J.E., 1997. Water vapour transfer in waterproof breathable fabrics: part 3: under rainy and windy conditions. Int. J. Cloth. Sci. Technol. 9 (2), 141–153.

Sayed, U., Dabhi, P., 2014. Finishing of textiles with fluorocarbons. Int. J. Adv. Sci. Eng. 1 (2), 1–7.

Schreuder-Gibson, H., Gibson, P., Senecal, K., 2002. Protective textile materials based on electrospun nanofibers. J. Adv. Mater. 34 (3), 44–55.

Shim, H., McCullough, E.A., Jones, B.W., 2001. Using phase change materials in clothing. Text. Res. J. 71 (6), 495–502.

Shishoo, R., 2005. Introduction. In: Shishoo, R. (Ed.), Textiles in Sports, first ed. Woodhead Publishing, Cambridge, pp. 1–14.

Shishoo, R., 2015. Introduction to textiles in sport. In: Shishoo, R. (Ed.), Textiles for Sportswear, first ed. Woodhead Publishing, Cambridge, pp. 3–16.

Slater, K., 1985. Human Comfort. CC Thomas, Springfield, IL.

Speedo, 2017. Speedo official website. Available from: https://speedo.com.au/product/fastskin-junior-lzr-racer-x-openback-kneeskin (accessed 11.06.17).

Sport England, 2016. Sport England official website. Available from: https://www.sportengland.org/research/about-our-research/active-people-survey/ (accessed 10.02.17).

Transworld Snowboarding, 2017. Transworld Snowboarding magazine official website. Available from: http://snowboarding.transworld.net/news/burton-debuts-2014-sochi-olympic-snowboarding-uniforms/#iEq6VboyUyCbCGY3.97 (accessed 11.06.17).

Uttam, D., 2013. Active sportswear fabrics. IJIEASR 2 (1), 34–40.

Ventile, 2017. Ventile official website. Available from: http://www.ventile.co.uk/ (accessed 22.05.17).

Wang, Z., Li, Y., Kwok, Y.L., Yeung, C.Y., 2003. Influence of waterproof fabrics on coupled heat and moisture transfer in a clothing system. Sen'i Gakkaishi 59 (5), 187–197.

Women's Tennis Blog, 2017. Women's Tennis blog. Available from: http://www.womenstennisblog.com/2012/07/24/fila-olympic-outfits-for-jelena-jankovic-and-vera-zvonareva/ (accessed 11.06.17).

Yoon, B., Lee, S., 2011. Designing waterproof breathable materials based on electrospun nanofibers and assessing the performance characteristics. Fibers Polym. 12 (1), 57–64.

Protective clothing

15

Jan Marek, Lenka Martinková
INOTEX Ltd, Dvůr Králové n.L., Czech Republic

15.1 PPE: A strategic commodity of the market

Personal protective equipment (PPE) is products the user can wear or hold, to be protected against hazards either at work, at home or while engaging in leisure activities. Most PPE is based on textiles, like work clothes protecting against foul weather and work place conditions; barrier clothing and hoods protecting against flame, heat stress and molten metal splashes; respiratory filters and masks, gloves, etc. (Fig. 15.1)

PPE as one of the EU LMI (lead market initiatives) including protective textiles plays an important role in prevention of accidents. They underline the importance of protection and prevention. PPE is defined by the European Commission (Council Directive 89/686/EEC, 1989) as, '…any device or appliance designed to be worn or held by an individual for protection against one or more health and safety hazard.' Modern protective clothing concepts unite the best possible protection from a variety of dangers with unique wear comfort. An example of prototype of a developed protective clothing against multiple hazard is demonstrated by Fig. 15.2. Reliable protection against heat and flame, cut and stabbing, UV-light, liquids, aggressive chemicals, static charge and microorganisms is generally required depending on end-use applications. It is often narrowed to the protection of workers, but it addresses all individuals, including sport practitioners.

Health and safety at work is one area where the EU and other countries focusing on industrialization have had the biggest impact with a solid legal framework covering the maximum levels of risk. Manufacturing of PPE uses around 10% of all clothing and technical textiles, about four million tonnes worldwide. Making materials water-, oil- and stain-repellent has an important impact on required safety and comfort parameters and reduces maintenance frequency and intensity. To avoid incorrect selection of efficient protective garments and/or improper use, both providers and users must have thorough knowledge about the conditions and limits of protection. From the selection of materials and optimized way of multifunctional finishing to adequate maintenance procedure, all these elements significantly influence the quality, lifespan and comfort of this category of products. Rising costs of high-quality protective garments can be balanced with their prolonged lifetime. Design, manufacturing and maintenance management contracted out as one package becomes the most efficient PPE business mode (Scheffer, 2011).

15.1.1 Societal background

Global development and increased production volume in the new millennium focuses not only on rapid manufacture and competition but also on quality improvement. Quality and prolongation of human life based on improved conditions is increasingly

Waterproof and Water Repellent Textiles and Clothing. https://doi.org/10.1016/B978-0-08-101212-3.00014-9

Fig. 15.1 Types of protective clothing (illustrative).

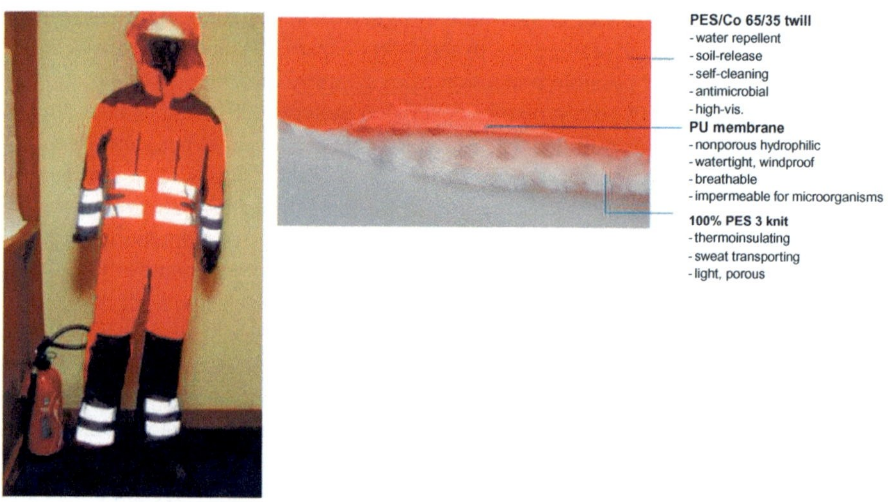

Fig. 15.2 Protective clothing for personnel exposed to extreme weather conditions (rescue teams). Prototype of multifunctional protective clothing for multiple hazards developed within the EU project SAFEPROTEX - High protective clothing for complex emergency operations (FP7-NMO-2008-SME-2, No. 228439).

emphasized as the result of increased levels of scientific background and higher living standards throughout globalized society, together with demographic and societal changes.

Rising awareness toward safety and security at workplaces is expected to steadily raise the demand for personal protective equipment and clothing. Workplace management must include occupational health and safety. The risk of workplace fatalities in extreme conditions is forcing employers to focus on employees' health and safety, which is anticipated to propel market growth. Companies can boost productivity, and employees can achieve better working conditions. Therefore, employers have prioritized worker safety and established safety guidelines to reduce workplace hazards that adversely impact operational costs in the form of compensation and arbitrated settlements. At the start of this era of smart economy and high demand for skilled people, workforce efficiency and full engagement play an important role. To increase efficiency, employers strive to reduce absenteeism caused by sickness and maintain the safety and wellbeing of workers through use of functional protective garments. These factors drive the need for more value-added protective textiles.

Continual growth of protective clothing production and use is expected. The economic potential of risk prevention is equally relevant as demographics change. Stringent government regulations regarding the safety of employees are forcing employers to comply with industry standards, thereby increasing protective clothing demand. Modern workwear with aesthetic appeal is becoming increasingly similar to fashionable outdoor clothing both in shape and colour. Functional requirements continue to be extremely important. Thanks to modern high-tech textiles and materials, professionals in strenuous fields, like emergency response, can wear special protective clothing that is extremely heat resistant and also breathable and water repellent. It is no longer enough just to give employees the right personal protection equipment. They are also equipped with relevant sensory and detecting systems monitoring the risk level and incidents in real time, often while connected with action commanders or control centre (Playing it Safe, 2016).

Technological innovations, along with changing consumer need for protective clothing that combines safety with fashion and attractiveness is further expected to augment market expansion. Growing awareness toward employee health and safety, coupled with increasing industrial fatalities owing to lack of protective gear, is anticipated to drive market growth, as described in market research reports (Grand View Research, 2015a, 2016).

15.1.2 Market situation and forecast

According to Textile Exchange Market Report (Textile Exchange, 2016) the global textile industry continues to grow thanks to the growth of new markets in developed countries and increasing international trade. It has generated sales of approximately \$750 billion during 2015, a growth of more than 5 percent for the year. By 2019, the value of the global textile industry is expected to reach values of \$910 billion, although weakening global economic conditions may slow down this growth. According to this report, the global mill consumption share of all major fibres for 2015 (PCI Wood Mackenzie) is as follows: polyester 55%, cotton 27%, cellulosics 6%, nylon 4.7%, polypropylene 4%, acrylic 2%, wool 1.3%.

As published in Market Research Report (Grand View Research, 2015b) the global technical textiles market is expected to generate \$193.16 billion in sales by 2022.

According to market research, the global personal protective equipment (PPE) market was valued at \$38.38 billion in 2015 (Grand View Research, 2016).

Stringent occupational health and safety regulations are expected to drive their global growth over the forecast period. As described in this report, Europe dominated the global PPE market in 2015 and is expected to remain the largest regional market for the future. Increasing occupational fatalities coupled with the requirement for highly durable mechanical wear resistance and high-utility protective gear in most of the core industries such as refining, metal manufacturing, oil and gas extraction, and automotive is anticipated to augment penetration of these market segments. Moderate growth of the PPE market in the near future is expected consequent to the rising automated manufacturing industry in advanced economies. North America accounted for over 29% of the global PPE demand in 2015. The North American market is characterized by stringent norms regarding occupational health and safety by regulatory bodies such as OSHA, ANSI and CSA enforcing employers in the region to adopt PPE to maintain worker safety in industries. The global PPE market is highly fragmented with the top five players accounting for just over 50% of the market share in 2015. Honeywell was the market leader with a total share of 24.5% of global revenue. This company has continually been involved in R&D to develop various application specific products. Other companies having a notable stake in the global PPE industry include Ansell, 3M, DuPont, MSA Safety, Lindstrom and Alpha Pro Tech Ltd. The majority of the companies were involved in product launches and development of innovative products.

The *global market for industrial protective clothing* is expected to reach $4.06 billion by 2020 (Grand View Research, 2015a). According to the study, increasing industrialization in emerging markets of India and China, rising importance of safety in industries, and the emergence of a blue-collar workforce are expected to remain key driving forces for the market. High-cost and complex manufacturing processes associated with industrial protective clothing are expected to be key challenges for market participants. The global industrial protective clothing market in 2014 achieved a significant rise in production and sales because of high requirements in end-use industries including chemical, oil and gas extraction and processing, refineries, petrochemical, pharmaceutical and industrial waste handling. The industry is marked by a high degree of integration at different levels of the value chain. Major industry players including Royal Ten Cate NV, DuPont, Bulwark FR, Australian Defense Apparel, Teijin, Benette Safetywear, PBI Performance Products and Lakeland Industries took part in this business.

New analysis of the European workwear and uniforms market from Frost & Sullivan (PR Newswire Europe Limited, 2017) found the market earned revenues of €6.77 billion in 2012 and estimates this to reach €7.43 billion in 2017. The research covers general workwear, corporate workwear and uniforms. Market revenues for both direct and rental channels were analysed. Concurrently, the healthcare textile sector is considered the PPE growth leader in the US from $9.40 billion in 2012 to $10.47 billion in 2017 (Lange-Chenier et al., 2014).

Protective clothing represented about 22% of the PPE market, as published by Ecorys, 2009. According to this report, the percentage of other commodities in the total PPE market was: 29% hand protection, 15% respiratory devices, 13% footwear,

6% eye protection, 6% gas detection, 4% fall prevention, 3% hearing protection, 2% head covering. PPE according to end-user segments is as follows: 11% automotive; 10% construction; 10% oil, gas and petrochem; 9% healthcare; 9% utilities; 8% metal fabrication; 8% fire service; 7% pulp and paper; 7% transportation; 21% others.

15.1.2.1 European export of textiles and clothing

As reported by Euratex (2016), total European exports of textiles and clothing amounted to €43 billion in 2014; in 2015, the overall size of the textile and clothing industry in the EU-28 represented a turnover of €169 billion and investments of around €4 billion. Thanks to a revival of EU activity, 174,000 textile and clothing companies employ over 1.7 million workers. EU external trade was more dynamic than the previous year with €45 billion of textile and clothing products exported and €109 billion imported from non-EU markets. According to the European Commission (EC) (2017c), textiles and clothing is a diverse sector that plays an important role in the European manufacturing industry. EU legislation on fibre names and labelling aims to ensure consumer protection and provide correct information to stakeholders. The EC is engaged in dialogues with non-EU countries on policy and regulatory issues that affect the textiles and clothing industry. As published by the market research company Macrom A+A 2015 (Playing it Safe 2016): Germany is the largest European market for outdoor clothing with market volume of €1.8 billion, increased by over 4% in sectors where PPE is vital.

15.1.3 Lead market initiative

Protective textiles belong to six sectors of high societal and economic value focused by Lead Market Initiative (LMI) established by the EC for identification and support of promising emerging markets, in which the EU has the potential to become world leader and where coordinated action is needed: eHealth; sustainable construction; technical textiles for intelligence personnel, protective clothing and equipment (PPE); innovative use of bio-based products; recycling; and renewable energy. These sectors are supported by actions lowering barriers to bring new products or services into the market (European Commission, 2016).

15.1.4 EU standards and legislation

European standards relating to textiles and clothing are developed through technical body CEN/TC 248 of the European Committee for Standardization (European Commission, 2016). The standards relate notably to the determination of size test methods, terminology and minimum performance requirements for certain types of textile products, and environmental aspects of textile products (European Commission, 2017a).

Standards are technical specifications defining requirements for products, production processes, services or test-methods. These specifications are voluntary. They are developed by industry and market actors following some basic principles such as consensus, openness, transparency and non-discrimination. Standards ensure interoperability and safety, reduce costs and facilitate companies' integration in the value chain and trade.

European Standards are under the responsibility of the European Standardization Organizations (CEN, CENELEC, ETSI) and can be used to support EU legislation and policies.

15.1.4.1 Textiles and clothing legislation

The EU has aligned laws in all member countries with Textile Regulation (EU) No 1007/2011 on fibre names and related labelling and marking of the fibre composition of textile products (EC, 2017d). This was done to protect consumer interests and eliminate potential obstacles to the proper functioning of the internal market. *European PPE Legislation*: The EU has issued a number of directives to improve health and safety at work and to ensure high-quality PPE (EC, 2017b).

The *PPE Council Directive 89/686/EEC* (1989) covers the *manufacture and marketing of PPE*. It defines legal obligations to ensure that PPE on the European market provides the highest level of protection against hazards. The CE marking affixed to PPE provides evidence of this protection.

As this is a new approach directive, manufacturers or their authorized representative in the EU can comply with the technical requirements directly or with European Harmonised Standards. The latter provide a presumption of conformity to the essential health and safety requirements. The PPE guidelines aim to facilitate a *common interpretation and application of the PPE Directive*. It should be noted, however, that only the national transposition of the directive is legally binding.

After April 21, 2018, Directive 89/686/EEC will be repealed by the new *Regulation (EU) 2016/425 of the European Parliament and of the Council of March 9, 2016 on personal protective equipment*.

The new PPE Regulation is aligned to the New Legislative Framework policy. In addition, it slightly modifies the scope and the risk categorization of products. It also clarifies the documentary obligations of economic operators.

15.2 Fluorocarbons and environmental issues

Many items of PPE are treated with a repellent finish to enhance their performance and reduce staining. Because of their extremely low surface tension, fluorocarbon resins exhibit unique attributes such as great water, oil and other liquid repellency, important for protective clothing in many end-use categories. C8-based fluorocarbons provide textiles with the highest and most durable water- (DWR) and oil-repellent surface treatment. Recent trends, however, boosted by actual eco-toxicological knowledge, led to their ban (Baurer et al., 2015).

Most efficient in terms of water and oil repellency, C8 fluorocarbons release the biopersistent and toxic component perfluorooctane sulfonate (PFOS) and perfluorooctanoic acid (PFOA). PFOS has very serious effects on health and the environment; consequently, PFOS and related substances are being phased out. Moreover, to improve water repellency and durability, formaldehyde containing melamine resins, or blocked isocyanates as additives for crosslinking, are often used. Unfortunately, these agents are toxic themselves (Atav and Bariş, 2016).

Fluoroacrylate polymers are highly resistant to biodegradation. The half-life of fluoroacrylate polymers is reported to be in the range of 1200–1700 years (Russell et al., 2008).

Because of the negative impact PFOS, PFOA and relevant related substances have on the environment and human health, the authorities, as well as the NGOs and many voluntary programmes, are focused on banning the use and placement of these substances on the market.

Around 2000, the US Environmental Protection Agency (EPA) became concerned over the presence of PFOA, which had been found in human blood in the general population (Holme, 2014a,b). In the EU, PFOA (EC 206-397-9, CAS 335-67-1) and its ammonium salt, ammonium pentadecafluorooctanoate (APFO) (EC 223-320-4, CAS 3825-26-1), as bioaccumulative, persistent and toxic for reproduction (PBT) substances, were added to ECHA's Candidate List for Authorisation (July 20, 2013) as part of the REACH legislation. Actually, the restriction of PF C8, 7 by REACH is under preparation and expected.

Greenpeace's DETOX campaign—Fluorocarbons and clothing

In 2011 Greenpeace started its Detox campaign to promote environmentally friendly and non-toxic clothing (Baurer et al., 2015). The objective of the campaign is to eliminate 11 specific substance groups from the textile industry. Among those substance groups are, e.g. alkyl phenol polyglycol ethers (APEOs) and per-or polyfluorinated chemicals (PFCs). This does not affect fluorocarbon resins themselves, but their toxic impurities, perfluorooctanoic sulphonic acid (PFOS) and perfluorooctanoic acid (PFOA) that can be formed during the manufacturing process. PFOS and PFOA are persistent, bioaccumulative and toxic (PBT) substances. Therefore these compounds accumulate in the environment and the human body and are furthermore carcinogenic, mutagenic and toxic for reproduction (CMR). In addition, fluorocarbon resins have been considered critically as well, because during their degradation PFOS and PFOA can be formed in our environment. In response to the Detox campaign, 19 major apparel and footwear brands as well as 7 associations made a shared commitment to eliminate 11 prioritized hazardous groups or substances (MRSL) from the textile supply chain by the end of 2020. Currently, many members of the ZDHC initiative have already stopped the application of fluorocarbon resins, whereas others are going to proscribe them in the near future after certain transition periods have ended. Based on this fact, leading producers of textile auxiliaries put great effort into the research and development of a new fluorocarbon-free system with a high acceptance by the consumer and markets.

One of the biggest initiatives today, especially in the outdoor sector, is the development of durable water repellent formulations that do not contain PFCs. As part of the US EPA's 2010/2015 PFOA Stewardship Program, certain chemical companies agreed to voluntarily stop manufacturing PFOAs by the end of 2015. In January 2006, the EPA requested the participation of the eight largest fluorocarbon producers in the 2010/2015 PFOA Stewardship Program (EPA, 2016), accessed January 2016. This asked for their commitment to reduce PFOA and related chemicals globally by 95% by 2010 and 100% by 2015.

The ZDHC initiative of international brands and retailers committed in 2011 to set forth a timeline for elimination of durable WR technologies that may contain or could degrade into long-chain PFAAs such as PFOA and PFOS.

The Oeko-Tex Association has revised its limit values for PFOA, and they are now much stricter (PFOA Crackdown, 2014). The Oeko-Tex Association has updated its valid test criteria and limit values for product certification in accordance with Oeko-Tex Standard 100, with the specification for perfluorooctanoic acid becoming much stricter. The values cannot exceed the limit of 1 $\mu g/m^2$ of Product class I–IV (Oeko-Tex, 2015).

Fruitful collaboration between the ZDHC brands, the Outdoor Industry Association (OIA), the European Outdoor Group (EOG), the German Sporting Goods Association (BSI) and representatives of the chemical industry ensured the elimination of DWR technologies associated with long-chain perfluoroalkyl acids (PFAAs).

Even if declared to be PFOA- and PFOS-free, many brands' products tested for PFCs in the membranes, coatings and finishes showed detectable concentrations of toxic PFOA or fluorotelomer alcohols that can break down into PFOA (Greenpeace Calls, 2013).

FLUOROCOUNCIL, a global organization representing the world's leading fluoro-technology companies (Fluorocouncil, 2014), has released a guidance document to help finishing textile mills to minimize waste and environmental releases, while keeping desirable fluorinated product performance in the textile. The document, 'Guidance for Best Environmental Practices (BEP) for the Global Apparel Industry: Including Focus on Fluorinated Repellent Products', outlines practical steps to implement environmental practices (Fluorochemical Advice, 2015).

15.3 Repellent finishing systems; C8 Fluorocarbons alternatives

C8 fluorocarbons provide textiles with high level water and oil repellency. Many eco-friendly alternatives, non-accumulative and safe to the environment and the human body, have been developed and used for textile and clothing production. These finishing products, as well as innovative technologies and processing approaches, are studied and transferred into the industrial production sphere (Baurer et al., 2015).

Alternatives to C8 fluorocarbons have been developed and commercialized based on:

- shorter (C6, C4) FC chains (Table 15.1)
- paraffin, waxes and fatty acids (Table 15.2)

Table 15.1 **Fluorocarbon-based; PFOA and PFOS-free**

Product	Producer	Characteristics
Eco release	Devan	Stain release, water- and stain-repellent technology, based on fluoro-based, but not in the form of C6 or C8; PFOA and PFOS-free, doesn't require high curing temperature, soft handle (Devan Chemicals, 2017)
Anthydrin NK 6	Zschimmer & Schwarz	C6- based fluorocarbon product for low temperature processes; WR, OR, no influence on handle (Zschimmer and Schwarz, 2017)
Unidyne TG-5546	Daikin	Novel short chain (carbon C6) fluoroproducts with improved properties (Alemdaroglu, 2016)
UNIDYNE Multi-Series	Daikin Ind. and Dow Corning Corp.	UNIDYNE TG-5601*,TG-5541*, TG-5543*, TG-5502, TG-811; PFOA-free* natural hand of treated substrates thanks to fluoro-silicone hybrid technology (Daikin Industries Ltd, 2017)
Asahi Guard E-SeriesAsahi Guard AG E550D	Asahi Kasei, AGC Chemicals Americas, subsidiary of Asahi Glass Company Ltd.	Fluorocarbon based short-chain patented C6 chemistry, PFOA-free Water and oil repellent. Nonflammable compatible with other finishing chemicals. Suitable for work clothing. AGC is a member of the FluoroCouncil (AGCCE, 2015)
Careguard 66 (New)	Sarex	Nonflammable C6-based; padding, foam, spraying. PES, Cotton, nylon, wool and blends (Productpilot, 2017)
Baygard NFC	Tanatex	C6 chemistry (Tanatex Chemicals, 2017a)
Ruco-Guard AFC6, AFR6, AFS6, AFT6	Rudolf	Aqueous C_6-based Fluorocarbon Polymeric Dispersions (Rudolf GmbH, 2017d)
Lurotex Duo	BASF	Combination of C6 fluorocarbon with a special Perapret booster to enhance SR
Scotchgard Protector Repellent Finish	3 M	Based on C4 and shorter fluoropolymers, suitable for active wear, durable WR; (3M, 2017)

(Continued)

Table 15.1 Continued

Product	Producer	Characteristics
Phobol CP, Phobol CP-CR	Huntsman and Chemours Company	C6-based sold under Teflon trademark; content of C6 and less FCs—Capstone (Chemours, 2017, and Huntsman International LLC, 2017a)
Tubiguard 6-F	CHT/Bezema	FC6 – water and oil repellency, stain release, spray application: RotaSpay ECO_2 – ROTO FINISHER rotating atomizer system (RotaSpray GmbH, 2013)
Fluorolink PFPE	Solvay	Product based on bifunctional perfluoropolyether structure with a high molecular weight. Doesn't contain and does not degradate to PFOA, PFOS or C8 telomere-like structure (Solvay, 2017)
Texfin C6-D	Texchem	C6-based. Oil and WR up to 50 washes. Requires tumble drying or hot pressing for reactivation after each washing (Texchem UK Ltd, 2017b)
NK Guard S Series	Nicca Chemical Company Ltd	Fluorocarbon-based water and oil repellent finishes, PFOA-free (below 5 ppb), formaldehyde-free (Nicca Chemical Co Ltd, 2017b)

- hyperbranched polymers/dendrimers (Table 15.3)
- microencapsulated systems and plant extracts (Table 15.4)
- hydrophobic polymers and F-free systems inspired by nature (Table 15.5)
- silicones and fluorosilicones (Table 15.6)
- alkoxysilanes [Table 15.7]

All the following tables contain manufacturer's claims on performance.

Because these alternative products show generally lower efficiency and durability, and slight oil-repellency (namely common silicones, waxes and paraffins) compared with C8 fluorocarbons, several approaches to boost the liquid repellency properties are studied and commercialized, e.g. newly developed and commercialized application technologies (spraying, plasma, hyperbaric thermal dry finishing (TDF), and Reactive Surface Treatment (RST) technologies are summarized in Tables 15.8 and 15.9).

15.3.1 Innovative approaches

In addition to these commercialized repellent products and technologies, numerous new alternative repellent systems, inspired by nature, are intensively studied and, optimized, e.g.

Table 15.2 Paraffin, wax and fatty acid-based repellent finishing systems

Product	Producer	Characteristics
Arkophob FFR	Archroma	F-free nature-liquid encapsulated wax, hydrophobic finishing for outdoor apparel, abrasion resistivity, handle, breathability, improved tear strength and sewability. Wash resistance of more than 20 washes, a performance close to C6 fluorochemicals and superior to other fluorine-free products, only water repellent. After laundering, ironing is not required (Archroma, 2017b)
EcoRepell	Schoeller Textil AG	Paraffin-based DWR and SR: Dispersion of long paraffin chain biodegradable (80–100% in accordance with OECD 302 B) paraffins and fatty acid modified melamine resin + dispersion of blocked polyisocyanates. The chains wrap around individual fibres to give a low-surface-energy film (Schoeller, 2017a)
EvoProtect DWA	DyStar	Modified fatty acids for wool and silk, soft and bulky handle. Bluesign approved
Phobotex range	Huntsman and Chemours Company FC	F-free hydropolymers for rain protection and stain management APK: paraffin with aluminium salt JVA, RHP, RSH, RHW: dispersion of paraffin oils and fat modified melamine resins RCO: dispersion of paraffin wax and acrylic kopolymerZAN: paraffin with a zirconium salt. 30 washings at 40°C. Quick drying effect, wicking properties (Huntsman International, 2017b)
Zero F1	CHT/Bezema	Paraffin-based (Marks and Spencer plc, 2017)
Itoguard NFC	LJ Specialities	Fatty acid derivative combined with a melamine containing paraffin wax (Technical Textiles, 2017)
Texfin HTF	Texchem	Modified wax dispersion for work wear, outdoor textiles (Texchem UK Ltd, 2017a)
Neoseed NR-158	Nicca Chemical Company Ltd	F-free, hydrocarbon based, that exhibits the same WR effect as fluorinated finishes (Nicca Chemical Co Ltd, 2017a)

15.3.1.1 *Novel short chain FCs*

A novel short chain (C6) perfluorinated acrylate polymer (NPFA) containing quaternary ammonium and fluorinated groups has been designed and synthesized for cotton finishing (Cattermole, 2016; Li et al., 2013; Holme, 2015a; Alemdaroglu, 2016).

Table 15.3 Hyperbranched (modified) polymers—dendrimers and 3D molecules

Product	Producer	Characteristics
Barrier Eco, Barrier RCF	HeiQ Materials Switzerland in coop with Rudolf Group	F-free organic chemistry for DWR inspired by nature: Nanoscale surface structure. Dendrimer and 3D hyperbranched technologies (Height, 2015; HeiQ Materials AG, 2017a)
CareGuard FF Ecoguard-EFd	Sarex	F-free similar effect to C6, not known composition, cotton, PES, blends, breathable, abrasion, tear resistivity, durable. Careguard FF is based on a hydrocarbon matrix with hyper-branched polymers, free from formaldehyde and paraffin wax and is suitable for LAD (Laundry Air Dry). Wash resistance more than 20 washings. No yellowing, improves stability (Sarex, 2017b)
HydroEco concept: Baygard WRC Baygard WRS	Tanatex	3D molecules, F-free, based on hydrocarbons (WRC—cotton, WRS—synthetics). Long lasting performance (Tanatex Chemicals, 2017b)
Bionic Finish ECO brand: Ruco-Dry ECO PLUS	Rudolf	Dendrimer (hyperbranched polymer) with hydrophobic end-groups connected to patented comb polymers (Rudolf GmbH, 2017b)
RUCOSTAR EEE6, EEN6, EEW6		C6 with integrated dendrimer component WR and OR; (Rudolf GmbH, 2017c)

15.3.1.2 UV curable FCs

Low-surface-energy coatings can act as a physical barrier limiting water penetration. Fluorinated compounds in UV-curable coatings offer fast cure and no volatile organic compounds (Baillié et al., 2015). Environmentally friendly UV curing technology may be successfully used in textile finishing at low total cost. UV-curable coatings and perfluoroalkyl acrylates have shown outstanding results;. more durable than heat-cured coatings. A urethane resin (specified), a photoinitiator (2-hydroxy-2-methyl-1-phenyl-1-propanone Darocur 173 from Ciba initiating free radical generation at 365 nm) and four fluorinated compounds: three perfluoroalkyl acrylates (PFAA)—hexafluoropropyl, heptafluoroproypyl and tridecafluorooctyl acrylates, and one commercial (i.e. Zonyl 8857A DuPont). The degree of cure is determined by IR spectroscopy, water contact angle (WCA) and spray test. UV-curing is economically advantageous compared to conventional thermal technology. It reduces large amounts of evaporated water in contrast with conventional thermo-curable water-based systems.

Table 15.4 **Microencapsulated systems and plant extracts**

Product	Producer	Characteristics
Smartrepel Hydro CMD	ARCHROMA	F-free, nature-friendly microencapsulated non-fluorine repellent. Excellent performance of over 20–30 wash-wear cycles, excellent abrasion resistance. CMD (cotton and blends, microencapsulated, durable) finish has a high resistance in Bundesmann tests with a high-repellency effect to water, coffee, red wine and ketchup (Lang, 2015; Archroma, 2017e)
Smartrepel Hydro PM liq		Microencapsulated for polyester and polyamide, softer handle. Better than wax, comparable with C6 fluorocarbon, stain repellency –no oil repellency, no chemical repellency, limited UV-stability –microencapsulated technology uses a repelling agent and anchoring agent that together create a perfect symbiosis of water protection, breathability and durability. High water repellence performance for cotton, polyester and polyamide(AATCC 22), non-yellowing (Lang, 2015; Archroma, 2017d)
Zelan N3	Huntsman and Chemours Company FC LLC	63% renewable non-GMO plant-based, non-fluorinated durable WR finish Claimed to be up to 3 times more durable than existing non-fluorinated repellents. Extender Phobol XAN is recommended to boost wash durability for at least 30 washes. (Huntsman International LLC, 2017c)
Itoguard NFC 90, 180	LJ Specialities	F-free Botanical extracts hydrocarbon-based polymer (Technical Textiles, 2017)

Table 15.5 **Hydrophobic polymers, F-free systems inspired by nature**

Product	Producer	Characteristics
H2O Repel	Devan	F-free inspired by nature, high curing temperature not needed (Devan, 2017)
Ecoperl Active	CHT/ Bezema	F-free inspired by nature, high curing temperature not needed (CHT, 2017)
OC-aquasil TexW and OC-aquasil TexN	OrganoClick	Fluorocarbon-free based on biodegradable compounds, creating covalent bond; inspired by lotus flowers' 3D structure (OrganoClick, 2017)
Altopel F^3	Bolger at O'Hear	F-free based on a hydrophobe-fortified polymer repellent is now Bluesign approved; wash durable (Bolger O' Hearn, Inc., 2017)

Table 15.6 Silicon-based repellent finishing systems

Product	Producer	Characteristics
DWR 7000 Soft Hydro Guard	Dow Corning	Silicon-based emulsion technology that provides water repellency, even after 30 wash cycles. Soft handle, WR and SR (EnsembleIQ, 2017)
Magnasoft NFR-A + Magnasoft NFR-B (catalyst)	Momentive Performance	Water repellent silicone emulsion + catalyst; handle, combination with fluorocarbons possible. Smooth handle, LAD effect. Cotton and blends (Momentive, 2011)
Wacker HC321 Wacker HC401	Wacker Chemie AG	Reactive silicone-based systems HC321: aqueous emulsion; no heat treatment required, restoring impregnation in outdoor clothing, spray; HC 401: solvent formulation, restoring by spraying in a textile washing machine (Wacker Chemie AG, 2017)

Table 15.7 Alkoxysilane-based systems—nanoscale surface modification (lotus effect)

Product	Producer	Characteristics
Mincor TX TT	BASF	Nanotechnology; particles with a diameter of less than 100 nm embedded in a carrier matrix; lotus effect (The Biomimicry Institute, 2016)
Dynasylan F 8815	Evonik	Fluoroalkyl functional water-borne oligosiloxane, which acts as a surface modification agent on oxidic, carboxy and hydroxyfunctional substrates (cotton) (Evonik Industries, 2017).

Table 15.8 Superhydrophobic spray coatings

Product	Producer	Characteristics
NeverWet	NeverWet	Superhydrophobic and superoleophobic spray coating, layer thickness 25–27 μm, transparent, handle not influenced. (NeverWet LLC, 2017)
Always Dry	Nanex Company	Superhydrophobic water repellent spray, which contains nanopolymers forming an invisible protection layer with highly hydrophobic properties without any change of colour, texture or breathability. The nanocoating provides also oil repellency, offering potential for application to workwear (Glassmark Solution, 2016)

Table 15.9 Innovative hydrophobic finishing systems and new finishing technologies

Product	Producer	Characteristics
Plasma Guard, Nanofics 110 coatings	Europlasma	DWR coating for outdoor apparel applied by low-pressure plasma technology won the WTiN award for Best Innovation in Sportswear and Outdoor Apparel category of ITMA Future Materials Awards 2015 in Milan. Nanoscaled functionalization. Water repellent according to ASTM D5964, Oil repellent ISO 14419. nanocoating PFOA, PFOS-free deposited by low-pressure plasma technology (Europlasma, 2017)
Encapsil Down	Patagonia	Silicon-based plasma-bonding process for down processing + outdoor apparel for cold environment (Ethan, 2013)
TDR	Green Theme Technologies	Hyperbaric thermal dry finishing. In situ free-radical polymerization dry finishing (TDF). A mixture of hydrocarbon-based monomers, cross-linkers, polymerization initiators, plus 'attributable' agents applied by gravure or spray coating and then cured by thermal curing. Released free radicals during TDF trigger free-radical polymerization. Because oxygen and air kill the free-radical polymerization process, the oxygen must be removed using steam, CO_2, He, Ar or nitrogen (Selwyn, 2016)
Cleanshell	Alexium	Advanced process for water- and oil-repellent treatment of textiles based on Reactive Surface Treatment (RST) technology, which uses microwave technology to apply the treatment rapidly and energy-efficiently. The process involves encapsulation of a fabric's fibres with a nanoscopic coating to protect the entire fabric (Nanowerk, 2011)

15.3.1.3 Hydrophobins

Spherical proteins with diameters of few nanometers with a hydrophobic and a hydrophilic part (Santhyia et al., 2010). They were found as thermally stable (up to 100°C) protein amphiphiles of filamentous fungi (mushrooms head). Orientation of hydrophilic and hydrophobic parts is decisive. These amphophilic proteins are suitable for synthetics hydrophobization having switchable effect (Opwis and Gutmann, 2011; Kaufmann et al., 2014).

15.3.1.4 CNTs

Cationized cotton was modified with carbon nanotube ion pairs synthesized by thermally fusing oxidized single-walled carbon nanotubes with octadecylamine (Soboyejo and Oki, 2013). The desorption of CNTs with linear structure onto materials can lead to highly ordered arrangements resulting in hydrophobic properties, also described as superhydrophobicity (SH), and a bionic approach because of the structural similarity to the surface of rice leaves. Modification of CNTs is necessary for a durable effect (Liu et al., 2007).

15.3.1.5 Dendrimers and comb polymers

Dendrimers are highly branched fractal-like macromolecules of well-defined, three-dimensional structures, shape and topology polymers with a regular structure and a high density of end groups (Bionic-Finish ECO, Rudolf; wax-based dendrimer containing fluorocarbons Ruco-dry DHN). There are two types of dendrimers: Dendrimer fluorocarbons containing fluorinated alkyl groups (WR, OR) and non-fluorinated dendrimers applied in combination with FCs. (Atav and Bariş, 2016; Fisher and Vögtle, 1999).

A recent approach to water repellency is based on the use of dendrimer nanoparticles (Namligoz et al., 2009). The repellency effect of dendrimers is based on their forming nano-sized crystalline structures that impart wash resistance through water repellency and high abrasion resistance. Minimizing the amount of FCs used in fabric treatments would be environmentally beneficial.

15.3.1.6 Sol–gel-based systems and nanolayers

Sol–gel materials are prepared from inorganic particles in sizes of up to 80 nm stabilized in a liquid. If applied onto a substrate, the particles aggregate to a 3D network and the inorganic coating is formed on the substrate. When hydrophobically modified, the surface gains hydrophobic properties (Takenori et al., 2009).

Organomodified alkoxysilane-nanosols-based coatings are prepared by acidic or alkaline hydrolysis of alkoxysilane precursors (including perfluorinated alkyl trialkoxy silanes, or silanes with a long hydrophobic alkyl chain (Holme, 2015a)). Alkoxylates, especially star polymers, can adhere to the substrate with anchor groups and exhibit hydrophobic groups to the surface. They can be crosslinked to increase washing fastness. Surface roughness induced by nanostructured materials results in biomimetic-made superhydrophobic surfaces: lotus leaf effect (Shirgholami et al., 2013). The surface of cotton and polyester fabrics was modified to create a water repellent finishing by depositing a modified silica-based film using the sol–gel technique (Montarsolo et al., 2013). Furthermore, low-temperature plasma pre-treatment was used to activate the fabrics to improve the sol–gel coating adhesion, resistance to abrasion and fastness to washing stresses. Plasma pre-treated samples showed significantly higher values of contact angles after the abrasion test because of the improved adhesion of the sol–gel coatings on fibres treated with oxygen plasma.

The Fraunhofer Institute for Silicate Research (ISC), Würzburg, Germany, developed a novel water-based refinement solution (Scrimshaw, 2015). The coating system InnoOlTex enables the combination of up to six functions in a single process: hydrophobicity/hydrophilicity, easy-care, anti-static, flame protection and antimicrobial properties. It allows combination of incompatible functions: abrasion-resistance, flame-resistance, hydrophobicity, anti-static and antimicrobial resistance and washability (60°C) in one solution based on Ormocers with controllable properties according to requirements. The water-based hybrid modified alkoxysilane precursor-based coating system InnoSolTex that offers the six combinable properties is curable at 170°C.

15.3.1.7 Nanoparticles

Nanoparticles of metallic oxides such as the oxides of titanium, zinc, aluminium and iron, and also silicates, have been used for stain repellency, self-cleaning fabrics and ultraviolet protective finishes (Holme, 2015b) Polymer nanocomposites utilize nanoparticles and/or nanotubes incorporated within a polymer matrix for use in military apparel, protective clothing and other technical-textile uses. Nanocomposites in protective clothing can detoxify biological and chemical weapons, e.g. nerve gases. Nanocoatings are very thin surface coatings on textiles that can incorporate nanoparticles or other functional additives. Nanocoatings may be produced using sol–gel treatments, or by physical vapour deposition (PVD) or chemical vapour deposition (CVD). Plasma technology could be used to functionalize the fibre surface and to introduce chemical groups that can improve the adhesion of the nanocoating. To functionalize the fibre surface and introduce chemical groups that can improve adhesion of the nanocoating, plasma technology could be used. Multilayered nanocoating could introduce multifunctionality. A wide variety of applications could be envisaged for nanocoatings and sol–gel treatments. It can also increase abrasion resistance. Functionalized nanoclays should extend the effect portfolio.

15.3.1.8 Silicone-based innovations

Sol–gel-based fluorocarbon products were generated via hydrolysis and condensation of an ormocer fluorocarbon monomer to yield a cross-linked perfluorocarbon (Holme, 2015c). Sol–gel C6 fluorocarbon treatments can be boosted with a simultaneous paraffin application or a cyclodextrin pretreatment. Silicone formulations for textile hydrophobization (INSA – Lyon in Tex-shield project): Lotus effect with WCA over 150° and self-cleaning behaviour with a water-sliding angle of less than 10° based on nano- and micro-sized surface roughness because of raspberry-shaped silica particles functionalized with a monoepoxy functionalized PDMS (polydimethylsiloxane). Silicone macroemulsion is crosslinked by heat curing. Precipitation of silica results in a double-scales surface roughness. Surface post-functionalization by grafting with a benzoxazine-functional silane or incorporation of CF_3 groups oriented toward the external surface improves oil repellency.

15.3.1.9 Plasma

Gas plasma treatments are now becoming well established commercially as a means of changing the surface properties of textiles without altering their bulk physical and mechanical properties (Commercializing Plasma, 2016; Holme, 2014a,b). The technique offers a clean, dry approach that uses considerably less energy than nearly all traditional wet treatments. Plasma treatments, depending on their nature, can render changes in fibre hydrophilicity or hydrophobicity (Coulson, 2007). Advances in plasma technologies for textiles have been reported in the groundbreaking textbook edited by professor Shishoo (2007). Plasma can be used for surface functionalization, e.g. for (permanent) introduction of chemical groups or surface coating/finish deposition. Atmospheric pressure plasma (APP) technology has been proven to be an effective method for improving the hydrophobicity of fabric surface. For APP treatment tetrafluoromethane (CF4) can be used as the polymerizing monomer (Kwong et al., 2013).

Plasmatreat, established in 1995, developed the atmospheric plasma processes to create tailored surface treatments. The process Openair Plasma pretreatment employs a reactive carrier gas for wettability improvement. Surfaces can be provided with hydrophobic and dirt-repellent finishes (e.g. CREPIM PlasmaPlus polymerization process for thin FR (flame retardant) coatings creation). Europlasma's ultra-thin coating solutions have been increasingly used in the textile production of outdoor, sporting, military and workwear applications. Europlasma's PlasmaGuard solution is a durable water repellent coating that can be applied to fabrics and finished products such as garments, shoes or accessories. The coatings are PFOA- and PFOS-free, have a strong antibacterial effect. Sportswear brand O'Neill wetsuits uses Europlasma technology.

15.3.1.10 MLSE technology

There is a new method of treating textiles that removes the use of chemicals, is greener than plasma, and is completely dry (New Dry Treatment, 2016). Developed by MTiX in partnership with and co-funded by the Eco-Innovation Initiative of the European Union, the new process, named Multiplexed Laser Surface Enhancement (MLSE) brings together plasma and laser technologies to make fibres or fabrics hydrophilic, hydrophobic, FR or antimicrobial without use of harmful chemicals or water and by 99.6% reduction of energy consumption. The dry process is carried out at atmospheric pressure, using safe inert gases: N_2, O_2, Ar, CO_2. Combination of plasma and photonic energy creates material synthesis on the fibre surface and reduces chemical usage by 94.8%, particularly fluorocarbons. Emission of greenhouse gases is reduced by 90.9%; water consumption by 75.5%. This process is suitable for natural and manmade fibres, including wool (MTIX LTD, 2017).

New, efficient repellent systems are developed under *Tex-shield project* (TEX-SHIELD, 2017). (University College Ghent) aimed to replace PFCs based upon C8 chemistry at the repellency sequence: C8 > C6 > C4. The project focuses on increasing nano surface roughness using inorganic nanoparticles (SiO_2, TiO_2, Al_2O_3) or organic nanoparticles. PMMA (polymethylmethacrylate) or SMI poly(styrene maleic anhydride) or beta-cyclodextrin increases the repellency of C6-based fluorocarbon finishes. Plasma etching using a Grinp atmospheric plasma treatment also increases fibre-surface roughness.

15.4 Protective clothing with multi-barrier properties

The fundamental function of PPE is reliable protection based on a range of functional barrier effects like hydrophobicity, oleophobicity, resistance against liquids and penetration of microorganisms, flame retardancy, resistance against heat and cold environments, antistatic properties, UV protection and antimicrobial/antiodour effect. High-visibility colouration combined with retro/reflective strips and soil repellent properties significantly support the wearer's safety in lowered visibility conditions. At the same time, the appropriate physiological parameters provide relevant wearing comfort without movement restriction, enabling the wearer to perform properly with adequate comfort during work, rescue or sport activities without excess fatigue or inconvenience. Multifunctional lightweight constructions with durable barrier effect and prolonged service life with moisture management properties (quick sweat transport and quick drying effect) supported by thermo-regulating characteristics and thermoinsulation effects protect wearers against thermal heat and stroke, mechanical injuries, burns and scalds. Pleasant touch, aesthetic look and easy donning and doffing are also very important attributes of protective clothing.

A combination of these properties is possible to achieve with an appropriate choice of textile fibres, construction of yarn, textile structures and combinations, and by modification of their properties in mass or by application of various surface finishing systems. Utilization of new functional fibres and modified natural materials and their blends in new special constructions and implementation of new technologies are becoming a growing trend of new high-performance protective clothing. These advancements with development and production ensure reliable functional and physiological parameters durable in end-use conditions and prescribed maintenance cycles. Workplace management must include occupational health and safety (Playing it Safe, 2016). Companies can boost their productivity, and employees can achieve better working conditions. The economic potential of preventative health measures is equally relevant as demographics change. Modern workwear needs to have also an aesthetic appeal similar to fashionable outdoor clothing both in shape and colour. Functional requirements, continue to be extremely important and thanks to modern high-tech textiles and materials, emergency response staff can wear special protective clothing that is extremely heat resistant and also breathable and water repellent. It is no longer enough just to give employees the right PPE. Safety specialists also need to be able to monitor their operations in real time, so that they are aware of an incident the moment it arises.

15.4.1 Thermoregulation and moisture management—an integral part of PPE functional and physiological performance

The primary role of textiles is to keep the human body in an appropriate thermal environment and to assist its own thermal balance and comfort under various circumstances of environmental conditions and physical activities. It helps to keep physical conditions around the body suitable for survival (Senthilkumar et al., 2012).

Thermal conductivity and heat transfer are involved in thermal management in addition to air permeability, moisture vapour permeability and moisture regain, influencing wearing comfort.

High thermal resistance is essential for protection from cold. New barrier properties like thermal stress protection, air permeability and water penetration resistance combined with breathability based on the rapid liquid moisture transport characteristics of textiles becomes more available on the market, requiring added value textiles with thermal regulation functions, i.e. materials with thermal management. Numerous alternative solutions based on new fibre qualities and manifold profiled fibres like hollow fibres for better insulation, microfibres capable of trapping air, channelled, hollow, crater and porous fibres with wicking effect, texturized filament yarns, etc. Fabric constructions like waffle and pocket weaves, fleeces (increased surface), double structures, irregular cross-sections, surface modifications and surface treatments like IR-reflective, metal-foil-covered, special colourations (including thermochromic dyes) widen the options to realize the multifunctional effects.

The moisture management performance of textiles significantly influences human perceptions of moisture sensation. If moisture released from the skin is not efficiently transported to the surrounding environment, it reduces comfort by generating a sensation of wetness that causes the garment to feel clammy, and by reducing thermal insulation as water condenses in fabric pores, produces a cold feeling during wear (Yoo and Barker, 2004).

15.4.2 Moisture management

By definition, clothing physiology is interaction of body, climate and clothing. Moisture management is understood to be the ability of a textile to transfer gaseous or liquid humidity from the skin, to transport it to its outer surface, and to release it into the surrounding air (TextInfo, 2012). Changing the liquid transfer in clothing significantly influences the wearer's perception of comfort declination. A key cooling mechanism of the human body is sweating and evaporation. Water vapour carries heat away from the body as it evaporates from the skin surface or the fabric surface. In the garment-skin microclimate, absorption of sweat by the garment and its transportation through and across the fabric accomplished by evaporation from its surface is related to clothing comfort perception (Hu et al., 2005). In hot conditions, trapped moisture may heat up and lead to fatigue or diminish performance, while in cold conditions, trapped moisture will drop in temperature and cause chilling and hypothermia. Excess moisture may also cause the garment to become heavy, as well as cause damage to the skin from chafing (Defence Update, 2017). The fabric should not retard the fast transmission of liquid sweat and moisture during body strain (when the body sweats actively).

Wear comfort of clothing is one of the most important properties demanded by consumers. Moisture management, that is, tactile and thermal comfort synergy, is very important.

15.4.3 Wicking effect

Often poor wear comfort of synthetics can be eliminated. Thermal insulation does not depend on the fibres only, but on fabric construction parameters: thickness, weight, porosity and airtightness. Combination of natural and synthetic fibres, or doublefaced textiles with proper constructions are other options. Textiles from customized synthetics could enable heat and moisture management comparable to natural fibres while cellulosic fibres absorbing water exposed by moisture should become heavy and sticky. They also they require longer drying.

Functional fibres with moisture management properties can be found on the market and are widely used. There is also growing interest in moisture management fabrics from the heat-resistant workwear and flame-retardant apparel market (Defence Update, 2017). Heat-resistant workwear garments are usually worn in hot and humid conditions. One of the most common causes of occupational death among firefighters is heart failure because of heat stress caused by loss of body fluid required to produce perspirations. Firefighters can lose up to 4 L of fluid per hour when in proximity to the fire (Holmes, 2000). Suitable fibres are used or a subsequent finishing is applied. Combinations of fibres and finishes can be employed (DR PETRY, 2008). A capillary effect of the customized constructions can be supported by use of microfibres. Their larger surface area promotes evaporation. New opportunities are offered by the irregular cross-sectional fibres (e.g. COOLMAX) or by the shell–core composite fibres with wicking performance.

During physical activity, the body provides cooling partly by producing perspiration. If the water vapour cannot escape to the surrounding atmosphere, the relative humidity of the microclimate inside the clothing increases, causing increased thermal conductivity of the insulating air, and clothing becomes uncomfortable. In extreme cases, hypothermia can result. As the body produces heat during exercise, which would otherwise cause the core temperature to rise above 37°C, the body tries to regulate the temperature via perspiration and evaporative cooling. Too high absorbency could lead to the negative effect, as often happened with use of cotton (Wallance, 2002). In cotton materials, particular (one-way) water repellent treatments can be applied creating 'wicking window' allowing the moisture to be wicked to the outside of the garment.

To develop MM fabrics the following aspects have to be optimized:

- Humidity has to be transported to the outer surface as quickly as possible. In order to evaporate, humidity has to reach the surface of the fabric first. Wicking isn't the same as wettability.
- Humidity has to evaporate as quickly as possible. The larger the surface, the faster evaporation
- Skin has to feel dry. A humid feel is unpleasant. Whereas cotton can absorb a certain volume of water without feeling humid, polyester feels wet and clammy even with small amounts of humidity stored in it. From this point of view, PES microfibre is better than common PES fibre.

Thick textiles absorb more humidity compared to thinner fabrics and their drying takes considerably longer. The moisture retention capacity can be influenced (reduced) by a chemical treatment, e.g. wrinkle-resistant finishes (Table 15.10).

The most efficient moisture management fabrics are high-tech synthetic fabrics made of polyamide or polyester microfibres. They can be lightweight, capable of transporting moisture effectively, and dry relatively quickly. Specific functional properties offered by bicomponent double-layer materials: usually a low-absorptive (non-absorbent) layer inside and high-absorptive layer outside.

Table 15.10 Examples of finishing systems for moisture management characteristics improvement based on fluoropolymers, silicones, waxes

Finishing systems	Producer	Characteristics
Wicking windows	Cotton Incorporated	Moisture management finishing reduces absorbent capacity and increases wicking performance of cotton accompanied by increased drying rate (Cotton Inc, 2017)
Ultraphil HCT	Huntsman Int.	Product of the High-IQ range based on silicone microemulsion, which provides cool comfort (Huntsman International LLC, 2017a)
Hydroperm RPU Liquid	Archroma	A thermoreactive polyurethane softener for soft hand feel and excellent hydrophilic properties of cellulose and polyamide (Archroma, 2017a)
Nano-Dry	Nano-Tex LLC	Durable hydrophilic finish determined for nylon and polyester fibres (California in the NanoEconomy, 2013)

Promising results of research aimed at nanocrystalline cellulose (NCC)-based finishing of polyester increasing moisture regain are published, e.g. hydrophilic modification of polyester fabric by applying nanocrystalline cellulose containing surface finish (Zaman et al., 2013) or preparation of polyester fibres with cellulosic surfaces by use of ionic liquids (Textor and Gutmann, 2014).

Polyester enzymatic hydrophilization is an alternative bio-based cleaner production system used instead of the harsh alkaline de-weighting process. It is significantly changing the weight and barrier properties of PES fabrics (Lee and Song, 2010; Martinkova and Marek, 2010).

15.4.4 Soil release

Natural fibres exhibit little soil repellency but a high level of soil releasability (by cleaning—laundry). Synthetics exhibit low level of soil repellency but also a low level of soil releasability.

Soil repellent – soil release finishes offer a different effect. Soil release finishes are hydrophilizing finishes, enabling better laundering (better wettability for soil removal). Oily soil can diffuse into the fibre. Oily residues may remain within the capillary structure making soil removal difficult in laundering.

Types of soiling mechanisms include:

- mechanical adhesion by direct contact
- rubbing of the garment against the skin
- picking up dirt from liquids or from the air

Different mechanisms of soil removal have been recognized: lotus effect based roll up and down of the soil and dirty particles, penetration of detergent into the soil-fibre interface followed by impurities solubilisation and emulsification, surface mechanical abrasion and swelling of the finish to reduce the inter-fibre spacing (Goldstein, 1993).

Ordinary laundering is sometimes not adequate for removal of intensive soil. to enhance the soil release properties, special finishes have been developed that are capable of facilitating removal of particulate soil by decreasing adhesion of particles and enhancing diffusion of water and detergents into the particle-fibre interface. This depends upon the ability to provide a hydrophilic surface during the laundry process. Finishing aimed to high-performance soil-release effect usually combines and utilizes more of the mechanisms mentioned above (Kissa, 1981) The first soil-release finishes were introduced in 1966. Soil-release agents are generally based on carboxy, hydroxyl, ethoxy (combination with DMDHEU crosslinkers) and fluorine chemistry. Soil-release fluorochemicals are usually modified by copolymerization with hydrophilic moieties to yield hybrid polymers. Hybrid fluorocarbons (Holme, 1993) are hydrophobic and oleophobic in air and hydrophilic and oil releasing during the laundering process. This is because hydrophilic blocks are shielded by fluorocarbon when dry, but swell after immersion. Examples of commercial products for SR finishing are described in Table 15.11 along with manufacturer's claims. US Patent 3,654,244: *Polymers for soil-release textile finishes* describes special polymers that preventing strongly prevent the staining of fabrics while simultaneously providing finished fabrics with high soil-release properties, i.e. the stains are readily washed out during laundry. The polymers are products of copolymerization of at least two different monomers: one with oleophobic properties based on (meth)acrylate with a terminal perfluoroalkyl group and the other with hydrophilic properties provided by alcohol groups in the acrylate or methacrylate structure.

Table 15.11 Examples of commercial products for SR finishing

Finishing systems	Producer	Characteristics
Ecoguard-SR 6	Sarex	C6-fluorocarbon based finishing for easy removal of stains under home laundering even at low temperatures (Sarex, 2017a)
DWR 7000 Soft Hydro Guard	Dow Corning	This silicon-based emulsion for soft handle, water repellency and soil release effect even after 30 washings (EnsembleIQ, 2017)
BIONIC-FINISH	Rudolf	A patented system for water, oil and soil repellent finish of textiles based on dendrimers (Rudolf GmbH, 2017a)
Smartrepel Hydro PM liq	Archroma	Microencapsulated F-free stain repellent/stain release agent for polyester and polyamide fabrics (Archroma, 2017c)

US Patent 3,639,156: *Siloxane polymers for soil-repellent and soil-release textile finishes* describes soil-repellent and soil-release polymers composed from an oleophobic silane monomer with perfluoroalkyl terminal group and a hydrophilic silane monomer.

15.4.5 PPE with thermoinsulating and thermoregulating effect

For protective clothing, water- and oil repellency are required simultaneously with high physiological comfort, thermoregulation/thermoinsulation and movement freedom ensuring safe and efficient working performance even in harsh environment conditions (firefighters, foundry workers, first responders or rescue team members and others working in cold outdoor conditions and/or wet environments).

Thermoinsulating effect is generally achieved by decreasing the heat transfer coefficient. This can be achieved by a combination of textile materials and constructions selection, functional fibre incorporation, application of barrier thermoreflective finishing systems and special apparel design (3D textiles with air gaps, multilayered systems and laminates, ventilation zones, etc.), as demonstrated by the following examples.

15.4.5.1 PPE for high-temperature environments

Heat stress and heat stroke

Heat stroke is characterized by body temperatures above 40.6°C. Heat stroke can begin when external temperatures are low as 32.22°C. Even with the best outer gear, the environment inside a firefighter's uniform can become dangerously hot. It can lead to heat stress, a condition where high temperatures disrupt the natural thermoregulation of the body, causing not only discomfort but potentially lethal conditions leading to circulatory collapse (Munsante, 2013a). According to ISO 7243, an essential requirement for work in hot environments in occupational settings is that the body core temperature may not exceed 38°C in a work period. Above that, the risk for hyperthermia and related heat disorders may occur and threaten workers safety and health (Zhao et al., 2013).

The garment creates the barrier between the person and the hazardous working environment. Nevertheless, it is also an obstacle for heat and moisture exchange between the body and the environment. The heat produced by the body during work cannot be fully transported to the environment, which leads to an imbalance between body heat production and heat loss. Then, a fast increase in the values of temperature and relative humidity of the undergarment microclimate occurs, and consequently causes an increase of the body and skin temperature. Working in impermeable protective clothing increases the thermal stress of its user and his or her cardiovascular system. Unfavourable changes in the values of the human physiological parameters (heart rate, core and skin temperature) and the physical parameters of the microclimate under impermeable protective clothing are often the reason for thermal stress of its user (Bartowiak et al., 2013).

Research has shown prolonged work in a hot environment leads to thermal stress and exhaustion and may cause disorders of the cardiovascular system. It has been reported the number of risky behaviours and accidents in the workplace increases with work intensity and ambient temperature. Studies concerning hazards resulting from prolonged exposure of workers to a hot microclimate emphasize that it is necessary to eliminate or decrease heat accumulation in the body. Additional heat stress can be caused by protective clothing characterized by high thermal resistance and low water vapour permeability (Bartowiak et al., 2015).

The right undergarment can help reduce that risk. Functional underwear that supports the body's natural cooling function is composed from two materials: hydrophobic on the inner skin contact side, and hydrophilic on the outer side. The outer layer is made from material designed to absorb sweat and move it away from the body. Non-melting materials need to be used.

The supporting effect of multiple air gaps on heat transfer of firefighters' protective clothing exposed to low thermal heat fluxes has been investigated. It works on the thermal heat fluxes—ventilation through the intermediate (3D distance fabric) layer (Fu et al., 2014).

Workers in high-heat environments (firefighters or foundry workers) are subjected to a daily exposure to potentially lethal conditions. Their jobs regularly put them at risk of being burned, or asphyxiated from fumes or heat inhalation. Adding to these dangers is the ever-present danger of heat stress, a condition where the human body's systems overheat and collapse, causing a risk of death (Munsante, 2013b). Various testing devices have been constructed to simulate these risk conditions.

PyroMan is a life-size manikin for testing of thermal protective performance of wildland firefighter clothing materials at North Carolina State University Thermal Protection Laboratory (NC State University, 2017). The fire test chamber simulates the same extreme conditions from heat and flame experienced in fires. PyroMan is regularly used to test the FR limits of protective gear used by military organizations and major manufacturers also used the lab to test a wide variety of new and existing FR technologies. *RadMan* (innovation in radiant heat lab) is equipped with multiple sensors. The big threat is radiant heat. Similar test manikins were developed in special testing laboratories (Scandic consorcial project modified also for the specific Nordic climate conditions—VTT-UNI Tampere; Hohensteiner Institutes).

Synthetics could melt against the skin. This can be effectively eliminated by using a flameproof blend of synthetics with FR finished cellulosic fibre (e.g. INDURA Ultrasoft). When exposed to flame, the charred cellulosic fibre is capable of absorbing the melted synthetic part, which prevents melting against the skin. The synthetic part (high-tenacity polyamide) ensures improved mechanical parameters. Oversaturated by moisture, the protective clothing exposed by heat could be waterlogged and scald the skin. Multilayer designed garment systems aspire to be an optimal solution regarding functional properties and wearing comfort. For the base layer, thin synthetic materials with low absorptivity capable of transporting sweat rapidly from skin to outer layer(s) can be recommended. Because the base layer is in a direct contact with the skin, the material should be heat resistant having a high melting point to avoid its melting against the skin. The intra-layer (if present) should be primarily thermoinsulative (heat or cold protective). The material should be low water-absorptive/sweat transporting and breathable as well (e.g. polyester 3D knit). For sweat transport acceleration from the skin and for a cooling effect, a superabsorbent layer can be used. The outer layer should be a basic barrier layer against flame and liquid penetration. This layer should be water impermeable but breathable, enabling rapid evaporation of the moisture transported from the inner side of the clothing. Combinations of inherent FR fibres of FR finished materials with a hydrophobic/repellent finishing or a watertight membrane are often used.

Besides thermoinsulating protective materials, thermoregulating fibrous materials are becoming increasingly used are being intensively developed. These thermoregulating systems are capable of balancing temperature differences to some extent, which can be very advantageous for protective clothing in risky (hot and cold) climate conditions, increasing thermal comfort of users and prolonging the time period needed for escape from a dangerous place (firefighters). This contributes to reduction of risk of injury (or death) and decreases the risk of heat stress and heat stroke. This is also important for workers as well as athletes in extreme temperature environments (climbers, skiers, bikers) in combination with high and/or intermittent physical load because of suppression of sweating (prevention of scald risk). These thermoregulating materials are used for functional clothing and for sleeping bags, bed linens, textile parts of shoes, protective blankets, etc.

For thermoregulation, special cross-sectioned fibres and textile multilayered constructions and finishing systems enabling quick sweat transport to the outer layer and evaporative cooling can be employed. These systems have developed and used innovative smart themoresponsive materials such as PCMs (Phase Changing Materials), which melt at higher temperatures, absorb heat and solidify at low temperatures, and release heat, or thermo-responsive polymers and active textile surface structures, which change shape and porosity in response to environment temperature change.

PCMs for PPE with smart thermoregulating properties

Phase Change Materials (PCMs) are substances with a high specific heat capacity and serve as thermal energy storage systems (Onder and Sarier, 2015). Inorganic or organic-based PCMs are capable of absorbing and releasing a great amount of latent heat during phase transition between their solid and liquid phase, over a narrow temperature range reacting to the environment temperature change (Kuru and Aksay, 2014). For textiles modification, mainly organic PCMs (poly(ethylene)glycols, fatty acids and their derivatives, polyalcohols) serve as the polymer in mass modification of fibres: Outlast PCM viscose, Outlast acrylic and Outlast polyester (Outlast Technologies, 2014a), Outlast PCM filling for the apparel market Spherix, a blend of PCM viscose fibres, and polyester fibre balls, designed to provide dynamic climate regulation by controlling the production of moisture before it begins to enhance comfort in apparel such as jackets (Outlast Technologies, 2014b) and Nanonic SmartCel clima fibre (Nanonic, Inc., 2014).

PCMs can be also applied on textiles as encapsulated systems in the finishing stage, e.g. Outlast Thermocules (Outlast Technologies, 2014c), or Devan THERMIC. Thermoregulation technology based on reactive microcapsules is capable of being firmly anchored to the fibre by a chemical bond (Devan Chemicals, 2013). Lundgren et al. (2012) indicated PCMs are effective in high-humidity environments, but not in a hot and dry environments. Sweat evaporation may be the only way for them to lose heat in hot environments, but the highly insulated protective clothing impedes sweat evaporation and contributes to heat stroke. The position of the PCM layer in relation to the skin plays a vital role in its efficiency.

Functional fibres

Specially shaped fibres and functional fibre blends including modified polymers incorporation are used for quick sweat transport from the skin, keeping the wearer comfortable in dry conditions as demonstrated by the following commercially-successful and newly launched product examples along with the manufacturer's claims (Table 15.12):

Flame-retarded and moisture management fibres

As described above, hot environment conditions and quick temperature changes combined with intensive physic activity are dangerous because of the risk both of injury (burns, scalds) and heat stress and stroke by overheating. Therefore utilization of protective clothing with a high level of barrier protection and also physiological properties providing relevant wearing comfort (thermoregulation, sweat transport) enabling high and long appropriate performance level (fatigue reduction, freedom of movement) complies with a crucial aspect of working conditions and successful actions.

Mental and physical exertion increases the body's core temperature. Increased blood circulation and the production of perspiration cool the body down. This cooling down process can be disrupted because of improper heat and moisture management, leading to problems including concentration fatigue, muscle cramps, difficulties breathing and ultimately, heat stroke. These properties are based mainly on the apparel material/textile fibres and clothing construction. A survey of commercial FR fibres is summarized in Table 15.13 including value of Limiting oxygen index (LOI) as a rating of flameproof properties.

Examples of commercial protective fabrics and garments from flame retarded fibres with enhanced wearing comfort are described in Table 15.14 along with manufacturer's performance claims:

Superabsorbents

In a normal, open environment, cooling can be provided by a fast-wicking, fast-drying fabric. Garments typically worn by personnel working in a high-heat environment tend to be heavy and bulky, inhibiting the ability to move quickly. Development is aimed at reducing the weight of these garments while maintaining the same protection levels, keeping the wearer cool and comfortable. Bulky enclosed protective gear can

Table 15.12 Examples of functional fibres

Fibre	Producer	Characteristics
Coolmax Air fibre	Invista	Fibre with a patented propeller-shaped cross section. The micro-channels are extremely effective at wicking moisture dispensing it to quickly evaporate (Invista, 2017a)
Cordura denim	Invista/ Artistic Milliners	Denim line from blend cotton/Invista T420 nylon 6.6 with a high-stretch and para-aramide properties, thermoregulation, moisture management, enhanced tear resistance (Cordura Fabric, 2017)
Nilit Breeze	Nilit	Flat nylon 6.6 yarns with a special cross-section, unique texturing and an inorganic additive for cooling effect, efficient ventilation capabilities, and UV protection (Nilit, 2017)
Corkshell	Schoeller	A technology based on natural FSC-certified cork granulate (by-product in the production of wine corks) to provide high thermal insulation, breathability and wearer comfort (Schoeller, 2017a)
Inotek fibre	MMT Textiles	A smart polymer imitating the botanical structure of pinecones, which open and close in response to moisture level. The bicomponent sheath-core staple cut fibre (MMT Textiles, 2017)
Coolcore	Coolcore	Cooling textile based on a patented chemical free technology for moisture transfer away from the skin. Hohenstein Quality Label for 'innovative technology—cooling power' (Coolcore, 2017)
Omni-Freeze ZERO	Columbia Sportswear	Thousands of little blue rings containing a special cooling polymer embedded in the fabric. When exposed to moisture the rings swell (like goose bumps) creating a cooling sensation (Columbia sportswear Company, 2017; Sweat Activated Cooling, 2013)
Thermolite Pro brand	Invista	Textiles based on yarns that absorb NIR rays to heat up the garment. The hollow fibre technology provides insulation and minimizes heat loss even in freezing temperatures (Invista, 2017b; Noll et al., 2016)
S. Café	Singtex	Fabrics with incorporated recycled coffee grounds (2%) ensuring odour control, UV-protection, quick drying characteristics, moisture management and wearing comfort (SINGTEX Industrial, 2015)
UV-blocking and cooling fabrics	Chieftex	Combination of a UV-blocking filament and full dull polyester filaments with a special cross-section and an integrated nano-sized cooling mineral. UV protection of UPF 50+ to 100+ (Chieftex Enterprise, 2016)

Table 15.13 Commercial FR fibres

FR fibre chemistry	Trade mark	Producer	LOI(%)
PBI (poly-benzimidazol)	CelazolePBI	PBI Performance Products Inc.	41
Meta-Aramide MPIA (poly-metaphenylene-isophthalamide)	Nomex, Teijinconex, New Star	DuPont Yantai Spandex Co. Teijin Limited	30
Meta-Aramide PPTA (poly-metaphenylene-isophthalamide)	Kevlar Twaron Technora	DuPont Teijin Limited Teijin Aramid BV	29
Aromatic poly-imide-amide	Kermel	Kermel S.A.	32
PBO (poly-p-phenylene-2,6-benzobisoxazole)	Zylon	Toyobo Co.	68
PREOX (oxidized PAN, carbon fibre)	Panox Pyron Zoltek	SGL Carbon Group Zoltek Companies, Inc.	55
FR Viscose (regenerated cotton)	Lenzing FR	Lenzing AG	27
Novoloid (phenol-aldehyde)	Kynol	American Kynol Inc.	33
Modacryl (PAN-vinylidenchloride, Sb)	Kanecaron Protex	Kaneka Corp Waxman Libres Ltd	33
FR modified acrylic (halogen- and Sb free)	PyroTex	PyroTex GmbH	43
LCP (liquid crystal polymer—HS polyacrylate)	Vectran	Kuraray	37
Melamine	Basofil	BASF	32
Aromaric polyamide $-SO_2-$	NewFire	New Fire Co. Ltd	31
PPS (polyphenylene sulphide)	Procon	Toyobo Evonic	34
Polyamide 66 with P/N- FR additive	Nexylon FR	Griltech	28

lead to heat stress and discomfort. These garments limit the ability of a wearer to lose heat through normal cooling methods such as evaporation. The alternative current solution includes the use of cooling gel or ice packs usually placed into pockets of specially constructed vests. Other more advanced systems use a pumped coolant (weight, bulk and cost are disadvantageous). Newest developments are focused on wickability and sweat absorption to help maintain a cool environment next to skin.

Finishing systems for heat reduction textiles

According to Georg Lang (Archroma) there are three primary textile-finishing cooling technologies on the market (Munsante, 2013c). Some commercial examples are described in Table 15.15.

Table 15.14 **Commercial FR fabrics and garments**

FR fabrics	Producer	Characteristics
Lenzing FR	Lenzing AG	Heat and flame resistant viscose fibre produced from beechwood protects against skin burning and heat stress. This danger is significantly reduced thanks to wicking moisture away, as proven by physiological tests (Lenzing AG, 2017a,b). Lenzing FR fabrics are available also in an aluminized form for protection from extreme radiant heat. Fibre blends of FR viscose and aramid are used for their excellent FR properties, improved physiological parameters, e.g. Comfort Plus (PGI Inc., 2011)
Nomex MHP	DuPont	Multi-hazard protection fabrics with high strength and comfort based on combination of Nomex with Kevlar. Protection of workers against heat and flame, arc flash and small molten metal splashes (DuPont, 2017b)
Kevlar XP 2104		FR and water repellent ballistic fabric (body armour) offers wearing comfort in wet or tropical climates (DuPont, 2017a)
Kevlar AS450X		Fibre for anti-stab, anti-spike and ballistic protection in armour applications with increased mobility and comfort. Suitable for lightweight laminates in difficult climate conditions (Kevlar fabric, 2014)
FR base layers	Protal	Seamless high-performance knitted workwear from Protal fibre (Waxman Fibres) maintains body temperature and offers freedom of movement and comfort even under heavy armour. Optimal heat distribution draws moisture away during physically demanding work in variable temperatures (Protal, 2017)
DriRelease	DriFire	Targeting military and industrial markets. Fortex: a fabric developed in partnership with DuPont – combination of Nomex and DriRelease is a more comfortable FR fabric compared to flat Nomex. Fortex combines rayon with the Nomex to provide a softer hand and a permanent wicking (Drifire, 2017)
Battleshield and Battleshield X	Massif in cooperation with W.L. Gore	Nylon-based FR and water-resistant laminates for military markets, oil, gas and electrical industries, first responders, general industrial workwear. Gore's FR stretch technology with increased breathability and water repellency (Massif, 2017a)
Pro-Fil	SSM	Warp and weft knits and wovens for PPE in category of no melt/no drip filaments. FR fibres and multi-fibre blends with increased high-tenacity have a soft handle and wicking properties suitable for next-to-skin garments (SSM Industries, 2016b)

Table 15.14 **Continued**

FR fabrics	Producer	Characteristics
Pro-CFR		A moisture management, antimicrobial FR cotton fabric for industrial, military, fire service and racing. Combined with TransDry fabric ensures a moisture management, fast drying and antimicrobial/antiodour effect (SSM Industries, 2016a; Moisture management antimicrobial, 2014)
Carbon Armour	National Safety Apparel - NSA	High-strength protective apparel for the welding and metals industries. Protection against burns caused by sparks and welding spatter, while providing comfort, breathability, moisture management and quick drying. High resistance to pin holes (National Safety Apparel, 2017; Carbon Armour, 2015)
GlenGuard FR	G&K Services	Vented flame-resistant uniforms and knit garments, providing cooler, comfortable and flame-resistant clothing. The uniforms feature better visibility striping to offer more protection for employees operating in limited light conditions (G&K Services, 2017)
Core range	Massif	FR fabrics and garments for the US military suitable for industrial markets, such as oil, gas and utilities. The garments with a strategic seam placement offer multi-dimensional stretch for a better range of motion, moisture management and breathability (Massif, 2017b)

- Finishes that wick moisture away from the skin

  Wicking technologies offer a first line of defense from overheating by pulling the perspiration away from the body. Wicking finish causes moisture to be drawn away from the skin, thus reducing bacteria growth. Odour, as a product of bacteria excrement, is also reduced.
- PCM that regulate temperatures and humidity between fibre and skin

  Temperature-regulating fabrics store and release heat in response to the environment.
- *Finishes that deflect the sun's rays* preventing fabrics from absorbing heat, and providing enhanced UV protection and antimicrobial properties: determined specially for dark fabrics that (when finished/dyed) reflect both visible and invisible sunrays providing a cooling effect for the wearer of up to 15°F. These heat reduction textiles are suitable for sport and outdoor activities (Schoeller, 2017b; Archroma, 2017e).

15.4.5.2 Textiles for low-temperature environment

Special textiles suitable for low temperature environment have been used for emergency clothing and shelters developed within the FT7 project S(P)EEDKITS

Table 15.15 **Commercial examples of thermoregulating functional garments and technologies**

Thermoregulating garments and technologies	Producer	Characteristics
Syngero	Klopman Int.	High-performance workwear fabrics including uniforms, military and police clothing. Stretch, soft-shell and laminated versions are available (Klopman Int, 2017). The functionality can be enhanced with Schoeller's finishing technologies such as 3XDry, Ecorepel, NanoSphere and Coldblack (Schoeller, 2017c,d). The Synergo clothing with moisture management properties and increased breathability is available in natural stretch with Invista's Lycra T400 fibre and two- or four-way stretch options. FR versions and high-visibility options are available (Launch pad, 2013).
Dryfast Adaptive	Pontetorto	Next-to-skin range of fabrics (Innovation In Textiles, 2017) reacting dynamically to changes in body temperature adjusting moisture evaporation to provide both cooling and thermal comfort. The functionality is based on a hydrofunctional polymer. At low temperatures the adaptive polymer (HeiQ) absorbs and stores moisture in a film surrounding each fibre. At high temperatures, the polymer changes its structure by releasing the stored moisture (Innovation in Textiles, 2017).
Glide	HeiQ	Low friction for wearer comfort, good handle and moisture management determined for workwear from polyester, polyamide, cotton (HeiQ Materials AG, 2017b)

coordinated by Centexbel for regions affected by natural disasters (floods, earthquakes). Kits also include clothing, blankets, bedding and semi-permanent windproofing insulation for the floor and roof (Wilson, 2015).

Bonded seams and details are without doubt an integral part of the modern look of today's outdoor clothing and sportswear, adding an aesthetic quality and performance (Wilson, 2016). Some commercial examples of textile products and technologies for production of outwear to cold environments are mentioned in Table 15.16 along with manufacturer's claims.

Table 15.16 **Commercial examples of textiles for cold environments**

Textiles for low temperatures	Producer	Characteristics
TENCEL R 100 with kaolin	Lenzing AG	Lyocell fibre modified with kaolin—a natural, white, crystalline clay material with a plated 2D sheet structure composed of units of one layer of silica tetrahedrons and one layer of alumina octahedrons: heat insulating, impermeable to heat and hot gases (Bisjak, 2015)
Black Yak outdoor jackets	Black Yak	Insulated jackets with specific, segmented insulation zones. Hybrid jacket contains three different insulation materials—Primaloft Gold, Polartech Alpha and goose down (MountainBlog Europe, 2017)
Pertex Quantum	Pertex	Fabric ensuring windproof and water-resistant protection. The zoned four-way stretch fabric ensures unhindered movement (Pertex, 2017)
Drirelease	Drirelease/ Mammut	Alpine, climbing and outdoor equipment; design and construction of active wear from Drirelease fabric blends (Drirelease, 2015) with high comfort and moisture-wicking properties combined with a soft feel and eco-friendly FreshGuard odour neutralizer (Haran, 2015)
Bicomponent polymers	Natick	Smart polymers that contract when cold, and straighten when warm, incorporated in textiles that offer warmth and thermoinsulation in cold climates (Munsante, 2013d)
Encapsil	Patagonia	A patented sustainable plasma-based coating to create water repellent down (Patagonia, 2017). Thermoinsulating, lightweight and loft. Encapsil Down belay Parka (Ethan, 2013) made from a water-resistant high-loft goose down treated with a plasma deposition of a silicone material
Polartech Alpha	Polartec	A synthetic insulation textile material that is warm, wind resistant, highly durable, quick drying and breathable. The thermoinsulating jackets are more open and provide breathable 'puffy'-style comfort garments that maintain the insulation effect while wet (Polartec, 2017a)
Polartec FR		A high-performance yarn blend that provides both flash fire and arc flash dual-hazard protection for workers in outdoor cold conditions. Optional high-visibility is available for triple protection (Polartec, 2017b)
Solar+	Schoeller Textil	A technology that absorbs the rays of the sun, no matter what their colour, thus warming clothes even in cold weather. Even thin fabrics can keep the wearer warm (Fibre2Fashion, 2017)

(Continued)

Table 15.16 Continued

Textiles for low temperatures	Producer	Characteristics
Octa	Teijin Fibres	Polyester fibre suitable for innerwear with highly modified cross-sections of eight projections aligned in a radial pattern around a hollow fibre (octopus-like appearance) with high air content. Rapid sweat absorption and drying, low weight, heat shielding/insulation (Teijin, 2017)
Realfleece Nano	Icebreaker	Merino insulated softshell jackets with Bluesign-accredited nanotechnology for water, oil and mud repellency and breathability. Based on nanoparticles with a rugged surface. Triple-layer construction for warm, windproof, WR washable garments (Icebreaker, 2017)
Aria	Thermore	'Light as air' hypoallergenic insulation as an alternative to real down feathers, consists from 98% air by volume, ultra-light and soft insulator that mimics the appearance of down (Thermore, 2017)
Hexoskin Arctic	Hexoskin	A textile inspired by nature (polar bears that trap heat between their skin and fur). While allowing natural evaporation of perspiration, it absorbs very little water (Hexoskin, 2017). It dries rapidly, keeping the body temperature constant. This reduces energy consumption to a minimum and improves wearers' performance (Haran, 2016)

15.4.5.3 Multifunctional barriers

Liquid-tight and breathable laminates

A combination of barrier properties and physiological comfort in protective clothing cannot often be achieved only by a choice of material or its finishing. Besides the functionalization methods based on customized construction of the textile itself and/or application of continuously extending functional finishing systems, multilayer constructions based on different textile layers eventually supported by multifunctional membranes significantly widen the range of attainable solutions on the way to multifunctional effects. There are numerous modern lamination procedures (Singha, 2012). In combination with actually available functional membranes, various lamination processes promote innovations of protective and barrier textiles. Various forms of laminating (scattering of powders, coating of pressure sensitive systems, hot melt and dot print adhesives, etc.) significantly extend variability of attainable effects. The outer layer of the laminate ensures liquid repellency, antimicrobial, self-cleaning and anti-static properties. The functional membrane can be used as an intra layer, creating a barrier against liquid- and microorganism-penetration with some level of breathability (moisture vapour transport) being porous or through hydrophilic groups (usually PES-, PBT- or PU based). The membrane also provides the laminate with a windproof-, and therefore some thermoinsulation, effect. An inner, close-to-skin layer ensures sweat transport from the skin toward the membrane, which transfers the moisture to the outer layer where it is distributed and evaporated. Some typical examples of highly functional performance laminates and the manufacturer's claims are mentioned in Table 15.17.

Table 15.17 Examples of functional laminates for protective clothing

Multifunctional textiles	Producer	Characteristics
Phaseable Adaptable laminates	Sympatex Technology	3D apparel technology designed to intelligently influence the body climate (Sympatex, 2017b). 2.5 laminates designed with a 3D half-layer on the inside of the laminate, which only touches the skin with foam points. When physical activity increases, moisture is produced, causing the compact hydrophilic Sympatex membrane to swell, the foam points disappear, and the laminate gets closer to the skin enabling the moisture transport
Climate Technology Seamless		Windproof, waterproof, breathable knitted laminates with Sympatex membrane for occupational safety including footwear (Sympatex, 2017a). High abrasion- and tear–resistance
GORE PYRAD	W. L. Gore & Associates GmbH	Protective fabric for arc-flash and foul weather protection combining comfort and durability for outdoor clothing membranes. Breathable and FR moisture barriers using hot melt technology and FR adhesives based on a membranes with thermostable PU or dual-component ePTFE (W. L. Gore and Associates GmbH, 2017a)
Gore-Tex 'Surround' technology		Laminates for hiking footwear, with all-round breathability and waterproofness. Sweat is transported through the laminate positioned on the underside of the shoe (W. L. Gore and Associates GmbH, 2017c)

(Continued)

Table 15.17 **Continued**

Multifunctional textiles	Producer	Characteristics
PARALLON		A layered structure of two membranes: one membrane is placed onto a thermal material under a selectable upper material (e.g. Nomex, PBI matrix, titanium) as a carrier. The second is glued to the inner lining facing toward the body (W. L. Gore and Associates GmbH, 2017b)
FR laminates with Trans Textil membranes	Trans Textil	FR laminates for moisture barriers in firefighters' clothing. The composites are produced using hot-melt technology, thermostable breathable membranes. The moisture barrier provides reliable protection against wind, rain, fire-extinguishing water and liquid chemicals (Hänsch and Bochmann, 2015)
High Performance Plus		Breathable tri-laminate for surgical gowns resistant against liquids and microorganisms penetration. Breathable membrane from hydrophilic PU laminated by Point-to-Point technology (Trans-Textil GmbH, 2013a)
Topaz Carbon Membrane		Water vapour permeable antistatic membrane as an alternative of antistatic fibre incorporated into the outer layer of the laminate (Trans-Textil GmbH, 2013b)
Arnitel VT technology	Toyota Tsusho/ DSM	High-performance membranes for clothing made from Royal DSM's Arnitel VT (DSM, 2017). Flexible thermoplastic polyester-based elastomer. No perfluorinated chemicals and is 100% recyclable, waterproof and breathable. It acts as a barrier to liquids, bacteria and viruses, making it also suitable for use in surgical gowns
Nexar	Kraton	Membranes from sulfonated polymers for microclimate controlled textiles with enhanced evaporative cooling for high-performance textiles based on Nexar technology (Kraton Performance Polymerc Inc, 2017)
Tecasystem Millenia 450	TenCate	The lightweight laminates for firefighters' protection. EN 469 level 2 - light, strong and comfortable yet protective (Koninklijke Ten Cate nv, 2013). Thermal 3D spun lace nonwoven trapping air barrier and PTFE-PU breathable membrane laminate is thermostable

Nanomembranes

A special category of functional laminates are textiles equipped with a nanofibrous membrane with engineered nanofibre alignment resulting in extremely high resistance against water penetration remaining highly breathable. High-performance materials using functional nanofibres for heat and flame protective clothing have been developed (Serbezeanu et al., 2015). Hybrid membranes prepared using Kevlar as a support textile carrier, and aromatic polyimide (FR) nanofibres as a protective coating, exhibit high thermal stability, high water penetration resistance, fire resistance and improved air permeability, retaining a sufficient water vapour transport. By application of FR nanofibres by electrospinning, the physiological properties of Kevlar fibres are said to be substantially improved bringing a high barrier (watertight) effect.

The Czech company *NANOMEMBRANE* (Nanomembrane, 2017) offers a comprehensive product line of laminates with nanofibrous membrane for the armed forces, police, civil forces and rescue teams.

15.5 Standards

Protective clothing has to ensure reliable barrier properties. These properties are determined and quantified according to end-use applications and have to be tested according to relevant standards. There are many testing methods and categorization standards used to characterize and evaluate the required barrier properties of protective fabrics in order to secure permission for the fabrics to be commercialized for planned applications. With a diverse range of needs in PPE from high visibility to flame retardancy and impact protection, the list of potential assessment methods is vast. General requirements for basic PPE groups are summarized in Table 15.18. Standards for repellent properties and resistivity against liquids penetration are contained in Table 15.19 (water), Table 15.20 (oil) and Table 15.21

Table 15.18 Protective clothing—General requirements

EN ISO 13688	Protective clothing—General requirements
EN 469	Protective clothing for firefighters—Performance requirements for protective clothing for firefighting
EN 470-1	Protective clothing for use in welding and allied processes. Part 1: General requirements
EN 13921	Personal protective equipment—Ergonomic principles
EN 13795	Surgical drapes, gowns and clean air suits, used as medical devices for patients, clinical staff and equipment
ISO 7243	Hot environments. Estimation of the heat stress on working man, based on the WBGT-index (wet bulb globe temperature)

Table 15.19 Hydrophobicity, water repellency, rain

EN 24920	Textiles. Determination of resistance to surface wetting (spray test) of fabrics
EN 20811	Textiles. Determination of resistance to water penetration: Hydrostatic pressure test
EN 29865 (ISO 9865)	Textiles; determination of water repellency of fabrics by the Bundesmann rain-shower test
EN 343+A1	Protective clothing: Protection against rain
EN 14360	Protective clothing against rain. Test method for ready-made garments. Impact from above with high-energy droplets
ISO 23232	Aqueous liquid repellency: Water/alcohol solution resistance test
AATCC 193	Aqueous Liquid Repellency: Water/alcohol solution resistance test (DuPont test)
AATCC TM 22	Water Repellency: Spray Test
AATCC TM 42	Water Resistance: Impact Penetration Test
AATCCTM 70	Water Repellency: Tumble Jar Dynamic Absorption Test
AATCC TM 35	Water Resistance: Rain Test
AATCC TM 127	Water Resistance: Hydrostatic Pressure Test
ASTM D7334	Standard Practice for Surface Wettability of Coatings, Substrates and Pigments by Advancing Contact Angle Measurement

Table 15.20 Oil-repellency

EN ISO 14419	Textiles, Oil repellency: Hydrocarbon resistance test
AATCC 118	Oil Repellency: Hydrocarbon Resistance Test
AATCC 130	Soil Release: Oily Stain Release Method

(chemicals). Resistivity against heat and flame, flame spreading, ignitability, heat transport and other relevant properties (burn prediction, emission, smoke evaluation) are evaluated according to standards in Table 15.22. Thermoinsulation properties testing (Table 15.23), antistatic properties (Table 15.24), high visibility (Table 15.25), protection against UV-radiation (Table 15.26) and antimicrobial properties (Table 15.27), are also often required for some types of protective clothing. Mechanical properties (Table 15.28), and physiological parameters (Table 15.29), weather resistance (Table 15.30) and soil release properties (Table 15.31), are also very important parameters necessary to be tested according to relevant standards. Durability of barrier effects in repeated washing is necessary to be tested after standardized washing procedures (Table 15.32).

Table 15.21 Protection against liquid, chemicals, gaseous chemicals

EN 464	Protective clothing—Protection against liquid and gaseous chemicals, including aerosols and solid particles. Test method: Determination of leak-tightness of gas-tight suits (Internal pressure test)
EN 465	Protection against liquid chemicals—Performance requirements for chemical protective clothing with spray-tight connections between different parts of the clothing (Type 4 Equipment)
EN 466	Protective clothing—Protection against liquid chemicals: Performance requirements for chemical protective clothing with spray-tight connections between different parts of the clothing (Type 3 Equipment)
EN 467	Protective clothing—Protection against liquid chemicals: Performance requirements for garments providing protection to parts of the body
EN 943-1,2	Protective clothing against liquid and gaseous chemicals, including liquid aerosols, and solid particles – Part 1: Performance requirements for ventilated and non-ventilated gas-tight (Type 1) and non-gas-tight (Type 2) chemical protective suits
EN ISO 6529	Protective clothing—Protection against chemicals: Determination of resistance of protective clothing materials to permeation by liquids and gases
EN ISO 6530	Protective clothing—Protection against liquid chemicals: Test method for resistance of materials to penetration by liquids
EN 13034	Protective clothing against liquid chemicals—Performance requirements for chemical protective clothing offering limited protective performance against liquid chemicals (Type 6 and Type PB[6] equipment)
EN 14325	Protective clothing against chemicals—Test methods and performance classification of chemical protective clothing materials, seams, joins and assemblages
EN 14605	Protective clothing against liquid chemicals—Performance requirements for clothing with liquid-tight (Type 3) or spray-tight (Type 44) connections, including items providing protection to parts of the body only (Type PB [3] and [4])
EN 14786	Protective clothing—Determination of resistance to penetration by sprayed liquid chemicals, emulsions and dispersions: Atomiser test
EN ISO 17491-3	Protective clothing—Test methods for clothing providing protection against chemicals—Part 3: Determination of resistance to penetration by a jet of liquid (jet test)
EN ISO 17491-4	Protective clothing—Test methods for clothing providing protection against chemicals—Part 4: Determination of resistance to penetration by a spray of liquid (spray test)

Table 15.22 **Flameproof, heat-resistant**

EN ISO 15025	Protective clothing—Method of test for limited flame spread
EN 1103	Fabrics for apparel—Detailed procedure to determine the burning behaviour
EN ISO 15831	Clothing—Physiological effects: Measurement of thermal insulation by means of a thermal manikin
EN ISO 14116	Protective clothing against limited flame spread materials
EN 348	Protective clothing. Determination of behaviour of materials on impact of small splashes of molten metal
ASTM D1230	Standard test Method for Flammability of Apparel Textiles 45° Flame Test
EN 367	Protective clothing. Test method: Heat transfer index
EN ISO 6942	Protective clothing. Test method: Radiant heat transfer indexes
ISO/TR 2801	Clothing for protection against heat and flame—General recommendations for selection, care and use of protective clothing
ISO 9151	Protective clothing against heat and flame—Determination of heat transmission on exposure to flame
ISO 9150	Protective clothing—Determination of behaviour of materials on impact of small splashes of molten metal
ISO 13506	Protective clothing against heat and flame—Test method for complete garments—prediction of burn injury using and instrumented manikin
ISO 14460	Protective clothing for automobile racing drivers—Protection against heat and flame—Performance requirements and test methods
ISO 17492	Clothing for protection against heat and flame—Determination of heat transmission on exposure to both flame and radiant heat
ISO 17493	Clothing and equipment for protection against heat—test method for convective heat resistance using a hot air circulation oven
EN ISO 11611	Protective clothing for use in welding and allied processes
EN ISO11612	Protective clothing—Clothing to protect against heat and flame: Minimum performance requirements
EN ISO 9185	Protective clothing—Assessment of resistance of materials to molten metal splash
EN 12127-2	Clothing for protection against heat and flame—Determination of contact heat transmission through protective clothing or constituent materials—Part 2: Test method using contact heat produced by dropping small cylinders
EN 61482-1-2	Live working—Protective clothing against the thermal hazards of an electric arc -Part 1–2: Test methods
ISO 17846	Welding and allied processes—Health and safety: Wordless precautionary labels for equipment and consumables used in arc welding and cutting
ISO 16852	Flame arresters - Performance requirements, test methods and limits for use

Table 15.22 Continued

ISO 15383	Protective clothing for firefighters—Laboratory test methods and performance requirements
ISO/FDIS 16073	Wildland firefighting personal protective equipment—Requirements and test methods
ISO 5660-1	Reaction-to-fire tests—Heat release, smoke production and mass loss rate—Part 1: Heat release rate (cone calorimeter method)
NFPA 2112	Standard on flame-resistant garments for protection of industrial personnel against flash fire

Table 15.23 Thermoinsulation

EN 14058	Protective clothing—Garments for protection against cool environments
EN 342	Protective clothing—Garments for protection against cold (−50°C)
EN ISO 15831	Clothing—Physiological effects: Measurement of thermal insulation by means of a thermal manikin
EN ISO 15027-2	Immersion suits—Part 2: Abandonment suits, requirements including safety
EN ISO 11079	Ergonomics of the thermal environment—Determination and interpretation of cold stress when using required clothing insulation (IREQ) and local cooling effects
EN 31 092	Determination of physiological properties—Measurement of thermal and water-vapour resistance under steady-state conditions (sweating guarded-hotplate test)

Table 15.24 Antistatic

EN 1149-1	Protective clothing—Electrostatic properties—Part 1: Test method for measurement of surface resistivity
EN 1149-2	Protective clothing—Electrostatic properties—Part 2: Test method for measurement of the electrical resistance through a material (vertical resistance)
EN 1149-3	Protective clothing—Electrostatic properties—Part 3: Test method for measurement of charge decay
EN 1149-5	Protective clothing—Electrostatic properties—Part 5: Material performance and design requirements
ENV 50354	Electrical arc test methods for material and garments, for use by workers at risk from exposure to an electrical arc
AATCC TM 76	Electrical resistivity of fabrics

Table 15.25 **High visibility**

EN ISO 20471	**High-visibility clothing—Test methods and requirements**
EN 1150	Protective clothing—Visibility clothing for non-professional use—Test methods and requirements
EN 1486	Protective clothing for firefighters—Test methods and requirements for reflective clothing for specialized firefighting

Table 15.26 **UV-protection**

EN 13758-1 +A1	Textiles—Solar UV protective properties—Part 1: Method of test for apparel fabrics
EN 13758-2 +A1	Textiles—Solar UV protective properties—Part 2: Classification and marking of apparel
AS/NZS 4399 (1996)	Sun protective clothing—Evaluation and classification
AATCC TM 183	Transmittance or blocking of erythemally weighted ultraviolet radiation through fabrics
ASTM D6544	Standard practice for preparation of textiles prior to ultraviolet (UV) transmission testing
ASTM D6603	ASTM D6603

Table 15.27 **Antimicrobial**

EN ISO 22612	Clothing for protection against infectious agents: Test method for resistance to dry microbial penetration
EN ISO 22610	Surgical drapes, gowns and clean air suits, used as medical devices, for patients, clinical staff and equipment: Test method to determine the resistance to wet bacterial penetration
AATCC TM 100	Antibacterial finishes on textile materials: Assessment of (quantitative)
AATCC TM 147	Antibacterial activity assessment of textile materials: Parallel streak method (qualitative)
AATCC TM 90	Antibacterial activity assessment of textile materials: Agar plate method
AATCC TM 30	Antifungal activity, assessment on textile materials: Mildew and rot resistance of textile materials
EN ISO 20743	Textiles: Determination of antibacterial activity of textile products
JIS L 1902	Testing antibacterial activity and efficacy on textile products
ASTM E 2149	Determining the antimicrobial activity of immobilized antimicrobial agents under dynamic contact conditions
EN 14119	Testing of textiles: Evaluation of the action of microfungi

Table 15.28 **Mechanical**

EN 530	Abrasion resistance of protective clothing material—Test methods
EN ISO 12947-2	Textiles: Determination of the abrasion resistance of fabrics by the Martindale method. Martindale abrasion testing apparatus
EN 863	Protective clothing—Mechanical properties—Test method: Puncture resistance
EN 388	Protective gloves against mechanical risks
EN 381	Protection against hand-held chain saws
EN ISO 14877	Protective clothing for abrasive blasting operations using granular abrasives
EN ISO 13995	Protective clothing – Mechanical properties – Test method for the determination of the resistance to puncture and dynamic tearing of materials
EN ISO 13997	Protective clothing – Mechanical properties – Determination of resistance to cutting by sharp objects

Table 15.29 **Physiological properties**

EN ISO 9237	Textiles—Determination of the permeability of fabrics to air
EN ISO 11092	Textiles—Determination of physiological properties—Measurement of thermal and water-vapour resistance under steady-state conditions (sweating guarded-hotplate test)
EN ISO 15496	Textiles—Measurement of water vapour permeability of textiles for the purpose of quality control
EN 30192 (ISO11092)	Textiles—Determination of physiological properties—Measurement of thermal and water-vapour resistance under steady-state conditions (sweating guarded-hotplate test)
AATCC TM 195	Method quantifies moisture management properties for performance fabrics
AATCC 201	Drying rate of fabrics: Heated plate method

Table 15.30 **Weather resistance**

AATCC TM 111	Weather resistance of textiles: Exposure to daylight and weather
AATCC TM 202	Relative hand value of textiles: Instrumental method
AATCC TM 204	Water vapour transmission of textiles
AATCC TM 169	Weather resistance of textiles: Xenon lamp exposure
AATCC TM 186	Weather resistance: UV light and moisture exposure

Table 15.31 Soil release and staining

AATCC TM 130	130, Soil release: Oily stain release method
AATCC TM 151	Soil redeposition: Launder-Ometer method
AATCC Grey Scale	Grey scale for staining evaluation

Table 15.32 Washing for testing

EN ISO 15797	Textiles—Industrial washing and finishing procedures for testing of workwear

15.6 Conclusions

Protective clothing can be considered a major segment of PPE. Initial simple protection against bad weather by water-repellence is extended to fulfil a combined effect protecting against water and/or penetration of harsh liquids and contaminants from the outside, as well as breathability, which is the permeability of evaporated moisture caused by physical exertion of users. Often, a number of next-additional-protection functionality needs are warranted. New processing technologies make this possible through proper knowledge of synergies between chemistry and supporting procedures of pre-activation, as well as post-treatment of fabrics. Often the single layer of the textile carrier is not sufficient to afford all requested functionalities hence the need for multifunctional multilayer structures. Laminated (functional) membranes and sandwich-barrier textile structures penetrate the steadily growing PPE markets. Typical visual fashion tools like design and colour are significantly encouraged and often combined with visually un-perceptible functionalities.

Within the last few decades, there has been a synergic impact on the growing market, as rising demand for safe, healthy and comfortable clothing comes from industries influenced by rising diversity of workers' activities in advanced economies. These industries, pushed by the influence of workforce efficiency and decreased absence of skilled people, and supported by a steadily hard-and-fast role of public procurement and consonant pressure from health insurance systems and unions increasingly partnering in specification, development and use of PPEs. The need for the efficiency of a fully qualified workforce sets up demand for more safety offered by the newly introduced multifunctional textiles, respecting their often logically higher prices. An existing wide range of solutions, as presented in this chapter, starts with the selection of materials and goes through customized multifunctional finishing. A thorough knowledge of the capabilities and limits of protective clothing is necessary to efficiently avoid risking user safety while maintaining user readiness. Adequate maintenance procedures and the skill of their providers significantly influences not only PPE quality but also the lifespan of this product category. This is the key to a balanced covering of the often-higher acquisition costs. Unified strategy of design, manufacturing and maintenance management becomes the most efficient business model for this category of products. Customized production and cleaner production, as well as circular economy implementation, supporting whole life cycle dematerialization, must penetrate the protective clothing strategy of production and use. That's why these aspects were included. Right selection and proper implementation of quality and safety standards is an integral part of the PPE business. Moreover, rising technological and material possibilities created in the area of PPE penetrate into other highly important segments of the textile market: Outdoor articles are in greater demand. Contrary to PPE with necessarily proportioned acquisition costs (acceptable for public tenders), in the case of outdoor activities as a steadily rising part of the lifestyle, value and costs of functionality and comfort in protection open doors for better margins. Consequent to social and societal trends, quality of life and rapid development of health care will bring new requirements and opportunities for functional materials with specific protective parameters. The same is true in the area of living standards for the elderly in developed countries. This aging demographic trends clearly toward functional textiles and comfortable customized garments, thus increasing market potential (supported by selection of PPE as one of Lead Market Initiatives of EC preferential programme of innovation). Also in new growing economies, a tremendous growth of industries and the developing consumer demand for functional wellbeing textiles that combine safety and comfort, shift market demand toward added-value functional textiles.

References

3M, 2017. 3M Unveils a "Next generation" Scotchgard™ protector repellent finish for outerwear fabrics. [Online]. Available from: http://news.3m.com/press-release/company/3m-unveils-next-generation-scotchgard-protector-repellent-finish-outerwear-fab (accessed 29.05.17).

AGCCE, 2015. AsahiGuard E-Series. [Online]. Available from: http://www.agcce.com/asahiguard-e-series/ (accessed 29.05.17).

Alemdaroglu, F., 2016. Novel short chain (carbon C6) fluoroproducts with improved properties. In: International Conference on Textile Coating and Laminating, Prague, 16–17 March.

Archroma, 2017a. Archroma's Moisture Management System. [Online]. Available from: http://www.bpt.archroma.com/products-services/finishing/moisture-management/ (accessed 01.03.17).

Archroma, 2017b. Arkophob® FFR. [Online]. Available from: http://www.bpt.archroma.com/arkophob-ffr/ (accessed 29.05.17).

Archroma, 2017c. Protection that keeps textiles dry and nature cleaner. [Online]. Available from: http://textiles.archroma.com/smartrepel-hydro/ (accessed 26.05.17).

Archroma, 2017d. Smartrepel Hydro. [Online]. Available from: http://www.bpt.archroma.com/smartrepel-hydro/ (accessed 29.05.17).

Archroma, 2017e. Sun reflector—UV protector. [Online]. Available from: http://www.bpt.archroma.com/about-brand-performance-textile-specialties/cooperations/schoeller-cold black/ (accessed 03.03.17).

Atav, R., Bariş, B., 2016. Dendrimer technology for water and oil repellent cotton textiles. AATCC J. Res. 2 (2), 16–24.

Baillié, A.E., Warner, S.B., Fan, Q., 2015. UV-curable coatings for textiles: water repellent finishes. AATCC Rev. 5 (11), 35–39.

Bartowiak, G., Dąmbrowska, A., Marszałek, A., 2013. Analysis of thermoregulation properties of PCM garments on the basis of ergonomic tests. Text. Res. J. 83 (2), 148–159.

Bartowiak, G., Dąmbrowska, A., Włodarzcyk, B., 2015. Construction of a garment for an integrated liquid cooling system. Text. Res. J. 85 (17), 1809–1816.

Baurer, M., Lohrey, L., Bachus, H., Höflich, A., Zyschka, R., Otterbach, A., 2015. Durable fluorocarbon-free water-repellents without CMR compounds. Melliand Textilber. 21 (4), 258–259.

Bisjak, C., 2015. Designing a novel fiber for fire protection applications. Tech. Text. Int. 58 (2), E47–E48.

Bolger O' Hearn, Inc., 2017. Advanced Fluorine Free™ Water Repellent Bluesign® Approved. [Online]. Available from: http://bolgerohearn.com/tag/altopel-f3/ (accessed 29.05.17).

California in the NanoEconomy, 2013. Website With Resource for the Nanotechnology. [Online]. Available from: http://californiananoeconomy.org/content/nano-tex-0 (accessed 01.03.17).

Carbon Armour, 2015. Futur. Mater. 2, 33.

Cattermole, A., 2016. Catalyst for change. AATCC Rev. 16 (2), 30–38.

Chemours, 2017. Capstone™ Surfactants & Repellents. [Online]. Available from: https://www.chemours.com/Capstone/en_US/uses_apps/textiles/oleophobol_cp_products.html (accessed 29.05.17).

Chieftex Enterprise Co., 2016. Most Extreme Care. [Online]. Available from: http://www.chieftex.com.tw (accessed 02.03.17).

CHT, 2017. [Online]. Ecoperl Active. Available from: https://www.cht.com/cht/web.nsf/id/pa_en_productdetail.html?Open&pID=00084756.100_EN (accessed 29.05.17).

Columbia Sportswear Company, 2017. Explore Technologies That Keep You Warm, Dry, Cool, Protected. [Online]. Available from: http://www.columbia.com/technology-landing.html (accessed 27.05.17).

Commercialising Plasma Treatments, 2016. Int. Dye. 201 (5), 23–24.

Coolcore, 2017. Technology—Powered by Coolcore. [Online]. Available from: http://coolcore.com/technology/ (accessed 27.05.17).

Cordura Fabric, 2017. CORDURA® Denim Fabric—Hard Working Comfort™. [Online]. Available from: http://www.cordura.com/en/fabric-technology/denim-fabric.html (accessed 27.05.17).

Cotton Incorporated, 2017. Wicking Windows™ Technology. [Online]. Available from: http://www.cottoninc.com/product/product-technology/moisture-management/wicking-windows/ (accessed 26.05.17).

Coulson, S., 2007. Plasma treatment of textiles for oil and water repellency. In: Shishoo, R. (Ed.), Plasma Technologies for Textiles. Woodhead Publishing, Cambridge, pp. 13–201.

Council Directive 89/686/EEC, December, 21 1989. On the approximation of the laws of the member states relating to personal protective equipment.

Daikin Industries Ltd, 2017. Water and Oil Repellents. [Online]. Available from: http://www.zschimmer-schwarz.com/8.html?typ=X&id=ANTHYDRIN%20NK%206 (accessed 29.05.17).

Defence Update, 2017. Moisture Management Fabrics for Performance Apparel. [Online]. Available from: http://defense-update.com/products/m/moisture_management_fabric.htm (accessed 26.05.17).

Devan Chemicals, 2017. H2O Repel. Durable Fluorocarbon-Free Technology. [Online]. Available from: http://www.devan.net/product.aspx?prodid=14 (accessed 29.05.17).

Devan Chemicals, 2013. Thermoregulation Technology. [Online]. Available from: http://www.devan.net/files/pdf/THERMIC_A4_flyer_EN.pdf (accessed 27.05.17).
Devan Chemicals, 2017. Eco Release. Stain Release Technology. [Online]. Available from: http://www.devan.net/product.aspx?prodid=5 (accessed 29.05.17).
DR PETRY, 2008. Textile Auxiliaries—Moisture Management. [Online]. Available from: http://www.drpetry.de/fileadmin/user_upload/petry/pdfs/Moisture_Management_e.pdf (accessed 26.05.17).
Drifire, 2017. Drifire Product Line. [Online]. Available from: http://www.drifire.com/flame-resistant-clothing/drifire-product-line (accessed 27.05.17).
Drirelease, 2015. Mammut Wins Drirelease® Innovation Award for Legacy of Design Ingenuity. [Online]. Available from: http://drirelease.com/news/mammut-wins-drirelease-innovation-award-for-legacy-of-design-ingenuity/ (accessed 28.05.17).
DSM, 2017. Toyota Tsusho Launches PFC-Free Waterproof Breathable Membranes in Outdoor Clothing Based on DSM Technology. [Online]. Available from: http://www.dsm.com/products/arnitel/en_US/home.html (accessed 28.05.17).
DuPont, 2017a. DuPont Announces Millionth Vest Made with DuPont™ Kevlar® XP™. [Online]. Available from: http://www.dupont.com/products-and-services/fabrics-fibers-non wovens/fibers/press-releases/dupont-millionth-vest-kevlar-xp.html (accessed 27.05.17).
DuPont, 2017b. DuPont Launches Next-Generation Multi-Hazard Protection. Available from: http://www.dupont.com/products-and-services/personal-protective-equipment/thermal-protective/press-releases/Next-Generation-Multi-Hazard-Protection-Dupont-MHP.html (accessed 27.05.17).
Ecorys, 2009. Study on the Competitiveness of the EU security industry – Final report. [Online]. Available form: https://ec.europa.eu/home-affairs/sites/homeaffairs/files/elibrary/documents/policies/security/pdf/study_on_the_competitiveness_of_the_eu_security_industry_en.pdf (assessed 09.05.17).
EnsembleIQ, 2017. Dow Corning's Soft Hydro Guard Finish Aimed at Sports Apparel. [Online]. Available from: http://apparel.edgl.com/news/Dow-Corning-s-Soft-Hydro-Guard-Finish-Aimed-at-Sports-Apparel79141 (accessed 26.05.17).
EPA, 2016. New Chemicals Program Review of Alternatives for PFOA and Related Chemicals. [Online]. Available from: https://www.epa.gov/assessing-and-managing-chemicals-under-tsca/new-chemicals-program-review-alternatives-pfoa-and (accessed 26.05.17).
Ethan, S., 2013. The Patagonia Encapsil Down Belay Parka: An Origin Story. [Online]. Available from: http://www.patagonia.com/blog/2013/03/the-patagonia-encapsil-down-belay-parka-an-origin-story/ (accessed 28.05.17).
Euratex, 2016. Key Figures. [Online]. Available from: http://euratex.eu/fileadmin/user_upload/documents/key_data/Euratex_Keyfigures_-_2015.pdf and http://euratex.eu/press/key-data (accessed 05.03.17).
European Commission, 2016. New Networks of Public Procurers to Help Stimulate Innovation. [Online]. Available from: http://ec.europa.eu/growth/tools-databases/newsroom/cf/itemdetail.cfm?item_type=254&lang=en&item_id=3663 (accessed 06.03.17).
European Commission, 2017a. European Standards. [Online]. Available from: https://ec.europa.eu/growth/single-market/european-standards_en (accessed 24.05.17).
European Commission, 2017b. Personal Protective Equipment (PPE). [Online]. Available from: http://ec.europa.eu/growth/sectors/mechanical-engineering/personal-protective-equipment_cs (accessed 24.05.17).
European Commission, 2017c. Textiles and Clothing Industries. [Online]. Available from: http://ec.europa.eu/growth/sectors/fashion/textiles-clothing/ (accessed 06.03.17).
European Commission, 2017d. Textiles and Clothing Legislation. Available from: https://ec.europa.eu/growth/sectors/fashion/textiles-clothing/legislation_en (accessed 24.05.17).

Europlasma, 2017. Combine Durable Water Repellency With Outstanding Breathability, Flexibility and Lightweight. [Online]. Available from: http://www.europlasma.be/Application/PlasmaGuard-Sporting-Goods (accessed 29.05.17).

Evonik Industries, 2017. Dynasylan® The Brand for Functional Silanes. Available from: https://www.dynasylan.com/lpa-productfinder/page/productsbytext/detail.html?pid=193&lang=en (accessed 29.05.17).

Fibre2Fashion, 2017. Soak up the Sun With Solar+to Stay Warm in Winters. [Online]. Available from: http://www.fibre2fashion.com/news/textiles-company-news/newsdetails.aspx?news_id=119150 (accessed 28.05.17).

Fisher, M., Vögtle, F., 1999. Dendrimers: from design to applications. Angew. Chem. Int. Ed. 38, 884–905.

Fluorochemical advice, 2015. Int. Dye. 200 (3), 33.

FluoroCouncil, 2014. Guidance for Best Environmental Practices (BEP) for the Global Apparel Industry, including focus on fluorinated repellent products. Available from: https://fluorocouncil.com/PDFs/Guidance-for-Best-Environmental-Practices-BEP-for-the-Global-Apparel-Industry.pdf (accessed 26.05.17).

Fu, M., Weng, W., Yuan, H., 2014. Effects of multiple air gaps on the thermal performance of firefighter protective clothing under low-level heat exposure. Text. Res. J. 84 (9), 968–978.

G&K Services, 2017. GlenGuard® FR Products. Available from: http://www.gkservices.com/search?keyword=glenguard+fabric&type=All&=Search (accessed 27.05.17).

Glassmark Solution, 2016. Always Dry. Hydrophobic Surface Protection. [Online]. Available from: http://www.alwaysdry.com.my/ (accessed 29.05.17).

Goldstein, H.B., 1993. Mechanical and chemical finishing of microfabrics. Text. Chem. Color. 25 (2), 16–21.

Grand View Research, 2015a. Market Research Report—Industrial Protective Clothing Market Analysis by Product and Segment Forecast to 2022. [Online]. Available from: http://www.grandviewresearch.com/industry-analysis/industrial-protective-clothing-market (accessed 05.03.17).

Grand View Research, 2015b. Market Research Report—Technical Textiles Market Analysis by Technology and Segment Forecast to 2022. [Online]. Available from: http://www.grandviewresearch.com/industry-analysis/technical-textiles-market (accessed 07.03.17).

Grand View Research, 2016. Market Research Report—Personal Protective Equipment (PPE) Market Analysis by Product, by End-Use and Segment Forecast to 2024. [Online]. Available from: http://www.grandviewresearch.com/industry-analysis/personal-protective-equipment-market (accessed 05.03.17).

Greenpeace Calls for PFC Ban, 2013. Int. Dyer 198 (1), 6.

Hänsch, M., Bochmann, R., 2015. Flammhemmende atmungsaktive Nässesperren für Schutzkleidung zur Brandbekämpfung. Tech. Text. Int. 58 (2), 88–89.

Haran, F., 2015. The great outdoors. Futur. Mater. 5, 13–15.

Haran, F., 2016. The future is now. Int. Dye. 201 (2), 25–26.

Height, M., 2015. Water repellency without fluorine–learning from nature. In: ITMA—Textile Colourant and Chemical Leaders Forum 16 November, Milano.

HeiQ Materials AG, 2017a. Full Steam Ahead for Fluorocarbon-Free Water Repellency Innovations by HeiQ. [Online]. Available from: http://heiq.com/2015/11/03/full-steam-ahead-for-fluorocarbon-free-water-repellency-innovations-by-heiq/ (accessed 29.05.17).

HeiQ Materials AG, 2017b. Move Without Limits–Effortlessly. [Online]. Available from: http://heiq.com/technologies/heiq-glide/ (accessed 27.05.17).

Hexoskin, 2017. Hexoskin Arctic Smart Shirt. [Online]. Available from: https://www.hexoskin.com/products/hexoskin-arctic-smart-shirt-mens-cold-weather (accessed 28.05.17).

Holme, I., 1993. New developments in the chemical finishing of textiles. J. Text. Inst. 94 (4), 520–533.
Holme, I., 2014a. Durable water repellents: alternatives to C8 fluorocarbons. Int. Dye. 199 (4), 14–16.
Holme, I., 2014b. Advancing technologies in plasma treatment. Int. Dye. 199 (5), 18–20.
Holme, I., 2015a. Advances in repellent finishes. Int. Dye. 200 (6), 18–21.
Holme, I., 2015b. Nanotechnology in textile coloration and finishing. Int. Dye. 200 (1), 14–17.
Holme, I., 2015c. The textile coating industry in transition. Int. Dye. 200 (7), 33–37.
Holmes, D.A., 2000. Waterproof breathable textiles. In: Horrocks, A.R., Anand, S.C. (Eds.), Handbook of Technical Textiles. Woodhead Publishing, Cambridge, pp. 282–315.
Hu, J., Li, Y., Yeung, K.W., Wong, A.S.W., Xu, W., 2005. Moisture management tester: a method to characterize fabric liquid moisture management properties. Text. Res. J. 75 (11), 57–62.
Huntsman International LLC, 2017a. PHOBOL® CP-CR. [Online]. Available from: http://www.huntsman.com/textile_effects/a/Solutions/Product%20Highlights/Chemicals/PHOBOL_R%20CP-CR (accessed 29.05.17).
Huntsman International LLC, 2017b. PHOBOTEX® Advanced Fluorine-Free Rain and Stain Management. [Online]. Available from: http://www.huntsman.com/textile_effects/Media%20Library/global/files/phobotex_brochure_LR.pdf (accessed 29.05.17).
Huntsman International LLC, 2017c. Zelan™R3. [Online]. Available from: http://www.huntsman.com/textile_effects/a/Solutions/Product%20Highlights/Chemicals/Zelan%20R3 (accessed 29.05.17).
Icebreaker, 2017. Realfleace®. [Online]. Available from: http://nz.icebreaker.com/en/realfleece (accessed 28.05.17).
Innovation in Textiles, 2017. Pontetortos Dryfast Adaptive Offers Dynamic Comfort. [Online]. Available from: http://www.innovationintextiles.com/pontetortos-dryfast-adaptive-offers-dynamic-comfort/ (accessed 27.05.17).
Invista, 2017a. Coolmax®Fibre. [Online]. Available from: http://www.invista.com/en/brands/coolmax.html (accessed 27.05.17).
Invista, 2017b. Thermolite®Fiber & Insulation. [Online]. Available from: http://www.invista.com/en/brands/thermolite.html (accessed 27.05.17).
Kaufmann, R., Böker, A., Siebgens, I., Monßen, S., Mahltig, B., 2014. Amphiphile Proteine zur hydrophilen Ausrüstung von Synthesefasern. Melliand Text. Int. 95 (4), 194–196.
Kevlar fabric for wet conditions, 2014. Futur. Mater. (1), 22.
Kissa, E., 1981. Mechanism of soil release. Text. Res. J. 51 (8), 508–519.
Klopman Int, 2017. Synergo™. [Online]. Available from: http://www.klopman.com/en/catalog/synergo/ (accessed 27.05.17).
Koninklijke Ten Cate nv, 2013. New: TenCate Tecasystem™—Millenia 450. [Online]. Available from: http://www.tencate.com/fr/emea/protective-fabrics/markets/emergency-response/news/new-system-millenia-450.aspx (accessed 28.05.17).
Kraton Performance Polymerc Inc., 2017. NEXAR® Polymers Next Generation Cooling Technology for Performance Materials. [Online]. Available from: http://www.kraton.com/products/nexar.php (accessed 28.05.17).
Kuru, A., Aksay, S.A., 2014. Cellulose-PEG grafts from cotton waste in thermo-regulating textiles. Text. Res. J. 84 (3), 337–346.
Kwong, C.H., Ng, S.P., Kan, C.W., Molina, R., 2013. Parametric study of CF_4-plasma on the hydrophobicity of polyester synthetic leather. Fibers Polym. 14 (10), 1608–1613.
Lang, G., 2015. Recent advantages in sustainable repellency solutions. In: ITMA—Textile Colourant and Chemical Leaders Forum, 16 November, Milano.
Lange-Chenier, H., Osborn, A., Ryall, J., 2014. A question of standards. Futur. Mater. 6, 8–10.
Launch pad. 2013. Futur. Mater. (4), pp. 24–25.

Lee, S.H., Song, W.S., 2010. Surface modification of polyester fabrics by enzyme treatment. Fibers Polym. 11 (1), 54–59.

Lenzing AG, 2017a. Lenzing™ FR. [Online]. Available from: http://www.lenzing-fibers.com/en/lenzing-fr/lenzing-frr/ (accessed 27.05.17).

Lenzing AG, 2017b. More performance in extreme situations—physiological tests. Available from: http://www.lenzing-fibers.com/en/lenzing-fr/performance-test/ (accessed 31.05.17).

Li, Z.-R., Tu, W.-P., Zou, J., Zhang, Z.-H., 2013. Synthesis of novel perfluorinated polymer containing quarternary ammonium group and its application on cotton fabric. AATCC Rev. 13 (3), 53–57.

Liu, Y., Tang, J., Wang, R., Lu, H., Li, L., Kong, Y., Qi, K., Xin, J., 2007. Artificial lotus leaf structures from assembling carbon nanotubes and their applications in hydrophobic textiles. J. Mater. Chem. 17, 1071–1078.

Lundgren, K., Kuklane, K., Gao, C., Holmér, I., 2012. Effect of heat stress on working populations when facing climate change. Ind. Health 51, 3–15.

Marks and Spencer plc, 2017. ECP Module 11 PCF Substitution Guidelines. [Online]. Available from: http://corporate.marksandspencer.com/documents/policy-documents/ecp-modules/module-11.pdf (accessed 29.05.17).

Martinkova, L., Marek, J., 2010. Multifunctional multilayer textiles for health care sector with protective barrier properties and enhanced wearing comfort using enzymes for PET modification for improved breathability and final treatment. In: 22nd IFATCC Int. Congress, 5–7 May, Stresa.

Massif, 2017a. Battleshield. [Online]. Available from: https://www.massif.com/catalogsearch/result/?q=battleshield (accessed 27.05.17).

Massif, 2017b. Field Tested. Combat Proven. [Online]. Available from: https://www.massif.com/ (accessed 27.05.17).

MMT Textiles, 2017. Biomimetics—The Pine Cone Effect. [Online]. Available from: http://www.inotektextiles.com/technology/ (accessed 27.05.17).

Moisture management antimicrobial FR cotton fabric. 2014. Futur. Mater. (1), 41.

Momentive, 2011. Magnasoft* NFR-A & NFR-B Textile Water Repellent Finish. [Online]. Available from: http://www.essentialingredients.com/pdf/magnasoftnfrpresentation.pdf (accessed 29.05.17).

Montarsolo, A., Periolatto, M., Zerbola, M., Mossotti, R., Ferrero, F., 2013. Hydrophobic sol-gel finishing for textiles: improvement by plasma pre-treatment. Text. Res. J. 83 (11), 1190–1200.

MountainBlog Europe, 2017. Black Yak. [Online]. Available from: http://www.mountainblog.eu/brand/blackyak/ (accessed 27.05.17).

MTIX LTD, 2017. Utilizing MSLE® Technology for the Treatment of Fabrics. [Online]. Available from: http://mti-x.com/ (accessed 26.05.17).

Munsante, G.B., 2013a. A firefighter's firewall. AATCC Rev. 13 (5/6), 38–39.

Munsante, G.B., 2013b. Feet to fire. AATCC Rev. 13 (7/8), 26–31.

Munsante, G.B., 2013c. Relief from the blistering heat. AATCC Rev. 13 (5/6), 32–33.

Munsante, G.B., 2013d. Weathering the arctic blast. AATCC Rev. 13 (5/6), 26–31.

Namligoz, E.S., Bahtiyari, M.I., Hosaf, E., Coban, S., 2009. Performance comparison of new (dendrimer, nanoproduct) and conventional water, oil and stain repellents. Fibres Text. East. Eur. 76 (5), 76–81.

Nanomembrane, 2017. Army. [Online]. Available from: http://www.nanomembrane.cz/en/our-solution/army (accessed 28.05.17).

Nanonic Inc., 2014. Smartcel TM Clima Functionality. [Online]. Available from: http://www.nanonic.us/smartcel-CLIMA.html (accessed 27.05.17).

Nanowerk, 2011. Alexium Rolls Out Cleanshell Process. [Online]. Available from: http://www.nanowerk.com/news/newsid=19879.php (accessed 29.05.17).
National Safety Apparel, 2017. Carbon Armour™ FR Jacket. [Online]. Available from: www.thinknsa.com/product/carbon-armour-jacket (accessed 27.05.17).
NC State University, 2017. Mannequin Men. [Online]. Available from: https://news.ncsu.edu/2011/11/mannequin-men/ (accessed 26.05.17).
NeverWet LLC, 2017. Rain Repellent. [Online]. Available from: http://www.neverwet.com/new/index.php (accessed 29.05.17).
New dry treatment method removes chemical usage, 2016. Int. Dye. 201 (5), 22.
Nicca Chemical Co Ltd, 2017a. NEOSEED NR series. Available from: http://www.nicca.co.jp/english/02productinfo/attention/04.html (accessed 29.05.17).
Nicca Chemical Co Ltd, 2017b. NK Guard S Series. Available from: http://www.nicca.co.jp/english/02productinfo/attention/04.html (accessed 29.05.17).
Nilit, 2017. Nitlit®Breeze//Performance Fibers. Available from: http://www.nilit.com/fibers/brands-nilit-breeze.asp (accessed 27.05.17).
Noll, I., Seide, G., Gries, T., 2016. The fibres of tomorrow. Futur. Text. 2, 30–31.
Oeko-Tex, 2015. OEKO-TEX® Standard 100 Test Criteria: New Regulations in 2015. [Online]. Available from: https://www.oeko-tex.com/en/ot_press/newsroom_1/pressdetailpage_89153.html (accessed 26.05.17).
Onder, E., Sarier, N., 2015. Phase change materials for thermal regulation. In: Paul, R. (Ed.), Functional Finishes of Textiles. Woodhead Publishing, Cambridge, pp. 17–78.
Opwis, K., Gutmann, J.S., 2011. Surface modification of textile materials with hydrophobins. Text. Res. J. 81 (15), 1594–1602.
OrganoClick, 2017. Water Repellent Additives for Textile and Nonwoven. Available from: http://organoclick.com/products/performance-textiles-nonwoven/water-repellent-additives/ (accessed 29.05.17).
Outlast Technologies, 2014a. Fibers. [Online]. Available from: http://www.outlast.com/en/applications/fiber/ (accessed 27.05.17).
Outlast Technologies, 2014b. Outlast Expands Into New Spheres. [Online]. Available from: http://www.outlast.com/uploads/media/1Outlast_Spherix_gb.pdf (accessed 27.05.17).
Outlast Technologies, 2014c. Technology. [Online]. Available from: http://www.outlast.com/en/technology/ (accessed 27.05.17).
Patagonia, 2017. Materials & Technology. [Online]. Available from: http://eu.patagonia.com/enCZ/patagonia.go?assetid=9052 (accessed 28.05.17).
Pertex, 2017. Pertex Quantum. [Online]. Available from: http://www.pertex.com/product/quantum/ (accessed 27.05.17).
PFOA crackdown in updated Oeko-Tex criteria, 2014. Int. Dye. 199 (2), 13.
PGI INC, 2011. COMFORT PLUS™—Nomex®(meta-Aramis)/Lenzing FR®featrures. [Online]. Available from: http://www.cobrahoods.com/index.php/materials/Nomex_Meta_Aramid_Lenzing_FR/),%20TenCate%20Coolderm%20TM%20fabric%20for%20firefighters%20 (http://www.tencate.com/emea/protective-fabrics/markets/emergency-response/default.aspx (accessed 31.05.17).
Playing it safe. 2016. Futur. Text. (4), 42–43.
Polartec, 2017a. Impossible Made Possible for Further Science of Fabric. [Online]. Available from: http://www.polartec.com/ (accessed 28.05.17).
Polartec, 2017b. Protection. [Online]. Available from: http://polartec.com/products (accessed 28.05.17).
PR Newswire Europe Limited, 2017. Frost & Sullivan: Emphasis on Fashion and Fabric Quality Promotes Development of Western European Workwear and Uniforms Market.

Available from: http://www.prnewswire.co.uk/news-releases/frost–sullivan-emphasis-on-fashion-and-fabric-quality-promotes-development-of-western-european-workwear-and-uniforms-market-215970161.html (accessed 06.03.17).

Productpilot, 2017. Careguard 66(New). [Online]. Available from: http://www.productpilot.com/en/supplier/sarex-chemicals/product/careguard-66new/ (accessed 29.05.17).

Protal, 2017. Table of Performance. [Online]. Available from: http://www.protal.co.uk/table-of-performance/ (accessed 27.05.17).

RotaSpray GmbH, 2013. Rotating Atomizer System ROTO-FINISHER®. [Online]. Available from: http://www.rotaspray.com/resources/ECO2_015_Fluorocarbon+C6$28e$29.pdf (accessed 29.05.17).

Rudolf GmbH, 2017a. Bionic-Finish®C6 Water-, Oil and Dirt-Repellent Textile Impregnation With Fluorine-Reduced c6 Technology. [Online]. Available from: http://www.rudolf.de/en/products/product-brochures-archive/details/bionic-finishr-c6/ (accessed 29.05.17).

Rudolf GmbH, 2017b. Bionic-Finish®Eco Impregnation for Ecologists. [Online]. Available from: http://www.rudolf.de/en/technology/bionic-finish-eco/ (accessed 29.05.17).

Rudolf GmbH, 2017c. Ecologically Optimized Finish. [Online]. Available from: http://www.rudolf.de/en/technology/bionic-finishr/ (accessed 26.05.17).

Rudolf GmbH, 2017d. Fluorocarbon Repellents & Booster. [Online]. Available from: http://www.rudolf.de/en/products/textile-auxiliaries/finishing/fluorocarbon-repellents-booster/ (accessed 29.05.17).

Russell, M.H., Berti, W.R., Szostek, B., Buck, R.C., 2008. Investigation of the biodegradation potential of a fluoroacrylate polymer product in aerobic soil. Environ. Sci. Technol. 42 (3), 800–807.

Santhyia, D., Burghard, Z., Greiner, C., Jeurgens, L.P.H., Subkowski, T., Bill, J., 2010. Bioinspired deposition of TiO2 thin films induced by hydrophobins. Langmuir 26 (9), 6494–6502.

Sarex, 2017a. Export Product List. Available from: http://www.sarex.com/textile/wp-content/uploads/2015/11/Export-Product-List-English.pdf (accessed 29.05.17).

Sarex, 2017b. List of Providing Products. Available from: http://www.sarex.com/textile/product_cat/solution-providing-products/ (accessed 26.05.17).

Scheffer, M., 2011. Results of the prometei initiative. In: PPE Conference 2011, 1–2 February 2011, Brussels.

Schoeller, 2017a. Environmentally Friendly Water and Oil Repellence. [Online]. Available from: https://www.schoeller-textiles.com/en/technologies/ecorepel (accessed 29.05.17).

Schoeller, 2017b. Protection From Heat and UV Rays. [Online]. Available from: https://www.schoeller-textiles.com/en/technologies/coldblack (accessed 27.05.17).

Schoeller, 2017c. Protective Textiles. [Online]. Available from: https://www.schoeller-textiles.com/en/textiles/protective-textiles (accessed 27.05.17).

Schoeller, 2017d. Three Times Dry. [Online]. Available from: https://www.schoeller-textiles.com/en/technologies/3xdry (accessed 27.05.17).

Scrimshaw, J., 2015. Six functions in one textile coating. Int. Dye. 200 (2), 22.

Selwyn, G.S., 2016. Hyperbaric dry finishing technology for sustainability combined with high-performance. In: International Conference on Textile Coating and Laminating, 16–17 March, Prague.

Senthilkumar, M., Sampath, M.B., Ramachandaran, T., 2012. Moisture management in an active sportswear: techniques and evaluation–a review article. J. Inst. Eng. E 93 (2), 61–68.

Serbezeanu, D., Popa, A.M., Stelzig, T., Sava, I., Rossi, R.M., Fortunato, G., 2015. Preparation and characterization of thermally stable polyimide membranes by electrospinning for protective clothing applications. Text. Res. J. 85 (17), 1763–1775.

Shirgholami, M.A., Shateri-Khalilabad, M., Yazdanshenas, M.E., 2013. Effect of reaction duration in the formation of superhydrophobic polymethylsilsesquioxane nanostructures on cotton fabric. Text. Res. J. 83 (1), 100–110.

Shishoo, R., 2007. Plasma Technologies for Textiles. Woodhead Publishing, Cambridge.

Singha, K., 2012. A review on coating & lamination in textiles. processes and applications. Am. J. Polit. Sci. 2 (3), 39–49.

SINGTEX Industrial Co., 2015. S.Café® Sustainable Performance. [Online]. Available from: http://www.scafefabrics.com (accessed 27.05.17).

Soboyejo, N., Oki, A., 2013. Functionalization of cationic cotton with octadecylammonium—swcnt carboxylate ion Pairs. AATCC Rev. 13 (4), 47–54.

Solvay, 2017. Fluorolink® PFPE. [Online]. Available from: http://www.solvay.com/en/markets-and-products/featured-products/Fluorolink.html (accessed 29.05.17).

SSM Industries, 2016a. A Durable, Flame Retardant Cotton for Professionals. [Online]. Available from: http://www.ssmind.com/pro_c_fr.html (accessed 27.05.17).

SSM Industries, 2016b. Flame Resistant Filament Fabrics for Professionals. [Online]. Available from: http://www.ssmind.com/24.html (accessed 27.05.17).

Sweat activated cooling for sportswear. 2013. Int. Dye. 198 (4), 27.

Sympatex, 2017a. Climate Technology by Sympatex®. [Online]. Available from: http://www.sympatex.com/en/technologies/221/climate-technology (accessed 28.05.17).

Sympatex, 2017b. Phaseable®. [Online]. Available from: http://www.sympatex.com/en/technologies/313/phaseable (accessed 28.05.17).

Takenori, T., Hiroyuki, N., Yoshie, K., Yuji, O., Kazufumi, O., 2009. Development of a water- and oil-repellent treatment for silk and cotton fabrics with fluoroalkyl-trimethoxysilane. J. Text. Eng. 55 (1), 13–21.

Tanatex Chemicals, 2017a. Finishing: Oil, Water, Soil Repellence. [Online]. Available from: http://www.tanatexchemicals.com/products/finishing/finishing-oil-water-soil-repellents (accessed 29.05.17).

Tanatex Chemicals, 2017b. HydrECO Concept. 3D-Equitized Hydrophobics. [Online]. Available from: http://www.spot.com.tr/uploads/1/4/4/6/14462940/hydreco_concept_2013.pdf (accessed 29.05.17).

Technical Textiles, 2017. Chemicals. LJ Specialities. [Online]. Available from: http://www.technical-textiles.net/terms/chemicals-0 (accessed 29.05.17).

Teijin, 2017. Teijin Frontier Develops Octa™ Neo Multilayer Fiber. [Online]. Available from: https://www.teijin.com/news/2015/ebd151217_24.html (accessed 28.05.17).

Texchem UK Ltd, 2017a. Fluorine Free Water Repellents. [Online]. Available from: https://www.texchem.co.uk/home-1/fluorine-free-repellents/ (accessed 29.05.17).

Texchem UK Ltd, 2017b. Fluoropolymers (C6). [Online]. Available from: https://www.texchem.co.uk/home-1/fluorocarbons/ (accessed 29.05.17).

TEX-SHIELD, 2017. Environmental Friendly and Durable Oil and Water Repellence Finish on Technical Textiles. [Online]. Available from: http://www.texshield-project.eu/ (accessed 26.05.17).

Textile Exchange, 2016. Preferred Fiber Market Report © 2016. [Online]. Available from: http://textileexchange.org/wp-content/uploads/2017/02/TE-Preferred-Fiber-Market-Report-Oct2016-1.pdf (accessed 05.03.17).

TextInfo, 2012. Moisture Management. [Online]. Available from: https://textlnfo.files.wordpress.com/2012/05/moisturemanagement_e.pdf (accessed 26.05.17).

Textor, T., Gutmann, J.S., 2014. Ionic liquids for finishing polyester fibers with cellulosic surfaces. Melliand Textilber. 20 (4), 203–204.

The Biomimicry Institute, 2016. Mincor TX TT BASF the Chemical Company. [Online]. Available from: https://asknature.org/idea/mincor-tx-tt-textile-coating/#.WSvuRevyiUk (accessed 29.05.17).

Thermore, 2017. THERMORE® ARIA—The Effective Down Substitute. Available from: http://www.thermore.com/en/insulation-for-outerwear-products/aria/ (accessed 28.05.17).

Trans-Textil GmbH, 2013a. Medical Applications. Available from: https://www.trans-textil.de/html/medical_appl/medical_applications.html (accessed 28.05.17).

Trans-Textil GmbH, 2013b. Membrane Systems. Available from: https://www.trans-textil.de/html/technologies/membrane_systems.html (accessed 28.05.17).

W.L. Gore & Associates GmbH, 2017a. What is the GORE-TEX® Fabric? [Online]. Available from: https://www.gore-tex.com/technology/what-is-gore-tex (accessed 28.05.17).

W.L. Gore & Associates GmbH, 2017b. GORE® PARALLON™ Liner System. [Online]. Available from: https://www.gore-tex.com/professional/fire/parallon (accessed 28.05.17).

W.L. Gore & Associates GmbH, 2017c. GORE-TEX® SURROUND® Footweardry and Comfortable Feet All-Around. [Online]. Available from: https://www.gore-tex.co.uk/technology/footwear/gore-tex-surround-footwear (accessed 28.05.17).

Wacker Chemie AG, 2017. SEPAWA 2014. Improved Fluorine-Free Impregnating Agents Based on Silicone Technology. Available from: https://www.wacker.com/cms/media/documents/markets_brands/consumer_care_1/SEPAWA_2014_Improved_Impregnation.pdf (accessed 29.05.17).

Wallance, M., 2002. 100% Cotton moisture management. JTATM 2 (3), 1–11.

Wilson, A., 2015. Beyond personal. Futur. Mat. 3, 10–12.

Wilson, A., 2016. Going for gold at ISPO. Futur. Text. 3, 16–19.

Yoo, S., Barker, R.L., 2004. Moisture management properties of heat-resistant workwear fabrics—effects of hydrophilic finishes and hygroscopic fiber blends. Text. Res. J. 74 (11), 995–1000.

Zaman, M., Liu, H., Chibante, F., Ni, Y., 2013. Hydrophilic modification of polyester fabric by applying nanocrystalline cellulose containing surface finish. Carbohydr. Polym. 91 (2), 560–567.

Zhao, M., Gao, C., Wang, F., Kuklane, K., Holmér, I., Li, J., 2013. The torso cooling of vests incorporated with phase change materials: a sweat evaporation perspective. Text. Res. J. 83 (4), 418–425.

Zschimmer, Schwarz, 2017. Anthydrin NK6. [Online]. Available from: http://www.zschimmer-schwarz.com/8.html?typ=X&id=ANTHYDRIN%20NK%206 (accessed 29.05.17).

Further reading

Advanced Textiles, 2017. TChIP provides chemical hazard and risk assessment data. [Online]. (accessed 25.05.17).

Huntsman International LLC, 2011. Textile Effects: Towards Zero Discharge of Hazardous Chemicals—Huntsman Positive List. [Online]. Available from: http://www.huntsman.com/textile_effects2/Media%20Library/global/files/Huntsman%20ZDHC%20positive%20list.pdf (accessed 26.05.17).

Kofler, P., Herten, A., Heinrich, D., Bottoni, G., et al., 2013. Viscose as an alternative to aramid in workwear: influence on endurance performance, cooling, and comfort. Text. Res. J. 83 (19), 2085–2092.

Munsante, G.B., 2014. Inside the crystal ball: a look at the textile industry's evolving future. AATCC Rev. 14 (1/2), 28–35.

Schoeller, 2017. Functional Textiles With Cork. [Online]. Available from: https://www.schoeller-textiles.com/en/technologies/corkshell (accessed 27.05.17).

Trans-Textil GmbH, 2013. Flame-Retardant Breathable Moisture Barriers for Firefighters' Protective Clothing. Available from: https://www.trans-textil.de/html/service_e/news.html (accessed 28.05.17).

TWD fibres polyester fiber with self-cleaning effect. 2013a. Tech. Text. Int. 56 (1), E16.

W.L. Gore & Associates GmbH, 2017. Enhanced Heat and Flame Protection Using Non-FR Textiles. [Online]. Available from: https://www.gore-tex.com/professional/technology/products/gore-pyrad-flame-retardant (accessed 28.05.17).

Healthcare textiles

16

Angela Davies
De Montfort University, Leicester, United Kingdom

16.1 Introduction: Key applications in healthcare textiles

Textiles used in healthcare (often referred to as medical textiles) can be seen as one of the most rapidly expanding sectors in the technical textile market (Infiniti Research Limited, 2015). Products defined under this category range from wound dressings and bandaging materials to scaffolds for tissue engineering and implantable prostheses. Textile use is prevalent and through innovation is continuously expanding within the healthcare and hygiene sector in both home and hospital environments.

The key market drivers for enhanced medical textile developments and applications can be identified as primarily due to a worldwide population growth, an ageing population (due to less dangerous professions, and better diagnosis and care of health conditions) and the modern lifestyle choices (poor diet and inactivity) of individuals (Infiniti Research Limited, 2015). With better awareness about hygiene, demands for an enhanced quality of life, and increased access to better healthcare facilities, the pressure on the healthcare sector is enormous. As the private and public health sectors attempt to cope with the demand and expectations of modern society, textiles are playing a central role. Innovations in fibrous materials are creating new capabilities in the drive to provide cost effective and quality healthcare to an expanding population while mitigating risks of infection to both patients and health workers. Materials can be engineered with a range of properties from softness and lightness and flexibility to absorption, filtration and the promotion of cell renewal (Qin, 2016). Figs 16.1 and 16.2 show examples of healthcare textiles.

To provide a general overview, healthcare textiles can be categorized into the following areas (Anand et al., 2010; Qin, 2016):
External Devices—these consist of materials that work on the outside of the body either directly against damaged skin or to support and control body functions such as movement and blood flow

- Wound care dressings, wadding and plasters
- Bandages
- Compression hosiery and pressure garments

Protection in the healthcare setting—materials used in the healthcare environment which serve the function of providing comfort, modesty, barrier properties (between patient and healthcare worker), cleanliness and sterility.

- Surgical clothing, wipes and facemasks
- Surgical covers for the operating table

Waterproof and Water Repellent Textiles and Clothing. https://doi.org/10.1016/B978-0-08-101212-3.00015-0

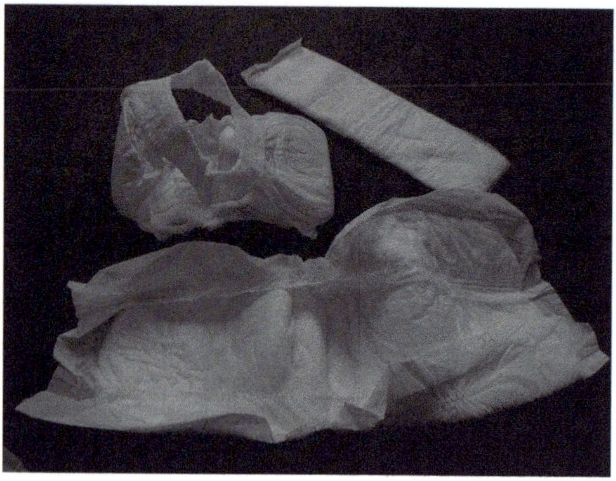

Fig. 16.1 Disposable incontinence products.

Fig. 16.2 Wound dressing.

- Staff uniforms
- Hospital bedding and curtains

Implantable materials—those inserted into the body either on a temporary or permanent basis. Biocompatibility is of paramount importance if materials are to be accepted by the body.

- Surgical sutures
- Cardiovascular implants such as vascular grafts (for redirection of blood flow)
- Artificial ligaments, tendons, skin, lumen, joints, bones
- Scaffolds for tissue engineering

Extracorporeal devices—materials used within the body which support the function of mechanical organs used in blood purification

- Artificial livers, lungs, kidneys
- Dialysis filters

Hygiene products—materials used primarily as an absorbent medium on the outside of the body in direct contact with the pubic area and vaginal tissue.

- Incontinence pads
- Nappies/diapers
- Feminine hygiene products

16.1.1 The impact of lifestyle choices and the ageing population on healthcare provision

Key focal points on the improved provision of healthcare over recent decades have centred around access to care, disease prevention and treatment, rehabilitation from trauma and improving the quality of life (Infiniti Research Limited, 2015). These improvements have seen advances including groundbreaking surgical procedures, improved hygiene and infection control and widespread immunization programmes. While healthcare today is a significant improvement on the past, new healthcare problems pose new challenges to the ever changing and dynamic responding nature of the healthcare system.

With advances in modern surgery it can be suggested that the demand on the future healthcare system will be in preventative applications for disease and in the treatment of and rehabilitation from trauma (due to extended life with a frail body), which is likely to be a significant cause of death.

There are now more than 3 million people aged 80 plus in the United Kingdom and more than 11.6 million over 65—almost 1 in 5 of the UK's total population of over 65 million (Office of National Statistics, 2015). On a global level just over 1% of the population in 2000 were over 80 years old, however, this is set to reach 4.1% by 2050 (United Nations, n.d.). It is predicted that by 2040, nearly one in four people in the United Kingdom (24.2%) will be aged 65 or over (Office of National Statistics, 2015). Life expectancy is predicted to increase with a baby girl born in 2011 having a one in three chance of living to 100 and a baby boy has a one in four chance (Department of Work and Pensions, 2011). The current good health life expectancy of people living in the United Kingdom is 72.5 years for women and 70.3 years for men with health deteriorating for the remaining years of life (WHO, 2015). Table 16.1 compares the UK healthy life expectancy with some of the best and worst performing across the world. Although healthy life expectancy exceeds the expectations of our predecessors this does imply many years of later life would require significant healthcare intervention. As a person ages the body becomes fragile with decreasing body functions contributing to greater healthcare considerations. In 2006, for example, 70,000 hip fractures occurred and it is predicted that by 2020 this number will exceed 101,000 (NICE, 2012). Weakened immunity, limited mobility, deterioration of organ and skin function, contributes to the vulnerability of the elderly to infection.

Table 16.1 **Healthy life expectancy across the world (WHO, 2015)**

Country	Female health	Male health
Japan (world best)	77.2	72.5
France (European best)	74.4	70.7
UK	72.5	70.3
USA	70.4	67.7
Sierra Leone (world worst)	44.8	43.9

While population across the world has increased, the birthrate has declined over recent years (Roser, 2016). *Although* birthrates have fallen, advances in science have increased the survival rates of pre-term births, one of the most vulnerable groups of humans. Similar to the healthcare issues seen in the elderly, pre-term infants have a weakened immune system and are at severe risk of infection as their premature system develops.

With the increases in the most vulnerable members of the population compounded by the over-use of antibiotic medication, recent advances in healthcare have focused upon the control and prevention of infections. HIV and AIDS created a global pandemic over latter part of the 20th century. Hepatitis B, C and the recent return of Tuberculosis and Whooping Cough all are threats to world health. Alongside infections within the hospital environment, Healthcare-Associated Infections (HCAIs), SARS, Ebola, Bird Flu, Norovirus and other viruses pose on-going threats along with new viruses constantly evolving across the world (Dye, 2014; Zins, 2011). Textiles are playing an increasing role in this battle against HCAIs, viruses, fungi and antibiotic resistant microbes therefore advances in both disposable and reusable healthcare products should address these challenges.

HCAIs can be defined as an infection that is acquired either in a hospital or other healthcare facility, which was not present or incubating at the time of admission (WHO, 2016). HCAIs present a very real risk to both staff and patients within the hospital setting. Significant research over recent years has focused on identifying areas of prevention (Collins, 2008; Gov.UK, 2016; James, 2011; WHO, 2016).

There are three key aspects to consider in the prevention of HCAIs:

- The sources of infection, e.g. patients with infectious organisms, environmental sources such as surfaces in wards, communal areas such as toilets where bodily fluids from infected patients may be found
- How transmission of infection take place, e.g. airborne, direct contact or percutaneous (e.g. via needle/sharps injuries)
- The role of the patient, e.g. compromised immune system, open surgical sites and healing wounds.

A greater emphasis on hygiene control of patients, staff and visitors within the healthcare setting, combined with advances in textile materials used within healthcare and their sterilability can contribute in the fight to prevent many HCAIs from developing. The increased textile performance and standardization of healthcare textiles can play a critical role within this arena.

16.2 Water repellent and waterproof healthcare textiles

With the increased movement of people between countries and the growing number of people accessing care, including the vulnerable such as the very young and old and those with impaired immune systems, the healthcare system needs to readdress preventative care. It has been identified (NHS, 2015; Llor and Bjerrum, 2014) that there has been a worldwide overuse of antibiotics either through misdiagnosis or preventative care. While new antibiotic developments are taking place, antibiotic resistant diseases are becoming a growing concern. This highlights the need of increased public hygiene and the development of materials that offer barrier properties including infection control. Patients in hospitals are at risk of infection and HCAIs are a contributory cause of death (WHO, 2016; Collins, 2008).

Microorganisms such as bacteria, viruses, and fungi can be found everywhere, some of which can enhance health and well-being. Many microorganisms, however, create and aggravate problems in hospitals and healthcare environments by transmitting diseases and infections through media such as clothing, medical covers, and bedding. Extensive research has been undertaken and guidelines produced over recent years to mitigate the impact of HCAIs, however, MRSA and MRSSA bacteria are still prevalent in hospitals (WHO, 2016; Royal College of Nursing, 2016; NICE, 2011). Incorrect or over-zealous use of antibiotics has led to strains of bacteria evolving increased resistance to, and reducing the effectiveness of antibiotics rather than enabled bacteria to evolve creating an environment where bacteria are no longer killed. Bacteria and fungi thrive in warm humid environments, temperatures in busy hospital wards can often be higher than the average domestic room in the region of 26°C whereas operating theatrers are often 22–26°C (Nursing Standard, 2014; Nocker, 2011). Fabrics therefore worn next to the skin in these warmer climatic conditions could be a contributory factor in the spread of HCAIs.

The ability of a material to repel fluids such as water, oil, and alcohol-based substances is an important attribute for medical textiles to possess. Preventing fluids from penetrating a material can be identified as a barrier property, while contributing to a material's ability rather than and the ability of a material to allow fluids to roll or pearl off the surface leaving the textile unchanged. These properties are advantageous in the hospital setting where bodily fluids can be a source of cross infection between patients and medical personnel.

The terms 'waterproof' and 'water-repellence' may be used when categorizing medical textiles based upon their effectiveness in acting as a barrier to fluids. A waterproof material can be defined as one that provides a continuous barrier to the passage of water (Slater, 1993). An undersheet used on the base of a hospital bed would be waterproof because it acts as a complete barrier to liquid penetrating through the material under a defined pressure. The undersheet, for example, would prevent fluids from leaking through onto the mattress below. Water repellent materials, on the other hand, are generally those not easily penetrated by water though water can penetrate if it strikes with sufficient force (Slater, 1993). Hospital uniforms would provide a degree of water repellency, ranging from minimal as seen in nurses' uniforms, to moderate in the case of that worn by surgeons for operations where there is low risk of fluid penetration. For surgeons working in higher risk surgery or where the presence of a large volume of bodily fluid is anticipated, superior or, in fact, waterproof materials would be used (BS EN, 13795:2011+A1:2013). Water repellent materials can be observed as those that will repel water but are not impervious to it. There is a level of water pressure the material can withstand before penetration starts to appear. The following sections will explore how these materials are used within the healthcare setting either as a complete barrier against liquids or microorganisms or used to support other more critical performance characteristics of the material.

16.2.1 Materials providing solely waterproof properties and their use

A vast majority of textiles used within the healthcare industry are required to possess barrier properties providing varying degrees of water repellence. These materials are discussed in more detail in Section 16.2.2. There are, however, some instances where it is required for a particular application that the material is impervious to liquids providing a total block against fluids of various forms under pressure for a short period of time. These are commonly static items, not worn against the body as clothing for extensive periods and may not require laundering. These materials may often require being wipe-clean only as part of a clinical area's wipe-clean programme. Hospital seating, mattress, and theatre cover material, as well as mattress protector covers are all instances where one of the key properties of the material is to block the passage of fluid under pressure. In the fight against HCAIs some protective covers, for example, seating in waiting rooms, requires functionality but should not permit the passage of liquids through the cover and into the cushioning or mechanism below. This allows for the surface to be cleaned with a suitable antiseptic cleaning agent on a routine basis as part of a standard hospital cleaning procedure (NHS, 2009). Commonly used materials include vinyl fabrics comprising polyurethane, PVC, or a blend of PVC with either cotton or polyester. The vinyl fabrics often possess a coating to resist bacteria and fungi on a long-term basis, alongside stain repellence, which may impede the longevity of the material to provide antibacterial protection over time.

The combination of a vinyl-based fabric with a waterproof finish prevents the passage of water or oil based solutions through the fabric. This provides a durable and effective surface for wipe clean applications. For some end uses such as hospital clothing the ability to block the passage of liquid alone is not sufficient. Providing a total barrier against fluids often comes at the cost of rendering the fabric impervious to the passage of moisture vapour from the inside to the outside. This means the fabric is not breathable and when used for wearable applications would lead to the wearer feeling sweaty and uncomfortable.

16.2.2 Materials offering barrier properties and their uses

Materials used in healthcare usually possess a set of core properties directly associated with their intended use. Those requiring water repellence may also require breathability, which means they must act as a barrier to prevent fluids passing from the outside through to the inside of the material, however, the material should also permit the passage of air and moisture through from the inside to the outside allowing sweat to evaporate.

Within the healthcare environment clothing items worn by professionals should provide a range of properties with regards to personal comfort and dignity, but more importantly as a form of protection between both the patient and medical professional from cross infection. Garments commonly worn include nursing uniforms, gowns, medical coats, face protection, protective footwear, and headwear.

16.2.2.1 Hospital clothing

Some forms of hospital clothing with the key purpose of providing barrier properties to resist penetration of blood, fluids, and microorganisms are not intended for long periods of use, due to the lack of comfort provided by the enhanced barrier properties. Surgical gowns are a prime example of a short-use, high-barrier product.

They routinely need to provide barrier protection, offer purity, low particle release rates, provide strength and, of course, a degree of comfort. Gowns are either required to have standard performance or, in cases where risk of infection and penetration from fluids is greater, superior performance. Gowns can be either reusable or disposable. Many tend to be single use and therefore do not require disinfection after use. Disposable gowns are usually a spunbonded, meltblown non-woven structure of a single layer, double-reinforced layers or multi-layers, depending on the level of protection required from liquid penetration. Materials used are commonly polyester, polypropylene, cotton or blends of the aforementioned. Spun-laced and spun-laid materials are often used with the former providing greater aesthetic and comfort properties to the wearer, and the latter providing greater barrier properties (Rajendran, 2006). These materials are then often treated with a fluorochemical finish, though it should be noted that with concerns over health and the environment, alternatives should be considered. Reusable equivalents are mostly comprised of PES filament woven fabrics treated with a fluorochemical finish. For greater performance a multi-layer composite might be used comprising PTFE, PES or PU membranes.

A key area highlighted as a cause for concern in the fight against HCAIs, are the uniforms worn in hospitals by medical professionals on a routine basis. These items tend to be reusable thus require laundering on a routine basis.

16.2.2.2 Hospital bedding and curtains

Primary areas for cross infection within the hospital setting are between patient-to-patient contact and patient and healthcare professional contact through bodily and clothing contact, and within the operating theatre. Although these are the prime locations for infection prevention, the importance of furniture and fittings within the environment should be duly considered. Hospital bedding and textiles used within the ward, such as privacy curtains, present secondary risk areas (NHS, 2010; Woodland et al., 2010). Bedding, although changed on a routine basis, is primarily in contact with the patient for long periods of time. Comfort, durability, and longevity are the core properties of bed linen, however, its use in infection control is not insignificant. Hygienically clean bed linen should be a standard in the hospital setting and with the use of commercial laundering services and associated disinfection protocol in place, hospitals can ensure effective laundering to minimize cross infection (NHS Executive, 1995). Barrier materials commonly accompany bed linen in hospital and care settings, used beneath the standard sheeting to protect the mattress. These may either be non-breathable vinyl or washable cotton/polyester nearest the skin with a polyurethane film base to provide the waterproof barrier.

Hospital cubicle curtains may be either wipe-clean only, or fully washable and sterilizable depending on the requirements of the hospital and the durability of the material and any associated finishes. Curtains should provide privacy but should also act as a barrier to germs to assist in infection prevention between patients. In terms of compliance, a key requirement of hospital curtains is to meet with flammability regulations. However, over recent years with curtains being identified as a potential source of cross contamination the ability of materials to provide infection control has been seen as an essential property (Woodland et al., 2010; Ohl et al., 2012).

The majority of curtains comprise a tightly woven polyester structure to provide strength and durability and have a fire retardant finish to comply with safety regulations (BS 5867-2:2008). Materials are commonly impregnated with an antibacterial finish or may have, for example, silver threads woven into the structure to provide a barrier against germs. Any finishes applied to a fully washable fabric should withstand repeated harsh temperature laundering to comply with hospital sterilization procedures. To restrict the accumulation of dust particles and soiling on the surface of the curtain the fabric may be also be composed of a stain-resistant or antistatic finish. Some modern finishes are designed to combine antimicrobial protection, fire retardancy, and stain resistance. Specific finishing of healthcare fabrics is discussed in more detail in Section 16.4.

The majority of clothing and bed linen used within the healthcare environment is reusable, thus the ability to repeatedly sterilize medical textiles means that the fibre composition must be able to withstand the harsh physical and chemical conditions of the sterilization process. With a focus on multifunctional fabrics and their ability to maintain barrier properties such as liquid repellence and antimicrobial properties following repeated harsh laundering, Sections 16.3.2 and 16.4.2 consider this aspect.

16.2.3 Materials providing absorbance properties and their uses

Many medical textile applications require that the material absorbs and retains fluids rather than acting as a barrier to repel. Incontinence products, wound dressings, swabs, and many maternity and operating theatre disposables are required to draw in and retain fluids and bodily products. Although the focus here is in repellence properties it is important to appreciate the materials requiring opposite properties.

A key burden on the healthcare sector contributed by the ageing population is the incidence of incontinence (Davies, 2009). There are two main types of products used within the management of incontinence; bed pads and body-worn pads. Body-worn pads, which are used in a similar way to babies' diapers, are complex, needing to be contoured around the body. Bed pads are often used as a back-up system on a patient's bed to absorb bulk liquid while body-worn pads may be used as the primary method of liquid absorption.

Placed on top of the waterproof mattress protector to accompany bed linen bed or under pads for incontinence or to prevent post surgery bodily fluid penetration may often be present in the hospital setting. These can be either disposable or reusable if laundered following standard hospital disinfection protocol (NHS Executive, 1995). The facing fabric on bed pads is usually cotton, polyester or a combination of the two. The absorbent cores are usually produced of viscose and/or polyester, which is either needle-punched felt or knitted, with a backing consisting of a polyurethane fabric. The fabrics are often quilted together to eliminate fabric shear, and waterproof backings can be integrated or separate. Separate backings are less easily handled but can easily be replaced if damaged. Bed pads, disposable in nature, are often non-woven bonded materials treated with an antiseptic finish and accompanied with a backing of polyurethane. The non-woven pad may contain fluff pulp as the absorbent medium. Composite fluff products are often used with two main additions to the fluff, these being hydrogel superabsorbent polymers or thermoplastic fibres, either individually or together (Davies, 2011).

16.3 Exploring specific properties in waterproof and repellent healthcare textiles

Healthcare textiles, whether they are designed to absorb or repel liquids, will possess a number of other properties critical for their specific application. In general the requirements of textile materials for medical applications are:

- Biocompatibility—therefore not harmful or toxic to living tissue
- Resistance to alkalis, acids, and micro-organisms—resistant to bodily fluids, sweat, deodorants, antimicrobial/antibacterial
- Dimensional stability—does not change shape/size in contact with water, steam, heat, light, liquids
- Elasticity—Free from contamination or impurities—not liable to cause allergies or irritation
- Absorption/Repellence—drawing in moisture or blocking penetration from moisture
- Air permeability—allowing air to move from the inside of the textile to the outside and vice versa

For materials inserted inside the body, biocompatibility will take precedence, whereas for materials used on the outside of the body, absorption or repellence may be the dominant factors.

16.3.1 General properties and the theory of fluid dynamics

The properties of a material designed to repel or absorb fluids are very different, however both rely on understanding fluid dynamics.

Improved materials for liquid management can only be developed by understanding fundamental interactions between liquids and fabrics. Understanding lateral wicking behaviour in textiles is necessary to improve the function of products designed to transport liquid away from undesired areas, such as restricting the liquid spreading within an absorbent pad. Pore structure is a fundamental property of fibrous substrates and governs liquid distribution and movement within such materials (Beckham et al., 2000). In materials designed to repel liquid, understanding lateral wicking can ensure that the material surface can be designed to block the passage of fluids from the face to the reverse.

Liquid transports through fabrics in two ways, by diffusion and wicking. Diffusion of liquid through the fabric is dependent on fibre type but is governed by the fabric structure. Wicking occurs by capillary transport. The size and number of capillary paths through a fabric is important, but these may be driven by other factors such as fibre size, yarn structure, and fabric structure. The capillary network of a fabric is dependent on the direction, demonstrating different wicking properties through the fabric thickness, to those along the wales and courses in a knitted fabric, or along the warp and weft in a woven fabric (Saville, 1999).

Water droplets are held together by cohesive forces, while adhesive forces hold together unlike molecules such as that of fabric and water. Cohesive forces between liquid molecules are responsible for surface tension, which is normally measured in N/m. These forces form a surface film over an object, in this case fabric, making it more difficult for the liquid to penetrate the surface. The greater this surface tension, the more difficult it is for the fabric to wet, and for liquid to advance into the fabric structure. Different liquid compositions and temperatures influence the surface tension, with lower levels created by heat and liquids such as detergents. The contact angle between the liquid and the fabric affects the fabric's wetting properties, with angles smaller than 90 degrees demonstrating fabric wetting and wicking into the material by capillary action.

Liquid movement in any porous media is driven by capillary action, which is governed by the liquid's properties of surface tension, viscosity, density; the liquid-media surface interaction and the geometric configuration of the pore structure in the medium, which is difficult to quantify (Hsieh, 1995).

Liquid must wet the fibre structure before being transported through the interfiber pores by means of capillary action. When water droplets are introduced perpendicular to the yarns, wetting and wicking occur. Initially the droplets will expand in the longitudinal direction of the yarns (wetting), and at the same time wick into the inner core of the yarn. Secondly, the liquid wicks in two opposite directions independently along the yarns, governed mainly by the contact angle and capillary system (Beckham et al., 2000). Capillary action plays an important role in liquid transport, with theory suggesting that smaller pore sizes produce higher capillary pressure, and thus enhance liquid spreading distance (Hsieh, 1995).

The amount of liquid a fabric absorbs depends on the total capillary volume, which is usually related to fabric thickness. Research by Crow and Osczevski (1998) has shown that water held in the intra-yarn spaces (the voids between the fibres) is as important as the liquid held in the inter-yarn spaces, particularly noticeable in fabrics with large voids.

To achieve effective repellence, the surface energy and roughness of the material needs to be considered. This includes the chemicals on the surface of the material such as those of any finish, and the properties of the liquids that come into contact with the surface (surface tension). The effectiveness of common finishes used in healthcare textiles are discussed in Section 16.4.

16.3.2 Repellence and barrier properties in hospital clothing

Staff uniforms within the hospital environment are a contributing factor in the transfer of bacteria and virus between patients and staff. Strict guidelines on the wear and washing of uniforms outside of the hospital environment are in place to support infection control (Royal College of Nursing, 2014). Uniforms that act as a barrier with protection against fluids and airborne pathogens are required. The barrier function is normally provided by the use of a fabric finish or applying a laminate to the material.

The main properties required in hospital clothing are that the materials and associated finishes are non-toxic, skin friendly, and provide comfort to the wearer such as ease of movement, and good handle. There should be a degree of temperature regulation since the climatic conditions within a hospital environment are generally high. One key property is that the clothing must provide protection against fluid penetration yet provide breathability for wearer comfort, which can be achieved by a microporous structure. An appropriate choice of fibre, fibre fineness, weave type, and fabric density all contribute to the porosity of the fabric. This combined with an appropriate finish can contribute to the creation of well-performing hospital clothing. It should be noted that for reusable healthcare textiles, high temperature laundering used in sterilization, in accordance with NHS Executive (1995), can decrease the effectiveness of barrier finishes thus re-application of the finish either periodically or after every wash cycle should occur (Riley et al., 2017).

Fluid resistance of uniforms can be achieved by using a combination of a very dense, tightly woven structure, comprising of piled yarns and a fluoropolymer finish. Blends of polyester/cotton usually 50/50 can be seen as commonplace for uniforms, surgical gowns, drapes, and wraps. Hydrostatic head (BS EN ISO 9073, 2008) is considered the amount of water pressure required to penetrate a given fabric. This fibre composition and associated finish can provide hydrostatic head in the region of 40–60 cm (Zins, 2011). Nursing staff generally wear polyester/cotton blends with a durable press, stain proof/soil release finish, and moisture-wicking capabilities to provide a uniform that is comfortable and durable. The use of multifunctional finishes can reduce the number of individual treatments applied to a fabric to provide an effective uniform with appropriate barrier properties.

Gowns worn by patients in a hospital environment are commonly 50/50 polyester/cotton lightweight woven fabrics as these are a cost-effective combination. Bedding used in hospitals is commonly either 50/50 polyester/cotton or 100% continuous filament polyester woven structures. These are often combined with a soil-release finish and moisture-wicking performance. If the polyester component of a material is greater than 50% this can also be problematic for high temperature laundering because the fibre may be prone to heat damage. Reducing the pH of the wash by careful choice of surface agents can prevent the degradation of finishes. For patients with sensitive and vulnerable skin, fibres such as those based on algae or silver can provide improvements in skin condition. Alginate is non-toxic, non allergic, and able to be sterilized easily. Zinc alginate and copper alginate offer high moisture absorbency and provide antibacterial properties, and have recently found a use in hospital linen (Sifri et al., 2016).

The use of a high-count multifilament yarn comprised of polyester woven more densely than a blend of spun yarns combined with a fluorochemical finish can provide a superior fluid resistance in the region of 60–90 cm hydrostatic head pressure. Controlling fluid penetration at high levels of pressure requires a greater level of textile engineering. PTFE microporous membranes sandwiched between two layers of polyester fabric offer enhanced barrier protection with a hydrostatic head pressure greater than 100 cm (Zins, 2011). The garment itself needs to be designed to provide a seal against the surgeon's skin. The use of polyurethane-coated sealing tapes or ultrasonic bonding can provide a barrier at fabric joins ensuring the garment is fully waterproof.

One of the most important aspects of operating-theatre clothing are the barrier properties that are provided. The comfort to the wearer is often seen as less important. However, during long procedures the lack of comfort to the surgeon may have a detrimental impact on performance thus this aspect should not be overlooked. Recent advances have allowed finishes to effectively provide hydrophobic properties to the outer surface but be hydrophilic inside, allowing the body to feel less clammy and the material to manage moisture effectively. A hydrophilic lining could be attached to the hydrophobic outer surface, however, the main disadvantage of this is the product cost to the health service (Bartels, 2011). Other properties alongside comfort and launderability include static control, which is an additional requirement for clothing worn in sensitive operating theatres. Where polyester is used, a conductive carbon-impregnated polyester yarn is often inserted into the weave periodically. Antistatic properties are often added to textiles used within the operating theatre to reduce the risk of injuries due to static shock and interference with sensitive equipment during surgery.

16.3.3 Repellence and requirements of internally used textiles

Advances in textile engineering have enabled many developments in the area of regenerative medicine, providing support either *in vitro* or *in vivo* due to their varied mechanical properties.

Regenerative medicine can be in the form of replacing, engineering or regenerating human cells, tissue or organs either to restore or provide normal function. Materials can provide flexibility, elasticity, as well as the ability to engineer materials with specific properties such as porosity or rate of degradation.

Materials can be used to:

1. replace damaged cells with biological substitutes grown in the lab for insertion into the body through tissue engineering or using textiles, which provide a scaffold for cell growth.
2. become biocarriers, cells or signalling molecules to guide regeneration to occur in the body (Qin, 2016).

Materials used within the body can either be permanent or break down in the body after a *period of time* and decompose (biodegradable). Where appropriate, there has been a shift away from the use of permanent materials to those required for a matter of months to serve as a scaffold supporting new growth before being reabsorbed by the body.

A key property of implantables is the appropriate porosity for the required function. This would include the ability of the device to facilitate human tissue growth and to encapsulate the implant through correct pore size selection. Devices are also required to function with biomolecules, be biocompatible, and be appropriately biodegradable or non-biodegradable depending on the longevity of the device. Implants should be non-toxic and free from contaminants (Qin, 2016).

Long-term implants such as some artificial bones, joints, and ligaments may be required to stay in the body for a period of years. Non-biodegradable materials required for longer-term use in patients include polyester, polypropylene, PTFE, carbon, nylon, and polyethylene along with metals and ceramics for permanent insertion. These may fail, however, due to long-term complications such as infections, immune reactions, mechanical failure, and problems with biocompatibility with surrounding tissue. If the implant is not completely biocompatible, at the site of implantation infection can occur and the body starts to reject the implant (Romling and Balsalobre, 2012). To increase the chances of acceptance by the body, for many procedures, cells from the host body are combined with the textile scaffold. This could be a mix of biological and synthetic material, combined *in vitro* and then introduced to the body (Qin, 2016). Hohenstein (2010) have worked on the tolerability of implants using an animal-free substitute method known as the chorioallantoic membrane assay. Implants colonized with a patient's stem cells could serve as biologic textile implants (e.g. hernia mesh implants) that can be integrated into the patient's body more quickly and without rejection, to regenerate damaged tissues. One modification used on non-degradables is the use of a titanium coating to provide enhanced biocompatibility reducing the side effects caused by inflammation.

Materials that are required to stay within the body for several years and become accepted require bio-stability. An artificial ligament, for example, is required to be permanent and able to react with blood cells and surrounding tissue. Artificial bones and joints are often made of textile structural composites as opposed to the historically commonplace metal implants. Non-fibrous mats comprised of graphite can be seen with Teflon around the implant to promote tissue growth (Swicofil, 2016).

Many fibres used in regenerative surgery are re-absorbable over time to mitigate some of the problems caused by non-degradables, such as the formation of scar tissue after any inflammation and also chronic pain caused by the implant.

Biotolerance in medical textile terms is the capability of the body to tolerate a textile material (Hohenstein, 2010). To achieve good biocompatibility according to Edwards and Goheen (2011) the following properties can help:

- Interfacial free energy-surface tension between material and contact fluids etc. in the body
- Hydrophilic surfaces and balance between hydrophobic and hydrophilic properties
- The charge of the material—charged surfaces have been found to promote bone formation
- Lower molecular weight and flexibility of the material
- Surface patterns and roughness may influence thrombogenicity

Considerable research has been undertaken in the area of biocompatible textiles in recent years with advances in regenerative medicine (Qin, 2016; Bartels, 2011; Pan and Sun, 2016; Ivanova et al., 2014; Zhong, 2013; Gokarneshan et al., 2015). Cellulose is the most commonly used natural polymer and forms the basis of many healthcare textiles. Keratin and fibroin (structural biopolymers) are commonly used due to their biodegradability and biocompatibility.

Regenerated cellulose has been seen as a replacement for traditional materials with examples including viscose and Lyocell, both of which are hydrophilic. Alginate, polylactic acid (PLA), and polyglycolic acid (PGA), have shown potential for healthcare applications. Alginate is non-toxic, non-carcinogenic, non-allergic, haemostatic biocompatible, able to be sterilized and easily processed. PGA polymerized directly from glycolic acid is strong, has a high melting temperature and low solubility in organic solvents. PGA is biocompatible and biodegradable thus is widely used in scaffolding. PLA made most commonly from corn is polymerized directly from lactic acid. It shows good biocompatibility, bio degradability, is hydrophobic, strong, and durable. It is commonly used as a carrier material for cells to grow during bone tissue engineering. A variation of PLA and PGA is poly(lactic-co-glycolic) acid (PLGA), which has a shorter absorption time by the body as it undergoes hydrolysis in the body to produce original monomers, lactic acid, and glycolic acid. This makes PLGA ideal for implants (Ivanova et al., 2014; Qin, 2016).

Chitin, found naturally and widely available from shells of crabs and crustaceans, has excellent bioactive properties. Chitosan (a form of chitin), partially deacetylated, is biocompatible, biodegradable, and dissolvable in many aqueous acidic solutions. It is ideal for use in drug delivery and tissue engineering. Collagen is a commonly found animal protein. It has a controlled biodegradation rate, is biocompatible, and used in scaffolds, tissue culture, and wound healing. Hybrid scaffolds, comprising collagen and chitin, have been used in tissue engineering with gelatin, which is often used as an adhesive in scaffold manufacture (Lee et al., 2004).

In terms of biocompatibility a balance between hydrophilic and hydrophobic surfaces are required depending on the requirements and nature of the implant or scaffold. A hydrophilic surface, and thus lower surface tension, is thought to be more biocompatible (Edwards and Goheen, 2011). Hydrophobic surfaces when generating a contact angle with water of more than 65 degrees tend to adsorb proteins present in blood plasma. Rapid protein adsorption at the implant site plays a pivotal role in how the body responds to an implanted biomaterial and the associated initial inflammatory response (Thevenot et al., 2008).

For tissue engineering, materials should be biodegradable, meaning they will dissolve safely into the body after the scaffold has formed. Cotton, viscose, alginate, collagen, chitin, and chitosan can all be absorbed by the body within 2–3 months (Qin, 2016). To understand better how biocompatible a material is, testing can be undertaken to establish suitability. Biocompatibility testing might involve cytotoxicity testing (biological reactivity), sensitisation testing and irritation, systematic toxicity testing, genotoxicity, implantation or hemocompatibility (Qin, 2016).

16.4 Finishes and materials with specialist features

Healthcare textiles possess a variety of properties dependent on end use, whether to absorb or repel bodily fluids, protect wounds, support injured limbs or facilitate tissue growth in regenerative medicine. One common property, present across many medical textile materials, is the ability of the surface to act as a barrier to microbes or biocompatibility within tissue engineering. This is often achieved by the presence of a surface finish. There have been many advancements in dyeing and finishing technology over recent years with the development of an array of finishes to improve comfort and protection of clothing, durability of technical textiles, and improved performance across a number of conventional and smart textile applications. Specific properties can be applied through the use of specialist polymers, coating a textile, lamination, advanced dyes and colourants, and finishing. These can be utilized by the healthcare sector to provide fluid repellence including hydrophobic and super-hydrophobic features, stain and soil release, self-cleaning materials, and bioactive finishes for protection against biological attack including antimicrobial finishes.

16.4.1 Finishing of fibres and fabrics for waterproofness and water repellence

Many advances have been made since the start of the millennium with regards to water repellent and waterproof finishes. The original water repellent finishes were simple coatings of paraffin or wax on the fabric surface, which over time would, in general, wash out.

Semi-permeable liners, coatings, and films comprising hydrophilic or microporous membranes are commonplace providing a waterproof yet breathable material for comfort next to the body. Microporous structures contain numerous very fine pores to allow water vapour through but not liquid to penetrate, whereas hydrophilic structures work by transporting water molecules from an area of high humidity (inside garment) to an area of lower humidity (outside surface). Structures are commonly polyester (PES) hydrophilic membranes such as SympaTex, and polyurethane (PU) or polyurethane microporous structures such as Porelle dry.

In terms of fabric finishing, the dominant family of chemicals used to repel water, oil, and other stain causing liquids, are fluorochemicals of which there are perfluorochemicals (PFCs). Compounds of fluorocarbons form a low-surface-energy film on the surface of the material providing high repellence to both water and oil. This prevents penetration through the materials but also provides stain repellence. The finish, however, can make the fabric feel harsh and rough to the touch.

Materials with a PFC finish have effective non-stick properties, which can be seen to out perform alternatives which, when subjected to oil, are inferior in durability.

There have, however, been recent concerns over the safety of some of the family of fluorochemicals, PFCs in particular PFOA (Perfluorooctanoic acid, also known as C8 and perfluorooctanoate) and PFOS (Perfluorooctanesulfonic acid or Perfluorooctane sulfonate). For healthcare applications where exposure to toxicants can increase the risk of diseases especially in the vulnerable there is a need to ensure the fabrics used in contact with patients and staff do not pose a risk to health. Both PFOA and PFOS have been shown to damage human health and the environment. In a study by Grandjean et al. (2012), the chemicals have been linked to serious damage to the immune system in children.

C8 fluorocarbons are banned and have been replaced by C6 fluorocarbons. The smaller the fluorocarbon the quicker it breaks down in the environment thus the safer it is for public health. C4 fluorocarbons have been investigated to reduce the health threats further, however, it should be noted that the water repellence goes down as the size of the Perfluorocarbon goes down providing a fabric with potentially inferior properties. C6 is closest chemically to C8, but it contains no PFOA however it does not repel water and oil as well as C8 (Oecotextiles, 2010). Although water repellence is the dominant factor, oil repellence alongside water repellence may reduce the need for a stain-resistant finish in some healthcare textiles.

With advances in technology there has been a drive towards the use of biomimetics (taking inspiration from nature) and nanotechnology to create superior and safer materials (Shishoo, 2002). The concept of how the lotus leaf repels liquids and self cleans has provided inspiration. Examining the leaf at a microscopic level it comprises a rough surface covered with a wax-like substance of low surface tension that causes water to roll off. Using fluorochemicals and nanotechnology, finishes have been created to achieve water- and stain-resistance that mimics the lotus leaf. To achieve a super-hydrophobic surface with a greater contact angle (over 150 degrees) using inspiration from the lotus leaf, the surface roughness of the fabric should be increased mechanically, creating a substrate with a multiplicity of microscale to nanoscale projections or cavities (Gulrajani, 2013). It should be noted that these advances in technology are still in relative infancy and surface roughness in relation to durability of use in healthcare textiles requires further exploration.

Traditional nanocoatings with superior water and oil repellency are derived from long molecular chains, which contain traces of PFOA and PFOS. Because these chemicals pose a threat to health, many developers have sought to engineer new coatings free from PFOA and PFOS. GreenShield (2016) for example, use amorphous silica as the basic nanoparticle in their finishes. Amorphous silica is well established for human consumption in the United States by Food and Drug Administration (FDA). The use of silica allows for a significant reduction in the use of fluorocarbons to achieve good repellency. This creates a multifunctional finish providing water and oil repellence alongside stain resistance.

Many scientists are exploring alternatives to fluorochemical-based technologies in an attempt to become environmentally friendly and provide materials posing minimal threat to human health. In general, finishes form a waterproof layer on the material surface without filling up the interstices. This allows the fabric to circulate air and moisture. Recent advances have seen dyes being utilized to both colour and protect fabric providing a waterproof barrier. Several dyes with inherent water repellent properties have been developed allowing the fabric to be dyed and protected in unison (Sekar and Rajule, 2004). Using microencapsulation of paraffin-based waxes, fluorine-free water repellents have been developed for polyester, nylon, and cotton-based textiles (Mowbray, 2015). Super-hydrophobic surfaces have been developed using a fluorine-free silica-based treatment with a modifiable core. These can be used to create multifunctional surfaces, which are, for example, water repellent, flame retardant, and anti-microbial. Many of these advances are new and only just coming onto the market for commercial use.

16.4.2 Multifunctional finishes to provide bacterial and fungal prevention

Hospital uniforms are a common source of bacterial transmission between patients because staff have contact with a variety of patients throughout a single shift. With uniforms now requiring water repellence, stain repellence, and antimicrobial protection, many finishes are required to be multifunctional. A number of antibacterial finishes have been produced and adopted providing varying degrees of antimicrobial activity by either slowing down the growth rate to killing germs on contact. One important aspect in the development of effective finishes is that they should be non-leaching, which means they do not pose a risk of penetrating through the skin nor change the normal microbial flora of the skin. It should also be noted that once bacteria is present on a uniform, the population multiplies rapidly. A multifunctional finish providing both antimicrobial protection and water/oil repellence could kill or slow down the rate of bacterial growth and, at the same time, repel and prevent adhesion to the uniform in the first place.

There are various ways of applying effective antimicrobial protection to textiles.

1. adding bio-functional additives at the fibre production (polymer formation) stage
2. coating the fibres or fabric post fibre production

Many antimicrobial technologies have been developed since the new millennium, some of the key established developments being:

Quaternary Ammonium Salts (QAS)
Chitosan
Silver and copper impregnated textiles
N-Halamine based finishes

Table 16.2 provides a brief overview of the four key established developments.

Due to their multifunctional properties, the use of silver and copper has been explored further over the past decade. Nanotechnology has allowed nanoparticles

Table 16.2 Major antimicrobial technologies

Technology	Comments	Benefits	Drawbacks
Quaternary ammonium salts (QAS)	Widely used to effectively attract bacteria and penetrate into the membrane to kill the bacteria	Not consumed by the attacking microbes therefore can provide long lasting bacteria control for the life of the garment Washing also reactivates the action enhancing their longevity	Kill speed is relatively slow
Chitosan	Can provide a positive kill of microorganisms such as fungi, bacteria and some viruses. Savard et al. (2002) have however demonstrated that chitosan is more effective to kill fungi and algae than bacteria	Not allergenic to humans, promotes healing and is biodegradable which makes it suitable for clothing applications next to the skin (Knittel and Schollmeyer, 2000)	Kill speed is relatively slow
Silver and copper	The properties of silver and its safety in contact with human skin are well established (Sun, 2011) The use of copper as an antimicrobial has also been well documented (Vincent et al., 2016)	Both metals are effective on a broad spectrum of bacteria	Silver requires a moist environment to be most effective
N-Halamine chemistry	Effective on textile materials (Sun and Worley, 2006)	Exhibit a near instant kill affect through oxidation or chlorine transfer mechanism (Sun, 2011) The biocidal function can be reactivated through washing to provide an effective treatment over time on reusable textiles	

and nanolayers of silver to be applied to the surface of textiles. Plasma treatment has also been used to coat silver onto cotton/polyester fabrics. Cupron have produced a copper containing antimicrobial fibre. In the melt spinning process nanoparticles of copper oxide are mixed and dispersed in the fibre. The copper component suppresses the growth of bacteria, viruses, and fungi but can also stimulate collagen formation in tissue promoting skin regeneration (Qin, 2016). Coating-textiles with antimicrobial metallic particles like silver can provide optimum antimicrobial performance but are not toxic to human cells. Over *recent years* nickel and gold have also been investigated for antimicrobial protection (Morent and Geyter, 2011).

Silver and copper have also been demonstrated to provide effective protection though the method of application and cover on the textile will impact the effectiveness of the antibacterial protection across the full breath of the material. Silver, for example, requires a moist environment to be most effective meaning in some circumstances it underperforms. Copper, however, is effective across a broad spectrum of temperatures and humidity conditions meaning greater reliability in the hospital setting (Codiţă et al., 2010; Michels et al., 2008).

Antimicrobial finishes can be seen as an effective way of protecting hospital textiles against an array of bacteria, fungi, and in many cases spores. One important aspect is that any finishes should withstand repeated harsh laundering over time. Gao and Cranston (2008) comment that in general, finishes are not durable over time and become ineffective after repeated laundering. Over the last decade this aspect has been considered by two different solutions. The first is to physically embed antimicrobial agents into fibres making them more durable with some demonstrating slow-release properties (Gao and Cranston, 2008; Sun, 2011). The second method would be to regenerate the biocide by chemically recharging or reactivating the surface with an activating agent. Chlorine or oxygen bleach can serve as a charging or activating agent (Sun and Worley, 2005).

One additional benefit many finishes can provide is combating odour at the same time as providing antimicrobial protection, either by removing the odour precursor molecules or covering up the odours with fragrances. This can be advantageous for busy hospital staff. Recent antimicrobial finishes also consider factors beyond odour control with concerns over HCAIs such as chemical and biological infection of emergency service providers who move between the hospital environment and the community. One worrying find by Wendt et al. (1998) is that some microorganisms can survive on textiles for up to 90 days meaning that HCAIs can easily be moved in and out of the hospital environment if staff leave the premises in their uniforms. The solution is that future antimicrobial textiles should completely inactivate upon contact with the surface. Another factor is to review the transport and laundering protocol of hospital uniforms.

An active area of research is in the engineering of inherently antimicrobial dyes. New developments include photo-induced agents like titanium dioxide and dyes. Photosensitive agents have shown effective antimicrobial activity when exposed to UVA and daylight. The colourant is an excited compound possessing bio-radicals in triplicate status. Titanium can be applied through use of a coating to the surface of textiles which, when exposed to specific light sources, can provide antimicrobial protection and/or self-cleaning (Alihosseini and Sun, 2011).

Some natural dyes have been shown to provide antimicrobial properties and recent advances have concentrated on the use of both natural and synthetic dyes in providing effective antimicrobial action. Incorporating an antimicrobial effect at the same time as dyeing the material could be more environmentally friendly. Synthetic colourants with antimicrobial action include acridines, sulfamide-based dyes, and Quaternary ammonium dyes (Gao and Cranston, 2008; Alihosseini and Sun, 2011).

The ideal antimicrobial textile would provide permanent antimicrobial properties that are not reduced during the useful life of the garment or during the laundering process. The material should provide protection against a wide range of microorganisms present in hospital settings and should be non-toxic. The active antimicrobial effect should be limited to the textile surface thus not interfering with skin bacteria nor leaching to surrounding areas. The effectiveness and longevity of the biocide and speed of kill alongside environmental and health considerations are the main driving forces behind many new developments.

16.4.2.1 Reducing the rejection of implants (bio tolerance for regenerative medicine)

The requirements of medical implants and regenerative medicine are discussed in Section 16.3.3 alongside the importance of biocompatibility.

A key criterion of both short- and long-term implants is in reducing their rejection. Biofilms can be defined as communities of microorganisms that attach to each other and to surfaces, for example by bacterial adherence. Biofilms consist of both the cells and the extracellular matrix produced by the cells and can be problematic on medical implants. The build up of microorganisms on implantables and their dispersal in the body can lead to chronic inflammation, infection, and damage to host tissue which over time necessitates removal of the infected device (Romling and Balsalobre, 2012).

For effective regenerative medicine, a number of factors should be considered to ensure the foreign object does not become rejected. In terms of the implant itself, the surface should attract tissue cells with which it wants to form a bond but also prevent bacteria growth and attachment. Bruellhoff et al. (2010) suggest three lines of defence, which should be used either in isolation or, to ensure an effective barrier, used in combination.

1. prevent bacteria sticking to the surface of the implant
2. preventing the survival of bacteria if they do attach to the implant quorum sensing and stopping biofilm formation
3. disrupt any biofilm formation before it is established

To block the adhesion of bacteria onto the implant in situ, the use of non-adhesive coatings is a viable method. A number of studies have demonstrated that the surface roughness influences bacteria adhesion, with rougher surfaces attracting adhesion. Modified polyethylene terephthalate (PET) was examined by (MacKintosh et al., 2006) and showed hydrophilic surfaces resisted bacteria adhesion better than anionic and cationic surfaces, which caused adhesion. Although surface properties of the implant will contribute to adhesion resistance of bacteria, chemical modifications to surfaces are required to prevent biofilm formation.

To prevent the survival of bacteria on implant surfaces, a suitable coating either leaching or non-leaching should provide an instant kill. Metal nanoparticles comprising copper or silver ions have been demonstrated to be effective coatings for implants

(Bruellhoff et al., 2010). Ren et al. (2009) showed that copper oxide nanoparticles provided a good response against different bacteria in solution however, if incorporated in block polymers were less effective. A fundamental drawback of metal ions is that they may be potentially cytotoxic and contribute to DNA damage, however, silver, for example, has been successfully used in wound healing and reducing infection rates (Karlsson et al., 2008; Hardes et al., 2010; Bruellhoff et al., 2010).

Silicone elastomers incorporated with Triclosan have shown to be effective against S. epidermidis and 99% *E. coli* survival (Lewis and Klibanov, 2005). Chitosan-based coatings, antibacterial peptides, and silicone rubber with covalently coupled with QAS have all been demonstrated to show decreased infection rates on implants (Bruellhoff et al., 2010). With all coatings, they should provide effective kill against bacteria, however, they should also support the adhesion of desirable cells for tissue growth.

Research into disrupting any biofilm formation before it is established on the surface of the implant is still in its infancy, however, NCO-sP(EO-stat-PO)-based surface coatings and laser technology have potential in this area supported by antibiotic theory. Bruellhoff et al. (2010), Connaughton et al. (2014) provide information for advances in this area.

16.5 Future trends

Advances in fibre and fabric development and associated surface finishes, allow for the possibilities of fabrics with improved surface properties to meet the ever-changing requirements of the medical industry.

With the development of waterproof and water repellent surfaces in healthcare, the increasing use of plasma technology, dyes and colourants, and nanotechnology are striving towards the creation of more environmentally friendly, safe, and durable treatments. If produced cost effectively, finishes to encourage self-healing, along with the incorporation of fibres such as alginate into materials that come in contact with the skin, could encourage wound healing and promote well-being to those confined to hospital. There is a current focus on biomimicry to source ways to clean, protect, and safeguard patients within the hospital setting.

From a development point of view the perfect biocide is still some way off but there are constant developments in the fight to combat HCAIs. N-Halamine chemistry has been shown to provide powerful biocidal properties, however, more cost effective and environmentally friendly technologies are being explored.

Reducing the rejection of implants can be achieved through a combination of blocking bacteria adhesion to the implant surface, effective killing of any bacteria to reach the surface of the implant, alongside disrupting any biofilm formation before it becomes established. Disrupting the biofilm formation is where future research is focused with promising advances in NCO-sP(EO-stat-PO)-based surface coatings and laser technology to support antibiotic theory.

References

Alihosseini, F., Sun, G., 2011. Antibacterial colorants for textiles. In: Pan, N., Sun, G. (Eds.), Functional Textiles for Improved Performance, Protection and Health. Woodhead Publishing, Cambridge, pp. 376–403.

Anand, S., et al., 2010. Medical Textiles and Biomaterials for Healthcare. Woodhead Publishing, Cambridge.

Bartels, V.T. (Ed.), 2011. Handbook of Medical Textiles. Woodhead Publishing, Cambridge.

Beckham, H., et al., 2000. Fundamentals of moisture transport in textiles. National Textile Center 1–11. https://pdfs.semanticscholar.org/6481/1cebc478fb7994bec4ecbd9e199a168d4ced.pdf.

Bruellhoff, K., Fiedler, J., Möller, M., Groll, J., Brenner, R.E., 2010. Surface coating strategies to prevent biofilm formation on implant surfaces. Int. J. Artif. Organs 33 (9), 646–653.

BS 5867–2:2008 Fabrics for curtains, drapes and window blinds: part 2 flammability requirements specification, BSI.

BS EN 13795:2011+A1:2013, Surgical drapes, gowns and clean air suits, used as medical devices for patients, clinical staff and equipment, BSI.

BS EN ISO 9073–16:2008 Textiles. Test methods for nonwovens. Determination of resistance to penetration by water (hydrostatic pressure), BSI.

Codiţă, I., Caplan, D.M., Drăgulescu, E.C., Lixandru, B.E., Coldea, I.L., Dragomirescu, C.C., Surdu-Bob, C., Bădulescu, M., 2010. Antimicrobial activity of copper and silver nanofilms on nosocomial bacterial species. Roum. Arch. Microbiol. Immunol. 69 (4), 204–212.

Collins, A.S., 2008. Preventing health care–associated infections. In: Hughes, R.G. (Ed.), Patient Safety and Quality: An Evidence-Based Handbook for Nurses. Agency for Healthcare Research and Quality (US), Rockville (MD). (Chapter 41). Retrieved from: https://www.ncbi.nlm.nih.gov/books/NBK2683/.

Connaughton, A., et al., 2014. Biofilm disrupting technology for orthopaedic implants: what's on the horizon? Front. Med. 1 (22), 1–4.

Crow, R.M., Osczevski, R.J., 1998. The interaction of water with fabrics. Text. Res. J. 68 (4), 280–288.

Davies, A.M., 2009. The behaviour of liquid movement in three dimensional weft knitted spacer fabrics (PhD). De Montfort University, Leicester.

Davies, A.M., 2011. Use of knitted spacer fabrics for hygiene applications. In: McCarthy, B.J. (Ed.), Textiles for Hygiene and Infection Control. Woodhead, Cambridge.

Department of Work and Pensions, 2011. Differences in life expectancy between those aged 20,50 and 80 –in 2011 and at birth, James Evans, DWP, ad hoc research series.

Dye, C., 2014. After 2015: infectious diseases in a new era of health and development. Philos. Trans. R. Soc. 369, 20130426.

Edwards, J.V., Goheen, S.C., 2011. New developments in functional medical textiles and their mechanism of action. In: Pan, N., Sun, G. (Eds.), Functional Textiles for Improved Performance, Protection and Health. Woodhead Publishing, Cambridge, pp. 293–319.

Gao, Y., Cranston, R., 2008. Recent advances in antimicrobial treatments of textiles. Text. Res. J. 78, 60–72.

Gokarneshan, N., Anitha Rachel, D., Rajendran, V., Lavanya, B., Ghoshal, A., 2015. Emerging Research Trends in Medical Textiles. Springer, London.

Gov.UK, 2016. Healthcare associated infections (HCAI): guidance, data and analysis. Published online at Public Health England. Retrieved from: https://www.gov.uk/government/collections/healthcare-associated-infections-hcai-guidance-data-and-analysis.

Grandjean, P., Anderson, E.W., Budtz-Jorgensen, E., 2012. Serum vaccine antibody concentrations in children exposed to perfluorinated compounds. JAMA 307, 391–397.
Greenshield, 2016. The first and only textile finish inspired by nature. Retrieved from: http://www.greenshieldfinish.com.
Gulrajani, M.L., 2013. The use of nanotechnology in the finishing of technical textiles. In: Gulrajani, M.L. (Ed.), In: Advances in the Dyeing and Finishing of Technical Textiles, Woodhead, Cambridge.
Hardes, J., Von Eiff, C., Streitbuerger, A., Balke, M., Budny, T., Henrichs, M.P., et al., 2010. Reduction of periprosthetic infection with silver-coated megaprostheses in patients with bone sarcoma. J. Surg. Oncol. 101, 389–395.
Hohenstein, 2010. Hohenstein researchers make progress on biotolerance of textile implants, Hohenstein Institute. Retrieved from: https://www.hohenstein.de/en/inline/pressrelease_7366.xhtml.
Hsieh, Y., 1995. Liquid transport in fabric structures. Text. Res. J. 65 (5), 299–307.
Infiniti Research Limited, 2015. Medical Textiles Market: Global Trends, Demand, and Forecast 2015-2019. Infiniti Research Limited, London.
Ivanova, E.P., Bazaka, K., Crawford, R.J., 2014. New Functional Biomaterials for Medicine and Healthcare. Woodhead Publishing, Cambridge.
James, R., 2011. Infection prevention and control and the role of medical textiles. In: Bartels, V.T. (Ed.), Handbook of Medical Textiles. Woodhead Publishing, Cambridge, pp. 297–315.
Karlsson, H.L., Cronholm, P., Gustafsson, J., Moller, L., 2008. Copper oxide nanoparticles are highly toxic: a comparison between metal oxide nanoparticles and carbon nanotubes. Chem. Res. Toxicol. 21, 1726–1732.
Knittel, D., Schollmeyer, E., 2000. Permanent modification of fibrous materials with biopolymers. Adv. Chitin Sci. 4, 143–147.
Lee, S.B., Kim, Y.H., Chong, M.S., Lee, Y.M., 2004. Preparation and characteristics of hybrid scaffolds composed of B-chitin and collagen. Biomaterials 25, 2309–2317.
Lewis, K., Klibanov, A.M., 2005. Surpassing nature: rational design of sterile-surface materials. Trends Biotechnol. 23, 343–348.
Llor, C., Bjerrum, L., 2014. Antimicrobial resistance: risk associated with antibiotic overuse and initiatives to reduce the problem. Ther. Adv. Drug Saf. 5 (6), 229–241. https://doi.org/10.1177/2042098614554919.
Mackintosh, E.E., Patel, J.D., Marchant, R.E., Anderson, J.M., 2006. Effects of biomaterial surface chemistry on the adhesion and biofilm formation of Staphylococcus epidermidis in vitro. J. Biomed. Mater. Res. A 78A, 836–842.
Michels, H., Moran, W., Michel, J., 2008. Antimicrobial properties of copper alloy surfaces, with a focus on hospital—acquired infections. Int. J. Met. 2, 47–56.
Morent, R., Geyter, N. De, 2011. Improved textile functionality through surface modifications. In: Pan, N., Sun, G. (Eds.), Functional Textiles for Improved Performance, Protection and Health. Woodhead Publishing, Cambridge, pp. 3–26.
Mowbray, J., 2015. Flexible, fluorine-free water repellent. Ecotextile News. Retrieved from: http://www.ecotextile.com/2015012821272/dyes-chemicals-news/flexible-fluorine-free-water-repellent.html.
National Institute for Health and Care Excellence, 2011. Healthcare-Associated Infections: Prevention and Control. NICE, London.
National Institute for Health and Care Excellence, 2012. Osteoporosis: Assessing the Risk of Fragility Fracture. NICE, London.
NHS, 2009. Revised Healthcare Cleaning Manual. National Patient Safety Agency, Hospital Cleaning, June. Retrieved from: http://www.npsa.nhs.uk/cleaning.

NHS, 2010. NHS Professionals Standard Infection Control Precautions. Clinical Governance, V3, November.

NHS, 2015. The Antibiotic Awareness Campaign. NHS. Retrieved from: http://www.nhs.uk/nhsengland/arc/pages/aboutarc.aspx.

NHS Executive, 1995. Hospital Laundry Arrangements for Used and Infected Linen, HSG (95) 18. NHS Executive, London.

Nocker, W., 2011. Evaluation of occupational clothing for surgeons: achieving comfort and avoiding physiological stress through suitable gowns. In: Bartels, V.T. (Ed.), Handbook of Medical Textiles. Woodhead Publishing, Cambridge, pp. 443–460.

Nursing Standard, 2014. Nurses welcome government report on extreme ward temperatures, 28 (46), 13.

Oecotextiles, 2010. What about soil resistant finishes like Scotchgard, GoreTex, NanoTex and GreenShield – are they safe? Retrieved from: https://oecotextiles.wordpress.com/tag/c6/.

Office of National Statistics, 2015. National Population Projections: 2014-Based Statistical Bulletin. ONS, London.

Ohl, M., Schweizer, M., Graham, M., Heilmann, K., Boyken, L., Diekema, D., 2012. Hospital privacy curtains are frequently and rapidly contaminated with potentially pathogenic bacteria. Am. J. Infect. Control 40 (10), 904–906.

Pan, N., Sun, G. (Eds.), 2016. Functional Textiles for Improved Performance, Protection and Health. Woodhead Publishing, Cambridge.

Qin, Y., 2016. Medical Textile Materials. Woodhead Publishing, Cambridge. pp. 13–22, 133–143, 191–201.

Rajendran, S., 2006. Infection control and barrier fabrics: an overview. In: Anand, S. et al., (Ed.), Medical Textiles and Biomaterials for Healthcare. Woodhead Publishing, Cambridge, pp. 131–135.

Ren, G.G., Hu, D.W., Cheng, E.W.C., Vargas-Reus, M.A., Reip, P., Al-Laker, R.P., 2009. Characterisation of copper oxide nanoparticles for antimicrobial applications. Int. J. Antimicrob. Agents 33, 587–590.

Riley, K., Williams, J., Davies, A., Shen, J., Laird, K., 2017. The effect of low temperature laundering and detergents on the survival of Escherichia coli and Staphylococcus aureus on textiles used in healthcare uniforms. J. Appl. Microbiol. 123, 280–286.

Romling, U., Balsalobre, C., 2012. Biofilm infections, their resilience to therapy and innovative treatment strategies. J. Intern. Med. 272, 541–561.

Roser, M., 2016. Future World Population Growth. Published online at OurWorldInData.org. Retrieved from: https://ourworldindata.org/future-world-population-growth/.

Royal College of Nursing, 2014. Wipe It Out-Guidance on Uniforms and Work Wear, third ed. Royal College of Nursing, London.

Royal College of Nursing, 2016. Infection Prevention and Control Commissioning Toolkit, third ed. Royal College of Nursing and Infection Prevention Society, London.

Savard, T., Beaulieu, C., Boucher, I., Champagne, C.P., 2002. Antimicrobial action of hydrolyzed chitosan against spoilage yeasts and lactic acid bacteria of fermented vegetables. J. Food Prot. 65, 828–883.

Saville, B.P., 1999. Physical Testing of Textiles. Woodhead, Cambridge.

Sekar, N., Rajule, R.N., 2004. Studies in fluorine-containing monoazo acid dyes and their comparison with non fluorine analogues. Adv. Colour Sci. Technol. 7, 48–54.

Shishoo, R., 2002. Recent development in materials for use in protective clothing. Int. J. Cloth. Sci. Technol. 14, 201–215.

Sifri, C.D., Burke, G.H., Enfield, K.B., 2016. Reduced health care-associated infections in an acute care community hospital using a combination of self-disinfecting copper-impregnated composite hard surfaces and linens. Am. J. Infect. Control 44, 1565–1571.

Slater, K., 1993. Physical testing and quality control: a critical appreciation of recent developments. Textile Institute, Manchester.

Sun, G., 2011. Antibacterial textile materials for medical applications. In: Pan, N., Sun, G. (Eds.), Functional textiles for improved performance, protection and health. Woodhead Publishing, Cambridge, pp. 360–375.

Sun, G., Worley, S.D., 2005. Chemistry of durable and regenerable biocidal textiles. J. Chem. Educ. 82 (1), 60–64.

Sun, G., Worley, S.D., 2006. Halamine chemistry and its applications in biocidal textiles and polymers. In: Edwards, V. et al., (Ed.), Modified Fibers With Medical and Speciality Applications. Springer, Netherlands, pp. 81–89.

Swicofil, 2016. Biomedical textiles: implantable materials, Swicofil. Retrieved from: http://www.swicofil.com/biomedical_textiles.html.

Thevenot, P., Hu, W., Tang, L., 2008. Surface chemistry influence implant biocompatibility. Curr. Top. Med. Chem. 8 (4), 270–280.

United Nations, n.d. World Population Ageing 1950-2050. Population Division Department of Economic and Social Affairs (DESA). Retrieved from: http://www.un.org/esa/population/publications/worldageing19502050/pdf/90chapteriv.pdf.

Vincent, M., Hartemann, P., Engels-Deutsch, M., 2016. Antimicrobial applications of copper. Int. J. Hyg. Environ. Health 219, 585–591.

Wendt, C., Wiesenthal, B., Dietz, E., Ruden, H., 1998. Survival of vancomycin-resistant and vancomycin-susceptible enterococci on dry surfaces. J. Clin. Microbiol. 36 (12), 3734–3736.

Woodland, R., Whitham, D., O'neil, B., Otter, B., 2010. Microbiological contamination of cubicle curtains in an out-patient podiatry clinic. J. Foot Ankle Res. 3, 26.

World Health Organisation, 2015. Healthy life expectancy (HALE) at birth. Published online at WHO. Retrieved from: http://www.who.int/gho/mortality_burden_disease/life_tables/hale/en/.

World Health Organisation, 2016. The burden of health care-associated infection worldwide. Published online at WHO. Retrieved from: http://www.who.int/gpsc/country_work/burden_hcai/en.

Zhong, W., 2013. An Introduction to Healthcare and Medical Textiles. DEStech, Pennsylvania, USA.

Zins, H.M., 2011. Reusable medical textiles. In: Bartels, V.T. (Ed.), Handbook of Medical Textiles. Woodhead Publishing, Cambridge, pp. 80–105.

Further reading

Age UK, 2016. Later Life in the United Kingdom. Age UK.

Military applications: Development of superomniphobic coatings, textiles and surfaces

17

Quoc T. Truong, Natalie Pomerantz
US Army Natick Soldier Research, Development and Engineering Center, Natick, MA, United States

17.1 Introduction

17.1.1 Army needs

Since 1954, the NSRDEC, also known as the Natick Labs, has been a part of the Army's Research, Development and Engineering Command, and was given the responsibility for the Soldier Systems Integration Domain in coordinating soldier-related efforts across the command and in addressing soldier's needs and technology gaps (NSRDEC Public Affairs Office, 2017). As such, this chapter shares work on developing superomniphobic coatings, textiles and surfaces with the goal to provide self-cleaning and enhanced chemical/biological (CB) protective clothing for the individual soldier.

On a battlefield, while carrying out their missions, soldiers cannot avoid getting their uniforms dirty. Activities such as manoeuvring through muddy terrains or dusty battlefields, and oil-contaminated environments make their clothing dirty (Fig. 17.1). To remove dirt and/or contaminants from the surface of their clothing's textile materials, enzymatic, surface active and/or oil-dissolving detergents are needed. Keeping the soldiers' uniforms clean becomes a lower priority when maintaining personal safety in a CB warfare agent contaminated environment becomes a higher priority; however, keeping their clothing clean remains a necessity.

Additionally, CB protective clothing systems rely on a fluorocarbon finished outershell fabric to minimize surface wetting by water and liquid chemicals. With minimal exposure to water, soldiers could minimize hypothermia risks when conducting extended missions in the cold and rain. Also, as the fluorocarbon finish helps to reduce attachment of toxic chemicals on the soldiers' outershell fabric, toxic chemical permeation through their clothing is drastically reduced; therefore, effectively keeping them safe when operating in a contaminated environment. The use of a liquid wicking/spreading outermost fabric layer was traditionally thought to spread out and enhance evaporation from the clothing/fabric surface. However, in the case of low volatility agents which do not evaporate quickly, NSRDEC's live agent data has shown that repellent surfaces offer as much as 90% reduction in permeation as compared to an untreated chemical protective fabric system that helps to wick and spread the toxic chemicals. This will be discussed further in Sections 17.8.2, and 17.8.3.

Waterproof and Water Repellent Textiles and Clothing. https://doi.org/10.1016/B978-0-08-101212-3.00016-2

Fig. 17.1 Soldiers cannot avoid getting their uniforms dirty.
Image credit: US Army.

Liquid repellent properties can be attributed to the very low interfacial energies in these protective clothing systems, and there have been many efforts to develop more effective and durable repellent chemical treatments for outershell fabrics. However, improving the outershell fabric's surface chemistry is only one of three fundamental design modifications that helps to improve a fabric's resistance to wetting by a wide range of liquids having high (e.g. water) to very low surface tensions (e.g. methanol and n-hexane). In general, lower surface tension liquids are difficult to repel. They tend to wet most surfaces, even textiles and solid surfaces that were treated with fluorinated chemicals. To repel low surface tension liquids, an additional layer of surface roughness is necessary, and this surface roughness can be engineered by the addition of micro and nanoparticles in the coating solution. EverShield (UltraTech International, Inc., 2017) Durable Omni Repellent (DOR) is an example of a commercial omniphobic coating containing dual hierarchical micro/nanoparticulates in its formulation. This coating was originally developed with and for the NSRDEC under SBIR contracts with Luna Innovations, Inc. from 2008 to 2013 with the objectives to provide soldiers with durable self-cleaning and CB agent protective clothing (Truong and Koene, 2013a,b).

In keeping the clothing clean and the soldiers safe in a dirty and/or contaminated battlefield/environment, NSRDEC has identified selected polar and nonpolar liquid chemicals, which include chemical warfare agents (CWAs), in developments of durable liquid repellent polymer coatings for textiles with the goal to effectively repel the majority of these chemicals (Table 17.1). United States Department of Homeland Security (DHS) has also selected toxic industrial chemicals (TICs) that are identified in the National Fire Protection Association (NFPA) 1991 standard to protect

Table 17.1 Surface tensions of selected polar and nonpolar liquids

Liquids (polar and nonpolar)	Surface tension (mN/m)	Notes
1. Water	72.8 (Wikimedia Foundation, 2016a,b)	Polar protic solvent
2. Methylene iodide	50.8 (Surface Tension, 2016)	Diiodomethane Nonpolar solvent
3. Ethylene glycol	48.4 (US FDA&EMEA, 2016)	Polar solvent
4. Dimethyl sulfoxide	43.54 (Wikimedia Foundation, 2016a,b)	DMSO. Lewisite (L) chemical agent simulant. Polar aprotic solvent
5. Mustard (HD)	43.2 (National Research Council of the National Academies, 2005)	Blister Chemical Agent. Polar solvent. @20°C
6. Methyl salicylate	41.84 (LookChem.com, 2008)	MeS. Sulphur mustard (blister) chemical agent simulant. Polar solvent
7. Dimethyl methyl phosphonate	36.7 (Keller and Simmons, 2005)	DMMP. Sarin Nerve chemical agent simulant. Polar solvent
8. Canola oil	35 (Gunstone, 2009)	Nonpolar solvent. @20°C.
9. VX	32.01 (National Research Council of the National Academies, 2005)	Nerve Chemical Agent. Polar solvent. @20°C
10. Triethyl phosphate	30.61 (Washburn, 1928)	TEP. Nerve chemical agent simulant. Polar solvent
11. 2-Chloroethyl ethyl sulphide	28.4 ± 3.0 (Royal Society of Chemistry, 2015)	2-CEES. Blister chemical agent simulant. Polar solvent
12. Tributyl phosphate	27.79 (Keller and Simmons, 2005)	TBP. Nerve chemical agent simulant. Polar solvent
13. Hexadecane	27.47 (Surface Tension, 2016)	Nonpolar solvent
14. Sarin (GB)	26.5 (National Research Council of the National Academies, 2005)	Nerve Chemical Agent. Polar solvent. @ 20°C
15. Dodecane	25.35 (Surface Tension, 2016)	Nonpolar solvent
16. Soman (GD)	24.5 (Gorzkowska-Sobas, 2013)	Nerve Chemical Agent. Polar solvent. @ 26.5°C
17. Decane	23.83 (Surface Tension, 2016)	Nonpolar solvent
18. 2-Propanol (isopropyl alcohol)	23.3 (Diversified Enterprises, 2014)	Polar protic solvent
19. Methanol	22.1 (Diversified Enterprises, 2014)	Polar protic solvent
20. Octane	21.62 (Surface Tension, 2016)	Nonpolar solvent
21. Heptane	20.14 (Surface Tension, 2016)	Nonpolar solvent
22. N-Hexane	18.40 (National Physical Laboratory, 2015)	Nonpolar solvent

Table 17.2 Toxic industrial chemicals (TICs)

TIC name	Surface tension (mN/m)	CAS registry number
Sodium hydroxide	101.05 (35.9 wt%)[c] (Weast, 1986)	CAS: 1310-73-2 (50%)
Sulfuric acid	51.7[c] (Haynes, 2014)	CAS: 7664-93-9
Nitrobenzene	43.90[b] (Weast, 1977)	CAS: 98-95-3
N,N-Dimethylformamide	35.74[a] (Haynes, 2014)	CAS: 68-12-2
Tetrachloroethylene (PCE)	31.74[b] (Weast, 1977)	CAS: 127-18-4
Acetonitrile	28.66[a] (Haynes, 2014)	CAS: 75-05-8
Toluene	27.73[a] (Haynes, 2014)	CAS: 108-88-3
Dichloromethane	27.20[a] (Haynes, 2014)	CAS: 75-09-2
Tetrahydrofuran	26.40[c] (Flick, 1991)	CAS: 109-99-9
Ethyl acetate	23.39[a] (Haynes, 2014)	CAS: 141-78-6
Acetone	22.72[a] (Haynes, 2014)	CAS: 67-64-1
Methanol (polar)	22.07[a] (Haynes, 2014)	CAS: 67-65-1
Diethylamine	19.85[a] (Haynes, 2014)	CAS: 109-87-7
n-Hexane (nonpolar)	17.89[a] (Haynes, 2014)	CAS: 100-54-3

CAS, chemical abstracts service.
[a]@25°C.
[b]@20°C.
[c]Unknown temperature.

emergency responders such as firemen and hazardous chemical materials (HazMat) response personnel handling chemical spills and terror incidences (Table 17.2).

17.1.2 Omniphobic coatings

Hydrophobic means, 'water (hydro) hating (phobic)', and refers to the repelling of water. Similarly, oleophobic or oil (oleo) hating (phobic) refers to the repelling of oils. Omni means 'all', so omniphobicity refers to a surface's ability to resist wetting by various liquids. Once applied onto a surface, an omniphobic coating alters the surface chemistry, which effectively helps to resist wetting by not only high surface tension liquids like water, which has a liquid–vapour surface tension (γ_{lv}) of 72.8 mN/m, but it will also resist wetting to low surface tension liquids such as octane ($\gamma_{lv}=21.62$ mN/m) and methanol ($\gamma_{lv}=22.1$ mN/m). In general, omniphobic coatings are engineered differently when they are applied to hard surfaces (e.g. metal, transparent glasses and plastics) as compared to textiles (e.g. clothing, gloves and boots), with textile surfaces having more reentrant (or architecture) levels with additional levels of liquid repellency. In addition, while an omniphobic surface can be created to enhance liquid repellency, maintaining its durability for practical applications can be a challenge. For example, surface roughness created by applying a coating solution containing micron-size particles onto a surface (to mimic the micron-size nubs (see Fig. 17.3) on the lotus leaf) could be easily abraded away if the coating chemistry is not properly selected to ensure durable bonding of particles to the applied surface. Therefore durability of a roughened surface to maintain its micro/nanosurface architectures depends on the coating material composition (e.g. polymer vs. elastomer) and how the roughened surface is created (e.g. mechanically roughened surface vs. coated surface containing particles that can be abraded off very easily). For example, if steel wool is used to scratch the surface of a softer metal such as aluminium, the roughened surface left behind will be very durable. Likewise if micro/nanosize particles are deposited on a surface, and not secured in place by using a durable adhesive, these particles can be abraded off very easily. On the other hand, if a coating formula solution that has low viscosity, and contains an elastomer and micro/nanoparticles is used, then the micro/nanoparticles will be coated by the elastomer, and cured in place on the textile fibres' surface. These elastomer-coated particles will take a much longer time to be abraded off when in contact with an abradant. This is the reason that an elastomer is used to produce 'rubber' tyres on the road.

17.1.2.1 Hard surfaces

Hard surface refers to many different types of surfaces. For military, examples include all-terrain vehicles, communication equipment, weapons, etc. Keeping them from being wetted by water and liquid chemicals will prevent corrosion and costly repairs or replacements; therefore, would increase their service life. In 1944, Cassie and Baxter, in studying ways to mimic natural surfaces, discovered surfaces created with micrometre- and/or nanometre-scale roughness behave similarly to a lotus leaf surface, where water would not wet its surface (Cassie and Baxter, 1944). They found that surface roughness dictates how liquids interact with a surface. When a liquid completely wets a roughened solid surface (i.e. in full contact), it's referred to as a Wenzel state (Wenzel, 1936). When a liquid is in contact with the peaks of a roughened solid surface, it's referred to as a Cassie–Baxter state, where air pockets within the surface features help to support the liquid; therefore, the surface stays nonwetting. Cassie and Baxter's discovery has far reaching military applications in creating liquid repellent hard surfaces such as metal surfaces of vehicles and equipment, preventing corrosion of metal components due to water contact (i.e. oxidation of hard metal surfaces by contacting oxygen in the water molecule).

Surface chemistry is one of three surface design criteria that helps to improve a hard surface's ability to resist wetting by low surface tension liquids (Fig. 17.2), where omniphobic materials are both hydrophobic and oleophobic, with contact angles between 90° and 150° (Rame Hart Instruments, Inc., 2017). The second design criterion is the micrometre and/or nanometre scale surface roughness. Surface roughness dictates how liquid droplets will behave—either wetting (i.e. Wenzel state) or nonwetting (i.e. Cassie–Baxter state) as represented by the graphic in Fig. 17.3 and their equations. These micro/nanosurface roughnesses (or features) are found in nature. For example, on a lotus leaf surface (Ensikat et al., 2011), water strider's feet (Hensel et al., 2013) or in the unique skin structure of a springtail insect (Tuteja et al., 2007, Fig. 17.4). The third design criterion is the geometrical (trapezoidal-shaped) reentrant surface architecture which MIT had discovered (Tuteja et al., 2008a,b; Chhatre et al., 2010a,b, Fig. 17.5). This third surface design criteria enables the surface to become nonwetting to extremely low surface tension liquids. MIT's 'microhoodoo' surface architecture demonstrated that man-made superomniphobic coating is possible, where the contact angle of liquids are >150°. The key finding here is that the

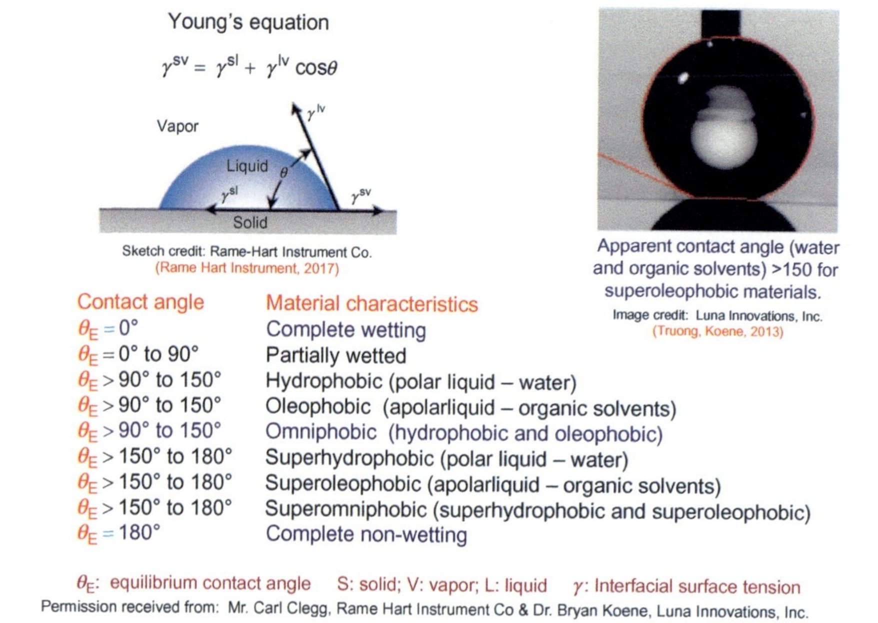

Fig. 17.2 Surface chemistry (high contact angle) (Rame Hart Instruments, Inc., 2017; Mabry et al., 2008).

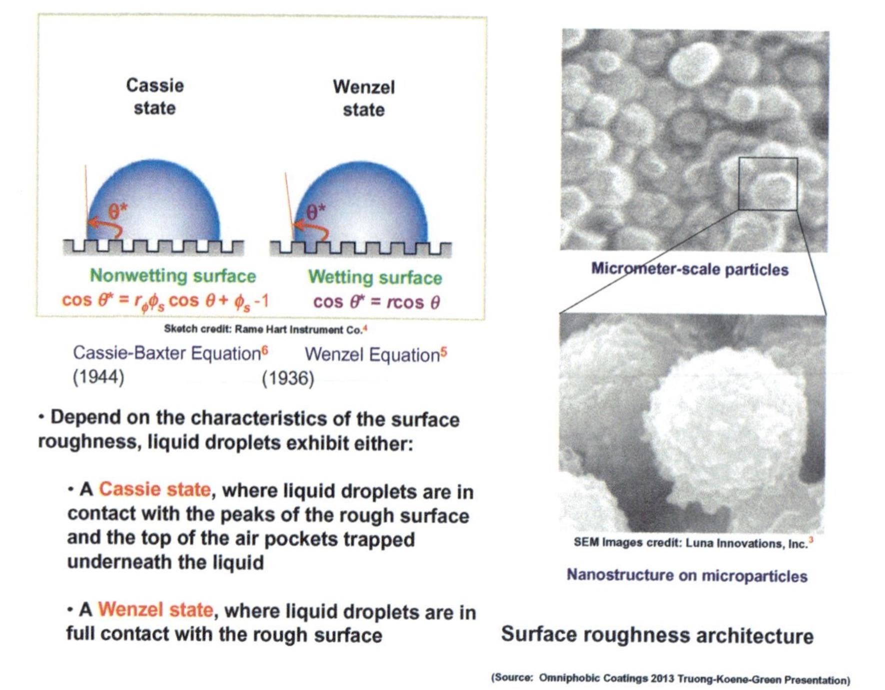

Fig. 17.3 Surface roughness (Truong and Koene, 2013a,b; Rame Hart Instruments, Inc., 2017; Wenzel, 1936; Cassie and Baxter, 1944).

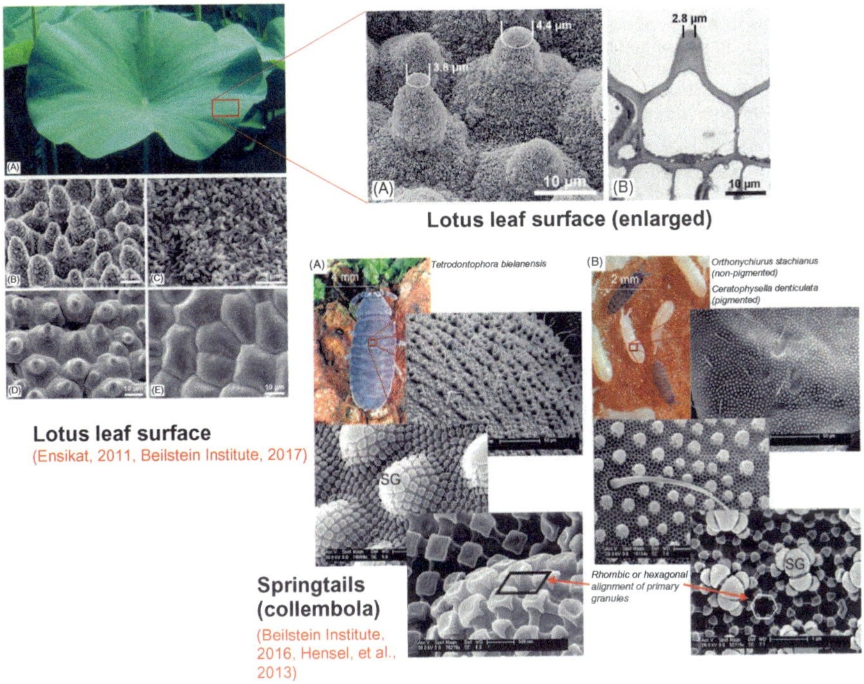

Fig. 17.4 Unique superhydrophobic surfaces and oleophobic (Springtails) surface structures of natural surfaces (Ensikat et al., 2011; Beilstein Institute, 2016, 2017; Hensel et al., 2013).

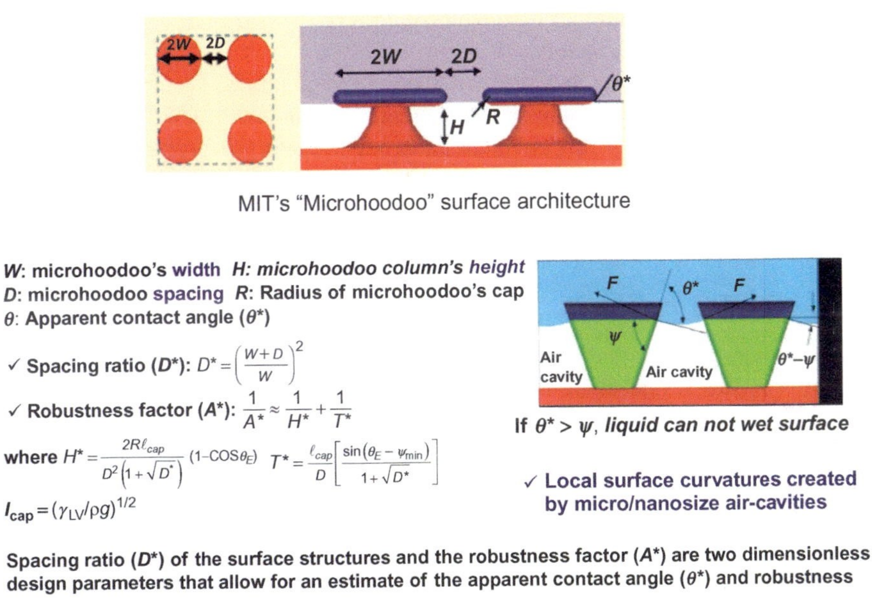

Fig. 17.5 Reentrant geometrical texture (Tuteja et al, 2008a,b; Chhatre et al., 2010a,b).

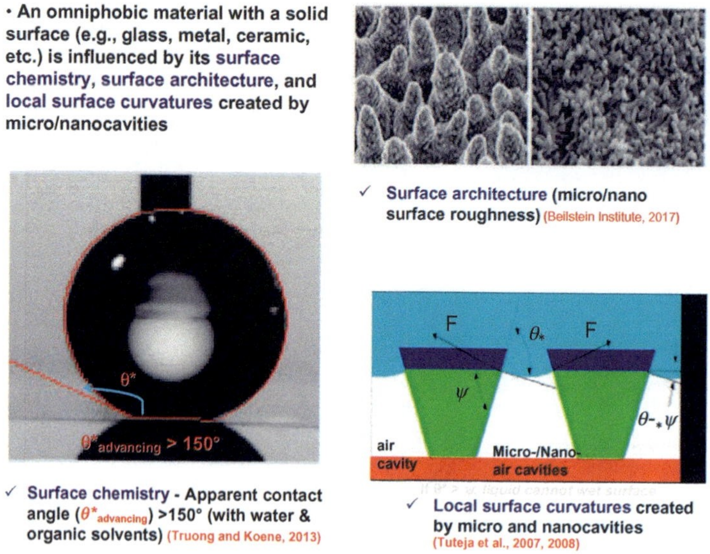

Fig. 17.6 Three important parameters in designing superomniphobic hard surfaces (Rame Hart Instruments, Inc., 2017; Hensel et al., 2013; Tuteja et al., 2007, 2008a,b; Chhatre et al., 2010a,b; Kleingartner et al., 2015).

ability to resist surface wetting (or liquid robustness) depends on the geometry of the textures. If the apparent contact angle (θ^*) is greater than that of the slanted angle of the side wall of the reentrant micro/nanocolumns (ψ) as shown in Fig. 17.5, a superomniphobic surface can be created.

To develop a material having a robust superomniphobic surface, the material must have all three key parameters: sufficiently low surface tension chemistry with a surface contact angle $>150°$, a micro/nanoroughness surface architecture and local surface curvatures created by micro/nanocavities in which liquid cannot wet the surface. (Fig. 17.6).

17.1.2.2 Textiles

Similar to designing liquid repellent hard surfaces, surface chemistry is an important criterion in the design of liquid repellent textiles. However, because textiles have more complex structures where liquid repellency can be progressively designed, NSRDEC had collaborated with MIT from 2008 to 2015 to study and develop a fundamental understanding of how fabric structures influence liquid repellency and liquid robustness. In 2010, MIT reported its findings that there are three key parameters in designing a superomniphobic (or supernonwetting) fabric. These include: fibres' surface chemistry, fabric weave structures or weave openings and the sizes of the individual fibres and fibre bundles (Chhatre et al., 2010a,b,

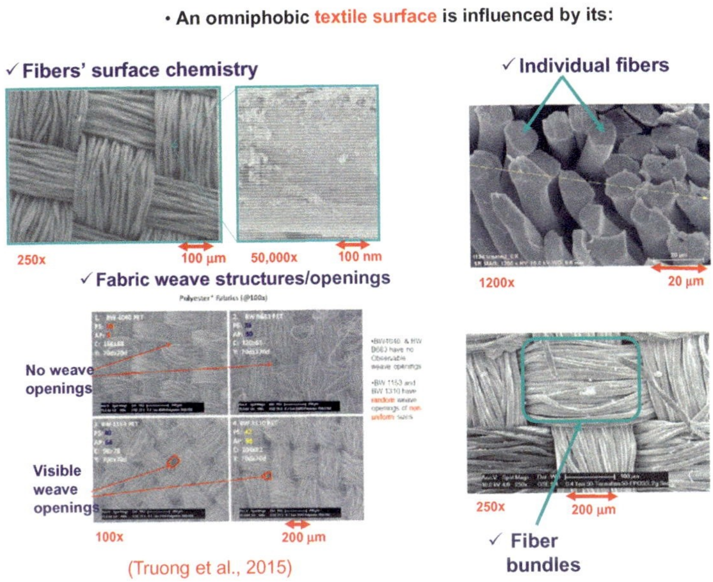

Fig. 17.7 Key parameters to achieving superomniphobic fibres and fabrics (Truong et al., 2015a,b).

Fig. 17.7). Based on the Cassie–Baxter equation, in changing the fabric's mesh size, the fabric's liquid robustness (ability to resist surface wetting) can be manipulated. For example, in a study of the liquid robustness of an idealized fabric structure where MIT used a stainless steel mesh structure (Fig. 17.8B), if the number of the openings between the stainless steel fibre structure (or the mesh size of this idealized stainless steel fabric) is reduced from 50 to 100, and to 325, then its robustness is increased accordingly. Likewise, changing fabric weave design will affect fabric's liquid robustness and its surface wetting behaviour, and increasing individual fibres' size and fibre bundles' sizes will decrease the fabric's liquid robustness.

Similarly, in changing the fibres' surface chemistry (i.e. Θ_E, or the equilibrium contact angle) and its surface texture (i.e. D^*, or the dimensionless distance or spacing between the fibres where $D^* = (R+D)/R$) (Truong et al., 2015a,b) one can design a textile surface (or a fabric) to become superomniphobic where its apparent contact angle (Θ^*) exceeds 150°. This can be seen as the line graphs shift upward from black ($\Theta^* = 0°$), to red ($\Theta^* = 90°$), to green ($\Theta^* = 120°$) and to blue ($\Theta^* = 150°$) [see Fig. 17.8A]. Reducing the mesh size (i.e. the opening space between the fibre bundles), will increase the number of openings per centimetre. By reducing the openings' size between the fibres of the idealized texture fabric, one could likewise increase the resistance to liquid penetration or liquid robustness (see Fig. 17.8B)—that is, an open structure lets more water through than a tight structure. Chhatre et al.'s referenced Langmuir paper (2010) contains much practical

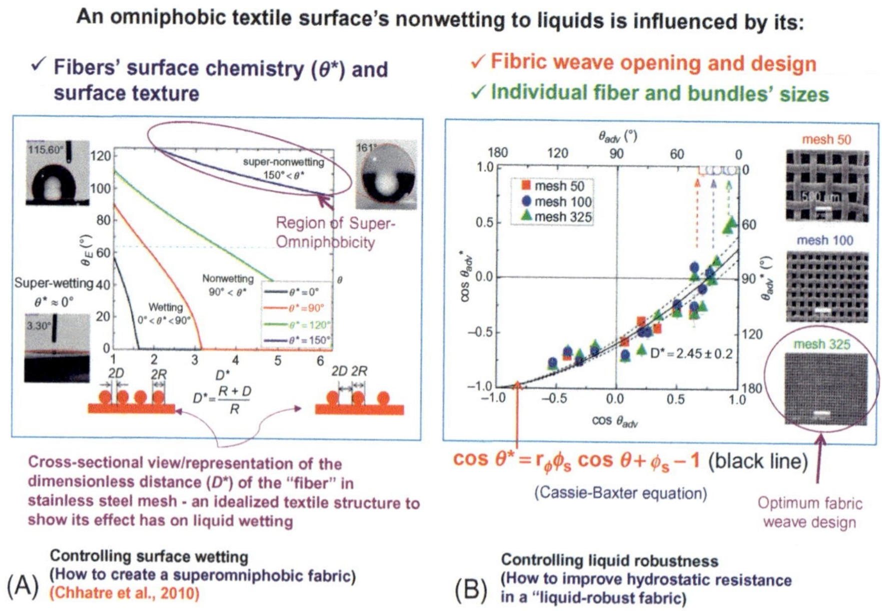

Fig. 17.8 Characteristics of omniphobic materials with textile surfaces (Chhatre et al., 2010a,b).

engineering design that would help material/fabric designers to design a super-omniphobic fabric, and/or a tightly woven fabric that has very good hydrostatic resistance that resists liquid penetration.

More recently, MIT has shown additional importance of applying hierarchical textured textiles, with the addition of micro and/or nanoparticulates, on the surfaces of the fibre (Tuteja et al., 2008a,b). In this NSRDEC funded study, MIT reported an additional reentrant textile structural level—micro/nanoparticulates added to the fibres' surface where each reentrant textile structural level contributes towards an increase in liquid nonwetting stage—namely: single fibre (1st level), fibre bundles or yarn size (2nd level-a), weave opening (i.e. how tight or open a fabric is) (2nd level-b) and fibre's surface texture (e.g. via deposition of micro and/or nanosize particulates) (3rd level) (Fig. 17.9).

In 2013, based on MIT's finding in 2007 (Tuteja et al.), NSRDEC has discovered yet another level of liquid nonwetting architecture (4th level) by creating a novel reentrant-shaped fibre (Truong et al., 2015a,b). This fibre is 30 μm in diameter (the current technological processing limit), and has repeating trapezoidal reentrant shape-channels along the length. On its surface (and inside the channels), there are nanosize surface features/roughness created by surface-bonded or surface-bloomed nanoparticles to mitigate the capillary effects, thus preventing liquid wicking along the channels. This new discovery has added an additional level to MIT's previously found three levels of liquid repellency (Fig. 17.10).

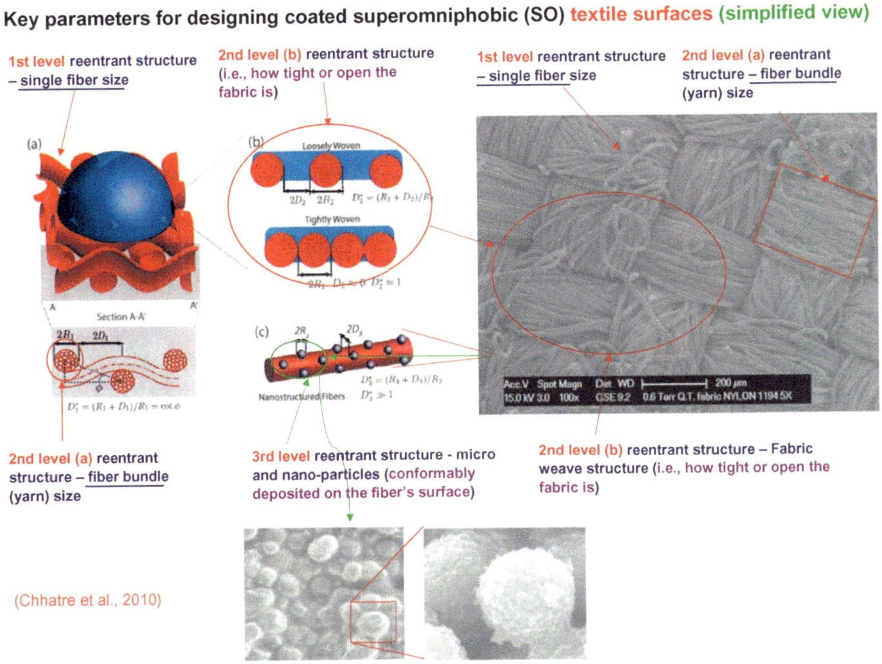

Fig. 17.9 Hierarchical reentrant textile structures and added particle-on-surface texture (Tuteja et al., 2008a,b; Truong and Koene, 2013a,b; Truong et al., 2015a,b).

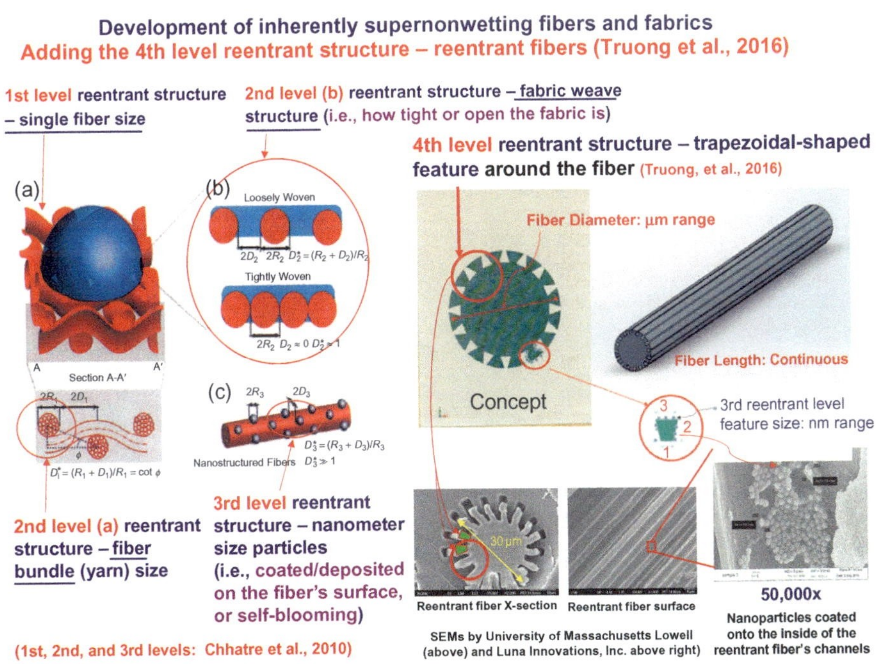

Fig. 17.10 Hierarchical reentrant textile structures with added inversed trapezoidal-shaped reentrant fibre structures (Tuteja et al., 2008a,b; Truong et al., 2015a,b).

17.1.3 C-8 (EPA banned) versus C-6 (environmentally friendly) chemistries

Perfluorooctanoic acid (PFOA) (Fig. 17.11A) and perfluorooctane sulfonic acid (PFOS) (Fig. 17.11B) are synthetic chemicals that do not occur naturally in the environment. Each molecule contains eight carbon atom units having fluorine atoms surrounding most of their molecular structures with oxygen and hydrogen atoms at one end, and are often referred to as C-8 chemical compounds. These C-8 chemicals have important applications in textile manufacturing because they are used in the process of producing liquid repellent textiles that are resistant to wetting by water, chemicals, oils and grease. C-8-treated surfaces have reduced surface friction; therefore exhibiting better surface lubricity. Although PFOA and PFOS are burned off during the textile manufacturing process and are not present in significant amounts in the final products (American Cancer Society, 2016), they are very persistent chemicals and remain in the environment and in living organisms. Furthermore, an investigation by Greenpeace has found a broad range of hazardous chemicals in children's clothing and footwear, which include the presence of nonylphenol ethoxylates (NPEs), phthalates, organotins and per/poly-fluorinated chemicals (PFCs) (Brigden et al., 2013). The Environmental Protection Agency (EPA) found these PFCs (C-8 chemicals) at very low levels in the environment and in the blood of the general U.S. population. They remain in people for a very long time and cause

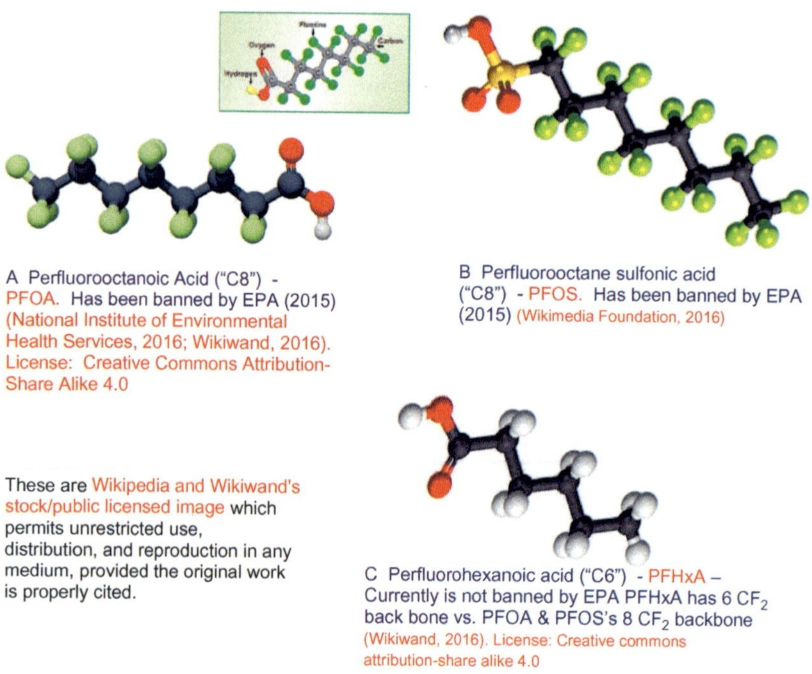

Fig. 17.11 Molecular structure differences between C-8 and C-6 perfluorinated chemicals (National Institute of Environmental Health Services, 2016; 123RF, 2017; Wikimedia Foundation, 2016a,b; Wikiwand, 2016).

developmental and other adverse effects in laboratory animals (United States Environmental Protection Agency, 2006a,b). 3M Company scientists submitted findings to the US government in 2001 that PFOA is present in the blood of 96 percent of 598 children tested in 23 states and the District of Columbia (Olsen et al., 2002.) The 2010–2015 PFOA Stewardship Program was set up in 2010 to phase out these chemical productions, and in 2015, EPA had finally banned their commercial uses (United States Environmental Protection Agency, 2006a,b).

Perfluorohexanoic acid (PFHxA), also known as C-6 chemicals (Fig. 17.11C), has been proposed as a green chemistry (i.e. environmentally friendly) replacement for C-8 compounds used in the production of omniphobic coatings. PFHxA has 2 less -CF2 backbone compared to PFOA and PFOS (see Fig. 17.11). However, it is still a fluorinated chemical; therefore it is very persistent in the environment (The Environmental Working Group, 2008). There are many current R&D efforts that aim to develop omniphobic coatings employing even shorter chain C-4 chemistry, and also nonfluorinated omniphobic coatings (e.g. silicone elastomer-based chemistry) (Jeevajothi et al., 2013; Hastbacka, 2015). However, currently, these alternative chemistries have been shown to have relatively less liquid repellency due to their shorter molecular chain (i.e. 6 vs. 8 carbons in their molecular backbones) (Truong, 2013; National Institute of Environmental Health Services, 2016; 123RF, 2017; Wikimedia Foundation, 2016a,b; Wikiwand, 2016). Therefore the Army's current research continues with development of alternative technologies to repelling water, toxic industrial chemicals and CWAs. Some of these alternative technologies have been identified and are being developed (see Section 17.11).

17.1.4 Commercially available products

While there are many web-based advertisements of 'commercial' omniphobic coatings such as the NeverWet coating, the EverShield Durable Omni Repellent (DOR) coating became the first commercial omniphobic textile-coating product that has been extensively researched, developed and field tested (Truong and Koene, 2013a,b). This product is now being produced and sold by UltraTech International, Inc. (UltraTech) under a licencing agreement with Luna Innovations Incorporated (Luna). Examples of commercial products that are using EverShield DOR coatings include: Occunomix, Huntech (Australasia's oldest and original hunting clothing brand), dust-proof clothing for field workers (Samsung Display Corp.), Raymond Luxury Cotton (India), G5 Textiles (Orlando, FL), Once treated, the EverShield DOR treated fabrics would repel water, oil and grease. They resist staining and are washable. A NSRDEC-Luna partnership, which began with a fluoro-alkyl poly dimethyl siloxane (PDMS) C-8 formulation, led to the successful commercialization of a durable, environmentally friendly (C-6) fluoroalkyl-based polyurethane omniphobic coating having dual micro/nanosurface features (Fig. 17.12A) due to the EPA ban of PFOA/PFOS chemical compounds. As applied to textile materials, EverShield coating's low viscosity has a consistency close to that of water; thus conformably coated around the textiles' yarns and fibres. The EverShield omniphobic coating does not interfere with the treated fabric's air permeability (Fig. 17.12B). This ultra-thin coating, once applied onto the textile substrate, cannot be felt by hand or visibly seen under high magnification (Fig. 17.12B).

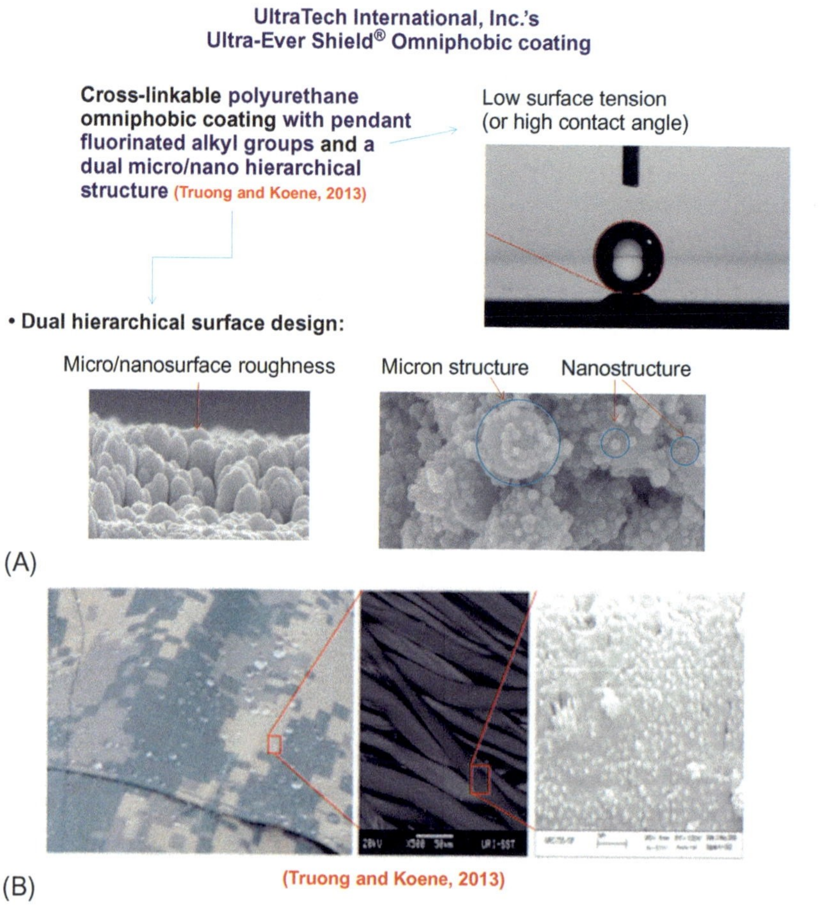

Fig. 17.12 (A) Dual hierarchical micro/nanosurface structures. (B) Conformal coating does not interfere with air permeability. Ultra-ever shield omniphobic coating and army combat uniform coated fabric.
SEMs and pictures by Luna Innovations, Inc.

However, its chemistry and nanometre-scale surface structure help to repel rain, oils and chemicals, while having minimal interference to the body's sweating process, which keeps the wearer comfortable and dry.

17.2 Material concept

17.2.1 Hard surfaces

Omniphobic coatings are usually based on a solution consisting of one or more low surface-free-energy (SFE) polymers and/or elastomers and the solvent that helps to solvate/dissolve it/them. The lower SFE the polymer has, the better omniphobicity the coating achieves. This is because once treated, the low SFE-polymer coating would alter (or replace) the surface chemistry of the treated/coated surface to take on its own SFE. Examples of selected polymers' SFEs are listed in Table 17.1. In this table, silicone elastomer has a relatively lower SFE than many polymers on this list, and because it also offers a good degree of abrasion resistance due to its rubbery elastomeric material structure, silicone is often a polymer choice in creating a durable omniphobic coating that would alter the surface chemistry of treated surfaces, and also be launderable. Another method for creating a durable omniphobic coating is through the co-dissolution of a low SFE polymer and a low SFE elastomer (Truong et al., 2015a,b) For example, Fluoro Polyhedral Oligomeric Silsesquioxane (Fluoro-POSS, Mabry et al., 2008; Iacono et al., 2010) and Tecnoflon (Solvay, 2017).

As discussed in Section 17.1, the degree in liquid wetting depends on a combination of the material's surface chemistry, surface's micro and/or nanoroughness and reentrant surface structures. Therefore, if an untreated hard surface such as glass or metal is treated with just a low SFE polymer or copolymer, its apparent contact angle will be lowered to that of the same coating with micro and/or nanosize particulates added to the coating formulation. These micro/nanoparticulates add an additional 'reentrant' layer (see Fig. 17.9, Tuteja et al., 2008a,b) which is similar to that of the sub-structure of the lotus leaf surface (see Fig. 17.4).

When a surface could be engineered with reentrant surface structure such as that found on the Springtail insect (Fig. 17.4), or the inverse trapezoidal surface features (Fig. 17.5), the hard surface would become superomniphobic (i.e. the engineered surface will likely have a contact angle consistently greater than 150°, and the surface would resist wetting by extremely low surface tension liquids such as methanol (22.1 mN/m), octane (21.6 mN/m, nonpolar) and N-hexane (18.4 mN/m). Current effort is ongoing at the HEROES Center at University of Massachusetts Lowell to develop a dual hierarchical FPOSS-Tecnoflon to create surface with inverse trapezoidal nanosurface features (Truong, 2017).

17.2.2 Textile surfaces

As discussed in Section 17.1.2.2, textile surfaces can be modified to have a much lower SFE by applying a low surface tension coating on their yarns or fibre bundles. Additionally, textile surfaces can be further engineered to have more reentrant levels as compared to hard surfaces by controlling the fibre and yarn sizes, weave openings, micro/nanosurface roughness and the fibre's reentrant structure (see Figs 17.7–17.10). These additional reentrant levels help to improve textile surfaces' repellency to low surface tension liquids. A liquid repellent finish helps to improve liquid repellency, but if its application coated over the weave openings of the textile surface, a soldier's comfort can be greatly reduced, and their mission time could be severely affected due to heat stress. It should be noted that an omniphobic (liquid repellent) finish can only help to shed water and low-surface-tension chemicals, and it can only offer a very small degree of liquid robustness (as measured in cm of water) when it is applied onto tightly woven fabric structures (Fig. 17.16). The smaller the yarn in a tightly woven fabric, the higher its resistance to low hydrostatic pressure. Liquid repellent finish cannot prevent liquid penetration through the fabric system. For prevention of liquid penetration under low hydrostatic pressure, a tightly woven fabric could be used to provide a degree of increased protection to liquid wet-through or liquid robustness; however, to prevent penetration of liquid wet-through at a much higher hydrostatic pressure while allowing a degree of evaporative cooling comfort, a waterproof/moisture vapour permeable (WP/MVP) membrane (i.e. a moisture vapour permeable liquid barrier) will need to be designed as an integral part of the clothing system which will employ a woven shell fabric, a liquid barrier and a comfort liner which will be worn next to the wearer's skin.

In 1959, NSRDEC (then known as the Quartermaster Research and Development Laboratories) had developed a durable water repellent (DWR) and oil-resistant finish to treat textile materials to replace rubber coating for taupe rainwear for all soldiers (GB Investments Inc., 2015). DWR finish was developed by combining two commercially available water repellent treatments and named Quarpel, which stands for Quartermaster-developed repellent finish. Quarpel was considered a technological breakthrough as the need for heavy and rubber coating on rainwear was no longer necessary. Quarpel DWR finish became the standard water repellent treatment in which the textile industry has been using for more than 50 years. However, overtime, it was found that the standard Quarpel DWR finish loses its effectiveness after repeated exposures to laundering, as well as harsh weather elements. Its effectiveness to staining and repellency to petroleum, oils, lubricants (POLs) and other low-surface-tension chemicals became greatly diminished after laundering (Fig. 17.13). Currently,

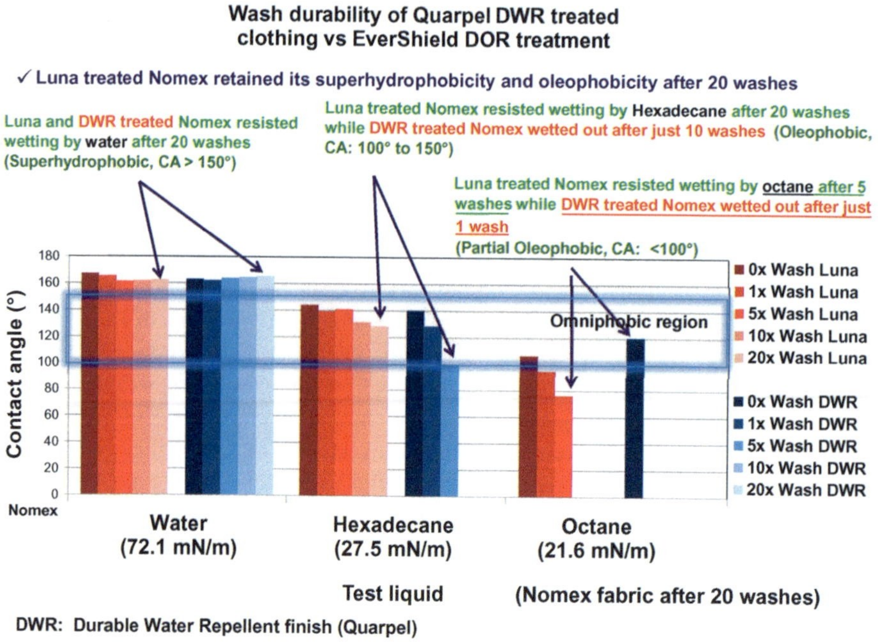

Fig. 17.13 Wash durability of Quarpel DWR treated clothing versus EverShield DOR treatment (Truong and Koene, 2013a,b).

the Army is working on replacing C-8 based Quarpel DWR finish with the C-6 based EverShield DOR finish since the use of C-8 chemistry was banned by the EPA in 2015 due to its persistence in humans, animals and in the environment (United States Environmental Protection Agency, 2006a,b). However, as discussed in Section 17.1.3, the use of C-6 chemistry is at risk due to similar health and environmental concerns as with C-8 based liquid repellent coatings.

Performance-wise, unlike the EverShield conformal coating, which shows no to minimal reduction in air permeability of the treated textile surface, the Quarpel DWR finish is a nonconformal coating, so when it is applied onto the surface of an air-permeable base fabric, e.g. Army Combat Uniform (ACU), it restricts air flow through the fabric, thus inhibiting convective air cooling that is required to keep the wearer cool, especially under strenuous activities. It must be noted that development of a omniphobic coating is to modify the very top layer of the fabric system to help enhance comfort and liquid protection via liquid shedding, which helps to minimize liquid (water and toxic chemicals) contact with the wearer's skin. In heavy driving rain and in extreme cold weather, an omniphobic coating will only help to minimize liquid wetting and wet-through of the outershell fabric layer. A WP/MVP membrane such as Sympatex or Gore-Tex membranes, an additional thermal insulative layer, for example, 3M Thinsulate, a faceshield (i.e. goggles or masks) and proper closures and interfaces will need to be used to prevent driving rain, wind chills and/or sub-zero weathers. Likewise, in a chemically contaminated environment, a CB protective barrier such as ChemPak selectively permeable membrane, or an activated carbon based air permeable layer such as Von Blucher carbon layer must be used underneath the omniphobic treated outershell fabric to provide protection from toxic vapour chemical molecules from coming in contact with the wearer's skin.

17.2.3 *Fibre structures*

Fig. 17.10 demonstrates that the fibre structures can be manipulated (i.e. fibre and yarn bundle sizes, and distance between yarn bundles (D^*)/weave openings) as well as engineered (i.e. changing the fibre core geometry/designs) to have better liquid repellency and liquid robustness. See discussions in Section 17.1.2.2.

17.3 Impact of surface chemistry, structures and polarity on liquid repellency

17.3.1 *Surface chemistry*

As discussed in Sections 17.1.2.1 and 17.1.2.2, surface chemistry plays a critical role in the control of surface nonwetting. If a surface contains or is treated with low surface free energy (SFE) materials such as those listed in Table 17.3, it will prevent liquids with higher surface tension than that of the surface's SFE value from wetting its surface. Surfaces that are wetted by water are referred to as hydrophilic surfaces,

Table 17.3 Solid surface-free-energy (SFE) of common polymers (Surface Tension, 2016)

Name	Surface free energy (SFE) at 20°C (mN/m)	Temp. coefficient SFE (mN/(m K))	Dispersive contrib. of SFE (mN/m)	Polar contrib. of SFE (mN/m)	CAS Ref.-No.
Fluorodecyl-POSS (Mabry et al., 2008; Iacono et al., 2010)	10	–	–	–	–
Tecnoflon BR9151 (Solvay, 2017)	18	–	–	–	–
Polydimethylsiloxane PDMS (Popular use in producing liquid repellent textiles – Fluorine-free)	19.8	−0.048	19	0.8	9016-00-6
Polytetrafluoroethylene PTFE (Popular use in producing repellent textiles – Fluorine-rich)	20	−0.058	18.4	1.6	9002-84-0
Polytrifluoroethylene P3FEt/PTrFE	23.9	–	19.8	4.1	24980-67-4
Polyhexylmethacrylate PHMA	30	−0.062	−27	−3	25087-17-6
Polypropylene-isotactic PP	30.1	−0.058	30.1	0	25085-53-4
Polyvinylidene fluoride PVDF	30.3	–	−23.3	−7	24937-79-9
Poly(t-butyl methacrylate) PtBMA	30.4	−0.059	26.7	3.7	–
Polychlorotrifluoroethylene PCTrFE	30.9	−0.067	22.3	8.6	25101-45-5
Polyisobutylmethacrylate PIBMA	30.9	−0.06	26.6	4.3	9011-15-8
Polybutylmethacrylate PBMA	31.2	−0.059	26.2	5	25608-33-7
Polytetramethylene oxide PTME (Polytetrahydrofuran PTHF)	31.9	−0.061	27.4	4.5	25190-06-1
Polyisobutylene PIB	33.6	−0.064	33.6	0	9003-27-4
Polycarbonate PC	34.2	−0.04	27.7	6.5	24936-68-3

Polyethylene-branched PE	35.3	−0.067	35.3	0	9002-88-4
Polyethylene-linear PE	35.7	−0.057	35.7	0	9002-88-4
Polyethylmethacrylate PEMA	35.9	−0.07	26.9	9	9003-42-3
Polyvinylacetate PVA	36.5	−0.066	25.1	11.4	9003-20-7
Polyvinyl fluoride PVF	36.7	–	−31.2	−5.5	24981-14-4
Polyethylacrylate PEA	37	−0.077	30.7	6.3	9003-32-1
Poly-*a*-methyl styrene PMS (Polyvinyltoluene PVT)	39	−0.058	−35	−4	9017-21-4
Polystyrene PS	40.7	−0.072	−34.5	−6.1	9003-53-6
Polyamide-12 PA-12 (Nylon – popular use as textile polymer)	40.7	–	35.9	4.9	24937-16-4
Polymethylacrylate (polymethacrylic acid) PMAA	41	−0.077	29.7	10.3	25087-26-7
Polymethylmethacrylate PMMA	41.1	−0.076	29.6	11.5	87210-32-0
Polyvinylchloride PVC	41.5	–	−39.5	−2	9002-86-2
Polyetheretherketone PEEK	42.1	–	36.2	5.9	31694-16-13
Polyethyleneoxide PEO	42.9	−0.076	30.9	12	25322-68-3
Polyethyleneterephthalate PET (polyester – popular as textile polymer)	44.6	−0.065	−35.6	−9	25038-59-9
Polyvinylidene chloride PVDC	45	–	−40.5	−4.5	9002-85-1
Polyamide-6,6 PA-66	46.5	−0.065	−32.5	−14	32131-17-2

and the higher their polar contribution of SFE (see 5th column, Table 17.3), the higher their affinity to water. If a surface is nonwetting to water, it is referred as hydrophobic and has a lower polar contribution of SFE accordingly. Surfaces that repel oils and chemicals are called oleophobic surfaces. Surfaces that repel water, oils and chemicals are called omniphobic surfaces. These surfaces have very low SFE. Fig. 17.2 lists different surface wetting definitions. Polarity of the contacting liquids and surfaces is also an important consideration in the design of nonwetting surfaces as it affects both the surface tension and solubility of the contact liquid on a surface. For example, the surface of a hydrophilic (polar) material will be easily wetted by polar liquids such as water, methanol, ethanol and other polar liquids such as those listed in Table 17.1 because they are miscible in water. They also allow adsorption over time into the hydrophilic surface due to the solubility of polar liquids. For hydrophilic surfaces with high polar contribution of SFE, polar chemicals such as CWAs will have higher solubility in water when compared to nonpolar chemicals; therefore polar chemicals have a higher tendency to be absorbed into and diffuse through the hydrophilic material along with water molecules when compared to hydrophobic and oleophobic materials. Materials having a surface with much lower SFE (e.g. Fluorodecyl POSS, Tecnoflon, PDMS, PTFE) and much lower polar contribution of SFE, will resist wetting by lower surface tension liquids, [e.g. methanol (polar) and octane (nonpolar)] more effectively than hydrophilic materials. These materials have minimal or no adsorption by polar liquids and thus have little to no polar liquid molecule diffusion through the polymer material due to the difference in polarity.

17.3.1.1 Fluorinated

A fluorine-rich or fluorinated surface effectively lowers the surface tension of the base surface. The addition of an ultrathin layer of fluorinated chemical will effectively modify the surface properties of the treated base material to become highly nonpolar due to fluorine's highest electronegativity rating (4.0 on the periodic table as compared to other atoms). Additionally, Zisman (1964) established that the SFE values of the following chemical groups decrease in the order of $—CH_2—$ (molecular backbone) > $—CH_3$ (side group) > $—CF_2—$ (molecular backbone) > $—CF_3$ (side group)—that is, with more surface (e.g. $—CF_2—$) and pendant fluorine atoms (e.g. Fluoro alkyl side groups), the fluorinated surface's SFE decreases—thus becoming effectively more liquid repellent. A surface with a high electronegativity has lowest polar contribution to SFE. The use of long chain C-8 fluorinated surface treatment has therefore been very popular in the liquid repellent textile industry since the late 1950s until EPA's C-8 chemistry ban in 2015. Currently, the use of a shorter C-6 fluorinated chemistry is considered as environmentally safe, and EverShield is the first commercial C-6 omniphobic fluoroalkyl polyurethane coating, that was developed and field evaluated for Army textile applications and commercially transitioned in December 2013. UltraTech's EverShield omniphobic coating, however, is among a growing number of C-6 omniphobic coatings that are being developed and commercialized.

17.3.1.2 Nonfluorinated

Polydimethyl siloxane (PDMS) has been the popular material for the development of nonfluorinated liquid repellent coatings because its molecular structure does not contain fluorine; they have essentially the same SFE with slightly less polar contribution to SFE as compared to PTFE (see Table 17.3). However, while PDMS based coatings are durable products and have excellent water repellency, they have poor repellency to oil and solvents (i.e. having minimal to no oleophobicity). With the EPA's ban on the use of C-8 chemistry, there have been many recent efforts on the development of environmentally friendly fluorine-free liquid repellent chemistry for both nontextile (Bittner et al., 2017; Nanex, 2016) and textile (Lassen et al., 2015; Davies, 2014; Nanex, 2016) applications.

17.3.2 Molecular and morphological structures

17.3.2.1 Molecular structures

Molecular structure has a significant effects on solid surface free energy (Moore et al., 2010; Chhatre et al., 2010a,b). As it can be seen in Fig. 17.14, maintaining the cage-like molecule (T8) in shortening the fluorinated fluoroalkyl chain length ('R') that is surrounding the T8 cage will result in the loss of oleophobic properties (the left bar graph shows the solid surface energy of the dispersion liquid (red bars) increases from about 10 mN/m to about 25 mN/m). A similar observation is seen with polar liquids

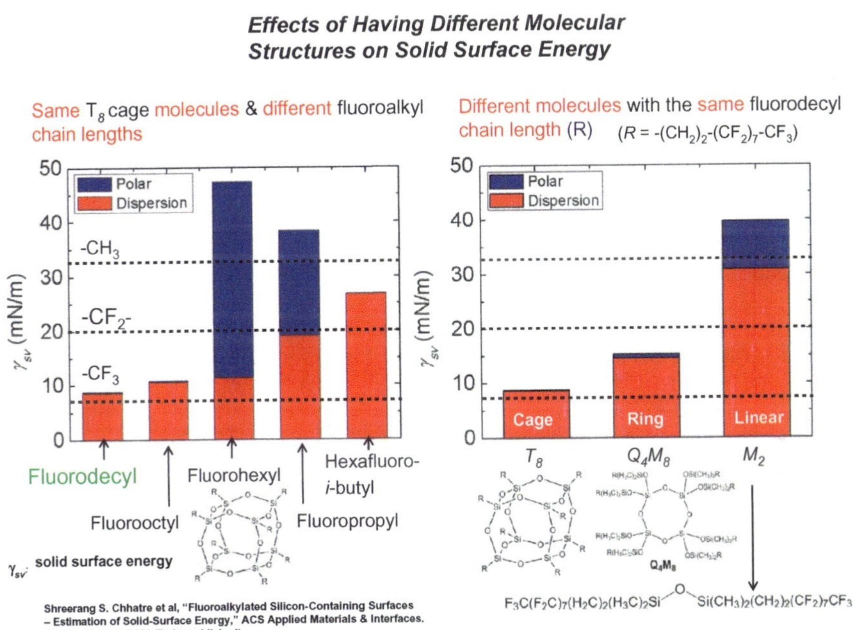

Fig. 17.14 Effects of having different molecular structures on solid surface energy (Moore et al., 2010; Chhatre et al., 2010a,b).

(water), but with a more drastic loss of omniphobic properties especially with fluorohexyl, then fluoropropyl and hexafluoro-ibutyl (blue bars).

In changing the molecules from being a linear chain (M2), to a ring structure (Q4M8), and then to a cage-like molecule (T8), the solid surface energy of both the dispersion and polar liquids decreases from about 30 mN/m to about 9 mN/m (see the red bars in the right bar graph). This shows the importance of using a cage-like structure omniphobic coating to create a surface with the lowest possible surface energies for increase liquid repellency to extremely low surface tension liquids.

17.3.2.2 Surface roughness

Micro- and/or nanosurface roughness can mimic natural surfaces (Sections 17.1.2.1 and 17.1.2.2, Fig. 17.4). Surface roughness can be added through coatings containing micro and nanometre size particles, where this surface modification process creates a dual hierarchical structure on solid and textile surfaces, which effectively adds an increased level of liquid repellency (Kleingartner et al., 2015).

17.3.2.3 Reentrant feature

NSRDEC is currently investigating dual hierarchical coatings where coating layers containing different sizes of nanoparticles are applied onto fabrics and cured in successive order to create a dual micro/nanosurface roughness where larger particles are deposited on top of smaller particles (Truong, 2017) to repel low polar and nonpolar surface tension liquids such as those listed in Tables 17.1 and 17.2. Currently, a random dual hierarchical micro/nanosurface roughness structure has been demonstrated on a tightly woven polyester fabric, where repellency to various polar and nonpolar low surface tension liquids which included dimethyl methyl phosphonate (DMMP, a polar nerve CWA simulant, 36.7 mN/m), tributyl phosphate (TBP, a polar nerve CWA simulant, 27.79 mN/m), hexadecane (a nonpolar solvent, 27.47 mN/m) and methanol (a polar solvent, 22.1 mN/m) was demonstrated, with N-hexane (a nonpolar solvent, 18.4 mN/m) showing partial wetting on the omniphobic treated fabric (Fig. 17.15). This 20:1 NP4 to NP2 ratio dual hierarchical coating has shown resistance to wetting of DMMP and TPB up to 4 min.

17.3.3 Fabric structure

17.3.3.1 Woven

As discussed in Section 17.1.2.2, woven fabric structure has an inherent reentrant level of liquid repellency due to fibre and yarn structures, and the size of the weave openings between the yarns. In a 2013 research study, the influence of inherently supernonwetting fibre and fabric designs on repellency and resistance to liquid penetration was examined. In this study, the effects of varying fabric weave structure has on liquid repellency and liquid robustness was studied. It was found that omniphobic coated fabrics constructed with a smaller yarn size (and tighter weave) yielded a higher resistance to wetting and also wet-through by the impinging liquids (Fig. 17.16); (Truong, 2013, 2014) different surface chemistries have different effects on liquid robustness (Fig. 17.17); and reducing yarn radii slightly increasing the liquid repellency (Fig. 17.18).

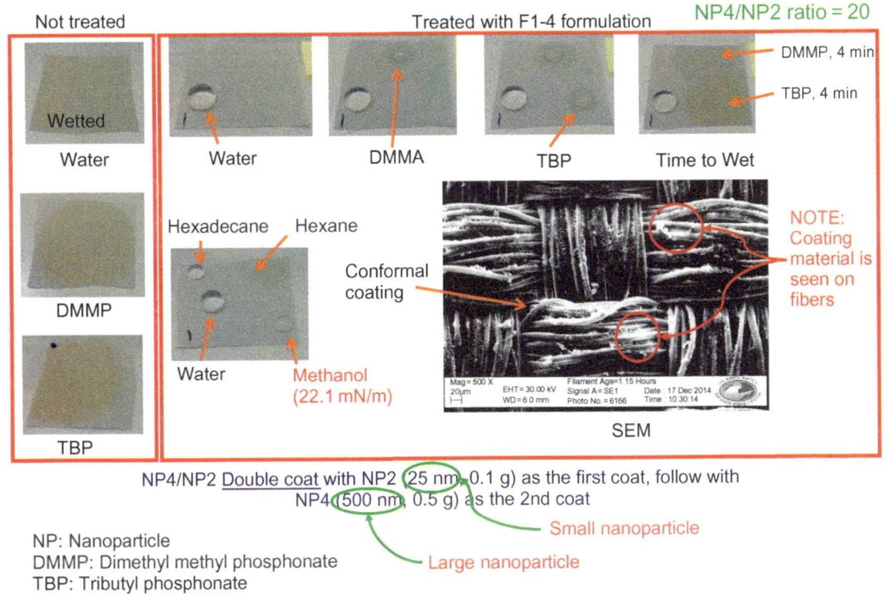

Fig. 17.15 Dual hierarchical coatings showing repellency to water, hexadecane and methanol. Pictures credit: NSRDEC (Truong, 2017).

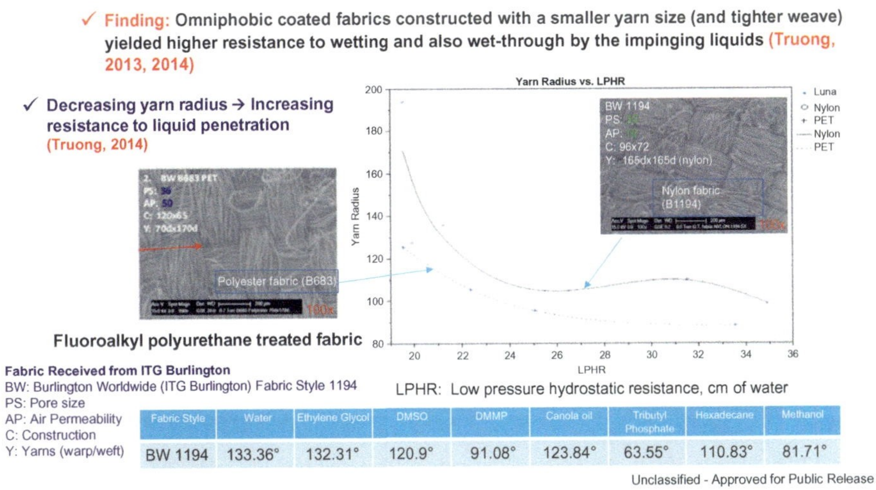

Fabric Style	Water	Ethylene Glycol	DMSO	DMMP	Canola oil	Tributyl Phosphate	Hexadecane	Methanol
BW 1194	133.36°	132.31°	120.9°	91.08°	123.84°	63.55°	110.83°	81.71°

Fig. 17.16 Yarn size, weave openings and resistance to wetting and wet-through by the impinging liquids (Truong, 2013, 2014).

17.3.3.2 Nonwoven

Similar to discussions in Section 17.1.2.2, nonwoven fabric structure has an inherent reentrant level of liquid repellency due to their spun-bond fibre size, nonwoven fibre structures and the size of the openings between the nonwoven fibres. In a comparative

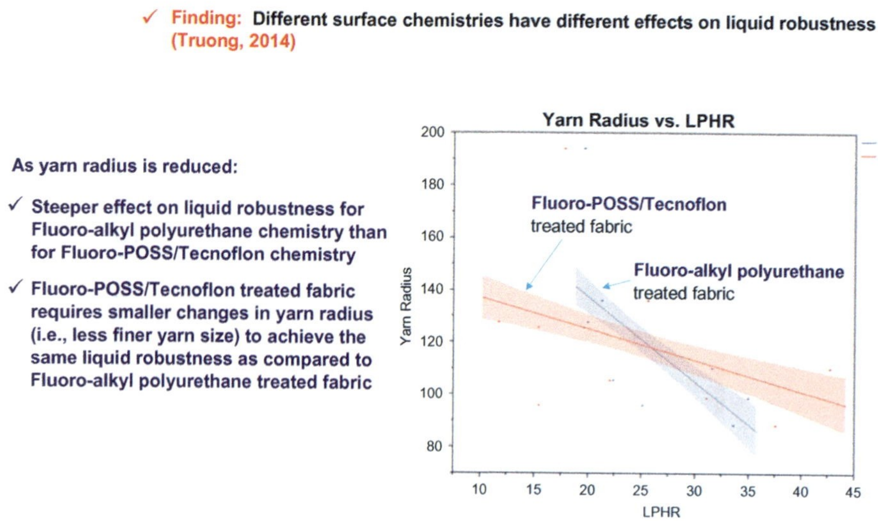

Fig. 17.17 Surface chemistry versus robustness (Truong, 2014).

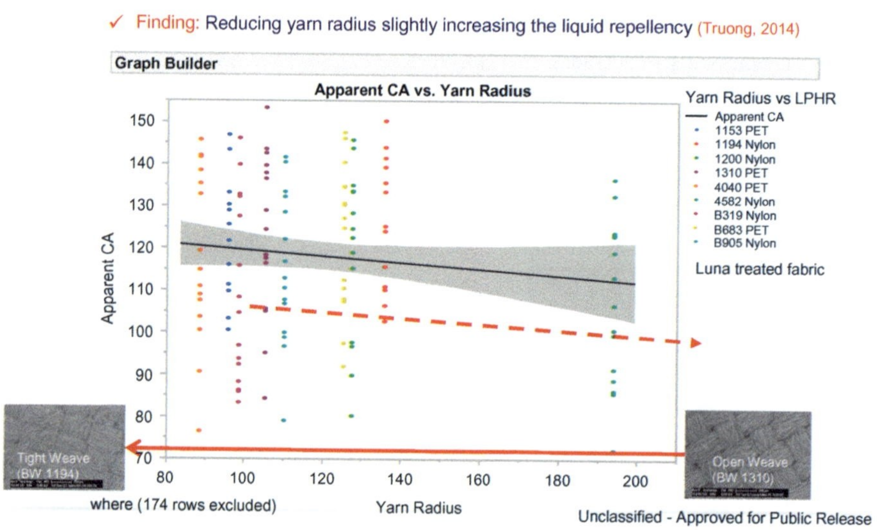

Fig. 17.18 Effects of varying fabric's individual fibre and bundles' sizes has on liquid repellency (Truong, 2014).

study of three methods for generating superomniphobic hydro-entangled nylon non-woven structures (Fig. 17.19) using three different methods (pulse plasma polymerization of 1H,1H,2H,2H perfluorodecyl acrylate (PFAC8), microwave-assisted fluorosiloxane (FS) condensation and fluorosiloxane condensation via wet processing), Although the three processes yielded different fibre surface

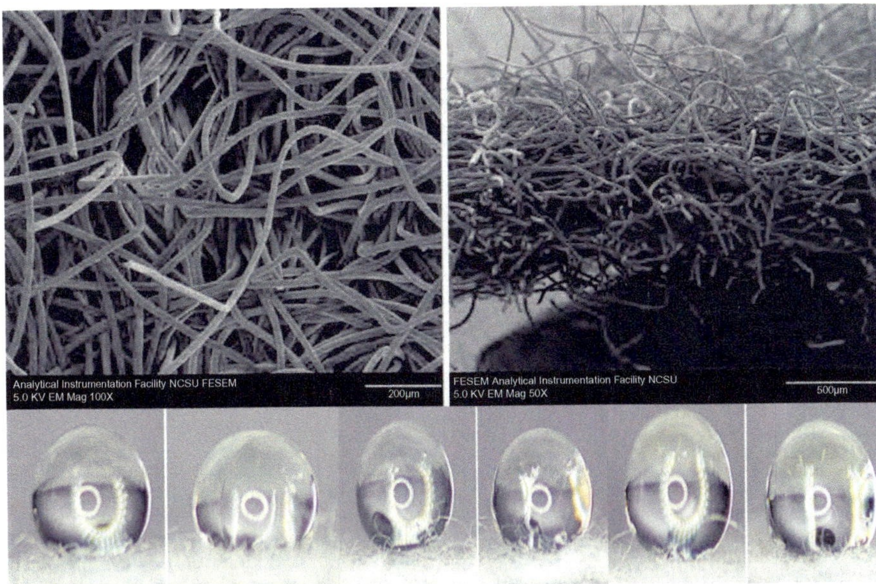

Fig. 17.19 Superomniphobic treated nylon nonwoven material (upper left: 200 μm). Water (lower left – left) and dodecane (lower left – right) droplets on plasma-polymerized PFAC8 on nylon nonwoven fabric; water (lower centre – left) and dodecane (lower centre – right) droplets on microwave-assisted FS condensation on hydroentangled nylon nonwoven fabric; and water (lower right – left) and dodecane (lower right – right) droplets on hydroentangled nylon nonwoven fabrics FS-condensed via wet processing.
Picture Credit: SEMs and Pictures by NCSU (Saraf et al., 2011).

morphologies, NSRDEC and its collaborators North Carolina State University (NCSU), US Air Force Research Laboratory (AFRL) and Defense Science and Technology Laboratory (Wiltshire, UK) found it is possible to create superomniphobic nonwoven fibre structures that has contact angle up to 174° for water and 160° for dodecane, where their roll-off angles were as low as 6° for water and 21° for dodecane for 50 μL droplet volume. Smaller droplets of water and dodecane (10 μL) did not roll off the treated nonwoven surfaces (Saraf et al., 2011) although liquid repellent nonwoven materials are not as durable when compared to woven textiles; however, their potential use could be realized in a one-time use disposable clothing applications.

17.3.4 Reentrant fibre structure

As discussed in Section 17.1.2.2, in 2013, NSRDEC had conceptualized and subsequently engineered a reentrant fibre structure (Fig. 17.20) in collaboration with its academic and industry partners which include MIT, Clemson, UMass Lowell and Luna Innovations, Inc. (Truong et al., 2015a,b) When polypropylene (PP) reentrant fibres were twisted into yarns and knitted into a knit fabric, the PP knit fabric without any repellent finish is superhydrophobic with a measured contact angle of 150°, but the knit fabric became wet when measured with ethylene glycol, hexadecane and octane with an

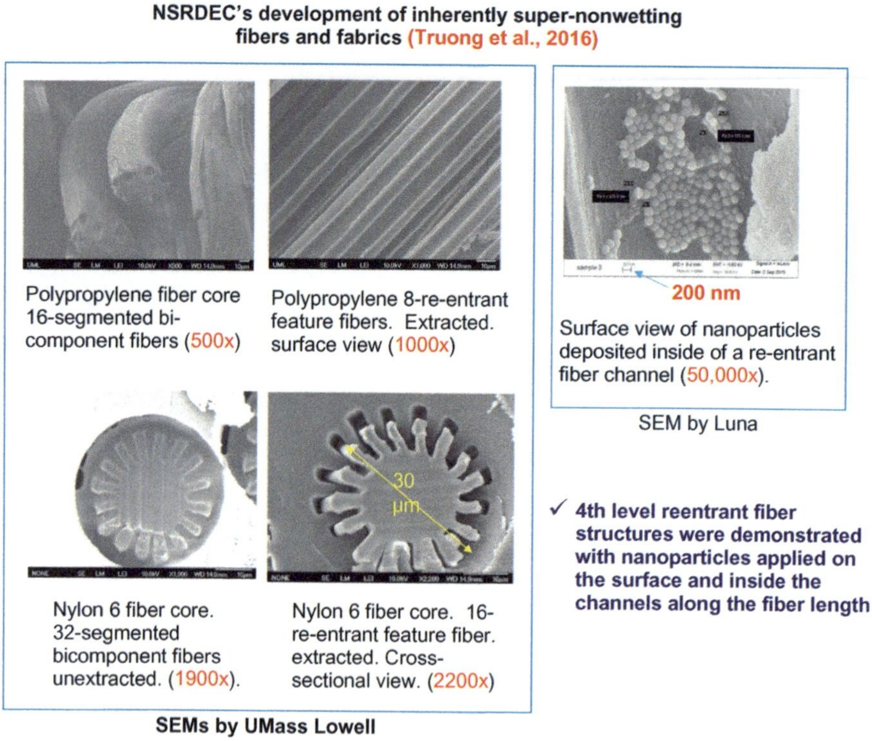

Fig. 17.20 16-segmented bi-component fibres (50 μm) were spun/produced (Truong et al., 2015a,b).

apparent contact angle of 0°. After the PP knit was treated with the EverShield omniphobic coating containing nanoparticles, the PP reentrant knit fabric's contact angle significantly increased. The water contact angle was increased from 150° to 157°, and contact angles for ethylene glycol, hexadecane and octane were increased from 0° to 131°, 120° and 80° respectively. Reentrant (extracted) new gear shape fibres have lower elongation, but better tenacity and higher modulus than round fibres. Whereas, (1) un-extracted and extracted fibres have essentially the same modulus, where the extracted/reentrant fibres have higher tenacity as compared to their un-extracted bi-component fibre structure; (2) extracted, reentrant fibres have higher tenacity and modulus than round fibres (Note: a high modulus fibre is a fibre that is able to better resist deformation, or the amount of stress applied on the fibre. And, high tenacity fibres mean fibres that are tough, and not easily broken.) (Figs 17.21 and 17.22).

17.4 Performance goals

17.4.1 Order of importance for the US military

The tests and measurements, test methods used and performance goals are listed in Table 17.4 below. These tests or measurements are ranked in the order of their importance for their liquid repellency, liquid robustness/protection, comfort and durability.

Reentrant fiber showed excellent liquid repellency (contact angles)

Untreated control (round) fiber polypropylene (PP) knit fabric

Untreated reentrant fiber PP fabric [1077-11A (F3)]

EverShield Treated reentrant fiber PP fabric (1077-11B)

ID	Samples produced	Treatment	Contact angle	Contact angle	Contact angle	Contact angle
			Water (°)	Ethylene Glycol (°)	Hexadecane (°)	Octane (°)
Control*	N/A	Untreated	126	105	0	0
1077-11A (F3)	Sep 2015	Untreated fabric – F3	150	0	0	0
1077-11B	Sep 2015	Treated** fabric – F3	157	131	120	80
1032-089 (F1)	Aug 2015	Untreated fabric	149	137	0	0
1032-089A1	Aug 2015	Treated** fabric – F1	156	145	114	0

*Regular polypropylene fabric knitted using conventional round-shape polypropylene fiber
** Treated with EverShield coating with dual size micro/nano particles

Unclassified - Approved for Public Release

Fig. 17.21 NSRDEC's reentrant fibre-based fabrics showed increased liquid repellency (Truong et al., 2015a,b).

Reentrant nylon fiber properties*

Sample	Nylon 6 fiber	Nylon 6/ G polymer fiber	Nylon 6 extracted fiber
Fiber	Single component	Bicomponent	Extracted
Polymers	Nylon 6	Nylon 6 and G polymer	Nylon 6
Polymer ratio	-	1:1 (N6:GPOLYMER)	-
Fiber shape	Round	New gear shape	New gear shape
Filaments	19	19	19
Denier	495.9	603	279
Denier/ filament	26.1	31.74	14.68
Fiber diameter	*~60 µm*	*~60 µm*	*~60 µm*
Elongation	520.04%	70.01%	236.99%
Tenacity	1.63 gpd	1.07 gpd	2.63 gpd
Modulus	5.75 gpd (SD: 0.41gpd)	10.12 gpd (SD: 0.65gpd)	9.26 gpd (SD: 0.56gpd)

The modulus for these two samples is not statistically different (due to standard deviation values of 0.65 vs. 0.56)

Higher tenacity because the reentrant features cause entanglements between the filaments

*Clemson University, NSRDEC Contract #W911QY-14-P-0413

Unclassified - Approved for Public Release

Fig. 17.22 Reentrant nylon fibre properties (Truong et al., 2015a,b).

Liquid repellency receives the highest importance due to the needs to repel a wide range of liquids ranging from high (e.g. water) to low surface tension liquids (oils, toxic industrial chemicals and warfare agents). This is to protect the soldiers from liquid exposure to enhance their survivability in the extreme environments such as hot/humid tropical conditions, cold and extremely cold climates and chemically

Table 17.4 **Performance goals for development of superomniphobic coated fabrics**

Ranking (#)	Tests and measurements	Test method	Performance goals	Test equipment
1. Liquid repellency performance	1. Surface tension 2. Apparent contact angle (CA) 3. Roll off CA 4. Timed droplet wetting	ASTM D5946 (water, oil and solvents), 20 μL drop	18 dynes/cm/ Apparent CA > 150° Roll Off CA ≤ 5	 Rame Hart Goniometer Model 300-00-115, with automatic tilt base
1. Liquid repellency performance	1. Surface tension 2. Apparent CA 3. Roll Off CA 4. Time droplet wetting	ASTM D5946 (after torsional flex and abrasion testing)	22 dynes/cm/ Apparent CA: 100° to 150° Roll Off CA ≤ 10°	 Rame Hart Goniometer Model 300-00-115, with automatic tilt base
1. Liquid repellency performance	5. Oil resistance	AATCC 118	7b to 8 ratings Oil rating scale: (higher rating indicates better repellency): 4 – tetradecane (26.4 dynes/cm) 5 – dodecane (24.7 dynes/cm) 6 – decane (23.5 dynes/cm) 7 – octane (21.4 dynes/cm) 8 – heptane (14.8 dynes/cm	 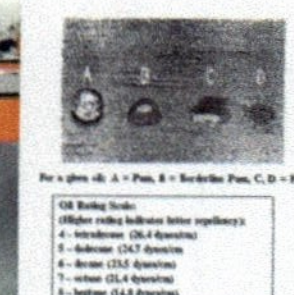 James Heal spray rating tester

1. Liquid repellency performance	6. Spray rating (water)	AATCC 22	100 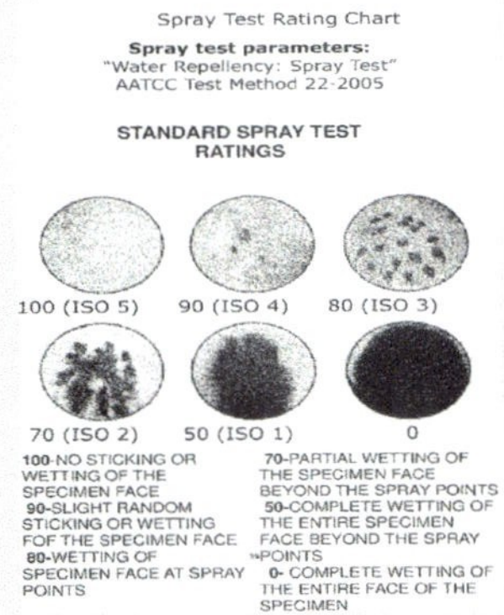	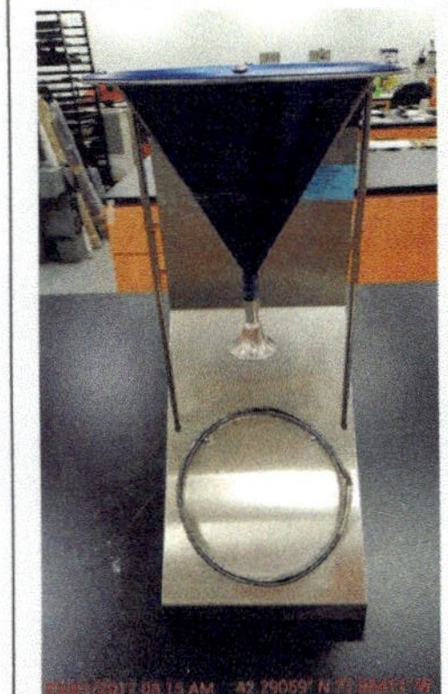 James Heal spray rating tester
1. Liquid repellency performance	7. Elemental analyses (surface chemistry)	Bruker Quantex 200 (EDS of X-ray microanalysis)	Weight % of fluorine and silicon, and others on textile fibres are analysed for their contribution to liquid repellency	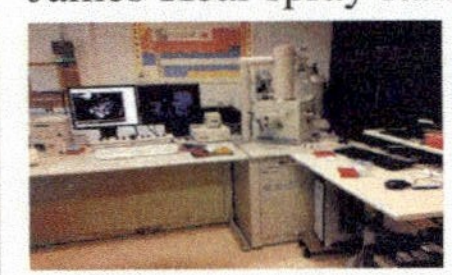 ESEM Model XL-30 by FEI
1. Liquid repellency performance	8. Surface topography (SEM)	NSRDEC's SEM (Surface and cross-section)	Examination of reentrant structures and surface roughness	 ESEM Model XL-30 by FEI
1. Liquid robustness/ protection	9. Hydrostatic resistance	ASTM D751, Proc. B	>32 cm of water (low pressure) >32 psi (high pressure)	 Hydrostatic Tester Model AH by Mullen Testers

Continued

Table 17.4 Continued

Ranking (#)	Tests and measurements	Test method	Performance goals	Test equipment
1. Liquid wicking and comfort	10. Vertical wicking (water)	AATCC TM 197	>30 min	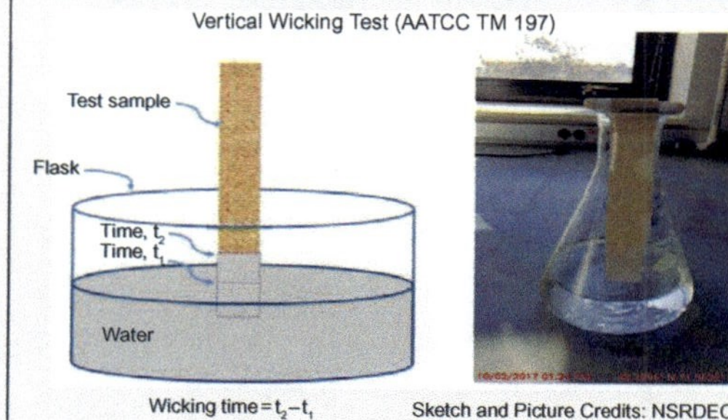 Vertical wicking test
1. Liquid chemical protection, and comfort	11. Liquid adsorption (immersion testing) (water and chemicals)	In-house method. CWA Simulants: DMSO, MeS, DMMP & TBP	Minimal to no weight uptake	 Sorption test using a Vestil wide mouth capped bottle
1. Liquid chemical protection	12. Drop roll-off CWA simulant permeation	In-house method, CWA Simulants: DMSO, MeS, DMMP & TBP	$<10\ g\ m^{-2}/24\ h$	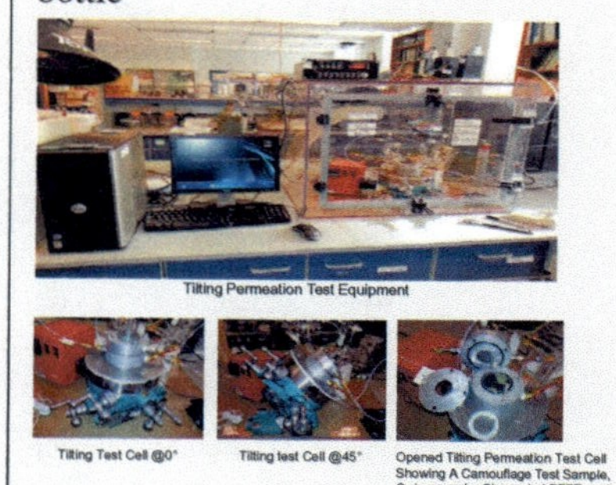 NSRDEC drop-roll off CWA permeation cell

2. Thermal comfort	13. Sweating guarded hot plate	ASTM D1518	Compare to untreated fabric: Clo ≤ 5% and $I_m \leq 10\%$	 NSRDEC Sweating guarded hotplate test chamber
2. Thermal comfort	14. Sweating manikin (US ARIEM: Army Research Institute of Environmental Medicine – co-located @NSRDEC)	ASTM F1291	Compare to untreated fabric: Clo ≤ 5% and $I_m \leq 10\%$	 US ARIEM thermal manikins
2. Thermal comfort	15. Moisture vapour transmission rate (MVTR)	ASTM E96-2007, Proc B	Air-permeable fabrics: Comparable ≥900 to 1,000 g m^2/24 h WP/B fabrics: ≥700 g m^2/24 h	 NSRDEC MVTR Wind tunnel
2. Thermal comfort	16. Air permeability	ASTM D737	<0.2 ft^3 of air/min/ft^2	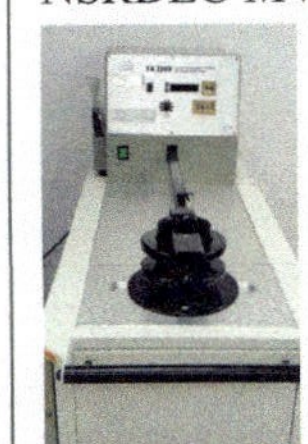 Testex Air Permeability Tester

Continued

Table 17.4 Continued

Ranking (#)	Tests and measurements	Test method	Performance goals	Test equipment
2. Thermal comfort	17. Drying time (water)	In-house method	<30 min	 Testex Padder
3. Durability	18. Abrasion resistance	ASTM D3884, No weight added	>2000 cycles. After abrasion, apparent CA $\geq$ of base fabric's	 Tabor Twin Abrader
3. Durability	19. Torsional flexibility	ASTM F392	>2000 cycles. After flexing, apparent CA $\geq$ of base fabric's	 NSRDEC Torsional Flex Tester
3. Durability	20. Launderability	AATCC 135	Pass after 5 launderings. (no contact angle reduction)	 NSRDEC Laundering Facility
3. Durability	21. Tensile strength	ASTM D5034; @break	Comparable to base fabric. Warp >200 lb; fill >125 lb; elongation: warp/fill >35%	 Instron Tester

3. Durability	22. Colour fastness: Foliage green Urban grey Desert sand	AATCC 61 2A (1x)	Minimal to no colour changes, after 1 laundering	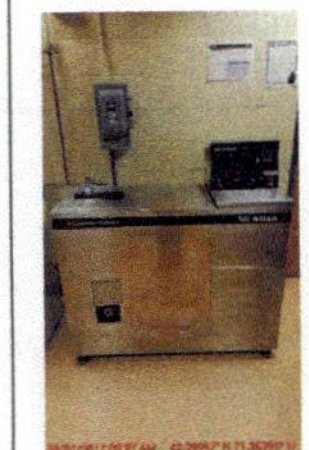 Launder-Ometer
4. Physical comfort	23. Dimensional stability	AATCC 135	<3% for both W and F directions, after 1 and 5 launderings	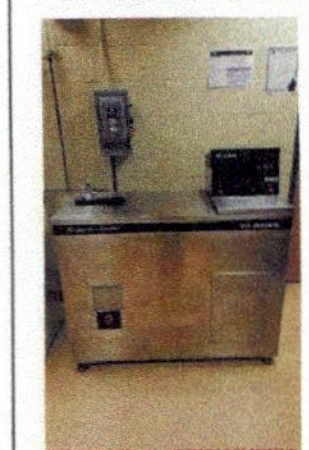 Launder-Ometer
4. Physical comfort	24. Weight	ASTM D3776; 0.0001 decimal	<0.1 oz/yd^2	 Analytical balance

Continued

Table 17.4 Continued

Ranking (#)	Tests and measurements	Test method	Performance goals	Test equipment
4. Physical comfort	25. Thickness	ASTM D1777	coating thickness range: nm to <1 μm	 Thickness gauge
4. Physical comfort	26. Stiffness	ASTM D747	Warp/Fill: <0.01 lb max	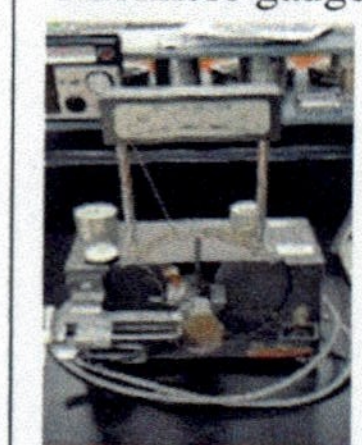 Tinius Olson Stiffness Tester

ASTM, American Society for Testing and Materials; *AATCC*, American Association of Textile Chemists and Colorists; *EDS*, energy-dispersive X-ray spectroscopy; *SEM*, scanning electron microscopy; *CWA*, chemical warfare agent; *DMSO*, dimethyl sulfoxide; *MeS*, methyl salicylate; *DMMP*, dimethyl methyl phosphonate; *TBP*, tributyl phosphate.

contaminated battlefield environments. A liquid barrier such as a WP/MVP membrane and effective closures and interfaces are needed when resistance to higher hydrostatic pressure is required in a protective clothing system. For example, for the soldiers who must carry out their missions in stormy weather with heavy/driving rain or when crossing a body of water. However, to minimize cold-water contact to the wearer's skin, it's important to prevent water from staying on the fabric. This is done through ensuring the omniphobic fabric would have high apparent contact angle and low roll-off contact angle to water. To minimize adsorption of water into the fabric, treated yarns or fibres would have low water affinity or uptake. To maximize liquid robustness, the fabric must be sufficiently designed with a tightly woven fabric structure. To maximize comfort, omniphobic coating must be conformably treated around the fibres and not blocking the interstitial openings (or weave openings) between the yarns to allow evaporative cooling to take place.

The performance goals listed in Table 17.4 help material developers in design and/or selecting clothing materials to minimize the soldiers' risk of hypothermia when they are exposed to the cold/extreme cold environments, or heat stress in a hot and humid environment. Likewise, to improve the soldiers' survival when carrying out their missions in a CB contaminated battlefield, omniphobic coated clothing must repel and prevent toxic industrial chemicals (TICs) and warfare agents from wetting the clothing and also from wetting through the protective clothing's barrier layer. Additionally, the omniphobic coating must be durable, hence the performance goals for abrasion resistance and durability testing which include launderability, torsional flexibility and tensile strength). The omniphobic coating must also have minimal effects on the base textile material's physical properties such as air permeability, weight, thickness and dimensional stability. Because blending in with the environment's background minimizes detection by the enemy force, camouflaging is critical for the soldiers' survival on a battlefield; therefore, colour fastness is very important, and the omniphobic coating must have minimal change to the clothing's dyed camouflaging patterns on its outershell fabric.

For the above reasons, NSRDEC has been using these tests and measurements to develop superomniphobic coated fabrics and fibres, and also to guide material developers in meeting the performance goals for military textile applications.

In regard to the above ranking, a material developer or user could decide on a different ranking based on end uses and needs. For example, if an air permeable chemical protective clothing system has a thick uncomfortable liner such as UK's no1mk4 activated carbon liner, one would prefer not to wear it due to heat stress in certain climates. In NSRDEC's R&D programmes and projects, the objective aims to achieve a balance of properties. In this chapter, for example, one of the objectives is to develop a conformal coating that would coat around the fibre and/or fibre bundles only and not across the weave or knit openings, which could inhibit the air and moisture vapour transfer from the body through the clothing into the environment for evaporative cooling. Furthermore, as NSRDEC's work also aims to provide enhanced CB protection to the soldiers operating in a contaminated environment. Repellency to deadly chemicals is considered more important than comfort because being alive is more important. If the soldiers prefer comfort and take off their clothing, they will be exposed to the deadly warfare agents. Liquid repellency of their clothing helps to shed off these dangerous warfare agents, and our aim is not to interfere with the protective layer underneath the shell fabric or affect the shell fabric's physical properties in any significant ways, but to help minimize the amount of CWA prior to contacting the protective barrier material layer. For these reasons, liquid repellency has our highest rating in group 1, thermal comfort in group 2, durability in group 3 and physical comfort in group 4.

17.4.2 Testing standards outside the US

Australia, the UK and Canada use similar methods of evaluating omniphobicity for R&D programmes, while relying on ISO test methods instead of ASTM. Repellency, durability, physical and thermal comfort are all taken into account and prioritized in accordance with the application the fabric is used for. In particular, the Oil Rating is commonly used to evaluate repellency as well as the isopropanol/water rating (AATCC 193-2007) in which water is mixed with increasing quantities of isopropanol to reduce the surface tension of the liquid applied. The higher the rating, the higher the ratio of isopropanol to water in the repelled droplet, and the lower the surface tension.

As discussed earlier, with regards to CB protection, there is a question as to whether or not it is preferable to repel CWAs, thus preventing the liquid agent from permeating through the fabric, or wick the agent as much as possible on the outer shell of the cover fabric. Wicking the agent would maximize the area of evaporation outward from the CB protective suit and lessen permeation through the sorptive layer of the fabric underneath. Ultimately, whether wicking or repellency is preferable depends on the properties of the agent and the properties of the fabric.

17.5 Test methods and assessment

Most of the testing listed above was conducted by the Textiles and Materials Evaluation Team (TMET), and the Chemical Science and Engineering Team (CSET) at NSRDEC. The exception was the sweating manikin test conducted by NCSU and the chemical agent testing done by Battelle Hazardous Materials Research Center (HMRC).

17.5.1 Chemical agent repellency

Superior CWA repellency is difficult to ascertain with the conventional methods detailed in Table 17.4. In Table 17.5, the contact angle of select agents is shown alongside the corresponding oil ratings for various fabrics. The oil rating varies between 5A and 7B for the fabrics, and yet the hydrocarbon HD has the same contact angle. Furthermore, the contact angles for GD, VX and Agent X—all organophosphates—are roughly the same despite the fact that GD wets all of the fabrics in less than a minute. Similarly, roll-off angle measurements did not differentiate between the fabrics either.

Table 17.5 CWA repellency of select fabrics. NSRDEC data (Pomerantz et al., in press)

		Average contact angle			
Material	**Oil rating**	**Agent X**	**HD**	**GD**	**VX**
A	6A	127	118	119	120
B	5A	129	121	118	120
C	5A	127	120	116	116
E	5A	121	118	114	114
F	7B	121	117	112	124
G	6A	119	117	109	114

Table 17.6 Weathering of CWAs on select fabrics. NSRDEC data (Pomerantz et al., in press)

	IPFS weathering test			
	Average weathering time (HH:MM)			
Material	**GD**	**HD**	**VX**	**Agent X**
A	<1 min	8:00	6:05	8:00
B	<1 min	5:28	0:27	8:00
C	<1 min	7:57	0:23	8:00
F	<1 min	0:55	0:05	7:29
E	<1 min	8:00	1:46	8:00
G	0:03	4:22	0:20	8:00

GD and VX did not roll off, and HD and Agent X did not roll off consistently despite using a heavier droplet size of 20 μL.

The test that did differentiate between the repellency of fabrics is the weathering test in which a 10 μL droplet of agent is placed on the fabric and the wetting of the fabric monitored and timed. The time of wetting was deemed the time until the agent had completely wet the fabric. Table 17.6 shows the weathering of four different CWAs on select fabrics. The test was performed up to 8 hours. As can be seen, little differences in GD wetting are observed. No differences in the weathering of Agent X were observed within the 8-hour time period with the exception of fabric F, although it is possible that a longer evaluation would have shown differences between the fabrics. The weathering of HD and VX clearly show differences in performance between the fabrics; and again, the oil rating shown in Table 17.4 does not predict the better performers.

Table 17.7 Weathering of VX and solvents compared to VX permeation (Units of permeation not shown). NSRDEC data (Pomerantz et al., 2015)

	Weathering time (MM:SS)			
Solvent	Fabric 1	Fabric 2	Fabric 3	Fabric 4
Isopropanol	1:30	0:20	12:50	15:00
Methanol	15:10	8:40	10:40	10:20
Tributyl phosphate	>4 h	>4 h	4:00	>4 days
VX	20:00	4.5 h	6 h	>24 h
VX permeation (units not shown)	1.7	2.5	1.3	0.7

While the droplet weathering evaluation clearly differentiates between the repellency of the fabrics, especially with VX where there is a clear best performer at a 6 hour wetting time, the test does not demonstrate the utility of agent repellency. Table 17.7 shows a clear correlation between VX permeation through fabrics with different coatings over the same sorptive liner, and the wetting time of VX. Clearly, it is of utmost importance to repel VX and prevent the droplet from wetting into the fabric over a long period of time. However, few facilities have the ability to evaluate the droplet wetting behaviour of VX in-house. For the sake of academic pursuits and quality control in textile mills, it is important to find a simulant, or series of simulants, whose performance can be correlated to that of VX.

Table 17.7 also shows the weathering behaviour of isopropanol and methanol, hydrophilic solvents with very low surface tensions, and tributyl phosphate, an organophosphate. Tributyl phosphate does not predict VX repellency. Fabric 3 had a 6 hour droplet wetting time for VX and only 4 minutes for tributyl phosphate, much lower than Fabrics 1 and 2, which had a faster wetting times for VX. Methanol and isopropanol do not appear to correlate well with VX repellency. However, both solvents evaporate after 12 minutes and the test cannot be performed much longer.

17.5.2 Surface properties

Surface properties are assessed using the test results or measurements from the tests and measurements: Surface tension, apparent and roll off contact angles, timed droplet wetting, oil resistance, spray rating, elemental analyses, surface topography, liquid adsorption and drop roll off CWA simulant permeation tests (see Table 17.4).

17.5.2.1 Vertical wicking

The vertical wicking test is performed by TMET in accordance to the Moisture Management Technical Supplement 06 (MM TS-06), or AATCC TM 197, Vertical Wicking of Textiles. The MM TS-06 is applicable to apparel, linens and soft goods. This test measures a 6 × 1 in (15 × 2.5 cm). vertically mounted fabric sample's ability to transport liquid along fibres and yarns. For the development of superomniphobic coated textiles and fibres, a less than 30 minutes wicking resistance goal is set per test standard, because that is the limit of the test time. The wicking rate offers the material/fabric designer with a means to quantify a fabric's ability to remove sweat from the body.

17.5.2.2 Thermal comfort

Thermal comfort is assessed using the test results from the tests: sweating hot plate, sweating/thermal manikin, MVTR, air permeability and drying time (see Table 17.4).

17.5.2.3 Sweating guarded hot plate

For EverShield omniphobic coating development, sweating guarded hot plate testing was not conducted due to the small differences in the air permeability and moisture vapour transport rate test results between untreated and the EverShield omniphobic treated clothing.

17.5.2.4 Sweating manikin

Thermal insulation and evaporative resistance of EverShield omniphobic treated and untreated ACU were measured at North Carolina State (NCSU) (Koene, 2012). The NCSU test simulates heat and sweat production to assess the influence of clothing on thermal comfort. The manikin is housed in a climate-controlled chamber and the surface is separated into 34 sections with independent sweating, heating and temperature measuring systems. The two ACUs were measured for thermal and evaporative resistance according to ASTM F 1291 Standard Method for Measuring the Thermal Insulation of Clothing using a Heated Manikin and ASTM F 2370 Standard Test Method for Measuring the Evaporative Resistance of Clothing using a Sweating Manikin. Three repetitions were completed for each garment. Test conditions are shown in Table 17.8.

Thermal and evaporative resistance measurements were taken from all sections. Thermal resistance values were converted to units of clo, and evaporative resistance values were converted to permeability index (I_m) values. The results show a slight decrease in the comfort of the Luna garment compared to the control ACU. Permeability index is slightly lower (with 1 representing totally permeability and 0 representing an impermeable fabric) in the Luna garment. The Luna garment also showed an 8% decrease

Table 17.8 Test conditions for sweating manikin test

NCSU test conditions	Thermal resistance (°C m^2/W)	Evaporative resistance (kPa·m^2/W)
Air temperature (°C)	20	35
RH (%)	40	40
Air speed (m/s)	0.4	0.4
Skin temperature (°C)	35	35

Table 17.9 **Average test values for ACU garments**

Sweating manikin test results	**Control ACU**	**Luna ACU**	**%**
R_t: total thermal resistance [°C m^2/W]	0.1557	0.1594	2
R_{et}: total evaporate resistance [kPa m^2/W]	0.02584	0.02903	11
R_{cl}: intrinsic thermal resistance, total thermal resistance provided by the garment only [°C m^2/W]	0.0970	0.1007	4
R_{ecl}: intrinsic evaporative resistance [kPa m^2/W]	0.01800	0.02119	15
I_t: total insulation value [clo]	1.00	1.03	3
I_m: permeability index, comfort parameter indicates effect of skin moisture on heat loss	0.368	0.335	−10
Q_p: total potential heat loss from dry and evaporated heat transfer [W/m^2]	224.33	206.23 (92%)	−9
$Q_{predicted,\ 25°C,\ 65RH}$: predicted total potential heat loss from dry and evaporate heat transfer at 25°C and 65% RH [W/m^2]	202.35	185.65	−9

in total potential heat loss in dry and evaporative heat transfer (Table 17.9). However, overall, these values are close to that of the untreated ACU's values.

17.5.2.5 Drying

The drying test is an NSRDEC in-house test method performed by TMET. In this test, the time to dry is set as the time required for wt% average of the test sample to become < 1%. Procedure: Samples were conditioned @22°C, 65% RH chamber, immersed in water for 30 minutes, then rung through the Testex padder @ 50 psi, and were returned to the conditioning chamber between weight measurements. Fig. 17.23 shows an example of the test results collected.

17.5.3 Durability

Durability is assessed using the test results from tests: abrasion resistance, torsional flexibility, launderability, tensile strength and colour fastness testing (Table 17.4).

17.5.4 Physical comfort

Physical comfort is assessed using the test results and measurements from tests: dimensional stability, weight, thickness and stiffness (Table 17.4).

✓ **Omniphobic treated Nomex fabric dried 4 times faster than untreated Nomex fabric. (15 min vs. 60 min)**

✓ **Omniphobic treated NYCO fabric dried 2.5 times faster than untreated NYCO fabric (30 min vs. 75 min)**

NYCO: 50% nylon/50% cotton
Nomex: 95% Nomex/5% Kevlar

Jaeger Nyco		Untreated 1	Untreated 2	Untreated 3	Average	Treated 1	Treated 2	Treated 3	Average
	Dry Weight (g)	0.58	0.6	0.558		0.572	0.58	0.547	
0 minutes	Weight (g)	0.777	0.755	0.705		0.618	0.638	0.571	
	wt% water	25.4%	20.5%	20.9%	22.2%	7.4%	9.1%	4.2%	6.9%
15 minutes	Weight (g)	0.696	0.674	0.633		0.584	0.587	0.561	
	wt% water	16.7%	11.0%	11.8%	13.2%	2.1%	1.2%	2.5%	1.9%
30 minutes	Weight (g)	0.651	0.633	0.598		0.577	0.586	0.552	
	wt% water	10.9%	5.2%	6.7%	7.6%	0.9%	1.0%	0.9%	0.9%
45 minutes	Weight (g)	0.6	0.611	0.572		0.577	0.583	0.552	
	wt% water	3.3%	1.8%	2.4%	2.5%	0.9%	0.5%	0.9%	0.8%
60 minutes	Weight (g)	0.587	0.603	0.566		0.575	0.583	0.552	
	wt% water	1.2%	0.5%	1.4%	1.0%	0.5%	0.5%	0.9%	0.6%
75 minutes	Weight (g)	0.581	0.602	0.561		0.574	0.583	0.551	
	wt% water	0.2%	0.3%	0.5%	0.3%	0.3%	0.5%	0.7%	0.5%

Nomex (S4)		Untreated 1	Untreated 2	Untreated 3	Average	Treated 1	Treated 2	Treated 3	Average
	Dry Weight (g)	0.431	0.417	0.397		0.385	0.409	0.443	
0 minutes	Weight (g)	0.563	0.537	0.504		0.413	0.418	0.456	
	wt% water	23.4%	22.3%	21.2%	22.3%	6.8%	2.2%	2.9%	3.9%
15 minutes	Weight (g)	0.486	0.464	0.438		0.391	0.411	0.446	
	wt% water	11.3%	10.1%	9.4%	10.3%	1.5%	0.5%	0.7%	0.9%
30 minutes	Weight (g)	0.438	0.425	0.4		0.388	0.41	0.445	
	wt% water	1.6%	1.9%	0.8%	1.4%	0.8%	0.2%	0.4%	0.5%
45 minutes	Weight (g)	0.436	0.424	0.4		0.388	0.411	0.444	
	wt% water	1.1%	1.7%	0.8%	1.2%	0.8%	0.5%	0.2%	0.5%
60 minutes	Weight (g)	0.434	0.421	0.399		0.388	0.411	0.445	
	wt% water	0.7%	1.0%	0.5%	0.7%	0.8%	0.5%	0.4%	0.6%
75 minutes	Weight (g)	0.435	0.422	0.399		0.388	0.41	0.445	
	wt% water	0.9%	1.2%	0.5%	0.9%	0.8%	0.2%	0.4%	0.5%

(Treated vs. Untreated)

***Time to dry = Time required for Wt% average < 1%.**

Procedure: Samples were conditioned @22°C, 65% RH chamber, then immersed in water for 30 min, then rung through padder @ 50 psi, and were returned to the conditioning chamber between weight measurements

Fig. 17.23 An example of NSRDEC (TMET)'s drying test method and data collection spreadsheet.

17.6 Limited field demonstration

In developing Luna Innovations Inc. (Luna)'s EverShield fluid-resistant coating based on a waterborne formulation via a simple padding process, many tests listed in Table 17.4 and discussed in Section 17.6 were conducted on test specimens. However, beyond these tests, as a final ultimate test, it was necessary to assess the omniphobic clothing's actual performance on the field. As an example, 20 of the EverShield omniphobic coated Army Combat Uniforms (ACU) were fabricated after receiving the toxicity clearance from the US Army Institute of Public Health. These EverShield-treated ACUs were evaluated by soldiers on the field at Fort Drum, New York in June 2011 during the Air Force Research Laboratory's Technology Warrior 2011 Program. Below are typical activities (Fig. 17.24 Left), and the average number of days and hours that the EverShield omniphobic treated ACUs were worn by the soldiers (Fig. 17.24 Right), and the selected findings (Fig. 17.25) after this limited field demonstration (Truong and Koene, 2012). At this limited field demonstration, NSRDEC's developed Luna omniphobic coating, now commercially known as UltraTech's EverShield product via a licencing agreement, had received the only Outstanding Warfighting Transition Award (Fig. 17.26).

17.7 Omniphobic coating technologies investigated

17.7.1 Fluorinated

Since 2008, there has been a growing number of research activities on the development and commercialization of environmentally friendly omniphobic coatings for textile and nontextile applications. Besides the successful development of the EverShield product, NSRDEC had evaluated and guided the developments of many more developmental and commercial liquid repellent coating technologies (Truong, 2013, 2014). These include experimental coatings from: Tiax LLC (Acrylate fluorinated coatings), Milliken & Co (ResQ Resist), Synthesized NanoCoatings/University of Oklahoma (Admicellar-polymerized fluorocarbon coating), Materials Modification, Inc., University of Virginia and University of Michigan; and from companies from Allegiance NanoSolutions, Inc. (Aqua Repulse), Triton Systems, Inc. (Invexus-LR), Ross Technology/NeverWet, LLC (NeverWet) and Innovative Surface Technology (product: ISurGuard) where they are offering their own commercial products. And, Aqua Repulse, Invexus LR and ResQ Resist showed promise in meeting NSRDEC's performance goals for self-cleaning military textile applications. These products are superhydrophobic; therefore have excellent water repellency for use in rainwear applications. Their physical properties are compatible for use with textiles. In that, as their coatings were applied onto the Army Combat Uniform (ACU) which is made of 50% nylon and 50% cotton, their coatings showed added durability to abrasion, did not cause the fabric to become stiff, added little weight to the base fabric, and did not interfere with the base fabric's air permeability and moisture vapour transport properties, which are essential for maintaining personal comfort while the soldiers carry out their regular

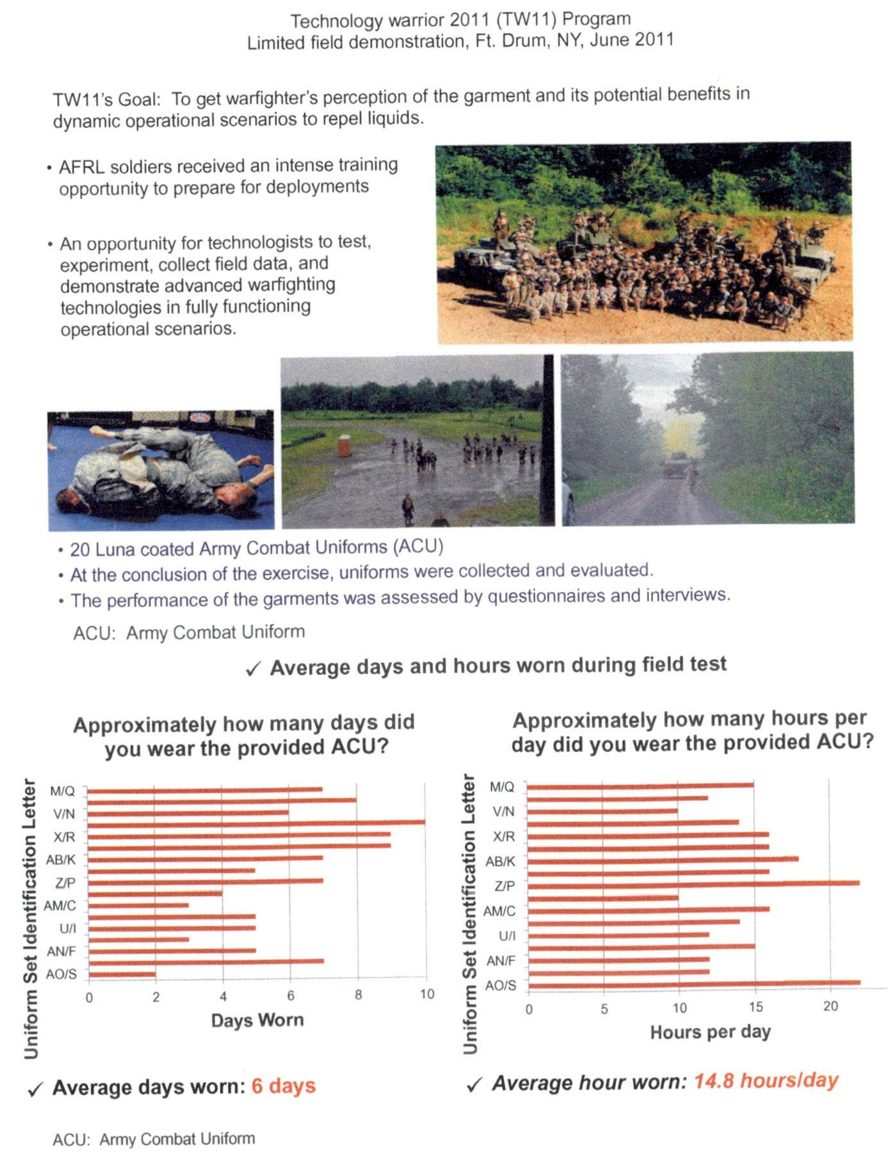

Fig. 17.24 Goal and various field training activities (left), and typical days and hours worn by the soldiers (right).
Pictures credit: Army.

duties. However, to be considered for potential use in enhancing Army's CB protective clothing, these products require further material development and/or optimization to effectively repel low surface tension nonpolar and polar chemicals. These chemicals include nonpolar chemicals such as hexadecane (27.47 mN/m,) and polar CWA

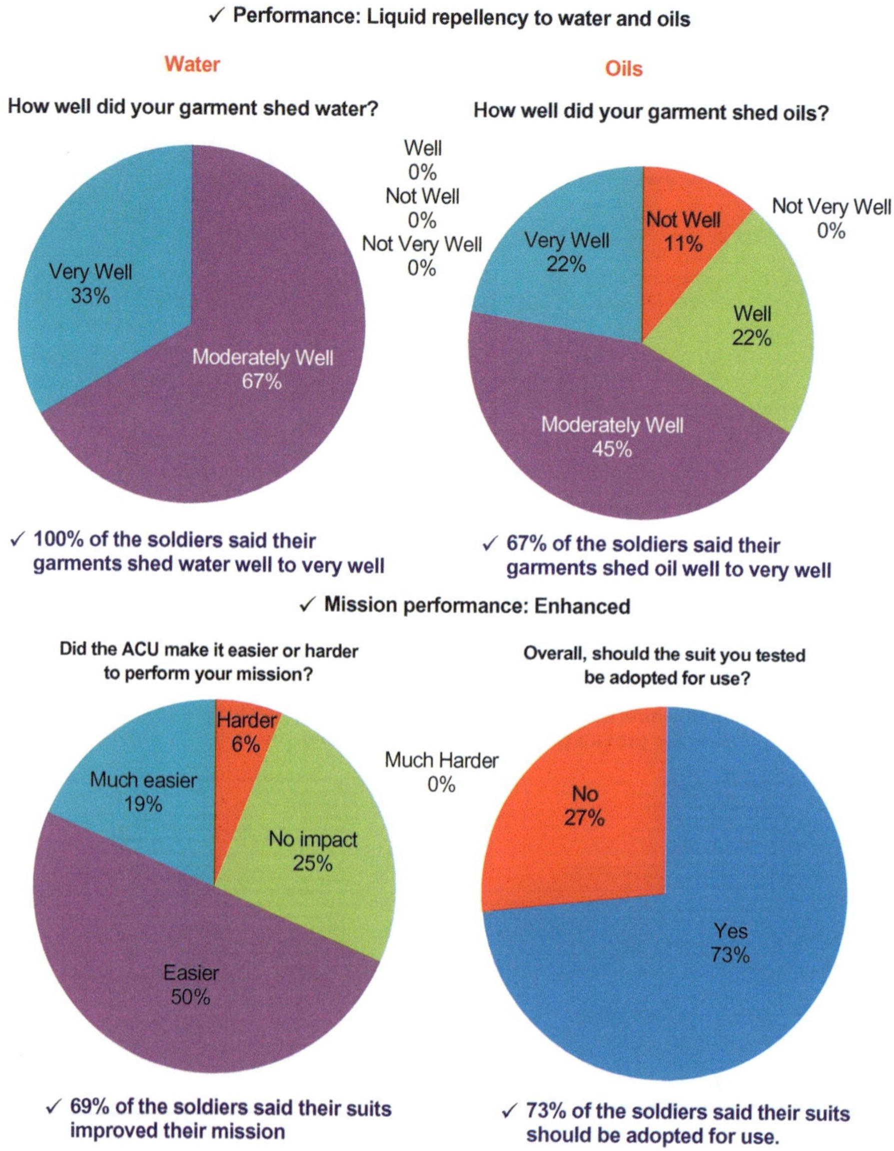

Fig. 17.25 Soldiers' assessment of EverShield omniphobic treated ACU (left), and how their mission performance was rated after the field training activities (right).

simulants such as tributyl phosphate (TBP) (27.79 mN/m), and dimethyl methyl phosphonate (DMMP) (36.7 mN/m).

The remaining omniphobic coatings NSRDEC had evaluated showed significant effects on the ACU's textile properties. For example, the coatings showed to be more

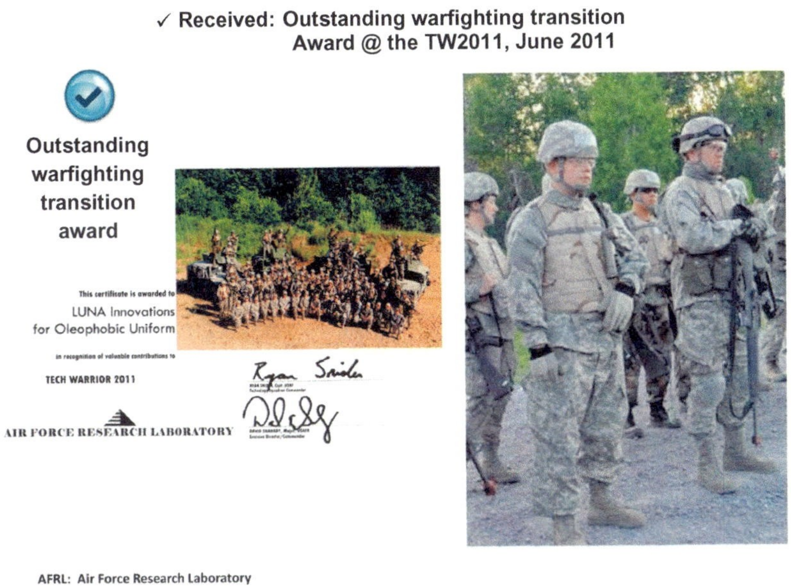

Fig. 17.26 EverShield omniphobic ACU received the only outstanding warfighting transition award at the Technology Warrior Demonstration 2011.
Pictures credit: Army.

durable than untreated fabric by 2 to almost 5 times in the resistance to abrasion, and were too thick, heavy and were not conformably coated only around the fibres and yarns, but their coating was applied across and over the yarns; therefore, had significantly reduced the ACU's air permeability and moisture vapour transport properties. The ACU also became stiffer, with added weight. However, some of these coatings such as the ISurGuard and NeverWet products show excellent potential for nontextile waterproofing applications for boots and hard surfaces such as concrete and metals. These two products require further material developments if the manufacturers would like to continue pursuing Army textile applications in which soldiers' personal comfort must be maintained.

17.7.2 Nonfluorinated

In the earlier development effort of EverShield omniphobic coating, it was found that while a nonfluorinated polymer coating based on polydimethyl siloxane (PDMS) showed to be superhydrophobic; however, it had a poor repellency to oils and solvents, thus is not oleophobic. In developing self-cleaning textiles, it is necessary for the textiles to have repellency to both water and organic solvents (Fig. 17.27).

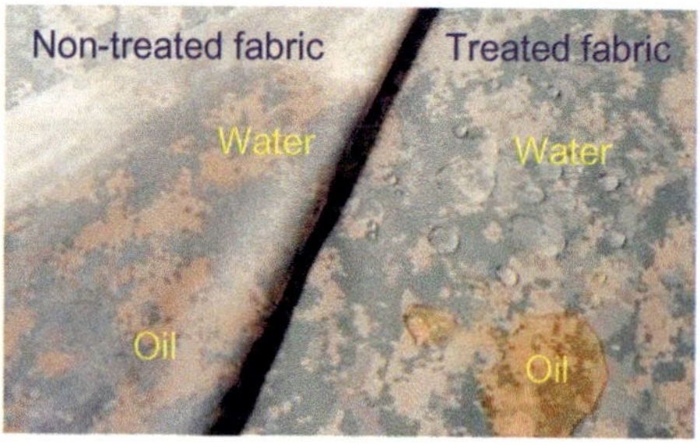

Fig. 17.27 Self-cleaning fabric is omniphobic, showing effective repellency to water and also to oil and solvents.
Picture Credit: NSRDEC.

Table 17.10 **Dirt weight % pick up of EverShield omniphobic treated fabric**

Weight (g)	Untreated	Treated
Before	5.943	5.996
After	6.235	6.034
Difference	0.292	0.038
Weight % pick up	4.91%	0.63%

17.8 Benefits to military textiles

17.8.1 Selfcleaning clothing

Omniphobic coated textiles pick up dirt about eight times less than untreated fabric (Table 17.10). It effectively sheds rain, repels oils and chemicals, and is compatible with a large range of textile materials including polypropylene, polyester, nylon, cotton, nylon/cotton, Nomex/Kevlar fabrics, wool, etc.; therefore, offers a wide range of military clothing applications (e.g. regular duty uniforms, combat uniform, rainwear/foul weather protective clothing, sleeping bags, back packs, tents, cold weather clothing, tropical clothing, body armour and CB agent protective clothing. Fig. 17.28).

17.8.2 Increased protection from low volatility agents

The significant, measurable advantages of omniphobic coatings to military textiles are in the area of CB protective clothing. Although increased protection against G and H agents has not yet been demonstrated, a significant reduction in permeation by LV (low volatility) agents was observed. The reason why repellent coatings reduce the permeation of LV agents is due to the fact that the vapour pressure of LV agents

Fig. 17.28 Selected omniphobic coated textile applications (left: operating in the rain; right: decontaminating vehicle). Pictures Credit: US Army.

is so low. Traditional agents such as those in the G and H series, even if in the form of sessile droplets on the surface of the fabric, still traverse through the material composite as a vapour whereas LV agents traverse primarily as a liquid. As such, the longer a low volatility agent remains a sessile droplet on the surface of the fabric, the longer the duration until the onset of liquid permeation, and the more protection the fabric will have. During the period prior to the onset of liquid permeation, only vapour will pass through the protective fabric, of which there is very little in the class of LV agents. A correlation has been observed between the retention of a sessile droplet, and reduced agent permeation in the case of VX (Table 17.7).

As can be seen in Fig. 17.29, in which several fabrics were paired with the same sorptive liner, the LV agent permeation through fabrics with a super repellent coatings (labelled IPFS C-8 and IPFS C-6) performed better than the standard Quarpel coating, which performed the same as or somewhat better than the uncoated fabric. A much larger reduction is seen between the uncoated baseline fabric and the fabrics treated with coatings developed within the IPFS (Integrated Protective Fabric Systems) programme. There is a 2 order of magnitude reduction in permeation for Agents W, X and Y, and an 85% reduction in VX permeation between the uncoated sample and IPFS C-6. The permeation values shown in the figure are averages with the error bars representing the maximum and minimum values measured.

A larger reduction is seen with Agents W, X and Y because VX is more difficult to repel. A 10 μL droplet of VX completely wets the IPFS C-6 coated fabric at 6 hours while Agent X remained a sessile droplet for at least 8 hours. The Quarpel coating did not perform as well as the super repellent coatings (IPFS C-6 and IPFS C-8 coatings) due to the lower repellency. A 10 μL droplet of VX soaked into the Quarpel coated

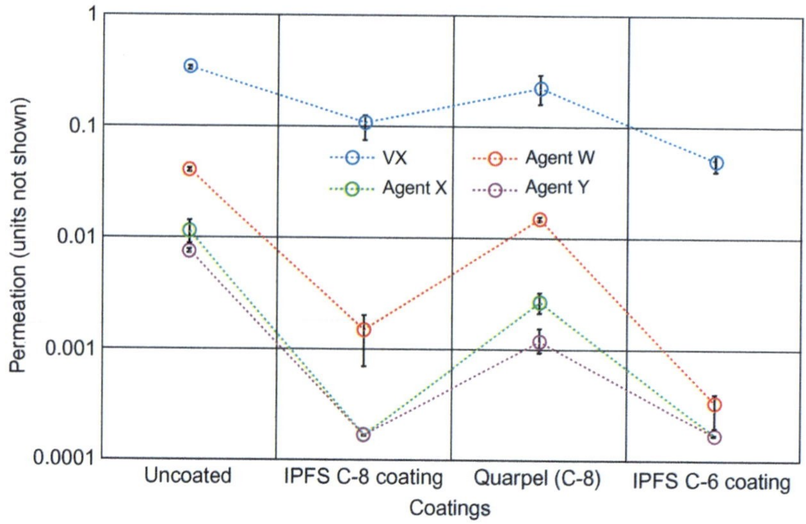

Fig. 17.29 Low volatility agent permeation through select protective fabric composites. NSRDEC data (Clay et al., 2015).

fabric after only 20 minutes. As such, the permeation of VX was only slightly reduced in comparison to a fabric without any coating.

The value of the super repellent coating can clearly be seen in the ability to reduce the permeation of low volatility agents by retaining a sessile droplet and delaying the passage of the liquid through the fabric composite. Repelling VX is more difficult than Agents W, X and Y and as such, protection to VX could be greatly improved by increasing the repellency towards VX in future R&D efforts.

17.8.3 Increased protection from agents after POL contamination

It is known that the exposure of CB protective fabrics to certain BFCs (Battlefield Contaminants) and/or POLs (petroleum, oils and lubricants) such as JP-8 (jet propellant-8) increases the permeation of both traditional and LV agents through the fabric composite via two mechanisms. The first mechanism is in which the JP-8 wets the fabric and carbon thus acting as a conduit for the CWA to penetrate the protective system in liquid form. The second mechanism is where the JP-8 vapours adsorb onto the carbon and block sorption sites for the agent, thus reducing the efficacy of the carbon as a sorbent material. The increase in permeation as a consequence of the first mechanism is much more drastic. Super repellent fabrics have the means to prevent the first mechanism from happening by repelling the JP-8 and preventing it from wicking into the fabric and acting as a conduit for the agent to bypass the carbon. The mitigation of agent permeation after contamination with JP-8 is demonstrated in Fig. 17.30 in which the permeation of HD and GD are shown for the same fabric composites evaluated in Fig. 17.29.

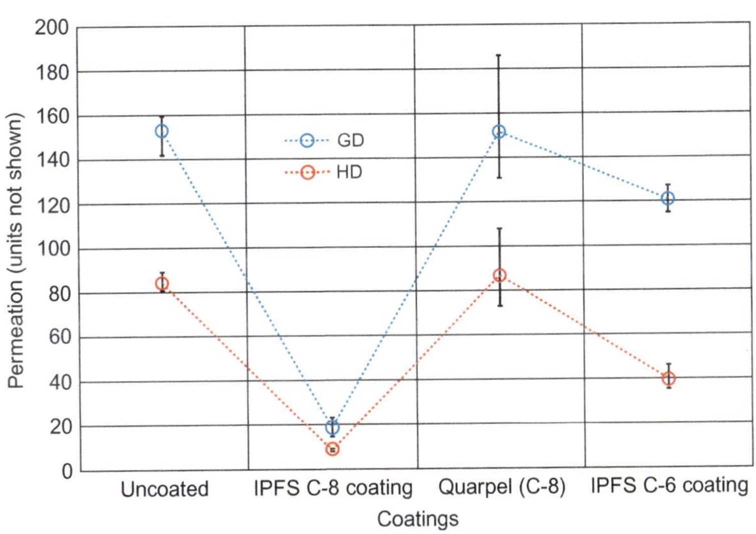

Fig. 17.30 Agent permeation through select protective fabric composites after contamination with JP-8. NSRDEC data (Clay et al., 2015).

As can be seen in Fig. 17.30, there is no difference in permeation between the treated and untreated fabrics. The repellency of the DWR Quarpel coating is not sufficient to prevent the JP-8 from wicking through. However, the IPFS C-6 coating resulted in a 20% permeation reduction for GD, and a 55% permeation reduction for HD. The IPFS C-8 based coating resulted in the largest reduction in permeation at 90% for both GD and HD. The oil rating was measured at 7B, a higher rating than for the IPFS C-6 coating, which had an oil rating of 6A. The higher oil rating of the IPFS C-8 fabric resulted in a much larger permeation reduction after the JP-8 contamination. For this reason, it is encouraged that any fabrics used with a repellent coating have an oil rating of 7B or higher to ensure protection from JP-8 contamination, and the subsequent increase in agent permeation. The same trend was seen with the low volatility agents. (Clay et al., 2015).

17.8.4 Reduction of agent transfer hazard from a contaminated surface

Soldiers in a contaminated environment may come into contact with contaminated surfaces. Whether the contact comes from holding a contaminated object or weapon, lying on a contaminated ground, pushing open a contaminated door, etc., contact is likely, despite taking measures to avoid it. This danger is even more prevalent with the LV agents, such as VX, which will stay for long periods of time after being disseminated because they do not evaporate quickly. However, if the CB uniform is super repellent, the agent will not wick into the garment and penetrate to the wearer, and will be more energetically stable remaining on the contaminated surface.

A test method currently in development in collaboration with Edgewood Chemical and Biological Center (ECBC) is designed to mimic the action of coming into contact with a contaminated surface for a brief period, and evaluating the resulting permeation. In the experiment, a protective material is placed face down on a specified amount of VX (or other LV agent). A weight of 1 psi is placed atop the material, removed and then material is transferred to a permeation cell.

Fig. 17.31 shows the permeation of VX after contamination for several fabrics atop the same sorptive material. As can be seen the fabric with a super repellent coating (IPFS C-6) reduced the VX permeation through a fabric as a result of contact transfer hazard by 60%. The baseline coated fabric produced a large error and no conclusions could be made. The results clearly indicate that super repellent coatings protect those that wear CB protective garments from contamination due to contact transfer hazard.

17.8.5 Retaining flame resistance after POL exposure

Flame resistant (FR) textiles are designed to turn fabric into char faster than a flame can spread, thus removing the source of fuel for fire, and to produce gases that displace oxygen at the base of the flame, thus suffocating the fire. When FR textiles become soaked with POLs such as the commonly used fuel, JP-8, they can lose their FR

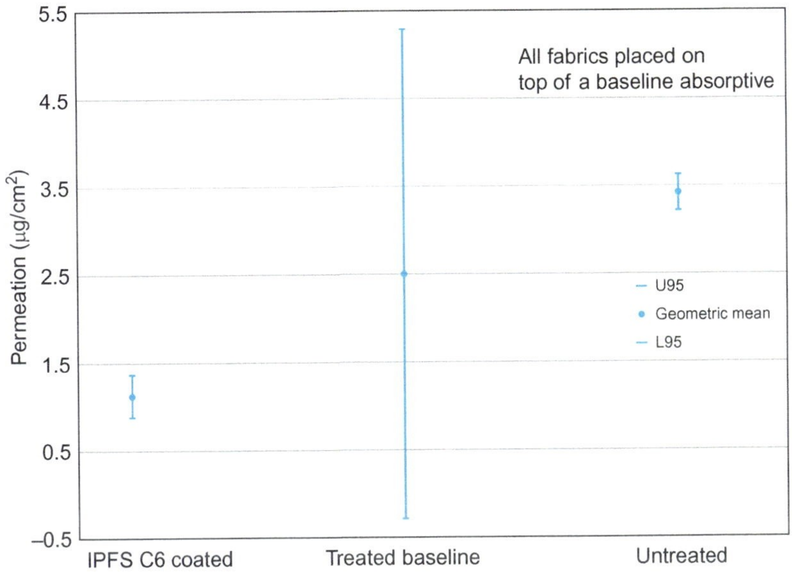

Fig. 17.31 VX permeation after exposure to a contaminated surface. NSRDEC data. (Pomerantz et al., in press).

capability. Although no standardized test methods have been used to evaluate this capability, the benefits of super repellency to FR have been seen. FR fabrics without any repellent treatment that are splashed with fuels such as JP-8 become saturated with the fuel as it soaks into the fabric. Upon exposure to fire the fabric is engulfed in flame, needing to be put out. Contrary to this, FR fabrics with repellent treatments do not become soaked with JP-8 because it beads up and rolls off for the most part. When exposed to fire, the residual JP-8 quickly catches fire and burns off, all superficially atop the fabric. However, repellent FR fabrics still self-extinguish, as would be expected of an FR material. Clearly, the super repellent coating enables FR materials to retain their capability after being exposed to large amounts of JP-8, as can happen with fuel handlers. The better the repellent coating, the more the JP-8 will roll off and not wick into the fabric. (Clay et al., 2015).

17.9 Future civilian applications

Omniphobic coated textiles offer a wide range of textile applications including protective clothing for emergency responders. Department of Homeland Defense (DHS) is responsible for the domestic preparedness for incidences involving the use of CB agents, and therefore must protect its emergency responders operating in contaminated environment in urban and city areas when responding to domestic threats involving weapons of mass destruction. Other civilian applications where minimal or no contact with water, oils, wines, blood, etc. is desired, include the following: chemicals and fuel handler garments, chef/cook uniforms, hospital gowns, regular workwear uniforms, rainwear, shipyards worker garments, sailor uniforms, sportswear (mountain climbing, skiing, swimming, sailing, etc.), awning, upholstery, convertible car tops, etc. For nontextile applications, omniphobic coatings can find their uses in waterproofing of concrete, electronic equipment (e.g. radios, cellphones), windows, buildings, ships, aircrafts, vehicles, etc.

17.10 Summary

- A robust superomniphobic surface must have three parameters: (1) A sufficiently low surface tension chemistry with a surface contact angle >150°, (2) a micro/nanoroughness surface architecture, and (3) local surface curvatures created by micro/nanocavities in which liquid cannot wet the surface.
- There are four reentrant textile structural levels that contribute towards an increase in liquid nonwetting stage: single fibre, fibre bundles or yarn size, weave opening (how tight or open a fabric is), fibre's surface texture and reentrant fibre structure.
- There are five factors that have strong impact on liquid repellency: (1) Surface chemistry, (2) surface roughness, (3) fabric structure, (4) reentrant fibre structures and (5) polarity of liquids and wetting surface.
- EverShield Durable Omni Repellent (DOR) coating became the first commercial omniphobic textile coating product extensively researched, developed and field tested by NSRDEC and Luna partnership. This product is being produced and sold by UltraTech under a licencing agreement with Luna Innovations Incorporated (Luna).
- EverShield omniphobic coating was proven to be more durable than Quarpel DWR, because it retains its oleophobic properties through repeated washings. It is being evaluated as the replacement for C-8 Quarpel DWR coating.
- Omniphobic coatings increase protection from low volatility agents and reduce agent permeation through fabrics contaminated with POLs. Repellent coatings also reduce the transfer hazard of coming into contact with a contaminated surface. Additionally, omniphobic coatings increase the flame resistance of FR fabrics after exposure to fuels.
- Currently, there is a growing number of developmental and commercial omniphobic coating technologies based on the greener C-6 chemistries. Omniphobic coatings, fabrics and fibres have many military and civilian applications.
- A novel reentrant fibre architecture was developed and demonstrated to offer an additional level of liquid nonwetting.

17.11 Follow-up work and remaining challenges

17.11.1 Repelling organophosphates and representative CWA simulants

As shown in Table 17.7, it is critical to be able to repel low volatility agents because the repellency will lead to increased protection, shown in Figs 17.29 and 17.30. However, live agent testing is expensive and few facilities have the capability to perform it. Academic institutions, national laboratories and industries in which many of these technologies are developed are forced to do iterative testing with solvents that do not adequately represent the CWA. Furthermore, textile manufactures do not have the capability to ensure that adequate VX repellency will be met in quality control evaluations. For these reasons, an adequate simulant, or series of simulants, should be found.

17.11.2 Durable one-sided repellency

Development of a fabric system that would prevent liquid wetting while wicking body sweat and vapour away from the body to evaporatively cool soldiers operating in jungle/tropical environments and also cold and extremely cold environments is being pursued. In this effort, the effects of using filaments versus staple and twist factor on liquid repellency will be examined to create a dual wicking and nonwetting yarn construction to improve soldiers' comfort in a tropical environments. Other work is currently being funded by the Australian military for repellency on the face side of the fabric, and wicking on the underside (Zeng et al., 2016) Additionally, laminating repellent fabrics to multiple layers, as is sometimes necessary for CB protection, lessens the adhesion and durability of the laminate (Pomerantz et al, 2017). Spray and foaming applications do not have the same coverage and durability of fabrics immersed in a trough in the pad-roll application. The ability to retain a durable omniphobic coating on one side only remains a challenge.

17.11.3 Nonfluorinated omniphobicity

As C-8 chemistry based liquid repellent coatings are no longer allowed, future non-fluorinated coatings are being developed for both nontextile (Moore et al., 2010; Chhatre et al., 2010a,b) and textile (Lassen et al., 2015; Davies, 2014) applications. Coating free superomniphobic fibres such as self-blooming reentrant super-omniphobic fibres meet this need as well (Fig. 17.32). Currently, no technology has been found to be as repellent as C-8 based chemistries, especially towards organ-ophosphates. Due to the demonstrated advantages of omniphobicity in CB protection towards LV agents such as VX, exemptions to the C-8 ban for CB protective suits have been discussed. However, there is no agreement at this time and an exemption cannot be relied upon. For this reason, it is critical to pursue R&D efforts towards achieving omniphobicity through other means.

17.11.4 Multifunctional superomniphobic coatings and fibres

There are only so many bonding sites on a fabric surface, and a repellent coating has to compete with other coatings and chemistries for the same real estate. The effect of repellent coatings on permethrin application and subsequent insect repellency has not yet been investigated. It is assumed that the coating would reduce the uptake

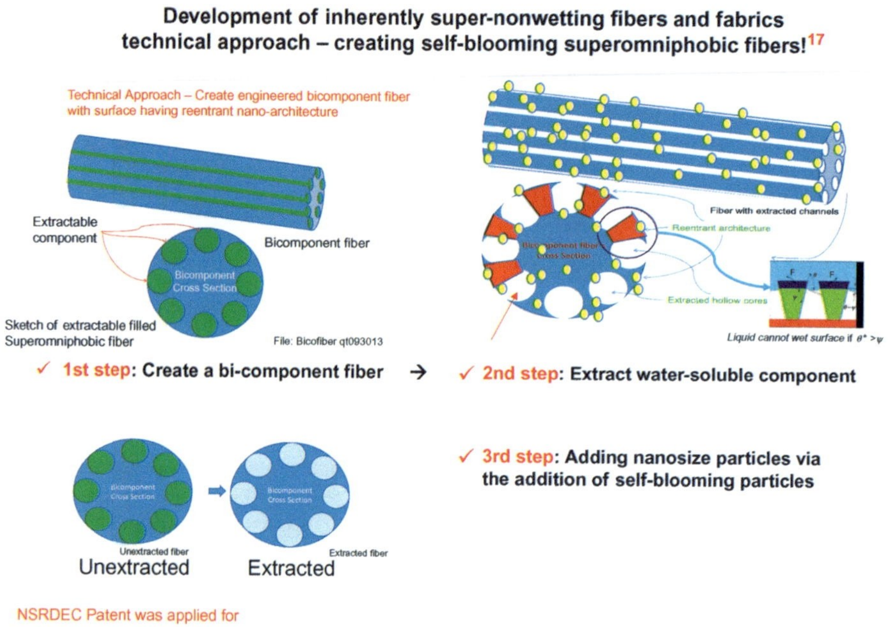

Fig. 17.32 NSRDEC's self-blooming superomniphobic reentrant fibres (Truong et al., 2015a,b).

of permethrin, which is applied after the garments are fabricated, and result in less insect repellency. Similarly, the IPFS programme has found that repellent fabrics have reduced biocidal properties as the microbes are repelled from the fabric and do not come into contact with the attached biocides (Doona, 2015). The effects of omniphobic coatings on properties that have not been evaluated yet include thermal signature, friend/foe identification, spectral reflectance and static discharge. Currently, there is an SBIR effort to incorporate broad spectrum reactivity against toxic chemicals and microbes into omniphobic coatings (Koene, 2016). Repellency to low surface tension liquids through a fluoro POSS/Tecnoflon coating by creating a random dual hierarchical coating where particles are stacked in a 20:1 ratio was demonstrated. Follow on work to this effort will focus on the use of charged nanoparticles and 3D printed nanosized surface features to develop an inverse trapezoidal-like nanosurface roughness (Fig. 17.33) which has been shown to create superoleophobic surfaces (Tuteja et al., 2007).

17.11.5 Aerosol and liquid repellent coatings

Current effort is focused on the development of CB aerosol and liquid repellent coatings. This effort aims to develop, assess and optimize CB aerosol and liquid repellent coatings for use on textiles and solid substrate surfaces. Presently, there is an SBIR topic calling for proposals to address this need (SBIR Source, 2016).

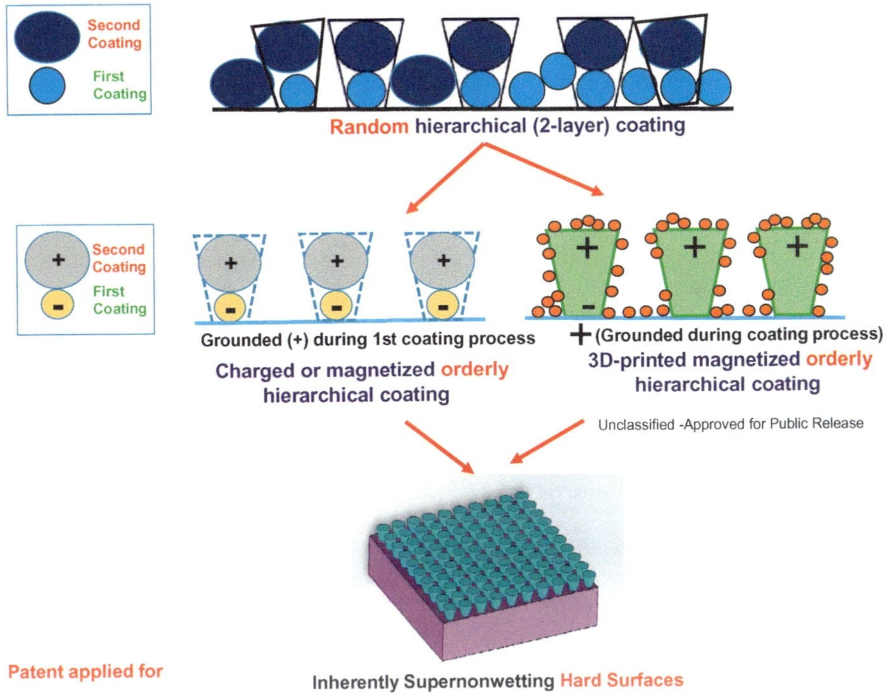

Fig. 17.33 Improving liquid repellency through manipulation of surface roughness.

Acknowledgments

The authors wish to thank the US Army Natick RDE Center and its scientists, engineers and technologists in CSET, TMET and the Molecular Science and Engineering Team (MSET) for their assistance in data analyses, electron scanning microscopy and testing and evaluation of omniphobic materials and fabrics, and for providing the research and development environment, management support, the Joint Science and Technology Office for Chemical and Biological Defense, and the US Congress and the SBRI Program Office for their generous funding support in which various exciting advanced and novel work in surface science and chemical/biological protection were possible since 2008. The results of these funded efforts are currently continuing at the present time.

Also, the authors would like to thank our technically competent and industrious technical collaborators from MIT, Clemson University, Hills Inc., Burlington Laboratories/ITG Group, University of Massachusetts Lowell, Luna Innovations, Inc., Ultratech, ITG Burlington Laboratory, US Air Force Research Laboratories, North Carolina State University, Defense Science and Technology Laboratory (Wiltshire, UK) and from many other academic and industry partners participated and collaborated in various R&D efforts with NSRDEC.

EverShield® coating was developed and commercially transitioned under NSRDEC's led SBIR effort (Truong and Koene, 2013a,b). This commercial product was a result of 2008 DoD Small Business Innovative Research (SBIR) Phase I topic on 'Superoleophobic/hydrophobic Coatings for Nonstick, Self-Cleaning Textiles', and follow-on SBIR funded efforts.

PAO #: U17-094.

References

123RF, 2017. Stock Photo—Perfluorooctanoic Acid (PFOA, C-8) Molecule. Important and Persistent Pollutant. Atoms are Represented as Spheres With Conventional Color Coding: Hydrogen (White), Carbon (Grey), Fluorine (Green), Oxygen (Red). http://www.123rf.com/photo_25357258_perfluorooctanoic-acid-pfoa-c8-molecule-important-and-persistent-pollutant-atoms-are-represented-as-.html (Accessed January 2017).

American Cancer Society, 2016. Teflon and Perfluorooctanoic Acid (PFOA). American Cancer Society, Atlanta, GA. https://www.cancer.org/cancer/cancer-causes/teflon-and-perfluorooctanoic-acid-pfoa.html.

Beilstein Institute, 2016. Springtails—Promising Templates for Robust Omniphobic and Anti-Fouling Surfaces. http://www.beilstein.tv/tvpost/springtails-promising-templates-for-robust-omniphobic-and-anti-fouling-surfaces (Accessed January 2017).

Beilstein Institut, 2017. Biomimetics. http://www.beilstein-institut.de/en/spotlight/biomimetics (Accessed January 2017).

Bittner, R., Bica, K., Hoffmann, H., 2017. Fluorine-free, liquid-repellent surfaces made from ionic liquid-infused nanostructured silicon. Monatsh. Chem. – Chemical Monthly 148 (1), 167–177.

Brigden, K., Hetherington, S., Wang, M., Santillo, D., Johnston, P., 2013. Hazardous Chemicals in Branded Textile Products on Sale in 25 Countries/Regions During 2013. Greenpeace Research Laboratories Technical Report. http://www.greenpeace.org/eastasia/Global/eastasia/publications/reports/toxics/2013/A%20Little%20Story%20About%20the%20Monsters%20In%20Your%20Closet%20-%20Technical%20Report.pdf.

Cassie, A., Baxter, S., 1944. Wettability of porous surfaces. Trans. Faraday Soc. 40, 546. 2.

Chhatre, S., Choi, W., Tuteja, A., Park, K., Mabry, J., McKinley, G., Cohen, R., 2010a. Scale dependence of omniphobic mesh surfaces. Langmuir 26 (6), 4027–4035.

Chhatre, S., Guardado, J., Moore, B., Haddad, T., Mabry, J., McKinley, G., Cohen, R., 2010b. Fluroalkylated silicon-containing surfaces – estimation of solid surface energy. ACS Appl. Mater. Interfaces 2 (12), 3544–3554.

Clay, M., Dugan, N., Gaffga, J., McVeety, B., Sieber, M., Soja, E., Stickel, G., 2015. Advantages of omniphobic coatings for chemical-biological suit protection. In: Presented at Nanotechnology for Defense Conference. City of Industry, CA, USA, November 16–20, 2015.

Davies, A., 2014. Durable Water Repellency – Study Phase I. An Evaluation of the Test Methods Used for Assessing Durable Water Repellent Fabrics Within the Outdoor Industry. De Monfort University, Leicester.

Diversified Enterprises, 2014. https://www.accudynetest.com/visc_table_print.html?sortby=sort_molecular_weight%20DESC (Accessed January 2017).

Doona, C., 2015. Integrated Protective Fabric Systems: Task 3b: Standardization of Advanced Biological Performance Testing, NSRDEC Technical Report, 2015.

Ensikat, H., Ditshe-Kuru, P., Neinhuis, C., Barthlott, W., 2011. Superhydrophobicity in perfection: the outstanding properties of the lotus leaf. J. Nanotechnol. 2 (2011), 152–161.

Flick, E.W. (Ed.), 1991. Industrial Solvents Handbook, fourth ed. Noyes Data Corporation, Park Ridge, NJ.

GB Investments Inc., 2015. Laundering field jacket? http://forums.gunbroker.com/topic.asp?TOPIC_ID=225248 (Accessed January 2017).

Gorzkowska-Sobas, A. A. (2013). Chemical warfare agents and their interactions with solid surfaces. Norwegian Defence Research Establishment (FFI). FFI-rapport 2013/00574.

Gunstone, F., 2009. Rapeseed and Canola Oil: Production, Processing, Properties and Uses. John Wiley & Sons, New York, NY. p. 90.

Hastbacka, M., 2015. Tiax, LLC, Non-fluorinated Omniphobic Coatings for Stain Resistant Textiles, EPA Contract No. EPD15029. SBIR Phase I. Sep. 1, 2015–Feb. 29, 2016.

Haynes, W.M. (Ed.), 2014. CRC Handbook of Chemistry and Physics. 95th ed. CRC Press LLC, Boca Raton, FL. pp. 6-186.

Hensel, R., Helbig, R., Aland, S., Braun, H., Voight, A., Neinhuis, C., Werner, C., 2013. Wetting resistance at its topographical limit: the benefit of mushroom and serif T structures. Langmuir 29, 1100–1112.

Iacono, S., Vij, A., Yandek, G., Smith, D., Mabry, J., Peloquin, A., 2010. Fluorinated Polyhedral Oligomeric Silsesquioxanes (F-POSS). Air Force Research Laboratory. Technical Paper No. AFRL-RZ-ED-TP-2010-070.

Jeevajothi, K., Subasri, R., Soma Raju, K.R.C., 2013. Transparent, non-fluorinated, hydrophobic silica coatings with improved mechanical properties. Ceram. Int. 39 (2), 2111–2116.

Keller, J.M., Simmons, C.S., 2005. The Influence of Selected Liquid and Soil Properties on the Propagation of Spills Over Flat Permeable Surfaces. Pacific Northwest National Laboratory, Richland, WA. PNNL-15058.

Kleingartner, J., Srinivasan, S., Truong, Q., Sieber, M., Cohen, R., McKinley, G., 2015. Designing robust hierarchically textured oleophobic fabrics. Langmuir 31 (48), 13201–13213 (American Chemical Society).

Koene, B., 2012. Superoleophobic Coatings for Easy Cleaning Textiles, US Army Soldier Systems Center, Natick, Massachusetts, USA. Contract #W911QY-10-C-0071, Luna Innovations, Inc. Phase II Final Report.

Koene, B., 2016. Self-Decontaminating Textiles, Contract Tyndall Air Force Base, Tyndall, Florida, USA. Contract FA8051-16-P-0014, Luna Innovations, Inc. Phase I Monthly report 5.

Lassen, C., Jensen, A., Warming, M. (Eds.), 2015. Alternatives to perfluoalkyl and polyfluoroalkyl substances (PFAS) in Textiles. In: Survey of chemical substances in consumer products No. 137, The Danish Environmental Protection Agency, Copenhagen, ISBN: 978-87-93352-16-2.

LookChem.com, 2008. Methyl salicylate. http://www.lookchem.com/Methyl-salicylate/ (Accessed January 2017).

Mabry, J., Vij, A., Iacono, S., Viers, B., 2008. Fluorinated polyhedral oligomeric silsesquioxanes (F-POSS). Angew. Chem. Int. Ed. 47, 4137–4140.

Moore, B., Mabry, J., Chhatre, S., Guardado, J., McKinley, G., Cohen, R. (2010). Fluroalkylated Silicon-Containing Surfaces – Estimation of Solid Surface Energy (Preprint). Air Force Research Laboratory Technical Report Number AFRL-RZ-ED-JA-2010-436.

Nanex, 2016. About Nanex. http://nanexcompany.com/ (Accessed January 2017).

National Institute of Environmental Health Services, 2016. Perfluorinated Chemicals (PFCs). https://www.niehs.nih.gov/health/topics/agents/pfc/index.cfm (Accessed January 2017).

National Physical Laboratory, 2015. Surface Tensions. http://www.kayelaby.npl.co.uk/general_physics/2_2/2_2_5.html (Accessed January 2017).

National Research Council of the National Academies, 2005. Interim Design Assessment for the Blue Grass Chemical Agent Destination Pilot Plant. The National Academies Press, Washington, DC. p. 9.

Olsen, G., Burris, J., Lundberg, J., Hansen, K., Mandel, J., Zobel, L., 2002. Identification of Fluorochemicals in Human Sera. III. Pediatric Participants in a Group A Streptococci Clinical Trial Investigation. Epidemiology, Medical Department, 3M Company, St. Paul, MN 55144. Technical Report EPI-0011. Web source: http://static.ewg.org/reports/2003/pfcs/kid_blood_full.pdf?_ga=1.156333821.773454566.1486228790.

Pomerantz, N., Dugan, N., El-Samaloty, N., Richards, M., Sieber, M., Sousa, C., Tew, S., in press. Integrated Protective Fabric System Program: Omniphobic Technologies Report. NSRDEC Technical Report.

Pomerantz, N., Dugan, N., Richards, M., 2015. Updates on omniphobicity and test methods. In: Presented at the CBR Enterprise Protect Working Group. Defense Science and Technology Laboratory, Porton Down, GBR. October 19–23, 2015.

Pomerantz, N., Dugan, N., Richards, M., Tew, S., Zukas, W., 2017. Integrated Protective Fabric Systems (IPFS) Program: 1QFY17 Quarterly Technical Report. NSRDEC Internal Report.

Public Affairs Office, 2017. US Army Natick Soldier Research, Development, and Engineering Center. NSRDEC Overview, http://nsrdec.natick.army.mil/media/fact/NSRDEC.pdf.

Rame Hart Instruments, Inc., 2017. Rame-hart Glossary of Surface Science Terms. http://www.ramehart.com/glossary.htm.

Royal Society of Chemistry, 2015. Sulfide, 2-chloroethyl ethyl. http://www.chemspider.com/Chemical-Structure.11215101.html (Accessed January 2017).

Saraf, R., Lee, H., Michielsen, S., Owens, J., Willis, C., Stone, C., Wilusz, E., 2011. Comparison of three methods for generating superhydrophobic, superoleophobic nylon Nonwoven surfaces. J. Mater. Sci. 46, 5751–5760.

SBIR Source, 2016. Development of Chemical and Biological Aerosol and Liquid Repellent Coatings. https://sbirsource.com/sbir/topics/93403 (Accessed in January 2017).

Solvay, 2017. Technical Data Sheet Tecnoflon BR 9151-216547. http://www.solvay.com/en/binaries/TDS_Tecnoflon_BR_9151-216547.pdf (Accessed January 2017).

Surface Tension, 2016. Solid Surface Energy Data (SFE) for Common Polymers. http://surface-tension.de/solid-surface-energy.htm (Accessed January 2017).

The Environmental Working Group, 2008. Teflon's Replacement: Still Toxic? EWG, Washington, DC. http://www.ewg.org/enviroblog/2008/06/teflons-replacement-still-Toxic.

Truong, Q., 2013. Task 6 "Omniphobic Fabrics with CWA Repellent Chemistries," Integrated Protective Fabric System Program, NSRDEC Internal Report (2012–2013).

Truong, Q., 2014. Superomniphobics. Integrated Protective Fabric System Program. Task 7. Internal Report. 1QFY13. NSRDEC.

Truong, Q., 2017. Dual Hierarchical Superomniphobic Coatings Based on Fluorodecyl Polyhedral Oligomeric Silsesquioxane (Fluoro-POSS) and a Fluoroelastomer. Smart Coatings 2017.

Truong, Q., Koene, B., 2012. Omniphobic coatings for enhanced chemical/biological agent protective clothing. In: Chemical, Biological, Nuclear, Radiological and/or Explosives (CBNRe) Conference.

Truong, Q., Koene, B., 2013a. Superoleophobic Coatings for Easy Cleaning Textiles. NSRDEC-Luna Innovations, Inc. SBIR Contracts W911QY-09-C-0016 -W911QY13P0175, Jan. 8, 2009–Dec. 31, 2013.

Truong, Q., Koene, B., 2013b. Omniphobic coatings for self-cleaning and enhanced chemical/biological (CB) agent protective clothing. Nanotechnology 1, 663–666 (Chapter 9).

Truong, Q., Pomerantz, N., Yip, P., Sieber, M., Mabry, J., Ramirez, S., 2015a. Pilot-scale coating of fabrics with fluorodecyl polyhedral oligomeric silsesquioxane/fluoroelastomer blends. J. Surf. Innov. 2 (2), 79–93.

Truong, Q., Zukas, W., Welsh, E., Stenhouse, P., Brown, P., Hoffman, N., Mead, J., Barry, C., Mooney, P., Panwar, A., Koene, B., Martin, R., Cohen, R., McKinley, G., Kleingartner, J., Chen, D., 2015b. Inherently superomniphobic fibers and fabrics. In: Advanced Materials TechConnect Briefs 2016, Coatings, Surfaces & Membranes, pp. 272–275 (Chapter 8).

Tuteja, A., Choi, W., Ma, M., Mabry, J., Mazzella, S., Rutledge, G., McKinley, G., Cohen, R., 2007. Designing superoleophobic surfaces. Science 318 (5856), 1618–1622.

Tuteja, A., Choi, W., Mabry, J., McKinley, G., Cohen, R., 2008a. Robust omniphobic surfaces. Proc. Natl. Acad. Sci. U. S. A. 105, 18200–18205.
Tuteja, A., Choi, W., McKinley, G., Cohen, R., Rubner, M., 2008b. Design parameters for superhydrophobicity and superoleophobicity. MRS Bull. 33, 757
UltraTech International, Inc., 2017. EverShield. Technical Info, http://www.evershield.com/technical-info/.
United States Environmental Protection Agency, 2006a. Assessing and Managing Chemicals under TSCA. http://www.epa.gov/oppt/pfoa/index.html.
United States Environmental Protection Agency, 2006b. Per- and Polyfluoroalkyl Substances (PFASx) under TSCA. Docket ID: EPA-HQ-OPPT-2006-0621. https://www.epa.gov/assessing-and-managing-chemicals-under-tsca/and-polyfluoroalkyl-substances-pfass-under-tsca#tab-3.
US FDA&EMEA, 2016. Ethylene Glycol. http://www.druglead.com/cds/ethylene-glycol.htm (Accessed January 2017).
Washburn, E.W., 1928. International Critical Tables of Numerical Data, Physics, Chemistry and Technology, first ed. Vol. IV. McGraw-Hill Book Company, Inc., New York, NY, p. 455.
Weast, R.C. (Ed.), 1977. CRC Handbook of Chemistry & Physics, 58th ed. CRC Press, Inc., West Palm Beach, FL. pp. F-46–F-48.
Weast, R.C. (Ed.), 1986. CRC Handbook of Chemistry and Physics, 67th ed. CRC Press, Inc., Boca Raton, FL. p. F-31.
Wenzel, R., 1936. Resistance of solid surfaces to wetting by water. Ind. Eng. Chem. 28 (8), 988–994.
Wikimedia Foundation, 2016a. File:Perfluorooctanesulfonic-acid-3D-balls.png. https://commons.wikimedia.org/wiki/File:Perfluorooctanesulfonic-acid-3D-balls.png (Accessed January 2017).
Wikimedia Foundation, 2016b. Surface Tension Values. https://en.wikipedia.org/wiki/Surface-tension_values (Accessed January 2017).
Wikiwand, 2016. Hexanoic Acid. http://www.wikiwand.com/en/Hexanoic_acid (Accessed January 2017).
Zeng, C., Wang, H., Zhou, H., Lin, T., 2016. Directional water transport fabrics with durable ultra-high one-way transport capacity. Adv. Mater. Interfaces. 3, 1–7.
Zisman, W., 1964. In: Fowkes, F.M. (Ed.), Contact Angle, Wettability, and Adhesion. Vol. 43. American Chemical Society, Washington, DC.

Further reading

Choi, W., Tuteja, A., Chhatre, S., Mabry, J., Cohen, R., McKinley, G., 2009. Fabrics with tunable oleophobicity, Adv. Mater. 21, 2190–2195.
Meuler, A., Chhatre, S., Nieves, R., Mabry, J., Cohen, R., McKinley, G., 2011. Examination of wettability and surface energy in fluorodecyl POSS/polymer blends. Soft Matter 7, 10122.
Truong, Q., 2015. Influence of inherently super-nonwetting fiber and fabric designs on repellency and resistance to liquid penetration. In: Advanced Textiles, 2015 Industrial Fabrics Association International 2015. http://www.ifai.com/wp-content/uploads/2016/03/at15_att_brochure_fnl.pdf.

Footwear

18

*Ameersing Luximon**, *Asimananda Khandual*†
*The Hong Kong Polytechnic University, China, †College of Engineering & Technology, Bhubaneswar, India

18.1 Introduction

Since the inception of human civilization, footwear has played a crucial role. Footwear is not only essential for protecting our feet and supporting mobility, but traditionally, it reflects wearer's lifestyle, demanding, therefore, aesthetic beauty along with quality, durability, comfort and fit. The world footwear market is becoming more vibrant now. In addition to the middle class population's increased spending capacity, there is widespread adaptation of healthy and secure lifestyles and a growing e-commerce retail market, increasing demand for comfortable, innovative footwear and customized product lines. Global Industry Analysts. Inc. (United States) projected that the global market for footwear is expected to reach $371.8 billion US dollars by 2020, at a CAGR of around 2% (2016–24). Regionally, Europe is anticipated to have the largest market share (as in 2015), followed by Asia Pacific and North America, whereas the Asia Pacific market is expected to have the highest CAGR, at about 8%. Sadly, their analysis reported that over 300 million children and 1.5 billion adults still don't have any shoes.

In 2015, the global athletic footwear market was $80 billion US dollars, and with a gross margin around 45%. Recently, increasing participation in sports, use of functional and lightweight athletic shoes for running & jogging for both women and men have become a driving force for innovative products and the demand for sustainable and faster growth, specifically for athletic footwear. Some of the leading key factors for the new generation shoe market are the availability of customized footwear, awareness of eco footwear, seasonal trends, and the availability and adaptability of e-commerce. In 2015, the largest market share was held by the men's footwear segment, approximately 60%, followed by women's footwear, at about 30% market share. The children's footwear segment is also growing significantly, as new generation schools have more sports facilities and infrastructure specifically designed for sports such as baseball, basketball, tennis, golf and soccer. Nike, VF Corporation, Adidas, New Balance and Asics are the five dominant major stakeholders, enjoying 80% of the total athletic footwear market share. In terms of revenue, Nike and Adidas accounted for around 45% and 35%, respectively, followed by Converse, Saucony, New Balance, Vans, Skechers, K-Swiss, Wolverine Worldwide Inc. and Deckers. In 2015–16, about 25% of the footwear purchased online was customized, amounting to 2 billion US dollars' worth of sales (BUSINESS WIRE News, August 23, 2016 01:43 PM Eastern Daylight Time). For customized shoes, retailers like Amazon are

Waterproof and Water Repellent Textiles and Clothing. https://doi.org/10.1016/B978-0-08-101212-3.00017-4

making use of their huge consumer base though big data; Reebok launched rbkcustom, and Nike launched NikeID in order to provide customize shoes with patriotic prints.

Today's most common key innovative indicators for diversified footwear applications are light weight, environmental friendliness, comfort with maximum support, lower maintenance, high durability, superior flexibility for muscles and water repellency.

18.2 Major raw materials and their constitution for shoe components

Ill-fitting footwear causes discomfort, pain and even foot problems such as calluses, corns, hallux valgus, ulcers and pressure sores (Rout et al., 2010; Luximon, 2001; Luximon et al., 2017; Xiong, 2008). There has been a long period of evolution in footwear material and design since the Stone Age, and the Age of Empire to the present era of technology. Diversified materials such as animal skin, textiles, leathers, advanced polymeric materials and composites are used for making shoes. Numerous designs and materials, as well as time-consuming production techniques are selected based on the demography, culture, age, sex, aesthetic gratification and functional needs of the footwear customer. Table 18.1 lists clusters of raw material and tool inputs involved in shoe production, Fig. 18.1 depicts the anatomy and dominant chemical constituents of a typical shoe, and Fig. 18.2 describes major raw materials for shoe components (Luximon, 2013).

18.2.1 Shoe soles

Materials that can be used for shoe uppers can also be used in shoe soles, that is, either chrome/vegetable-tanned leather or polyurethane/neoprene as a solid foam. These polyurethanes tend to be highly cross-linked, closely packed foams. For manufacturing soles, different types of rubber can be used, depending upon the application. Natural rubber and synthetic polybutadiene are popular rubbers for soles and heels. Neoprene is a widely used synthetic for oil-resistant work soles, utilizing dibenzothiazyl disulphide as an accelerator.

In general, a sole's material and design must have the following essential properties:

- Appealing and attractive to the buyer. Design and comfortable fit are the dominant essential factors.
- Able to absorb impact for comfort while walking or running.
- Good abrasion resistance for durability and maintenance of structural integrity.
- Good resistance to continuous flexion while running, walking or exercising.
- Good skid resistance on various types of surfaces, floors.
- Ability to provide specialized protection against hostile environments: resistant to stains, oils, solvents, hydrolysis, fungus, microbes, static charges, etc., for example, in safety boots.

Commonly available materials of various categories of shoes are shown in Fig. 18.3 (Kyriacos and GEM-Chem, 2009).

Table 18.1 **Inputs (raw materials and tools) for footwear production**

Shoes	Materials	Connection	Machines	Tools	Footwear chemicals
Cloth footwear	Textiles	Velcro	Groovers	Steel shanks	Adhesives
Leather footwear	Leather	Tapes	Eyeleting machines	Cutters	Solvents
Plastic footwear	Synthetic leathers	Cotton thread	Rotary mops	Punches	Stiffeners
Athletic footwear	Rubber	Nylon thread	Skiving machines	Needles	Dyes
Children's footwear	Plastics	Tacks and nails	Footwear moulds	Grindstones	Polish
Rubbers and boots			Sewing machine	Roughing brushes	Other chemicals
			Lasting machines	Cementing brushes	
			Ovens		

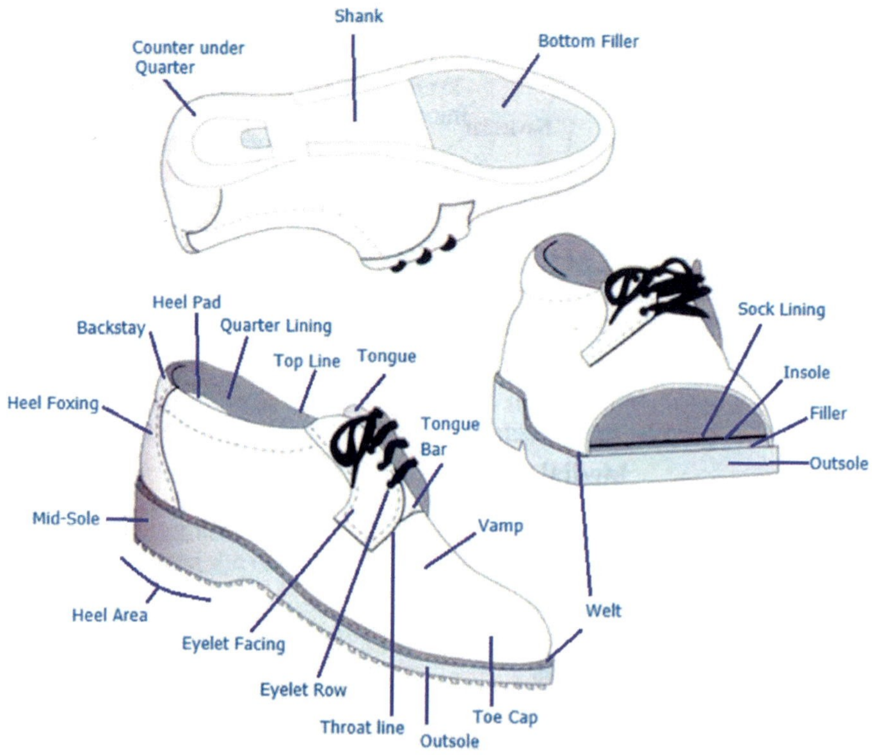

Fig. 18.1 Anatomy and dominant components of a typical shoe.

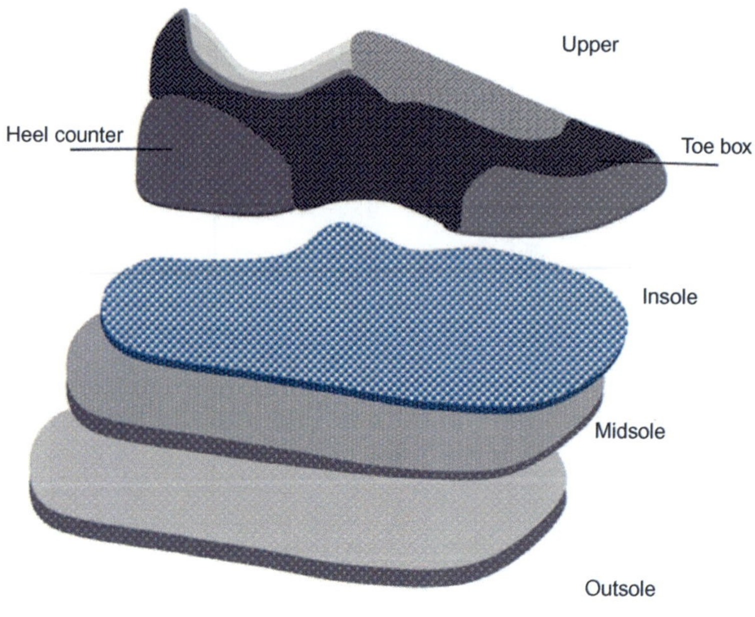

Fig. 18.2 Major components of a typical shoe.

Fig. 18.3 Examples of various categories of shoes and their raw material. (A) Street/fashion shoes: major constituents, (B) casual shoes soles

(Continued)

18.2.2 Shoe insoles

Once again, these can be made up of vegetable-tanned/leather-chrome or polyurethane or neoprene foam. Fibre-board material is one of the cheapest sources to produce. This is prepared by composite material usually made of paper fibres, occasionally wood or even leather fibres, which are reinforced in a glue matrix.

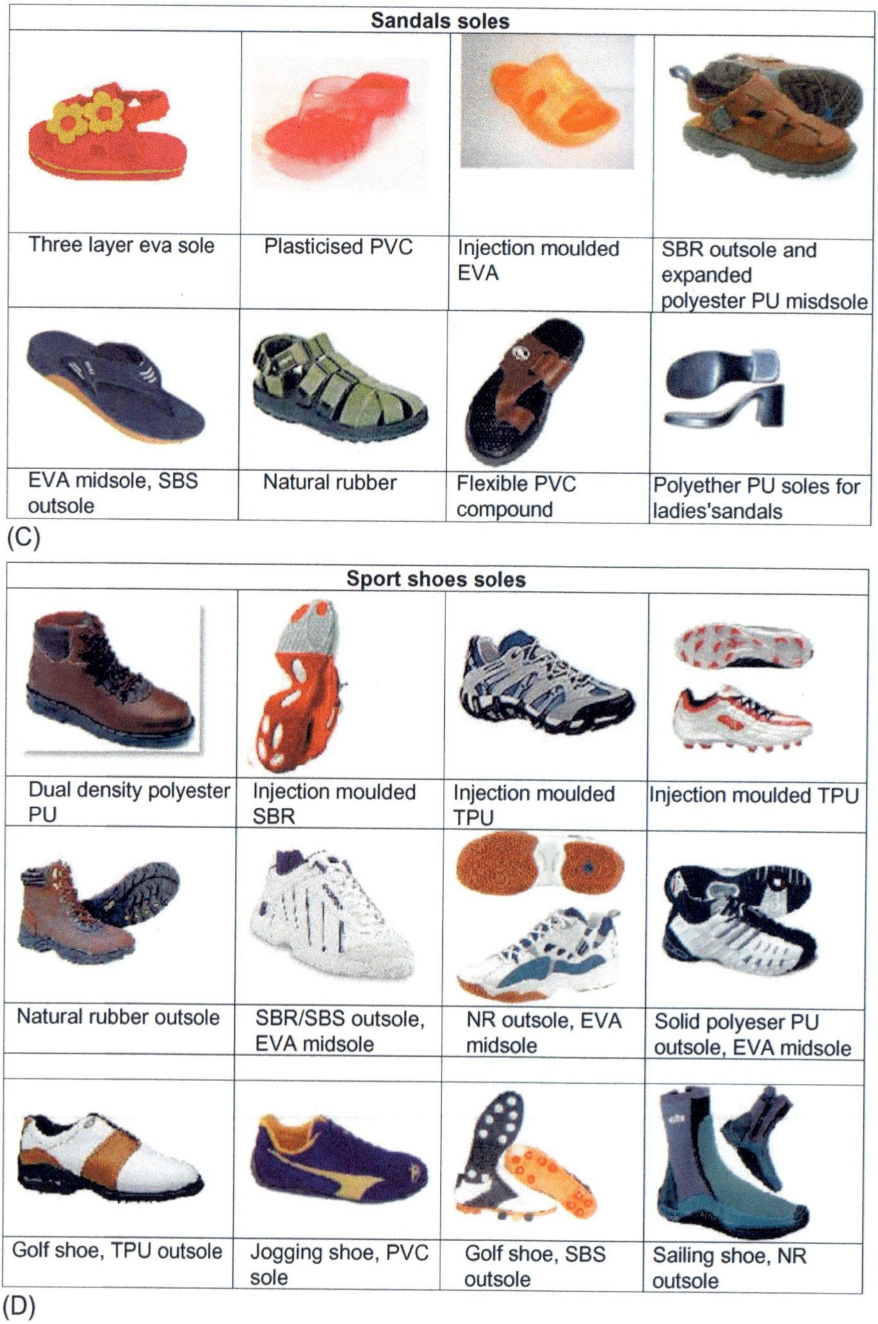

Fig. 18.3, Cont'd (C) sandals soles, (D) sport shoes soles and

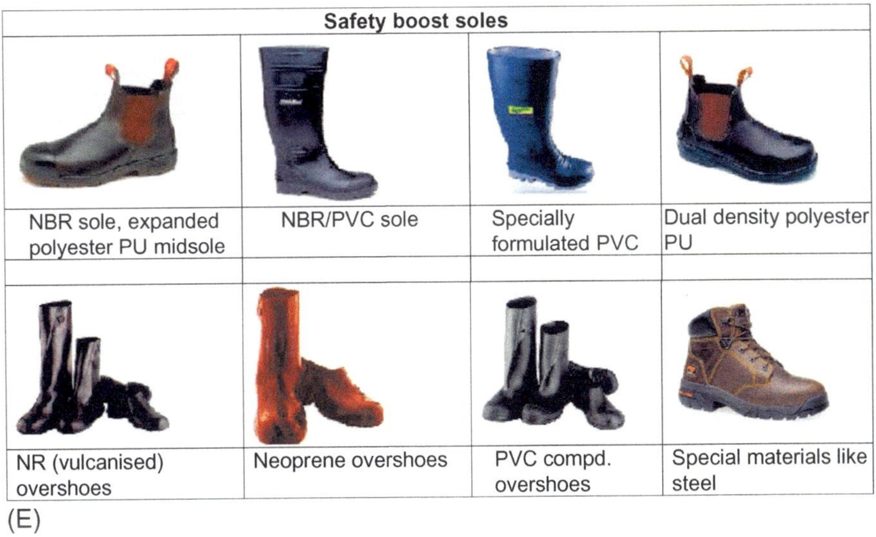

Fig. 18.3, Cont'd (E) safety boots soles. http://www.gem-chem.net/artpubsoles.html.

This biocide-preserved matrix emulsion consists of various rubber resins, such as styrene-butadiene, neoprene or acrylics. Copper-8-quinolinolate or Dowicide A are used as biocides. This mixture is conglomerated in sheets, and the insoles are cut out of these sheets. To make cheaper shoes, fibreboard can also be used as a lasting board or foundation. A second, cheaper choice is thick vegetable-tanned leather. However, fibreboard exposes wearers to a wide range of allergenic problems including plantar foot dermatitis, including rubber chemicals, leather chemicals, adhesives and biocides.

18.2.3 Midsoles

EVA (ethylene vinyl acetate) and polyurethanes are the most-used materials for midsoles. They are particularly well-suited as shoe materials due to their low cost and light weight, as well as their pronounced cushioning effect. The two base copolymers are ethylene and vinyl acetate (Fig. 18.4), which is expanded by incorporating azo compounds or carbon dioxide. The weight percent of ethylene may be from 60% to 90%, and remainder is vinyl acetate. EVA can be formed by injection moulding or compression moulding. Extrusion is tailored for flexibility, softness, strength, colours and desired thickness, and then the material is cut to shape. Both materials can be finished or flame laminated, depending upon the functional needs.

These elastomeric polymers are waterproof, UV-resistant, glossy, tough at low temperatures, resistant to stress-cracks and are hot-melt adhesive. EVA is well known as expanded rubber or foam rubber, and used extensively in sports gear applications like ski boots, corks, bicycle saddles, hockey pads, boxing and martial arts gloves, helmets, wakeboard boots, waterski boots, fishing rods and fishing reel handles. EVA slippers and sandals are also popular. EVA is low in cost, lightweight, easily

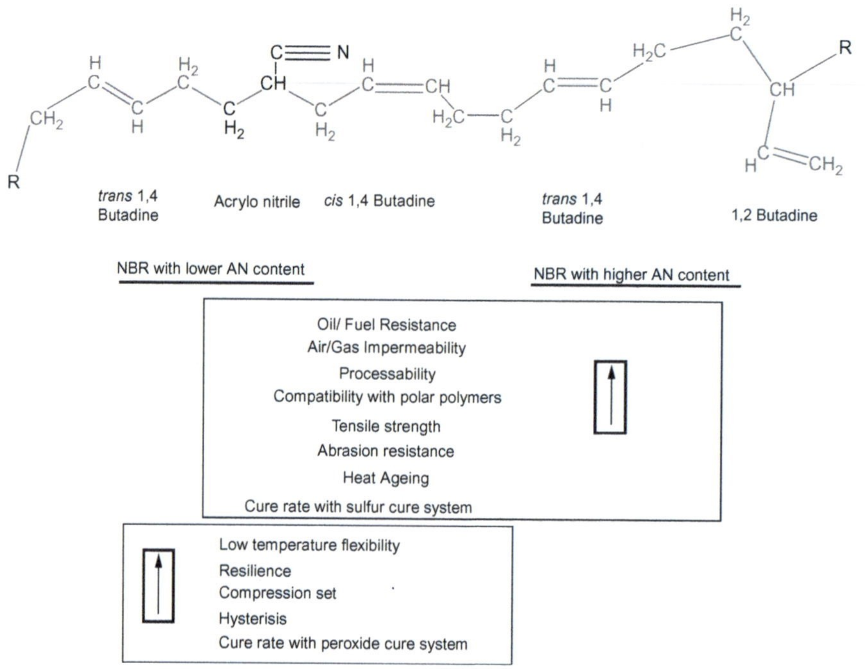

Fig. 18.4 NBR polymer and effect of AN content on various property. https://www.surplussales.com/Rubber-Plastic/pdf/rps-footrub-33_properties.pdf.

moulded, odourless, glossy and cheaper compared to natural rubber material, and because of its pronounced cushioning effect, it is often used as a shock absorber in sports shoes. Typically, polyurethanes have two-component system: an isocyanate prepolymer with polyester polyol, along with a chain extender like butanediol or ethylene glycol, water and catalysts (Fig. 18.5). The resulting products are open-celled.

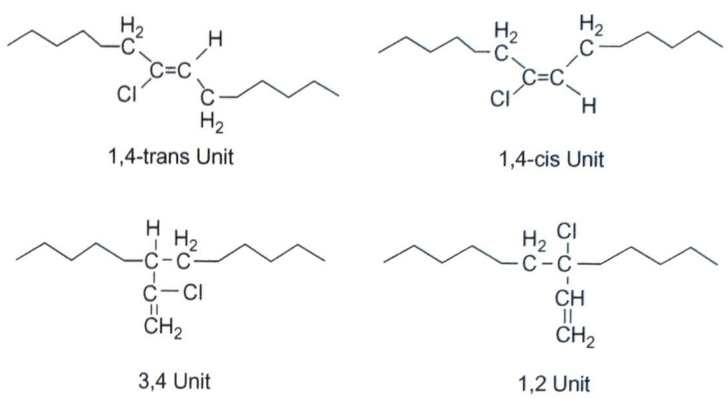

Fig. 18.5 Polymer units.

However, they are costlier and heavier and have less of a cushioning effect than EVA. However, they have a durability advantage (as they can be compression set) over EVA.

Due to severe toxicity issues associated with isocyanates, recent efforts have been made to reduce the use of isocyanates in synthesizing polyurethanes. A new class of polyurethane polymers with the desired functionalities, nonisocyanate based polyurethanes (NIPUs), are being developed to address health and ecological concerns (Zhang et al., 2016; Delebecq et al., 2012; Helou et al., 2011; Javni et al., 2013; Diakoumakos and Kotzev, 2004). Thermoplastic grade polyurethanes are also available now.

18.2.4 Outsoles

The major industrial sole materials are made up of leather, thermoplastics, vulcanized rubbers, thermoplastic elastomers (TPE) and polyether polyurethanes.

18.2.4.1 Leather

Despite several ecological and social issues, leathers have been popular since ancient times as the preferred soling material for fashionable shoes. Leathers are derived from the skin of many animals, including cattle and goats. Cattle skins, with their compact and coarse fibre structure, are treated with a tanning agent to stabilize the collagen fibres in their full thickness to make sole leather. Tanning agents should have a strong affinity and act as cementing agents for collagen and bind the fibrils together. Vegetable-based tanning agents are derived from wood, bark or leaf extracts containing either a derivative of the esters of phenolic carboxylic acids or derivatives of hydroxybenzophenones. Sole leather should also be water resistant and is sometimes impregnated with pore sealers and water repellents.

18.2.4.2 Thermoplastics

Among many thermoplastic materials, PVC is the most widely used, low-cost material for solid as well as expanded soles for street shoes, casual shoes, sandals and slippers. Typically, PVC is compounded with thermoplastic polyurethanes (TPU) to acquire required flexibility and resistance to abrasion, along with some additional components to enhance further oil and water repellency. PVC already has inherently good protection against animal fats, as well as many acids, alcohols, alkalis, bases, oils and petroleum hydrocarbons. For example, dry blending with nitrile rubber powder (NBR) imparts to the PVC polymer additional oil resistance, making it possible for the compound to be used in the moulding of soles for safety boots. Acrylonitrile butadiene styrene (ABS) is also becoming popular in the production of heels for ladies' fashion shoes. However, manufacturing the whole sole and heel results in limited flexibility.

18.2.4.3 Vulcanized rubbers

Natural rubber (NR)—The basic monomer for NR is *cis*-1,4, polyisoprene (synthetic counterpart is isoprene rubber). It has outstanding stretchability and excellent low-temperature properties, as well as superior resistance to slip, puncture and cutting, compared to PVC. It has good resistance to acids, bases, alcohols and diluted water solutions of most chemicals that are water-soluble. However, being a hydrocarbon, NR will be harmed by constant exposure to petroleum and oil-based solvents. Safety boots of NR are widely used in common industries such as food and agriculture, transportation and construction. municipalities.

Styrene butadiene rubber (SBR) is low in cost and most commonly used among synthetic rubbers. It has better abrasion resistance but worse fatigue resistance than that of natural rubber. Reinforcing fillers must be incorporated to enhance SBR's strength.

Nitrile rubber (NBR)—The copolymerization of butadiene and acrylonitrile produces these elastomers, which have excellent resistance to petroleum oils, gasoline, mineral and vegetable oils. Their properties can be well manipulated by acrylonitrile (AN) content and processing conditions as mentioned in Fig. 18.4.

Neoprene (Chloroprene, CR) rubber—Among the specialty elastomers, polychloroprene [poly(2-chloro-1,3-butadiene)] is one of the most important, with an annual consumption above 300,000 tonnes worldwide. First produced in 1932 by DuPont ('Duprene', later 'Neoprene') CR has since gained an outstanding position due to its favourable combination of technical properties. CR is used in different technical areas, mainly in the rubber industry (more than 60%), but is also important as a raw material for adhesives (both solvent-based and water-based, above 33%), and has different latex applications such as the production of dipped articles (e.g. gloves) and moulded foam, and in the improvement of bitumen. Neoprene has better resistance to oils, ozone, oxidation and heat exposure than natural rubber, but does not have natural rubber's low-temperature flexibility, preventing its use in low-temperature shock or impact applications. Neoprene is resistant to a broad range of animal fats and blood, oils, certain acids, alcohols, alkalies, caustics and certain solvents. It is less resistant to cuts, punctures and abrasion than rubber. Safety boots with neoprene outsoles are used in food processing, vegetable oil refining, chemical production and hazardous waste cleanup.

CR is not characterized by any one outstanding property, but its balance of properties is unique among the synthetic elastomers. It also demonstrates strong adhesion to many substrates and can be vulcanized by using various accelerator systems over a wide temperature range.

The high amount of trans-1,4-units in the polymer (about 90% under standard polymerization conditions) leads to synthetic rubber, which has crystallization as an inherent property (Fig. 18.5). Structure and properties of elastomers depend on the types of polychloroprene involved, such as normal linear grades (general purpose grades), precross-linked grades, sulphur-modified grades and slow crystallizing grades. The physical, chemical and rheological properties of the different grades of commercial polychloroprene are dependent on its molecular structure, which can be changed by changing the polymerization conditions, for example, polymerization temperature or monomer conversion, polymerization aids (co-monomers, type and amount of molecular weight modifier and emulsifier) and conditions during finishing.

18.2.4.4 TPE and TPU

This class consists of various classical grades and is formulated according to end use. In TPU, the mechanical performances are attributed to the regular arrangement of crystallites that consist of MDI/chain extender agglomerates, whereas in thermoplastic

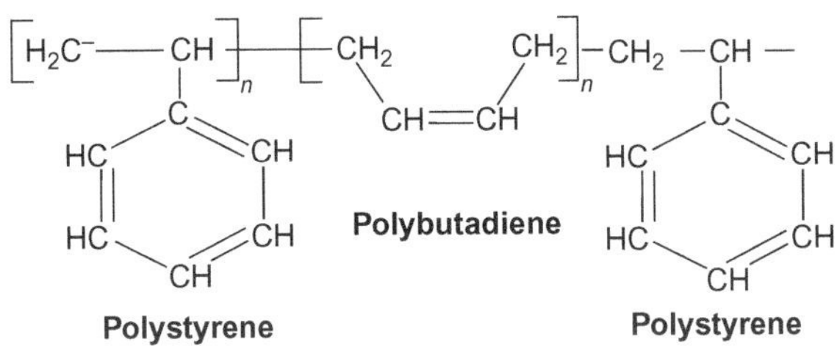

Fig. 18.6 Styrene-butadiene-styrene.
Polymer Science Learning Center, http:/pslc.ws/macrog/sbs.htm (accessed 05.02.17).

rubbers, the crystallites result from the precipitation of one component (polystyrene) from the polymer (styrene butadiene) matrix.

Poly(caprolactone) is a polyester-based and butane diol extended MDI polyurethane elastomer with superior abrasion resistance. They are the preferred material for tennis, running and soccer shoe soles. The polytetramethylene ether glycol based TP-polyurethanes have a better resistance to hydrolytic degradation with lower abrasion resistance. Another important class is SBS synthetic rubbers (styrene-butadiene-styrene), which are formed of polystyrene and polybutadiene blocks (Fig. 18.6). The styrene (29% and 55%) component is responsible for strength and hardness, whereas the polybutadiene blocks provide adherence.

18.2.4.5 Liquid polyurethane systems (polyester systems and polyether systems)

Polyester polyol or polyether polyol and butane diol extended systems with polyester MDI prepolymer are also used for producing lightweight soles with good mechanical properties and water/oil resistance.

18.2.5 Adhesives

Shoes are in constant intimate contact with the foot, and adhesives are a key factor in keeping the bond between the shoe lining and insole intact during walking. Adhesives' polymer properties (like inherent water repellency, bendability, torsional rigidity, elasticity and stain resistance) under the conditions of a broader range of temperature and higher moisture retention levels (due to sweat) play key roles in ensuring better performance and comfort. Manufacturers' use of adhesives which result in shoe allergies can be a major drawback for wearers with sensitive skin. The main adhesives used for gluing in shoe manufacturing are urethane, hot-melt, neoprene and natural rubber.

Hot-melt adhesives are improbable causes of shoe allergy. They are inert, high molecular weight polymers of EVA, having polyamides or polyesters in the rod, pellet or block form in the main polymeric chain. After melting, they must be kept hot. usually insole and heel attachments. However, use of the term is much broader in the adhesive industry, especially referring to water-based lattices such as natural or synthetic rubber latex. Urethane adhesives are polyurethanes in solution, modified to give the composition tack or adhesion. To obtain crease bond strength in special situations, isocyanates are added, along with other adhesives, and epoxy resins are required. In the case of neoprene, to apply EVA polymers, tackifiers such as acrylic or phenol-formaldehyde resins are used. In 80% of footwear, urethane-based adhesives are widely used as a substitute for rubber (which can trigger allergies) whereas, neoprene accounts for less than 20%.

Table 18.2 outlines a typical neoprene formulation. Strong neoprene adhesives contain isocyanates, which cause *p-tert*-butyl phenol-formaldehyde (PTBP-F) resin to undergo premature cure. Therefore, these strong adhesives either use terpene phenolic resins as tackifiers or consist of two components that are mixed immediately prior to use. Rubber latex adhesives tend to be made of natural rubber, as synthetic rubber is expensive (Table 18.3). Since they are water-based, they are biocide-preserved, and tackifying resins such as colophony are added.

18.2.6 Heel and toe counters

If a shoe is to retain its shape, it is necessary to stiffen it at the toe and heel; this leads to the addition of heel and toe stiffeners, or counters (Table 18.4). There are five main types, and all serve to strengthen the toes and heels of shoes. The simplest and least problematic types are those found in athletic shoes: a layer of hot melt adhesive. However, the other types are more elaborate, consisting of a web polyester or cotton material impregnated with a variety of resins to give stiffness and support.

Table 18.2 Neoprene formulation (rubber-based)

Chlorinated rubber (neoprene)	• Phenolic resin (*p-tert*-butylphenol-formaldehyde resin or terpene phenolic resin) • Magnesium oxide • Zinc oxide • Ethylene thiourea • Dioctyl-4-phenylenediamine • Fillers: sodium/calcium silicate • Tetramethylthiuram disulphide • Diorthotolyl guanidine • Sodium dibutyldithiocarbamate

Table 18.3 Water-based resins

Polyoxyethylene Rosin esters Polymerized wood resin Triethanolamine Diallyl phenyl sulphone	Phenol formaldehyde Parachlorometacresol (Collatone) Fatty acid Caustic potash

Table 18.4 Heel and toe counters

Types of counters	Composition
Pressed	–Fibreboard coated with urea or phenol formaldehyde resin, glue –Natural rubber latex 95% ethyl vinyl acetate polymer, hot melt
Thermoplastic	PVC impregnated with rubber resins
Solid styrene	Ethylene methacrylic acid (Surlyn) styrene plasticized with methyl ketone
Thermal	Woven fabric coated with rubber or melamine formaldehyde resins, glue
Three-part counters	Ethyl vinyl acetate hot melts 1. Woven fabrics impregnated with urea-formaldehyde resin and natural rubber latex 2. Ammonium chloride or sodium acetate 3. Pine oil, sodium-*N*-methyl-*N*-oleoyllaurate, dioctyl sodium sulphosuccinate

Counters may contain a number of potential allergens: natural rubber latex, EVA, hot-melt adhesive, phenol-formaldehyde resin, melamine-formaldehyde resin, urea-formaldehyde resin, pine oil and various biocides (Kronberg and Lindman, 2003).

Tables 18.4–18.6 summarize typical material composition of shoe components.

Discussion above in Section 18.2 and Tables 18.2–18.6 summarize typical material composition of shoe components. It is evident that most of the synthetic components to be chosen as raw material should have shown good water and oil repellency and must be able to be tailored well to provide other required functional properties such as strength, toughness, flexibility and durability, etc. Their repellency can further be enhanced by adding some amount of co-polymer during moulding or curing and also by various additives, fillings, adhesives, etc. From Fig. 18.2, it is clear that the greater concern is for other material like textiles and leather to be processed and finished to ensure repellency, along with other desirable properties like breathability, durability and comfort.

Table 18.5 Rubber polymers (some formulations may contain a mixture of natural and synthetic rubber)

Polymers	Monomers
Natural rubber	Isoprene
Synthetic rubber	
Styrene-butadiene rubber	Styrene, butadiene
Synthetic polyisoprene	Synthetic isoprene
Neoprene	Chloroprene
Thermoplastic elastomers	Styrene, isoprene, butadiene
High-styrene resins	Styrene, butadiene
	Isoprene, chloroprene

Table 18.6 Some less-familiar rubber chemicals

Chemical category	Chemical
Vulcanizers	Diorthotolyl guanidine
	Dicurnyl peroxide
Accelerators	Ethylene thiourea
Antioxidants	Octylated diphenylamines
	2,2,4-Trimethyl-1,2-dihydroquinoline
	Butylated hydroxytoluene
	Diphenylamines
Pigment	Titanium dioxide
Resins	Coumarone indene resins
	Terpene phenolic resins
	p-tert-Butyl phenol-formaldehyde resin
Blowing agents	Azodicarbonamide
	4,4-Oxybis (benzenesulphonylhydrazide)

18.3 Upper shoe material processing and finishing

Shoe uppers can be designed from both modern and traditional materials such as fabrics, foam leather and original leather or synthetic materials, including neoprene foam or polyurethane. Traditionally, leather was chrome-tanned, and exposed the wearer to the allergen potassium dichromate. In 95% of cases, primary tanning is done with the use of chrome; however, re-tanning may be done using other methods that include synthetic tanning, vegetable tanning, alum tanning and formaldehyde tanning. Leather's applications, like athletic shoe uppers, increase after it is tanned and shaved to a uniform thickness (Luximon, 2013).

Chrome tanning is the fastest process, taking only a few days, compared to the three weeks involved in vegetable tanning. Vegetable tanning uses all plant or fruit extracts, with myrobalans, quebracho, wattle and chestnut extracts being the major ones in use today. Vegetable tanning is not widely used but yields a highly resilient leather, and 60% of their composition is extracted from wood only. Hydrophilic leather that is ideal for soles, linings and insoles, is less malleable, harder and more difficult to work with than either synthetic- or chrome-tanned leathers.

Alum tanning (or formaldehyde) consumes formaldehyde or glutaraldehyde, which yields a soft waterproof leather often used for work wear footwear. Synthetic tanning is usually used as a final tannage, and utilizes chemicals such as sulphonic acid, naphthalene, phenolic resins, dicyandiamide, dimethylol urea and melamine-formaldehyde resin.

Tanning is followed by a fat liquoring (oiling) process, which replaces the natural oils leached out by the tanning process. The oils include sulphonated neat foot oil, cod liver oil and synthetic moellon oil (Cao et al., 2008).

After tanning, dyeing is the next process which may persist in the finished product as impurities and possible allergens through the byproducts of multiple dye production steps. Usually, acid dyes are used for chrome-tanned leather, and basic dyes for vegetable-tanned leather (Tomasino, 1992).

The next step is finishing, which is designed to give the leather a tough, wear-resistant outer surface and to make the leather to look more attractive. It consists of five main steps:

1. To improve the colour, spray the leather with a stain of metal complex pigments.
2. Add a base coat of acrylic, acrylonitrile or styrene-butadiene resins.
3. Add a protein paste of egg albumin, milk casein protein and vegetable wax solids to give a depth to the appearance.
4. Dry coat with a finish of nitrocellulose containing di-isobutyl ketone, or in less expensive finish coats, a combination of nitrocellulose and phthalate ester.
5. Add a water repellent coat of alkenyl succinic acid complex, fluorinated acid chromium complex, stereochromic chloride complex and silicones.

Many isomer mixers of long chain alkenyl compound, fluorinated metal complexes and silicones were being patented and commercialized to produce water repellent coats on textiles and leathers though lamination, transfer printing and pad dry cure techniques much earlier (US2741597 A, 1952; US2872428 A, 1959; US3028690 A, 1962; US3546151 A, 1970). Over the last decade, the direction of innovative formulation has been focused on developing ecofriendly formulations with better breathability for improved comfort. This has been discussed in more detail in Sections 18.2 and 18.3. For example, US 7485729 B2: Organometallic compositions and coating compositions describes novel organometallic complexes as catalysts for the reaction of compounds with isocyanate and hydroxyl functional groups with zinc metal for the production of urethanes and polyurethanes, which are important in many industrial applications. In fact, silicone-based micro-, macro- and nano-formulations have a wide range of flexibility in containing functional group, depending upon whether the requirement is hydrophobicity or hydrophilicity, oleophilicity or oleophobicity, along with moisture management properties, enabling them as potential candidates. Singha (2012) reviewed the processes and applications of coating and lamination in textiles well.

All of these steps may involve chemical compounds that can cause footwear dermatitis and sensitization. Biocides, a number of which contain allergens, are important additives in both the fat liquoring (oiling) and finishing stages of the process. Table 18.7 lists selected biocides which have been used in shoe manufacturing (Taylor et al., 2006).

18.4 Manufacturing comfortable waterproof/ water repellent shoes

'Waterproof' means water is not allowed at all to pass through a given material. To make a material entirely waterproof, the surface must be completely impermeable to water and provide a high degree of sustained water protection during extreme conditions. For instance, if a footwear product is marked waterproof, that means no water

Table 18.7 Biocides which may be used in shoe manufacturing

Shoe part	Biocide
Leather upper	*N*-Octyl-4-isothiazoln-3-one 8% in propylene glycol92% (Kathon LP) Di-iodomethyl-*p*-tolylsulphone (Amical48) 2Thiocyanomethyl benzothiazole (Busan 30 L) trichlorophenol in Taiwan and Korea Pentachlorophenol
Heel and toe counter	Sodium-*o*-phenylphenate (Dowicide A) 1,2-Benzisothiazolin-3-one 10% in an aqueous solution of propylene glycol (Proxel X L2)
Water-based adhesives	*N*-Trichloro-methyl-thiophthalimide (Fungitrol11) Copper-8-quinolinolate
Lasting board	Sodium-*o*-phenylphenate (Dowicide A) Copper-8-quinolinolate

can get in through the fabric, leather, seams or soles, because not only is the fabric waterproof, the seams are sealed and eyelets are covered.

Most of the basic mechanisms of water repellency and waterproofing, along with recent research and developments are well-documented in the context of textile materials (May, 1987; Tomasino, 1992; Shuhua, 2003; Zhang et al., 2008, 2013; Cao et al., 2008; Chen et al., 2009; Bagherzadeh et al., 2012; Artus et al., 2012; Joshi et al., 2012; Singha, 2012). The manufacture of comfortable waterproof shoes especially for autumn, spring and winter, involves the use of watertight fabrics which satisfy the tensile, hygienic and aesthetic requirements for the materials of which the shoe uppers are made.

It is essential to know the typical foot skin behaviour (Fig. 18.7) regarding heat and moisture build-up before choosing raw materials for shoes and socks, with design suitable to the functional needs of various activities. Repellency and wicking are opposing functions.

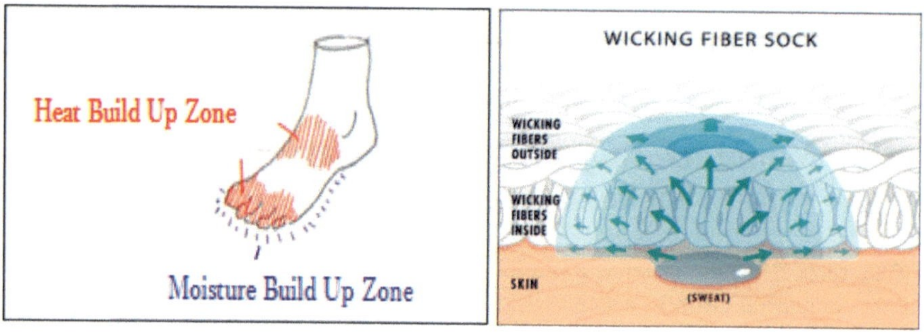

Fig. 18.7 Heat and moisture build-up zone on foot skin and sock material wicking.

Thus, the material close to foot skin needs to have (Wang et al., 2013):

- optimum heat and moisture regulation (thermo-regularity effect)
- absorption of surplus heat
- no/minimal water absorption in the layer of clothing positioned next to the skin
- rapid drying to prevent catching cold
- good air and water vapour permeability
- rapid moisture absorption and conveyance capacity
- easy care/launderability
- dimensional stability even when in contact with water
- soft feel
- light weight
- durability
- cost effectiveness

Waterproof or water-resistant components should be covered with a material that resists or does not allow water passage and remains relatively uninfluenced by water or resisting water passage. In fact, their applications depend on the degree of finishes applied with proper treatment. It is highly recommended in waterproof fabrics which contain a breathable membrane stitched within the fabric, making them both breathable and waterproof. In sports activities, such breathable fabric helps to regulate or dissipate heat and release moisture from the fabrics.

Durable water repellent (DWR) finishes are frequently used within the outdoor footwear and apparel industry to provide fabrics with water and oil repellent properties. For this purpose, fluorochemical-based finishes have long been favoured, but have recently come under scrutiny for their potentially hazardous properties. Legislative and voluntary restrictions are now being implemented across the industry (Herzke et al., 2009; Kähkönen, 2011; Le Marechal et al., 2012; Vestergren et al., 2015; Khandual, 2016). Secondary research unveiled that these finishes' environmental and health credentials are constantly an alarming concern in evaluation of their performance levels. Alternative technologies, even though not widely available and popular, are almost exclusively discussed by the chemical suppliers themselves or referred (Poulsen et al., 2005; Jönsson et al., 2018; Holmquist et al., 2016). DWR finishes are used in many sectors of the textile industry to imbue fabrics with water- and, occasionally, oil-repellent properties. The finish should cause water droplets to roll away instead of spreading onto the fabric surface and wetting the fabric; water drops should bead up on the surface, as shown in Fig. 18.8. Liquid repellency can be propagated when the critical surface tension of a fabric/leather/PU film or laminated surface is lower than that of the liquid; hence, DWRs achieve their properties by reducing the surface tension of the fabric to which they are applied.

They are still considered favourable by the outdoor footwear and apparel industry because they provide the best overall performance in terms of both water and oil repellency. PFC-based DWRs and other fluorosurfactants act as an electronegative atom to reduce the surface tension of the fabric beneath that of oil and water. The popularity of PFC-based finishes has increased due to their unique ability to repel oils as well as water. PFC-free finishes can create a surface tension on a fabric. Whereas silicones

Fig. 18.8 Suspended water droplets after DWR finish.

(24–30 mN m^{-1}) can only reliably provide water repellency, PFC (10–20 mN m^{-1}) can imbue a fabric with a finish that allows them to easily repel a variety of oils and still be durable (Holmquist et al., 2016).

18.4.1 Some recent developments

DWR finishes quickly became favoured by the textile industry when they were introduced in the 1950s with long-chain perfluoroalkyl functionalities, and by 1990, fluorochemical finishes represented 90% of the DWR industry. The desired performances of DWR finishes for textiles were initially achieved following their modification with hydrophobic polymers (Dechant, 1985), dendrimers or nanoparticles (Wong et al., 2006). Fig. 18.9 typically depicts the silicone-, fluorocarbon- and dendrimer-based bionic, durable, water repellent finish which forms a durable bonding with fibre surfaces with their polar or functional groups.

Water repellency is achieved using different product groups, but oil repellency is attained only with fluorocarbon polymers. In practice, this is mostly achieved with coatings, which have the disadvantages of stiff handle, lack of air and vapour permeability, and, consequently, poor wear comfort. It is imperative to apply a thin polymeric layer to its surface (Sarkar, 1981; Sundar et al., 2006; Nashy et al., 2012; Mohamed et al., 2009; Wang et al., 2013) or the use of a polymer composition containing nanoscale particles (Holmberg et al., 2004; Coulson, 2007; Joshi et al., 2012; Wong et al., 2006) at the finishing step of processing to make them waterproof; otherwise, they may be retired early. To prevent damaging shoes with rain, snow and salt, different finishes need to be applied. Functional concepts of recent developments in DWR textiles with comfort and feel through better transpiration are provided in Fig. 18.10.

The major practices can be grouped into four categories as follows:

1. Mechanical incorporation of water repellent products in or on the fibre and fabric surface (in the fibre pores and in the spacing between the fibres and the yarns). (Examples of these are paraffin emulsions.)
2. Chemical reaction of the repellent material with the fibre surface. (Examples of these are fatty acid resins.)

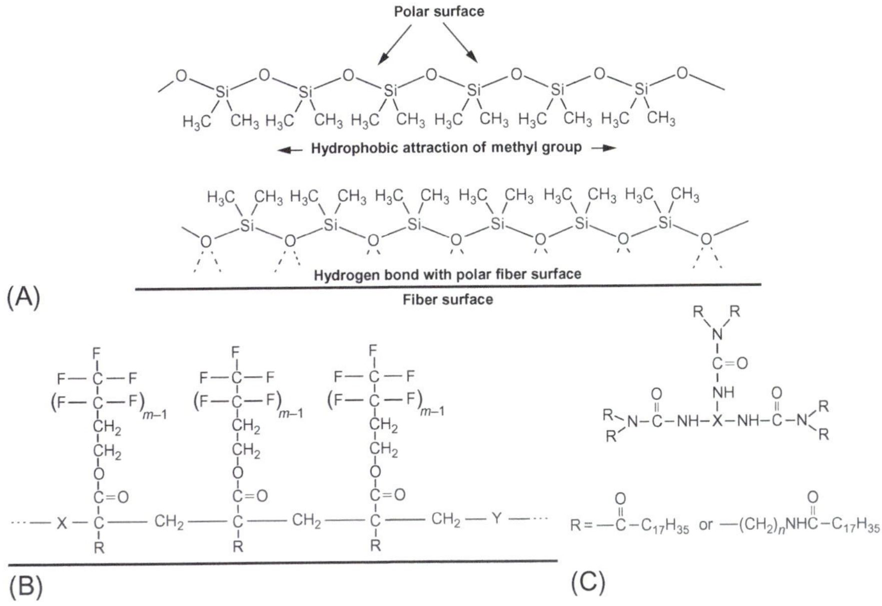

Fig. 18.9 Silicone, fluorocarbon and dendrimer based bionic DWR finish: (A) silicone double layer fabric, (B) fluorocarbon repellent on fabric and (C) dendrimers bionic finish.

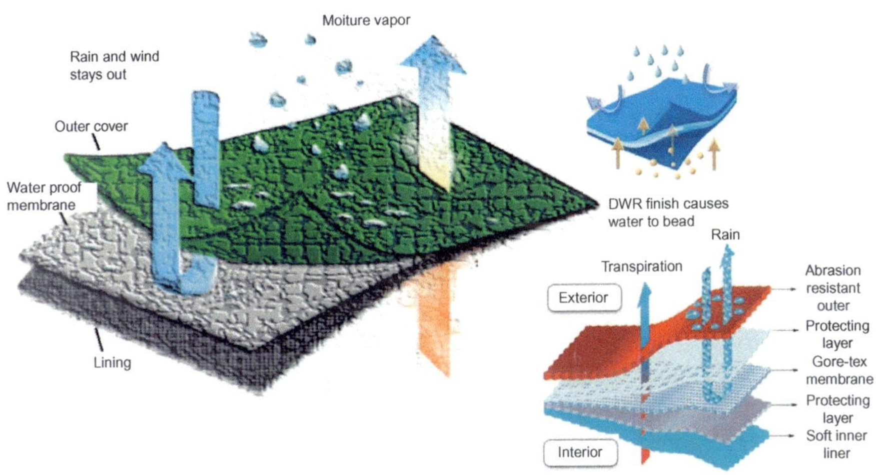

Fig. 18.10 Concepts of recent developments in durable water repellent textiles.

3. The formation of a repellent film on the fibre surface. (Examples of these are silicone and fluorocarbon products.)
4. Special fabric constructions (like stretched polytetrafluoroethylene films (Gore-Tex), films of hydrophilic polyester (Sympatex) and microporous coatings (hydrophilic modified polyurethanes)).

As discussed earlier, long-chain per- and polyfluoroalkyl substances (PFASs) have already been phased out for use as DWR chemistries, and four kinds of DWRs are now available: side-chain fluorinated polymers, silicones, hydrocarbons and other special products from dendrimer and inorganic nanoparticle compounds. Holmquist et al. (2016) gave a lucid critical review of the properties, performance and associated hazards of state-of-the-art DWR chemistries for textile finishing. They clearly found the lack of oil repellence of nonfluorinated alternatives and information on the hazards associated with DWRs, in particular, for the dendrimer and inorganic nanoparticle chemistries, with reference to the performance benefits balanced against the risks to the environment and human health.

To achieve a hydrophobic surface, the surface free energy must be lower than 72 mN/m, which is the surface tension of water. The surface free energy must be even lower, because most oils have a surface tension around 20–30 mN/m, in the case of an oleophobic surface. Only fluorocarbons demonstrated a low enough surface energy to be used for this purpose (Tomasino, 1992). Few companies claim that surfaces coated with their material have a contact angle of around 170 degrees, making the material ultra-hydrophobic. Therefore, it would make the surface perfectly nonwetting to achieve a contact angle of 180 degrees or more. Manufacturers generally measure DWR effectiveness using a spray test. Water is sprayed onto a textile, and the amount that sticks is visually assessed. A score of 90 points indicates that roughly 90% of the fabric has no water sticking to it. The higher the number, the better the performance. The test is then repeated after a number of washings to determine durability. Test scores and the number of washings are combined to create a rating. For example, a 90/10 rating means the spray test achieved a total of 90 points after 10 washes.

Not all manufacturers publish DWR performance ratings, but here is a basic guide for interpreting them:

- Good—80 points after 10 washes. This is a basic outerwear finish.
- Excellent—80 points after 20 washes. Marmot, for example, uses this as its minimum rating for outerwear.
- Superior—80 points after 50–100 washes (Recreational Equipment, Inc., 2017).

Microporous membranes and coatings are becoming popular as a component layer of water repellent fabric/polymers to achieve good air and vapour permeability and cushioning effect, which in turn consequently designed for good wear comfort. The concepts and application are well described by Mukhopadhyay and Midha (2008a,b, 2016), and the advantages and disadvantages of various raw materials and techniques are well addressed. Various methods for generating microporous membranes and coatings can be briefly outlined

18.4.1.1 Microporous membrane

- *Mechanical fibrillation*—for example, polytetrafluoroethylene (PTFE) films having high crystallinity over 98%, biaxial stretched, negative Poisson's ratio of range 50–400 μm with 5.8064 million pores/square metre goretex.
- *Wet coagulation*—for example, ultrafine pores <1 μm by salt leaching.

- *Thermocoagulation*—for example, volatile solvent—PU dissolved in methyl ethyl ketone, toluene and water, having 15%–20% solid content.
- *Foam coating*—fluorocarbon (FC) polymer, PU coating—soft and flexible—abrasion.
- *Radio frequency (RF)/ion/UV or E beaming*—Sputtering of Teflon by Ar gas.

18.4.1.2 Polyurethane coating

- Possible pore diameter of 0.0004 μm.
- PU-based film designed (Permatex) by J.B. Broadley coated on fabrics offers a vapour permeability of at least 70%.
- Grabotter membrane (Grabo Ltd.) used in waterproof shoes is a PU-based film.
- Microporous PU films by Acordis (Tarka) are applied by a transfer process from the release paper, and it can be applied to almost any type of substrate.

A safety shoe for high-end application with material of required functionalities is depicted in Fig. 18.11.

To overcome environmental issues and toxic potential, there is much research to reduce/eliminate fluorochemicals or find alternate material formulations through nanotechnology (Sheng et al., 2016a,b; Horrocks and Anand, 2016; Lee et al., 2016; Sadighzadeh et al., 2016; Li et al., 2016a,b; Duan et al., 2016). A novel fluoride-free functional membrane of electrospun polyacrylonitrile (PAN) nanofibres modified with polydimethylsiloxane (PDMS) has been reported recently by

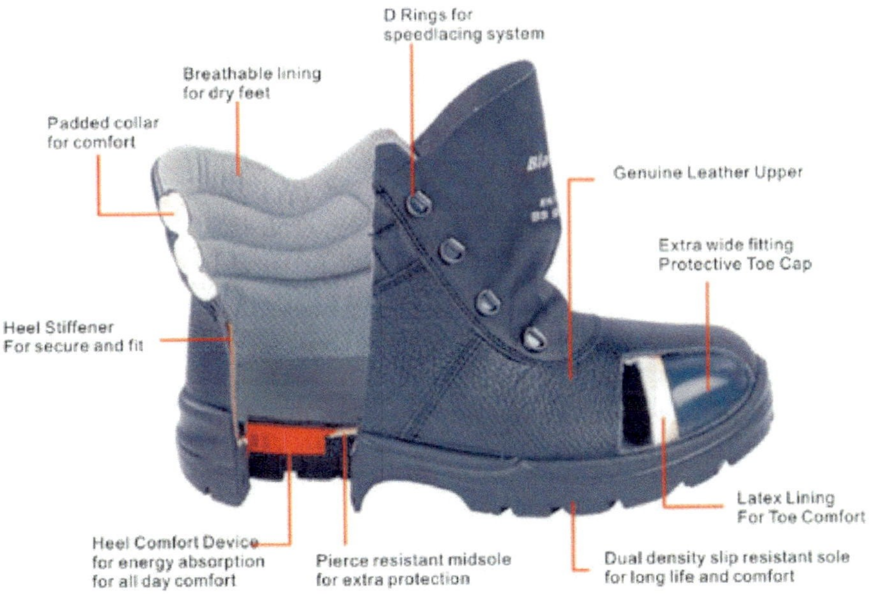

Fig. 18.11 A typical high-end safety shoe: material and functionality.
Wang, J., Safety Shoes, Safety Boots, Anbu Safety Industrial Co., Ltd. http:/safetyshoes-china.com (accessed 08.06.17).

Sheng et al. (2016c), which can be tailored in balancing between the water-resistance and breathability as well as mechanical properties.

In addition, a deluge of companies manufacture waterproofing sprays that are becoming popular due to easy use and maintenance as and when required. Companies such as Kiwi, Meltonian, Tarrago (High-Tech-Nano-Protector-Spray), Nikwax, Vizeodeal, Woly, Vans, etc., are making water and stain protector sprays, which are claimed to be safe to use on leather and suede, with quick-drying capabilities. These waterproofing sprays should be used liberally, as they are not meant to penetrate the leather, but rather form a protective coat on top of it.

18.4.2 Additional issues regarding shoe allergies

Apart from certain issues in water repellent raw polymer and finishes for environmental concern, it is quite important to discuss shoe allergy. An extensive list of potential allergens associated with shoe raw material, processing and composition have been investigated by Podmore (1995) and Podmore (2006). The most common causes of shoe allergy are chemicals found in leather, rubber, adhesives, biocides, dyes, trim, accelerators, plasticizers, tanning agents, antioxidants, UV stabilizers and exposure to them.

Important allergens include:

- Rubber chemicals: thiurams, benzothiazoles, carbamates and para phenylenediamine derivatives in rubber or adhesives.
- Chromates in leather tanning.
- *p-tert*-butyl phenol-formaldehyde (PTBP-F) resin in neoprene adhesives.
- Colophony in adhesives.
- Formaldehyde in leather tanning.
- Nickel in decoration and trim.
- Lanolin in some polishes.
- Para-phenylenediamine in some dyes.

Storrs (1986) reported the following allergens in 55 cases of shoe allergy: potassium dichromate (24%), rubber chemicals (44%), *p-tert*-butyl phenol-formaldehyde resin (4%) and 'unknown' (15%) (patients reacted as allergy-positive to shoe pieces). Two major chemicals (mercapto mix, followed by thiuram mix) were the most common rubber allergens found in shoe lining adhesives. Other sources were elastic in shoe uppers and rubber insoles. Patch testing may help diagnose additional cases of shoe allergy with a series of other shoe chemicals (Cronin et al., 1970; Babbit, 1978; Malten and Seutter, 1985). Podmore (1995) and Podmore (2006) assembled a list of chemicals by identifying all possible culprits and breaking down all of the constituents of shoes. The complexity of identifying potential allergens in shoe components is exacerbated by the fact that today's shoes are manufactured all over the world, often in countries where there is a lack of information on manufacturing components. Components' country of origin may not even be recorded when they are derived from different countries (Cronin et al., 1970). Several other potentially allergenic chemicals have been identified by other authors: p-aminoazobenzene, a dye, chloroacetamide, biocides thimerosal and phenyl mercuric nitrate, hydroquinone in rubber, vegetable tannins, glutaraldehyde (a rare sensitizer used as a leather tannin), other dyes, polyurethane chemicals (Desmodur, Desmocoll 400, Desmodur R and RF and dodecyl mercaptan), urethane adhesives and other adhesive components (1,2-benzisothiazolin-3-one, BIT) (Malten et al., 1982; Foussereau et al., 1976). Testing with additional shoe chemicals may be helpful in confirming that an allergen is present in patch test positive shoe pieces. In addition, it may be useful to identify patients who are allergic to rubber, adhesives or chrome among those having false-negative reactions to shoe pieces (Mathias and Maibach, 1979).

18.5 Conclusion

Footwear raw material selection, processing and design encompasses a diversified domain and moving parallel to the new era of development of materials, polymers, composites along with styles, fashion and functionalities. It should be designed to ensure the life of the footwear by restoring them from hostile exposures such as stains, rain and spillages along with the preservation of aesthetic and comfort properties. The production of water repellent and waterproof shoes without affecting adversely the comfort and breathability along with ecology is gaining much momentum in research and development of novel material and techniques. Currently the development in new generation breathable DWR finishes, composite layers of polymers, textiles and leathers, substitutional candidates for hazardous and toxic chemical through nano material and finishes, easy care spay finishes has been evolving in a big way that completely changing the concept of designing and assembling of shoes components for multifunctional purposes.

References

Artus, G.R., Zimmermann, J., Reifler, F.A., Brewer, S.A., Seeger, S., 2012. A superoleophobic textile repellent towards impacting drops of alkanes. Appl. Surf. Sci. 258 (8), 3835–3840.

Babbit, R.O. (Ed.), 1978. The Vanderbilt Rubber Handbook. RT Vanderbilt Company, Norwalk, CT.

Bagherzadeh, R., Latifi, M., Najar, S.S., Tehran, M.A., Gorji, M., Kong, L., 2012. Transport properties of multi-layer fabric based on electrospun nanofiber mats as a breathable barrier textile material. Text. Res. J. 82 (1), 70–76.

Cao, L., Price, T.P., Weiss, M., Gao, D., 2008. Super water-and oil-repellent surfaces on intrinsically hydrophilic and oleophilic porous silicon films. Langmuir 24 (5), 1640–1643.

Chen, W.H., Chen, P.C., Wang, S.C., Yeh, J.T., Huang, C.Y., Chen, K.N., 2009. UV-curable PDMS-containing PU system for hydrophobic textile surface treatment. J. Polym. Res. 16 (5), 601–610.

Coulson, S., 2007. Plasma treatment of textiles for water and oil repellency. Plasma Technologies for Textiles. Woodhead Publishing, p. 183.

Cronin, E., Bandmann, H.J., Calnan, C.D., Fregert, S., Hjorth, N., Magnusson, B., Maibach, H.I., Malten, K., Meneghini, C.L., Pirilä, V., Wilkinson, D.S., 1970. Contact dermatitis in the atopic. Acta Derm. Venereol. 50 (3), 183–187.

Delebecq, E., Pascault, J.P., Boutevin, B., Ganachaud, F., 2012. On the versatility of urethane/urea bonds: reversibility, blocked isocyanate, and non-isocyanate polyurethane. Chem. Rev. 113 (1), 80–118.

Diakoumakos, C.D., Kotzev, D.L., 2004. Non-isocyanate-based polyurethanes derived upon the reaction of amines with cyclocarbonate resins. In: Macromolecular Symposia. vol. 216. Wiley-VCH Verlag, pp. 37–46.

Duan, P., Yan, C., Luo, W., Zhou, W., 2016. A novel surface waterproof geopolymer derived from metakaolin by hydrophobic modification. Mater. Lett. 164, 172–175.

Foussereau, J., Cavelier, C., Selig, D., 1976. Occupational eczema from para-tertiary-butylphenol formaldehyde resins: a review of the sensitizing resins. Contact Dermatitis 2 (5), 254–258.

Helou, M., Carpentier, J.F., Guillaume, S.M., 2011. Poly (carbonate-urethane): an isocyanate-free procedure from α, ω-di (cyclic carbonate) telechelic poly (trimethylene carbonate) s. Green Chem. 13 (2), 266–271.

Herzke, D., Posner, S., Olsson, E., 2009. Survey, screening and analyses of PFCs in consumer products. Swerea IVF AB, Project Report, 9, p. 47.

Holmberg, K., Jönsson, B., Kronberg, B., Lindman, B., 2004. Surfactants and polymers in aqueous solution. J. Synth. Lubr. 20 (4), 367–370.

Holmquist, H., Schellenberger, S., van der Veen, I., Peters, G.M., Leonards, P.E.G., Cousins, I.T., 2016. Properties, performance and associated hazards of state-of-the-art durable water repellent (DWR) chemistry for textile finishing. Environ. Int. 91, 251–264.

Horrocks, A.R., Anand, S.C. (Eds.), 2016. Handbook of Technical Textiles: Technical Textile Applications, vol. 2. Woodhead Publishing.

Javni, I., Hong, D.P., Petrović, Z.S., 2013. Polyurethanes from soybean oil, aromatic, and cycloaliphatic diamines by nonisocyanate route. J. Appl. Polym. Sci. 128 (1), 566–571.

Jönsson, C., Posner, S., Roos, S., 2018. Sustainable chemicals: a model for practical substitution. In: Detox Fashion. Springer, Singapore, pp. 1–36.

Joshi, M., Bhattacharyya, A., Agarwal, N., Parmar, S., 2012. Nanostructured coatings for super hydrophobic textiles. Bull. Mater. Sci. 35 (6), 933–938.

Kähkönen, E., 2011. Regulatory (EU) guidance towards reduced chemical risks: a study on unintended outcomes in biocide applications.

Khandual, A., 2016. Green flame retardants for textiles. In: Green Fashion. Springer, Singapore, pp. 171–227.

Kronberg, B., Lindman, B., 2003. Surfactants and Polymers in Aqueous Solution. John Wiley & Sons Ltd., Chichester.

Le Marechal, A.M., Križanec, B., Vajnhandl, S., Valh, J.V., 2012. Textile finishing industry as an important source of organic pollutants. In: Organic Pollutants Ten Years After the Stockholm Convention-Environmental and Analytical Update. InTech.

Lee, Y., You, E.A., Ha, Y.G., 2016. Transparent, self-cleaning and waterproof surfaces with tunable micro/nano dual-scale structures. Nanotechnology 27 (35), 355701.

Li, Y., Ge, B., Men, X., Zhang, Z., Xue, Q., 2016a. A facile and fast approach to mechanically stable and rapid self-healing waterproof fabrics. Compos. Sci. Technol. 125, 55–61.

Li, Y., Yang, F., Yu, J., Ding, B., 2016b. Hydrophobic fibrous membranes with tunable porous structure for equilibrium of breathable and waterproof performance. Adv. Mater. Interfaces. 3(19).

Luximon, A., 2001. Foot shape evaluation for footwear fitting (Doctoral dissertation).

Luximon, A. (Ed.), 2013. Handbook of Footwear Design and Manufacture. Elsevier.

Luximon, A., Ganesan, B., Younus, A., 2017. Effects of socks and shoes on normal foot skin temperature. In: International Conference on Applied Human Factors and Ergonomics. Springer, Cham, pp. 485–492.

Malten, K.E., Seutter, E., 1985. Allergenic degradation products of para-tertiary butylphenolformaldehyde plastic. Contact Dermatitis 12 (4), 222–224.

Malten, K.E., Rath, R., Pastors, P.H., 1982. *p-tert*-Butylphenol formaldehyde and other causes of shoe dermatitis. Derm. Beruf Umwelt (Occup. Environ.) 31 (5), 149–153.

Mathias, C.G., Maibach, H.I., 1979. Polyvinyl chloride work boots in the management of shoe dermatitis in industrial workers. Contact Dermatitis 5 (4), 249–250.

May, C. (Ed.), 1987. Epoxy Resins: Chemistry and Technology. CRC Press.

Mohamed, O.A., Moustafa, A.B., Mehawed, M.A., El-Sayed, N.H., 2009. Styrene and butyl methacrylate copolymers and their application in leather finishing. J. Appl. Polym. Sci. 111 (3), 1488–1495.

Mukhopadhyay, A., Midha, V.K., 2008a. A review on designing the waterproof breathable fabrics part I: fundamental principles and designing aspects of breathable fabrics. J. Ind. Text. 37 (3), 225–262.

Mukhopadhyay, A., Midha, V.K., 2008b. A review on designing the waterproof breathable fabrics part II: construction and suitability of breathable fabrics for different uses. J. Ind. Text. 38 (1), 17–41.

Mukhopadhyay, A., Midha, V.K., 2016. Waterproof breathable fabrics. In: Handbook of Technical Textiles: Technical Textile Applications, vol. 2, p. 27.

Nashy, E.L., Essa, M.M., Hussain, A.I., 2012. Synthesis and application of methyl methacrylate/butyl acrylate copolymer nanoemulsions as efficient retanning and lubricating agents for chrome-tanned leather. J. Appl. Polym. Sci. 124 (4), 3293–3301.

Podmore, P., 1995. Shoes. In: Textbook of Contact Dermatitis. Springer, Berlin, Heidelberg, pp. 516–526.

Poulsen, P.B., Jensen, A.A., Wallström, E., Aps, E.N.P.R.O., 2005. More environmentally friendly alternatives to PFOS-compounds and PFOA. Environmental Project No. 1013, p. 2005.

Recreational Equipment, Inc., 2017. Rainwear: Durable Water Repellent (DWR) Care, viewed 5 Feb 2017. https:/www.rei.com/learn/expert-advice/rainwear-dwr.html.

Rout, N., Zhang, Y.F., Khandual, A., Luximon, A., 2010. 3D foot scan to custom shoe last. Int. J. Comput. Commun. Technol. 1 (2–4), 14–18 (Special issue).

Sadighzadeh, A., Valinejad, M., Gazmeh, A., Rezaiefard, B., 2016. Synthesis of polymeric electrospun nanofibers for application in waterproof-breathable fabrics. Polym. Eng. Sci. 56 (2), 143–149.

Sarkar, K.T., 1981. Theory and Practice of Leather Manufacture. Ajoy Sorcar.

Sheng, J., Li, Y., Wang, X., Si, Y., Yu, J., Ding, B., 2016a. Thermal inter-fiber adhesion of the polyacrylonitrile/fluorinated polyurethane nanofibrous membranes with enhanced waterproof-breathable performance. Sep. Purif. Technol. 158, 53–61.

Sheng, J., Zhang, M., Luo, W., Yu, J., Ding, B., 2016b. Thermally induced chemical cross-linking reinforced fluorinated polyurethane/polyacrylonitrile/polyvinyl butyral nanofibers for waterproof-breathable application. RSC Adv. 6 (35), 29629–29637.

Sheng, J., Zhang, M., Xu, Y., Yu, J., Ding, B., 2016c. Tailoring water-resistant and breathable performance of polyacrylonitrile nanofibrous membranes modified by polydimethylsiloxane. ACS Appl. Mater. Interfaces 8 (40), 27218–27226.

Shuhua, L., 2003. The developing prospect of waterproofed moisture permeable and water & oil repellent finishes for PET fabrics. J. Text. Res. 24 (5), 118–120.

Singha, K., 2012. A review on coating & lamination in textiles: processes and applications. Am. J. Polym. Sci. 2 (3), 39–49.

Storrs, F.J., 1986. Dermatitis from clothing and shoes. In: Contact Dermatitis. third ed. 283, Lea & Febiger, Philadelphia, PA, p. 337.

Sundar, S., Vijayalakshmi, N., Gupta, S., Rajaram, R., Radhakrishnan, G., 2006. Aqueous dispersions of polyurethane–polyvinyl pyridine cationomers and their application as binder in base coat for leather finishing. Prog. Org. Coat. 56 (2), 178–184.

Taylor, J.S., Erkek, E., Podmore, P., 2006. Shoes. In: Contact Dermatitis. Springer, Berlin, Heidelberg, pp. 703–716.

Tomasino, C., 1992. Chemistry & Technology of Fabric Preparation & Finishing. North Carolina State University, Raleigh, NC, pp. 13–30.

Vestergren, R., Herzke, D., Wang, T., Cousins, I.T., 2015. Are imported consumer products an important diffuse source of PFASs to the Norwegian environment? Environ. Pollut. 198, 223–230.

Wang, Y., Qiu, F., Xu, B., Xu, J., Jiang, Y., Yang, D., Li, P., 2013. Preparation, mechanical properties and surface morphologies of waterborne fluorinated polyurethane-acrylate. Prog. Org. Coat. 76 (5), 876–883.

Wong, Y.W.H., Yuen, C.W.M., Leung, M.Y.S., Ku, S.K.A., Lam, H.L.I., 2006. Selected applications of nanotechnology in textiles. AUTEX Res. J. 6 (1), 1–8.

Xiong, S., 2008. Pressure Perception on the Foot and the Mechanical Properties of Foot Tissue During Constrained Standing Among Chinese. Hong Kong University of Science and Technology, Hong Kong.

Zhang, B.T., Liu, B.L., Deng, X.B., Cao, S.S., Hou, X.H., Chen, H.L., 2008. Fabricating superhydrophobic surfaces by molecular accumulation of polysiloxane on the wool textile finishing. Colloid Polym. Sci. 286 (4), 453–457.

Zhang, M., Wang, C., Wang, S., Li, J., 2013. Fabrication of superhydrophobic cotton textiles for water–oil separation based on drop-coating route. Carbohydr. Polym. 97 (1), 59–64.

Zhang, K., Nelson, A.M., Talley, S.J., Chen, M., Margaretta, E., Hudson, A.G., Moore, R.B., Long, T.E., 2016. Non-isocyanate poly (amide-hydroxyurethane)s from sustainable resources. Green Chem. 18 (17), 4667–4681.

Further reading

International Institute of Synthetic Rubber Producers, Inc. (IISRP), Acrylonitrile-Butadiene Rubber (NBR), viewed 23 April 2017. http://www.iisrp.com/webpolymers/07nbr-18feb2002.pdf.

Lewin, M., Sello, S.B., 1984. Handbook of Fiber Science and Technology: Volume II. Chemical Processing of Fibers and Fabrics. Functional Finishes. Part B. CRC Press.

Polymer Science Learning Center, http://pslc.ws/macrog/sbs.htm (accessed 05.02.17).

Index

Note: Page numbers followed by *f* indicate figures, and *t* indicate tables.

G

9780081012123